DAMAGED

# The Law and Economics of a Sustainable Energy Trade Agreement

The widely accepted need to reduce the world's dependence on fossil fuels and move instead to low-carbon, renewable alternatives faces a host of challenges. Whilst the greatest challenges remain in engineering, political and public policy issues continue to play a very important role. This volume, which consists of contributions from leading figures in the field, presents the case for a Sustainable Energy Trade Agreement (SETA). It shows that by addressing barriers to trade in goods and services relevant for the supply of clean energy, such an Agreement would foster the crucial scaling-up of clean energy supply and promote a shift away from fossil fuels. In doing so, it illustrates how the Agreement would help to address a number of overarching sustainable development priorities, including the urgent threat of climate change, enhanced energy access and improved energy security. The book will appeal to academics and policy makers working at the interface of trade and energy policy.

**Gary C. Hufbauer** is the Reginald Jones Senior Fellow with the Peterson Institute of International Economics (PIIE), a position he has held since 1992. He was formerly the Maurice Greenberg Chair and Director of Studies at the Council on Foreign Relations (1996–1998). He has authored or co-authored over thirty publications on International Trade, Investment and Tax Issues. As a senior official in the US Treasury Department, he negotiated trade, tax and financial agreements.

**Ricardo Meléndez-Ortiz** is co-founding Chief Executive of the International Centre for Trade and Sustainable Development (ICTSD). He is a former negotiator in bilateral and multilateral trade, environment and development negotiations and processes, and served as Principal Advisor to the Colombian Minister of Economic Development and as Chief of Administration in the Office of the President of Colombia. He is currently the Convener of the E15Initiative, a joint ICTSD–WEF undertaking looking into the future of the global trade and investment system, and an Advisor to the Global Commission on Business and Sustainable Development.

**Richard Samans** is Managing Director and Member of the Managing Board of the World Economic Forum (WEF). From 2011–2013, he was Director-General of the Global Green Growth Institute (GGGI). He has previously served in the US White House as Special Assistant to the President for International Economic Policy, and as National Security Council Senior Director for International Economic Affairs.

# The Law and Economics of a Sustainable Energy Trade Agreement

*Edited by*
Gary C. Hufbauer, Ricardo Meléndez-Ortiz
and Richard Samans

# CAMBRIDGE
UNIVERSITY PRESS

University Printing House, Cambridge CB2 8BS, United Kingdom

Cambridge University Press is part of the University of Cambridge.

It furthers the University's mission by disseminating knowledge in the pursuit of education, learning and research at the highest international levels of excellence.

www.cambridge.org
Information on this title: www.cambridge.org/9781107092860

© International Centre for Trade and Sustainable Development (ICTSD) 2016

This publication is in copyright. Subject to statutory exception and to the provisions of relevant collective licensing agreements, no reproduction of any part may take place without the written permission of Cambridge University Press.

First published 2016

*A catalogue record for this publication is available from the British Library*

ISBN 978-1-107-09286-0 Hardback

Cambridge University Press has no responsibility for the persistence or accuracy of URLs for external or third-party internet websites referred to in this publication, and does not guarantee that any content on such websites is, or will remain, accurate or appropriate.

# Contents

| | |
|---|---|
| List of Figures | *page* vii |
| List of Tables | viii |
| List of Boxes | x |
| List of Contributors | xi |
| List of Selected Abbreviations and Acronyms | xvi |
| Foreword: Trade Policy Can Help Combat Climate Change | |
| H.E. MOGENS JENSEN | xxi |
| Acknowledgements | xxiii |

1 Introduction: Setting the Horse Before the Cart to Preserve a Viable World
GARY C. HUFBAUER, RICARDO MELÉNDEZ-ORTIZ AND RICHARD SAMANS — 1

2 Issues and Considerations for Negotiating a Sustainable Energy Trade Agreement: May 2012
GARY C. HUFBAUER AND JISUN KIM — 20

3 Trade in Sustainable Energy Services: October 2013
JOACHIM MONKELBAAN — 85

4 Governing Clean Energy Subsidies: What, Why, and How Legal?: August 2012
ARUNABHA GHOSH WITH HIMANI GANGANIA — 140

5 Trade Law Implications of Procurement Practices in Sustainable Energy Goods and Services: September 2012
ALAN HERVÉ AND DAVID LUFF — 196

6 Selling the Sun Safely and Effectively: Solar Photovoltaic Standards, Certification Testing and Implications for Trade Policy: December 2013
SUNNY RAI AND TETYANA PAYOSOVA — 242

7 Addressing Local Content Requirements in a Sustainable
  Energy Trade Agreement: June 2013
  SHERRY M. STEPHENSON                                          316

8 International Technology Diffusion in a Sustainable
  Energy Trade Agreement: September 2012
  THOMAS L. BREWER                                              349

9 Legal options for a Sustainable Energy Trade Agreement:
  July 2012
  MATTHEW KENNEDY                                               390

  *Index*                                                       456

# Figures

3.1　The Relation Between EGS, CFGS and SEGS　　　*page* 89
3.2　Contribution of the Wind Energy Sector in the EU to the
　　　Direct and Indirect Employment Forecast, 2020 and 2030　　122
3.3　Comparison of Job-Years Across Technologies　　123
3.4　Trade Restrictiveness in Services Trade (Tariff ad valorem
　　　Equivalents) by Sector Weighted by Trade Volumes　　125
4.1　Global Financial New Investment in RE, Quarterly Trend,
　　　2002–2011　　142
4.2　Worldwide Share of Many RE Sources in Total Electricity
　　　Generation Remains Small　　144
4.3　RE's Share in Germany's Electricity Consumption has Risen
　　　Steadily　　169
4.4　RE's Share in Total Electricity Capacity in the United States　　175
4.5　Clean Energy Subsidies Have Multiple Rationales, Impacts
　　　and Counteractive Reactions　　186
6.1　Stakeholders and Rationale for Standards　　244
6.2　The Role of National Certifying Bodies　　254
6.3　Cumulative Installed PV Capacity in the Top Eight Countries　　256

# Tables

| | | |
|---|---|---|
| 2.1 | List of SETA Products, Trade Values and Tariff Rates | page 23 |
| 2.2 | Tariff Rates for SETA Products by Selected Countries, Tariff Year 2010 | 31 |
| 2.3a | Trade with Other Core and Candidate Countries: NAFTA, 2010 | 44 |
| 2.3b | Trade with Other Core and Candidate Countries: EU-27, 2010 | 50 |
| 2.3c | Trade with Other Core and Candidate Countries: China, 2010 | 56 |
| 2.4 | Trade of All 39 SETA Products between Selected Countries, 2010 | 64 |
| 2.5 | Some Cases of LCRs in Place, as of January 2012 | 67 |
| 2.6 | Top 10 Solar PV Cell/Wind Turbine Manufacturers, 2010 | 76 |
| 2.7 | Trade of APEC 50 EGs, 2010 | 78 |
| 2.8 | Comparison of SETA, World Bank and APEC Lists | 79 |
| 3.1 | Key Sectoral Mitigation Technologies, and Corresponding Services Categories | 98 |
| 3.2 | Major Exporters and Importers of Architectural, Engineering and Other Technical Services | 100 |
| 3.3 | Sectoral Commitments on Other Professional, Technical and Business Services | 102 |
| 3.4 | Snapshot of GATS Commitments in Services Related to Wind Energy | 104 |
| 3.5 | Market Access and National Treatment Limitations on Mode 1: Other Professional, Technical and Business Services | 105 |
| 3.6 | Market Access and National Treatment Limitations on Mode 3: Other Professional, Technical and Business Services | 107 |

| | | |
|---|---|---|
| 3.7 | Major Exporters and Importers of Construction Services | 112 |
| 3.8 | Market Access and National Treatment Limitations on Mode 3: Construction Services | 114 |
| 3.9 | The Clean Development Mechanism and Services | 119 |
| 3.10 | Distribution of Jobs in the PV Sector | 123 |
| 4.1 | Clean Energy Subsidies Have Supporters and Opponents | 150 |
| 4.2 | A Typology of Clean Energy Subsidies | 152 |
| 5.1 | Main Measures Promoting SEGS in EU Procurement Policies and Relevant WTO Provisions | 227 |
| 5.2 | SEGS Prescriptions Based on PPMs and Gaps in Current Multilateral Trade Law | 235 |
| 5.3 | Possible Options for Procurement-Related Provisions of a SETA | 237 |
| 6.1 | PV Testing Standards | 252 |
| 6.2 | PV Performance Rating Tests | 272 |
| 6A.1.1 | Annual Installed Solar PV Capacity in Selected Countries and the World, 1998–2010 | 303 |
| 6A.2.1–6A.2.7 | Concordance of Various Standards for PV Modules | 304 |
| 7A.1 | LCR Statistics | 338 |
| 8.1 | Energy Efficiency and Other Sources of Energy-Related $CO_2$ Abatement in an IEA 450 Scenario | 356 |
| 8.2 | 'Key Groups' of Technologies in the IEA BLUE Scenario | 357 |
| 8.3 | Approximate Ranges of Technical Potentials of World Sustainable Energy Production Relative to Recent Production | 358 |

# Boxes

3.1 Services Involved in Solar and Wind Power Projects  *page* 96
3.2 The Four Modes of Services Supply  101
5.1 Sectors for Sustainable Energy Goods and Services
   Procurements Potentially Covered by the Revised GPA  203
8.1 Internationalisation of Renewable Energy Industries  361

# Contributors

THOMAS L. BREWER is a Senior Fellow at the International Centre for Trade and Sustainable Development (ICTSD), Geneva. He is also a Lead Author of the chapter on International Cooperation in the IPCC report (AR5). He has had various short-term appointments as a Visiting Senior Research Fellow at Oxford University in the Smith School for Enterprise and the Environment, and he has been a Schöller Foundation Senior Research Fellow at Friedrich-Alexander University in Nuremberg, Germany. He is a visiting scholar at MIT'S Center for Energy and Environmental Policy Research (CEEPR) and an emeritus faculty member of Georgetown University. His research focuses on the intersection of climate change issues with international trade, investment and technology transfer issues. He is the editor and author of the introduction to a symposium on those issues in the refereed journal *The World Economy*. His other publications include articles in the refereed journals *Climate Policy* and *Energy Policy*, as well as chapters in books published by the Brookings Institution, Cambridge University Press, Oxford University Press and other leading publishers. He has been a consultant to the World Bank, the United Nations Conference on Trade and Development (UNCTAD), and the Organisation for Economic Cooperation and Development (OECD).

HIMANI GANGANIA is a Research Analyst, at the Council on Energy, Environment and Water (CEEW), New Delhi. She holds a Bachelor of Engineering (BE) degree in Bio-Technology from Delhi College of Engineering (DCE), University of Delhi, India (2011). Her interest areas include: climate change and governance issues; sustainable global governance; bio-technology and sustainability; and new fossil fuel technologies. Himani's work domain includes sustainability finance and technology; technology horizons; and resource and energy efficiency development. Previously, Himani has undertaken academic research projects at the Defence Research and Development Organization (DRDO), Ministry of Defence, India to study the critical environment and climate-related effects on living organisms, in the near and distant future.

ARUNABHA GHOSH is CEO of the Council on Energy, Environment and Water (CEEW), consistently ranked (third year running) as one of South Asia's leading policy research institutions. With work experience in thirty-seven countries and having previously worked at Princeton, Oxford, the UNDP (New York) and the WTO (Geneva), Arunabha advises governments, industry, civil society and international organisations around the world. His recent books include *The Palgrave Handbook of the International Political Economy of Energy* (2016); *Energising India: Towards a Resilient and Equitable Energy System* (2016); *Human Development and Global Institutions* (2016); and *Climate Change: A Risk Assessment* (2015). Arunabha's essay 'Rethink India's energy strategy', in Nature, was selected as one of 2015's ten most influential essays. Dr Ghosh has advised the Prime Minister's Office and several ministries; was invited by France to advise on the COP21 climate negotiations; has been actively involved with the International Solar Alliance; and serves on the Executive Committee of the India–US PACEsetter Fund. He is a World Economic Forum Young Global Leader. He holds a DPhil from Oxford.

ALAN HERVÉ has been full professor of public law at the University of Bretagne Occidentale since 2014. Previously, he was lecturer at the Sorbonne Law School (University of Paris 1) for two years (2012–2014) and was a visiting professor at the College of Europe (Warsaw, Natolin Campus). He worked as a teaching assistant at the University of Rennes 1 (France) (2004–2010) and at the College of Europe (2010–2012). He obtained a PhD at the Université de Rennes 1 (France) in 2010, on The European Union and the Judicialization of the Dispute Settlement Mechanism of the World Trade Organization (WTO). Professor Hervé regularly publishes articles and contributions on the EU trade and investment policy, the WTO and free trade agreements and more recently on the legal aspects of the European and Monetary Union. He is one of the associate editors of the *European Papers* (www.europeanpapers.eu/en), an e-journal dedicated to the EU law and integration process.

H. E. MOGENS JENSEN is the Danish Minister for Trade and Development Cooperation. He is a member of the Social Democratic Party and has previously held the position as Culture and Media Spokesman as well as serving on several committees of the Danish Parliament including the Foreign Policy Committee, the Foreign Affairs Committee and the

Cultural Affairs Committee. He is the former Chairman of the Danish Delegation to the Parliamentary Assembly of the Council of Europe as well as of the Parliamentary Group of the Social Democratic Party. A long-standing and active member of the Social Democratic Party, Mogens Jensen has worked on a number of political issues prior to his parliamentary career, including development, education and culture. Since 2012 he has been the Vice-Chairman of the Social Democratic Party.

MATTHEW KENNEDY is a visiting professor at the Faculty of Law of the University of International Business and Economics, Beijing. Previously, he worked in the WTO Secretariat in Geneva for twelve years where he served as a senior lawyer in the Legal Affairs Division. In that capacity he was responsible for advising dispute settlement panels and providing institutional advice on the different means to give legal effect to changes in WTO law, including market access improvements and treaty amendment.

JISUN KIM has been a principal business analyst at the POSCO Research Institute, Seoul, since April 2010. Before that, she worked as a research analyst at the Peterson Institute for International Economics (PIIE), Washington DC. Previously, she worked as a tax consultant (US CPA) at PricewaterhouseCoopers in Seoul, Korea. She is co-author of *Global Warming and the World Trading System* (2009) and has co-authored several papers, including 'Climate Change and Trade: Searching for Ways to Avoid a Train Wreck' (2010), 'Controlling Climate Change: Challenges for Latin America and the Caribbean' (2010), 'After the Flop in Copenhagen' (2010), 'Reaching a Global Agreement on Climate Change: What are the Obstacles?' (2010), 'Prospects for International Climate Negotiations: Copenhagen and Beyond' (2009), 'WTO and Climate Change: Challenges and Options' (2009), 'Policy Options and the World Trade Organization' (2009), 'A View from Washington: Climate Policy under Obama – Implications for Canada and Trade' (2008), 'International Competition Policy and the WTO' (2008). She received her MA degree in international relations from the Maxwell School of Syracuse University.

DAVID LUFF is an international trade lawyer and a founding partner at the law firm Appleton Luff, with offices in Brussels, Geneva, Singapore, Washington and Warsaw. He has practised international trade law at both EC and WTO levels for seventeen years. Luff is also a professor of international trade law at the University of Liège in Belgium and at the College of Europe. He obtained hid PhD at the University of Brussels:

his thesis examined the regulatory powers of States under WTO law to pursue domestic policy objectives. Luff has published several articles in international reviews on subjects related to WTO law and has held visiting appointments in India, China, Vietnam, Switzerland and Italy. He has trained trade officials of the EC, FYR Macedonia, Montenegro, Kazakhstan, Ukraine, Uzbekistan, West African countries and China on WTO law and trade policy, including on SPS and TBT rules.

JOACHIM MONKELBAAN is an independent advisor on the Sustainable Development Goals (SDGs) and in particular on trade, climate change and the water–energy–food nexus. He also lectures at the University of Geneva (Institute for Environmental Sciences). He earned a doctorate in sustainable governance from the University of Geneva, a Master of International Law and Economics at the World Trade Institute (WTI) and an LLB and LLM in European Law from Maastricht University. He previously was a Senior Advisor at the United Nations Environment Programme (UNEP) and a Senior Programme Officer at the International Centre for Trade and Sustainable Development (ICTSD). He also worked at the International Center for Small Hydro Power (IC-SHP) in China, the International Union for the Conservation of Nature (IUCN), the Dutch Ministry of Economic Affairs and global law firms. He frequently publishes in the areas of trade, climate change and the green economy.

TETYANA PAYOSOVA graduated from the National University Odessa Law Academy, Ukraine, in 2007 and has a Master of Law degree (2009) from the University of Bern, Switzerland. Payosova has been a research fellow at the Institute for European and International Trade Law, University of Bern, since 2008, and is closely cooperating with Work Package 5 on Trade and Climate Change. The topic of her PhD research is the analysis of framework conditions for the promotion of renewable energy, with a focus on solar energy, and international trade regulation. Her research interests include the status of energy in the WTO, regulation of electricity markets, promotion of renewables, climate change mitigation, intellectual property law and EU law. She has co-authored *Climate Change and International Law: Exploring the Linkages between Human Rights, Environment, Trade and Investment* (2010) and *Challenges in International Monetary Law* (2011).

SUNNY RAI is the regional vice president for renewable energy at Intertek, providing strategic direction for the photovoltaic (PV), wind, smart grid and semiconductor businesses. With more than twenty-five years with Intertek, Sunny has helped several solar, semiconductor and FPD equipment manufacturers and users establish product safety and global regulatory compliance programmes. He was also instrumental in establishing Intertek as a key player in the PV industry, with laboratories in Lake Forest and San Francisco, CA; and Shanghai, China. He is recognised as an international expert in solar testing and certification, with several published articles and industry presentations to his credit.

SHERRY M. STEPHENSON is a Senior Fellow at the International Centre for Trade and Sustainable Development (ICTSD), Geneva. Currently she is Senior Advisor for Services Trade at the Organization of American States (OAS) in Washington, DC, where she has previously served as Director of the Department of Trade and Deputy Director of the Trade Unit during the ten years of the Free Trade Area of the Americas (FTAA) negotiations. She received a PhD in International Economics from the Graduate Institute of International and Development Studies/ University of Geneva and an MA in Economics from New York University. She has served as an Advisor to the Minister of Trade in Indonesia and held positions with the GATT and UNCTAD Secretariats and with the OECD. Her publications address in particular services trade, regional economic integration, labour mobility and non-tariff barriers (NTBs). She has edited two volumes on services trade and reform and published more than sixty articles and chapters in journals and books.

# Selected Abbreviations and Acronyms

| | |
|---|---|
| 3GF | Global Green Growth Forum (Denmark) |
| ABCs | Solar America Board for Codes and Standards |
| ANSI | American National Standards Institute |
| APEC | Asia-Pacific Economic Cooperation |
| ASEAN | Association of Southeast Asian Nations |
| ASTM | American Society for Testing and Materials |
| BIPV | building-integrated photovoltaics |
| CBD | Convention on Biological Diversity |
| CCS | carbon capture and storage |
| CDM | clean development mechanism |
| CECET | Center for Evaluation of Clean Energy Technology |
| CEEW | Council on Energy, Environment and Water (India) |
| CFGS | climate-friendly goods and services |
| CHP | combined heat and power |
| COP | Conference of the Parties (UNFCCC) |
| CPC | Committee for Progress and Coordination (UN) |
| CSi | crystalline silicon |
| CSP | concentrated solar power |
| CTCN | Climate Technology Centre and Network (Technology Mechanism) |
| CTESS | Committee on Trade and Environment Special Session |
| DSB | Dispute Settlement Body (WTO) |
| DSM | Dispute Settlement Mechanism (WTO) |
| DSU | Dispute Settlement Understanding (WTO) |
| EG | environmental good |
| EGS | environmental goods and services |
| EIA | Energy Information Administration (US) |
| FIT | feed-in tariff |
| FPD | flat panel display |
| FTA | free-trade agreement |
| G2A2 | Green Growth Action Alliance |

List of Selected Abbreviations and Acronyms xvii

| | |
|---|---|
| GATS | General Agreement on Trade in Services |
| GATT | General Agreement on Tariffs and Trade |
| GDP | gross domestic product |
| GGGI | Global Green Growth Institute (Korea) |
| GPA | Government Procurement Agreement (WTO) |
| GSP | Generalized System of Preferences |
| HS | Harmonized System (product coverage) |
| ICLEI | Local Governments for Sustainability |
| ICT | information and communications technology |
| ICTSD | International Centre for Trade and Sustainable Development |
| IEA | International Energy Agency |
| IEC | International Electrotechnical Commission |
| IECEE | IEC System of Conformity Assessment Schemes for Electrotechnical Equipment and Components |
| IEEE | Institute of Electrical and Electronics Engineers |
| ILO | International Labour Organization (UN) |
| IMF | International Monetary Fund |
| IMO | International Maritime Organisation |
| IP | intellectual property |
| IPCC | Intergovernmental Panel on Climate Change |
| IPR | intellectual property right |
| IRENA | International Renewable Energy Agency |
| ISO | International Organization for Standardization |
| IT | information technology |
| ITA | Information Technology Agreement (WTO) |
| LCR | local content requirement |
| LDC | least-developed country |
| LEAP | Local Authority Environment Management Systems and Procurement |
| MEA | multilateral environmental agreement |
| MEF | Major Economies Forum on Energy and Climate |
| MFN | most favoured nation |
| NAFTA | North American Free Trade Agreement |
| NCB | National Certification Body (IECEE) |
| NPR | non-product-related |
| NPR–PPM | non-product-related production and processing methods |
| NRTL | Nationally Recognised Testing Laboratory |
| NT | national treatment (GATT) |
| NTB | non-tariff barrier |

| | |
|---|---|
| OECD | Organization for Economic Cooperation and Development |
| OEM | original equipment manufacturer |
| PPM | production and processing methods |
| PPP | public–private partnership |
| PTA | preferential trade agreement |
| PTC | production tax credit |
| PV | photovoltaic |
| RCR | regional content requirement |
| RE | renewable energy |
| RFP | request-for-proposal |
| RTA | regional trade agreement |
| SCC | Standards Council of Canada |
| SCM | (Agreement on) Subsidies and Countervailing Measures (WTO) |
| SE4All | Sustainable Energy for All (UN) |
| SEGS | sustainable energy goods and services |
| SETA | Sustainable Energy Trade Agreement |
| SETI | Sustainable Energy Trade Initiative |
| SME | small and medium-sized enterprise |
| SOE | state-owned enterprise |
| SPS | Sanitary and Phytosanitary (Agreement) (WTO) |
| STC | standard testing condition |
| T&D | transmission and distribution |
| TBT | Technical Barriers to Trade (Agreement) (WTO) |
| TCA | Trade in Civil Aircraft (Agreement) |
| TEC | Technology Executive Committee (Technology Mechanism) |
| TFA | Trade Facilitation Agreement (WTO) |
| TiSA | Trade in Services Agreement (WTO) |
| TPP | Trans-Pacific Partnership |
| TPR | Trade Policy Review (WTO) |
| TRIMs | (Agreement on) Trade-Related Investment Measures (WTO) |
| TRIPS | (Agreement on) Trade-Related Aspects of Intellectual Property Rights (WTO) |
| TTIP | Transatlantic Trade and Investment Partnership |
| UNCED | United Nations Conference on Environment and Development |
| UNCITRAL | United Nations Commission on International Trade Law |
| UNCLOS | United Nations Convention on the Law of the Sea |

| | |
|---|---|
| UNEP | United Nations Environment Programme |
| UNFCCC | United Nations Framework Convention on Climate Change |
| UNIDO | United Nations Industrial Development Organisation |
| USITC | United States International Trade Commission |
| WBCSD | World Business Council for Sustainable Development |
| WEF | World Economic Forum |
| WIPO | World Intellectual Property Organization |
| WRI | World Resources Institute |
| WTI | World Trade Institute (Berne) |
| WTO | World Trade Organization |

# Foreword: Trade Policy Can Help Combat Climate Change

*H. E. Mogens Jensen, Minister for Trade and Development Cooperation, Denmark*

Climate change is one of the most pressing problems we face today. Trade policy is a crucial, but often overlooked, tool that can help us address this challenge. A key factor in a transition to sustainable energy alternatives, such as wind turbines and solar panels, is our ability to trade these technologies. Through a liberalisation of green trade, we can make it easier and cheaper to export and import sustainable energy products and services. This can benefit the climate and the environment and also create growth and employment.

Promoting green trade ought to be a main element in discussions on sustainable energy. The potential gains from trade in sustainable energy products and services are substantial but remain unharvested. This is in large part due to the high number of trade barriers that hinder the free flow of sustainable energy solutions across borders. The barriers stem from a 'governance gap' in the global architecture of green trade. Due to this, we are far from having an integrated, international approach to green trade. We need to find a framework to get rid of these barriers that stymie our efforts to achieve progress in this area.

Denmark plays a leading role in the EU in advancing green trade liberalisation. During the Danish EU Presidency in 2012, the issue was reintroduced on to the EU's agenda for the first time since COP15 in Copenhagen in December 2009.

The Danish government has also worked for green trade through the Global Green Growth Forum (3GF), a Danish government initiative to promote public–private cooperation to scale-up green solutions. 3GF has played a key role in supporting the development of the Sustainable Energy Trade Initiative (SETI) and consequently the SETI alliance, hosting meetings and facilitating the network to bring together public and private players. The alliance advocates negotiations to reduce tariffs and non-tariff barriers (NTBs) on sustainable energy products and services.

The good news is that we currently see an increasing international focus on green trade liberalisation. The historical agreement reached at the 9th World Trade Organization (WTO) Ministerial in Bali in December 2013 has given us the possibility to take a fresh look at the WTO, including the possibility to start up initiatives on green trade liberalisation.

This has resulted in the ground-breaking statement made by a number of WTO countries in Davos in January 2014 to pledge to launch negotiations in the WTO on liberalising trade in green goods. This agreement represents an important step forward in the efforts to promote trade in sustainable energy products. The negotiations will build upon the Asia-Pacific Economic Cooperation (APEC) list of green goods agreed on by the APEC countries in September 2012. The broad group of Pacific countries decided to reduce tariffs to 5 per cent or less on fifty-four climate and environmentally friendly products by the end of 2015.

It is now important that the Davos statement results in negotiations on an agreement on green trade liberalisation with a critical mass of participating countries. In the longer run we also, however, need to aim at an even more ambitious agreement. The goal should be a comprehensive international agreement on green trade that goes beyond the APEC Agreement. It should, if possible, also include services. Several options for such a comprehensive international agreement on green trade in goods and services are suggested in this book.

This book seeks to address existing misconceptions about green trade liberalisation, while sparking discussion and debate within the research and political communities. Hopefully, it will contribute to a much-needed impetus to agree on an ambitious international agreement on sustainable energy trade, to the benefit of our climate, environment, economic growth and development, energy security and public health.

# Acknowledgements

This book is the result of collaboration between the three institutes: the Global Green Growth Institute (GGGI), the Peterson Institute for International Economics (PIIE) and the International Centre for Trade and Sustainable Development (ICTSD). It is part of a broader project that has also involved a series of policy dialogues in Washington, Seoul and Geneva, as well as other locations.

The book has been possible thanks to the valuable contributions made by the authors of the respective chapters, as well as by reviewers; by experts, policy makers and private sector participants; by programme and editorial staff in the three institutes, as well as by the institutes' respective funders.

In particular, we would like to acknowledge Michael Liebreich and James Bacchus, both of whom have been great sources of inspiration throughout the work. We also wish to recognise the former Danish Minister of Trade, Ms Pia Olsen Dyhr, for her commitment and her determination to take the sustainable energy trade agreement (SETA) to the next level.

We want to take this opportunity to thank all the reviewers: Gabriel Barta; Chris Beaton; Francisco Boshell; Helen de Coninck; Mireille Cossy; Thomas Cottier; Devin McDaniels; Jane Drake-Brockman; Muge Dolun; Gary Horlick; Veena Jha; Peter Kleen; Markus Krajewski; Jan-Christoph Kuntze; Tom Moerenhout; Joost Pauwelyn; Amy Porges; Luca Rubini; Susan Shafi-Brown; Pierre Sauvé; Mike Sitkowski and Manuel Teehankee.

In addition, we would like to thank Ahmed Abdel Latif, Christophe Bellmann, Johannes Bernabe, Ingrid Jegou, Jisun Kim, Joachim Monkelbaan, Julia Muir, Pedro Roffe, Malena Sell, Sherry M. Stephenson, Mahesh Sugathan and Marie Wilke for their reviews, comments and guidance in producing the chapters. Anja Halle has provided valuable editorial support.

# 1 Introduction: Setting the Horse Before the Cart to Preserve a Viable World

*Gary C. Hufbauer, Ricardo Meléndez-Ortiz and Richard Samans*

## 1   The Key Message

2014 has been confirmed as the warmest year on record[1] whilst in its Fifth Assessment Report (AR5) the UN Intergovernmental Panel on Climate Change (IPCC) – the world's largest body of physical and social scientists dedicated to understanding the phenomenon – confirmed the unequivocal warming of the planet beyond its sustainability, and as a consequence of human activity. The main source of such warming – up to 80 per cent by some accounts – is emissions of so-called greenhouse gases (GHGs), most of which happen in the form of carbon dioxide ($CO_2$), stemming from energy use and supply. As a response, the international community has agreed to introduce remedial action and change course such that warming is kept at a safe 2° above normal. The twenty-first-century economy is energised to a very large extent by fossil fuels. But for the climate stabilisation goal to be reached, one-third of the world's oil, 50 per cent of its natural gas and up to 80 per cent of its coal reserves should remain *unburned*.[2] A dramatic scale-up of alternative sources of energy, as well as of massive energy efficient gains, is then imperative and most urgent. Towards resolving this predicament, the good news is that the low-carbon or carbon-neutral technologies currently on offer include a wide choice of mature innovations from

---

Special thanks are due to Ingrid Jegou (ICTSD) for her overall management of this project, as well as her critical editorial contributions to this chapter and the whole book. The introductory chapter is also built on substantive contributions from Jisun Kim (formerly at PIIE) and Mahesh Sugathan (ICTSD), and reflect discussions over various years of cooperation between the authors and their colleagues.

[1] US National Oceanographic and Atmospheric Administration (NOAA), National Climatic Data Center, Global Analysis, Annual 2014, www.ncdc.noaa.gov/sotc/global/2014/13, accessed 2 February 2015.

[2] Mark Carney, Governor, Bank of England, quoted in 'Mark Carney: most fossil fuel reserves can't be burned', *Guardian*, 13 October 2014, www.theguardian.com/environment/2014/oct/13/mark-carney-fossil-fuel-reserves-burned-carbon-bubble. See also McGlade and Ekins (2015).

solar photovoltaic (PV) to geothermal and biomass, at commercially competitive terms or in grid parity vis-à-vis fossil fuels.[3] The bad news is that the policy frameworks that would enable such an effort in a globalised economy characterised by international fragmentation of production and services are lagging behind, and in many instances run against this purpose. This book is a proposal to set things right, through international cooperation, and establish rules and conditions for international trade and investment to support the fundamental transformation in energy markets so vitally called for. In doing so, the world will also avoid the waste of time and resources involved in the current legal, economic and political quarrelling around the evolution of the clean energy technologies sector[4].

## 2   The Context

Multiple dangers cloud the global outlook in this century's second decade, including slow growth, persistent and new geopolitical tensions and overwhelming evidence and consequences of climate change. Amidst these troubles, innovative policies that align environmental objectives with economic development and energy security hold tremendous appeal. A scale-up in the use of renewable energy (RE) is right up that alley. Progress is being made, but not at the speed and magnitude required. Stable sources of finance and sustained investment, on the one hand, and enabling, predictable, rules-based policy frameworks, on the other, are essential. The latter include the use of trade and investment policy to proactively stimulate the growth of the clean energy sector, as well as ensuring the rollback of trade barriers increasingly being erected, targeting the sector.

Many countries have embarked on proactive policies to expand the deployment of renewable and sustainable energy. According to REN21, by early 2013, energy RE support policies had been identified in 127 countries. The share of RE in the global power generation mix stood at 22 per cent in 2013 (8.5 per cent excluding large-hydro)[5] and is expected to account for nearly half of the global increase in power generation to

---

[3] The Investment bank Deutsche Bank has suggested that solar power systems will be at grid parity in up to 80 per cent of global markets by the end of 2017, see 'Solar at grid parity in most of the world by 2017', *REneweconomy*, 12 January 2015, http://reneweconomy.com.au/2015/solar-grid-parity-world-2017.

[4] A. Gosh and R. Meléndez-Ortiz, 'Want clean energy? Avoid trade disputes', *Business Standard*, April 14, 2013, www.business-standard.com/article/opinion/want-clean-energy-avoid-trade-disputes-113041500023_1.html.

[5] Renewable Energy Policy and Network for the 21st Century (REN21), *Renewables 2014: Global Status Report*, Paris, June 2014.

2040, overtaking coal as the leading source of electricity.[6] In countries like Germany, the share of renewable sources in the electricity mix grew from about 9.4 per cent in 2004[7] to 31 per cent during the first half of 2014 (27 per cent excluding hydro).[8] Global investment in renewables soared to US$ 310 billion in 2014.[9]

Yet, efforts to scale-up sustainable energy more rapidly and at the scale needed continue to be constrained by insufficiency and inadequacy in current international regulatory frameworks as well as new and unstable, competing and conflicting policies and measures introduced all over the planet, at national and sub-national levels, in an uncoordinated fashion among countries. In addition, countries, companies and communities with a stake in fossil fuels are in no hurry to change their ways. Indicative of this is how, despite falling prices, cost differentials between sustainable energy sources and fossil fuels continue to be exacerbated by massive subsidies to fossil-fuel use.[10]

In the twenty-first-century global economy, RE technologies are made up of packages of goods, services and embedded intangibles (such as software) that come together as a result of multiple transactions involving the providers of supply chains operating across several jurisdictions. In this manner, the goods, services and intellectual property (IP) involved in a wind energy park or a solar PV installation have usually crossed several borders, some times more than once. Any such equipment would generally also include locally produced components and services. An efficient functioning international market is essential for these value chains to deliver the final technologies.

Persistent barriers to global commerce, possible incentives and innovative policies to foster more rapid growth in RE, as well as the current rules governing such commerce, are the focus of this book. This includes – on the impediments side – tariffs (though seldom the most significant),

---

[6] International Energy Agency (IEA), *World Energy Outlook 2014*, Paris, November 2014 and 'International Energy Agency global outlook report points to continued strength of fossil fuels while low emissions nuclear and renewable energy grows', *ABC Rural*, 13 November 2014, www.abc.net.au/news/2014-11-14/international-energy-agency-global-outlook-2014/5889474.

[7] European Commission EuroStat, 'Energy from renewable sources,' http://ec.europa.eu/eurostat/statistics-explained/index.php/File:T_RENEWABLES_RES_E_2012.png.

[8] 'German power sector 27 percent non-hydro renewable in 2014', *Renewables International*, 3 July 2014, www.renewablesinternational.net/german-power-sector-27-percent-non-hydro-renewable-in-2014/150/537/80072/.

[9] Bloomberg New Energy Finance.

[10] According to the IEA *World Energy Outlook 2014*, global fossil fuel subsidies in 2013 amounted to US$ 548 billion. This figure does not reflect the costly impact of $CO_2$ emissions on global warming. It also does not reflect subsidy reform sparked by the dramatic fall in oil prices in the second half of 2014.

restrictive technical standards, local content requirements (LCRs), other so-called 'buy national' policies and restrictions on the many services necessary to deploy clean technologies and deliver clean energy. On the positive side, the growth of clean energy has been driven to large extent by policy. National governments and regions as well as local authorities have sponsored research, subsidised the development of new technologies and supported producers as well as consumers through loans, guarantees, feed-in tariffs (FITs) for electricity, tax breaks and a range of other tools and instruments. Ensuring that all such measures do not work against each other and rather support the delivery of social, private and public goods at the pace and scale required is essential for the survival of the planet.

The rationale for reducing barriers to sustainable energy goods and services (SEGS), and engendering non-trade distortive policy incentives, is straightforward and strong. Prices for SEGS would fall sharply, facilitating faster deployment at lower cost and making a significant contribution to the control of $CO_2$ emissions. Moreover, reducing the dependence on fluctuating oil and coal prices would greatly benefit energy security in developed and developing economies, particularly those that are net importers of fossil fuels. Net coal import dependency in India, for instance, rose from practically nothing in 1990 to nearly 23 per cent in 2012.[11] In comparison to fossil fuels that have exhibited a high degree of volatility, electricity prices generated from renewable sources, particularly solar PV, have shown a declining trend.

A shift to clean energy is also associated with multiple, additional sustainable development gains. Importantly, clean energy offers interesting opportunities for enhancing access to modern forms of energy, in particular in developing countries. This is associated with multiplier effects that it would be foolish not to recognise: with access to electricity, as opposed to traditional forms of heating and lighting such as wood, economic activity can be carried out once the sun has set. Children can study at night, and women will no longer have to spend hours collecting lumber and can use their time more productively. Air pollution would be dramatically reduced, including in homes as households could shift from burning wood or kerosene, with important health improvements as an immediate consequence.

---

[11] US Energy Information Administration (EIA), 'India is increasingly dependent on imported fossil fuels as demand continues to rise', August 14, 2014, www.eia.gov/todayinenergy/detail.cfm?id=17551.

## 3 Origins of the Sustainable Energy Trade Agreement (SETA) Concept

In 2011, three institutes, the Global Green Growth Institute (GGGI), the International Centre for Trade and Sustainable Development (ICTSD) and the Peterson Institute for International Economics (PIIE), entered into a research partnership for the purpose of examining salient trade policy issues and options related to the potential negotiation of a framework of rules and mutual obligations on trade and investment policies and an eventual agreement on the technologies, goods and services required for sustainable energy. This initiative evolved out of the fusion of two streams of work: on the one hand, ICTSD's undertaking since 2008 to generate a model sustainable energy agreement to be discussed and presented in the contexts of the WTO and the UN Framework Convention on Climate Change (UNFCCC) and, on the other, a proposal for a Sustainable Energy Free Trade Area (SEFTA) led by Michael Liebreich of Bloomberg New Energy Finance and made by the World Economic Forum (WEF) Global Agenda Council on Sustainable Energy, as part of the Forum's Global Redesign Initiative in 2010.[12] The result, a Sustainable Energy Trade Agreement (SETA), was first outlined in the ICTSD research paper, *Fostering Low Carbon Growth: The Case of a Sustainable Energy Trade Agreement*.[13] The initiative to promote a SETA was formally launched at, and underwritten by, the inaugural meeting of the Global Green Growth Forum (3GF), convened by the Prime Minister of Denmark, and the governments of Mexico and the Republic of Korea in October 2011.[14]

In this book, the SETA addresses a number of critical policy issues and tools, including the elimination of tariffs, the disciplining of non-tariff barriers (NTBs) and alignment and coherence concerning standards with respect to certain RE goods and services, as well as the documentation and eventual elimination of fossil-fuel subsidies. Its presumed modality would be a negotiation driven by a like-minded global coalition of countries, similar to the way that other successful endeavours have been initiated in the past, including the way in which the Information Technology Agreement (ITA) was achieved in the 1990s. The ITA, which recognised the urgency to act on a critical technology, was steered and put together by a geographically diverse group of twenty-nine countries and customs

---

[12] See 'Sustainable energy free trade area: global agenda council on sustainable energy', in Samans, Schwab and Malloch-Brown (2010).

[13] ICTSD, *Fostering Low Carbon Growth: The Case for a Sustainable Energy Trade Agreement*, International Centre for Trade and Sustainable Development, Geneva, 2011, 2011/12/fostering-low-carbon-growth-the-case-for-a-sustainable-energy-trade-agreement.pdf.

[14] 3GF, http://3gf.dk/en/issues/trade/ and 3GF, *Global Green Growth Forum, Report*, Ministry of Foreign Affairs of Denmark, Copenhagen, 2011.

unions at the WTO Singapore Ministerial Conference in December 1996. The number of participants has since grown to seventy, representing about 97 per cent of world trade in the information technology (IT) products that underpin the digitalisation of today's economy.

The SETA aims to bring countries together for the imperative of a fast and massive scale-up of innovation, deployment and use of the critical RE sources required to confront the challenges of mitigation of GHGs, energy security and energy access. It would provide a global framework for production and trade in the technologies required for sustainable energy, by ensuring robust markets for SEGS, and introducing international governance where particular gaps exist.

## 4   International Progress Towards the SETA Concept

### The Momentum for a SETA

Since 2010, interest in a SETA has grown, spawning extensive discussion among various stakeholders including governments, academia, private firms and international organisations. Dialogue has led to the endorsement or inclusion of the SETA proposal at several high-level fora. In addition to an incremental understanding of the role of energy in climate change, and the ensuing urgency of reform of energy systems, in the context of the UNFCCC, this has also been facilitated by the global priority given to issues related to access to energy and energy security, as expressed by the UN's Sustainable Energy for All (SE4All) initiative and the 2012 Rio + 20 Conference, among others.

At the same time, recent trade tensions have contributed to creating a momentum for trade initiatives in the area of RE. The limits of existing trade rules are being tested as countries draft policies to promote the shift to a cleaner energy mix, often at the same time to trying to stimulate domestic economies and create jobs. Over the past few years this has led to an increase in energy-related spats among countries trading along the various supply chains of RE. An increasing number of these differences has been brought to the dispute settlement system of the WTO, notably regarding trade-distorting LCRs, as well as cases of alleged subsidisation and dumping of RE products.

### Broad Support from Industry, Civil Society and Intergovernmental Organisations

Seeking to improve market access and transparency, as well as to clarify rules, the private sector has voiced its enthusiasm for the SETA's potential to scale-up the production and application of sustainable energy. At the Business 20 (B20) Summit held in Mexico in June 2012, the Green Growth Task Force called for freer trade in green goods and services and endorsed the SETA as a meaningful vehicle to combat climate change while promoting economic growth and better access to energy.[15] As the Task Force explicitly acknowledged, 'the Sustainable Energy Trade Agreement (SETA), launched by a group of NGOs and research institutes and supported by a number of major corporations, shares many of these objectives'. This was followed by the launch of the 'Green Growth Action Alliance' (G2A2), a public–private partnership (PPP) with the objective of leveraging investment in green infrastructure projects, hosted by the World Economic Forum (WEF). The G2A2 put in place a working group on trade, which explicitly promoted the SETA.

Another example is the 'Friends of Rio + 20', an extraordinary coalition of business, scientists and civil society, which suggested practical actions to meet pressing sustainable development challenges. Among the coalition's recommendations, the SETA is highlighted as a significant multi-stakeholder collaboration that is capable of creating impact at scale.

More recently, the Alliance of the Sustainable Energy Trade Initiative (SETI Alliance) was launched in late 2012 during the Conference of the Parties of the UNFCCC in Doha. The Alliance is a PPP and a stakeholder support mechanism that works towards supporting and developing initiatives to promote a scale-up of innovation, production and deployment of SEGS through trade. Its wider aim is to contribute to climate change mitigation, to improve access to energy and enhance energy security. Among its members are companies from the wind and solar industries, as well as like-minded governments.

Among intergovernmental organisations, the UN Economic Commission for Latin America and the Caribbean (ECLAC), the Inter-American Development Bank (IADB), the United Nations Industrial Development Organization (UNIDO) and the International Renewable Energy Agency (IRENA) have all expressed interest in exploring the analytical concept of a SETA and its further dissemination.

Several private sector firms and research institutions such as Trina Solar, Vestas Wind Systems A/S, Siemens AG, Yingli Solar, Suzlon, Canadian Solar, Hanwha Solar, Grundfos, Danfoss, Solar Clarity, the Chinese, Danish and European Wind Energy Associations, the Organisation for Economic Cooperation and Development (OECD),

---

[15] The Business 20 Summit (B20) is an international forum that aims to foster dialogue between governments and the global business community. The B20's main objective is to provide leaders of the G20 with meaningful recommendations from the private sector.

the Governments of Denmark, Norway and the United Kingdom, Bloomberg New Energy Finance; the WEF, the World Resources Institute (WRI) and the MIT Energy Initiative are among members or key supporters of the initiative.

### Governments are Taking Action on Trade Initiatives on Sustainable Energy

On the government side, notable progress has also been made on trade initiatives in the area of sustainable energy. Indeed, a few countries, notably in Europe and Latin America, as well as China, have expressed their explicit support for a SETA.

In 2012 the European Commission Trade Policy Committee and Foreign Affairs Council (Trade) adopted an options paper on 'Trade and green growth', which acknowledges SETA's value from a commercial, environmental and climate viewpoint. Subsequently, in 2013, EU Member States agreed in the European Council that '[f]urther progress is required towards liberalisation of trade in environmental goods and services as a positive contribution to moving towards a resource-efficient, greener and more competitive economy'.

China has also expressed concern over the inadequacy of existing frameworks and ensuing tensions, and has called for dialogue on sustainable energy trade. For example, in the context of the 2012 and 2013 3GF meetings in Copenhagen, the Chinese Vice Minister of the National Energy Administration, Liu Qi, called for trade initiatives in the field of RE.

In June 2013, US President Barack Obama stated in his Climate Action Plan that:

> The US will work with trading partners to launch negotiations at the World Trade Organization towards global free trade in environmental goods, including clean energy technologies such as solar, wind, hydro and geothermal. The US will build on the consensus it recently forged among the 21 Asia-Pacific Economic Cooperation (APEC) economies in this area. In 2011, APEC economies agreed to reduce tariffs to 5 per cent or less by 2015 on a negotiated list of 54 environmental goods. The APEC list will serve as a foundation for a global agreement in the WTO, with participating countries expanding the scope by adding products of interest. Over the next year, we will work toward securing participation of countries which account for 90 per cent of global trade in environmental goods, representing roughly $481 billion in annual environmental goods trade. We will also work in the Trade in Services Agreement negotiations towards achieving free trade in environmental services.

This statement appears to refer to a plurilateral approach similar to the SETA.

*Intergovernmental Progress*

An increasing number of positive developments at intergovernmental level towards a framework for trade in SEGS have followed this growing multi-stakeholder support.

Most notable is the commitment by the Asia-Pacific Economic Cooperation (APEC) to define a list of environmental goods (EGs) and to address related applied tariffs as well as NTBs. The APEC leaders agreed during their Honolulu summit in November 2011 to develop a common list of EG during 2012. Eventually, in Vladivostok in September 2012, they managed to define such a list, comprising fifty-four EGs. They also committed to reducing applied tariffs to 5 per cent or less by 2015, and to subsequently eliminating NTBs, including LCRs that distort environmental goods and services trade.[16] The agreement is ground-breaking in the field of *environmental* goods negotiations, particularly in contrast to over ten years of acrimonious talks under the Doha Development Round at the WTO. It also prompted many, from numerous advocates in think tanks and civil society to clean tech firms to the Director General of the WTO and President Barack Obama, to propose to build upon the Agreement and to take it to the WTO.

Following the APEC Agreement, discussions took place in Geneva among a group of countries self-defined as *friends* of environmental goods and services (EGS) on how to build upon it. Energy-related goods are at the heart of those discussions. At the time of writing this introduction, at the end of 2014, this has resulted in concrete moves. A group of forty-one WTO members[17], including the twenty-eight European Community countries, made a joint statement in Davos during the WEF Annual Meeting in January 2014 to announce their commitment to achieve global free trade in environmental goods and pledge to work together, and with other WTO Members, to prepare for negotiations in the area. Although not specific to sustainable energy, the APEC list of goods that would serve as a basis for this process includes many of the key goods in that area.[18] In July 2014 this same group of countries formally launched negotiations towards an Environmental Goods Agreement (EGA) 'within the WTO', seeking an ambitious outcome to be implemented on a 'most favoured nation' (MFN) principle[19].

---

[16] The full text of the APEC 2011 Leaders' Declaration is available on the APEC website.
[17] Australia; Canada; China; Costa Rica; the EU; Hong Kong, China; Japan; Korea; New Zealand; Norway; Singapore; Switzerland; Chinese Taipei; and the United States.
[18] '"Green goods" trade initiative kicks off in Davos', ICTSD, *Bridges Trade BioRes*, 28 January 2014, www.ictsd.org/bridges-news/biores/news/%E2%80%9Cgreen-goods%E2%80%9D-trade-initiative-kicks-off-in-davos.
[19] '"Green goods" trade talks kick off in Geneva', ICTSD, *Bridges Trade BioRes*, 10 July 2014, www.ictsd.org/bridges-news/bridges/news/green-goods-trade-talks-kick-off-in-geneva.

The negotiations, of a plurilateral nature (involving only a selection of WTO Members), are set from the outset to treat other WTO Members in a non-discriminatory manner, such that ensuing benefits in terms of tariff liberalisation (for instance) will automatically accrue to all.

### The EGA Prospects of Advancing a SETA

ICTSD and its partners continue to advocate a result along the lines of the proposed SETA, and for an early harvest on energy to be linked to the climate change talks under the UN at its summit in Paris in December 2015. For the moment, the EGA talks have focused on reduction of tariffs at the border; however, most players agree that the inclusion of NTBs, and other aspects of trade policy, is desirable. Moreover, this emerging understanding would also include the notion advanced in the SETA concept that trade in services would be crucial to make the resulting more liberal frameworks effective in expanding the markets for clean energy technologies. An expansion of the negotiations with respect to thematic scope as well as to membership requires a better understanding of incentives for current EGA non-Members to join: both thematic scope and such incentives go hand in hand. Many of the current non-Members are rapidly developing policies to scale-up RE in their energy supply mix, and doing so successfully. It would make no sense for them to remain outside an enhanced EGA or an eventual purposeful SETA. Likewise, other negotiations at the regional and mega-regional level are now including provisions in a SETA-like direction. All these efforts will probably end up in convergence in a few years. The sooner they do, the better the world will be.

In most recent papers under this initiative we have argued along the following lines for the possibilities of the EGA to effectively contribute to the SETA objectives:

The APEC list provides a reasonably good coverage of certain clean energy supply products particularly in the solar PV and wind-power sectors. The list also includes products that may contribute to enhancing access to clean energy, for example small-hydro, ocean, geothermal and biomass gas turbine generating sets. On the other hand, some clean energy sectors are not included. For example, equipment used in hydropower applications does not make the cut for the APEC list. While the list does include both clean energy equipment and parts – which may be useful for a value-chain approach to reducing costs – certain segments of value chains are missing. For example a range of downstream components used in solar PV systems, such as solar inverters, are also not included, perhaps because the relevant [Harmonized System] HS sub-heading includes products that are principally applied for other uses.

Drawing on earlier submissions made in the WTO and those identified in work carried out under the SETA initiative by ICTSD and its partners, additional products and components relevant for renewable energy and access to clean energy could also be considered. These include, for example, Fresnel mirrors and reflector modules used in concentrated solar power (CSP) applications, heat pumps, as well as parts and components used in RE supply technologies. Some of these products are multiple-use with relatively large values of trade, such as switchboard and control panels, gearboxes, and ball bearings used in RE installations. Some other papers [included in this book] have also highlighted products that contribute to improved access to clean energy, in particular, off-grid solar appliances. Opportunities for including such products in the EGA could also be explored although lack of data may be a problem. For example, off-grid markets and trade flows for solar home-systems, mini-grids, solar pumps, solar cooking stoves, and solar lighting appliances are difficult to trace. Certain products that may be required in off-grid solar applications such as batteries, charge controllers, and energy converters have been included in earlier submissions to the WTO, but not in the APEC list, and they could be part of possible additions for the EGA.

The selection of additional products for the EGA will likely face similar challenges as those detailed in the APEC list. Some criteria may therefore be developed to guide the possible inclusion of additional products in the EGA, which should be driven primarily by environmental considerations. Possible impacts of tariff elimination and practical factors, such as the ease of implementing tariff cuts taking into account HS classifications and existing national tariff schedules, including the costs and benefits of creating new tariff lines, might also be considered.

Moving forward there is also a need to explore ways to arrive at a more accurate picture of trade in more narrowly defined EGs. Analysis of available information on trade in national tariff lines of key trading partners may give some insights around how to interpret trade flows estimated at the level of certain HS subheadings that include unrelated products. Additional indicators and business surveys of markets for environmental goods and services could be useful for that purpose.[20]

## 5 The Chapters in this Book

This book represents a collaborative effort by the GGGI, the ICTSD and the PIIE to examine more closely the trade policy issues presented by the potential plurilateral negotiation of a SETA. We have invited a number of leading experts to investigate many of the key design considerations for a SETA in order to facilitate a more informed debate among interested parties. This book contains eight chapters that explore the major issues surrounding the SETA proposal. The chapters are based on papers prepared between 2011 and 2014. In addition, the volume contains a foreword

---

[20] R. Vossenaar, 'Securing climate benefits in the Environmental Goods Agreement', ICTSD, *Bridges Trade BioRes*, 8 (10), December 2014, www.ictsd.org/bridges-news/biores/news/securing-climate-benefits%E2%80%A8-in-the-environmental-goods-agreement.

by H. E. Mogens Jensen, Minister for Trade and Development Cooperation of Denmark, responsible for trade in 2013–2014.

In Chapter 2, 'Issues and Considerations for Negotiating a Sustainable Energy Trade Agreement', by Gary C. Hufbauer and Jisun Kim, after an overview of issues that need to be addressed in the SETI discussions, the authors focus on tariff barriers. They examine in detail the tariff rates and trade flows for thirty-nine narrowly defined sustainable energy goods, both for the world as a whole and for selected countries. They then survey several non-tariff measures by way of introduction to the deeper analysis in subsequent chapters. Finally, they sketch key issues relating to negotiating approaches and the institutional framework.

While tariff rates range widely among candidate countries for launching a SETA, tariffs are not the biggest impediment to trade in SETA products. In fact, many EGs that already enjoy low tariff rates still face severe NTBs, such as restrictive standards, LCRs and preferential subsidies to national firms (these NTBs are usually associated with government procurement or state-owned enterprises, SOEs).

The authors suggest that tariffs should be eliminated first. After that, NTBs should be phased out over a period of five or ten years in the following order of priority: LCRs, services, investment, subsidies and standards. In their view, the plausible model for a SETA is the WTO Government Procurement Agreement (GPA), not the Information Technology Agreement (ITA), since benefits under the former would be extended on a conditional MFN basis, but the agreement would be open to all WTO Members that subscribe to the obligations.

While sustainable energy services are a key component of a SETA, their inclusion in any agreement faces several challenges. In Chapter 3, 'Trade in Sustainable Energy Services', by Joachim Monkelbaan, the author addresses these challenges by identifying 'core' services that are directly linked to the diffusion of sustainable energy technologies. He analyses the specific commitments made by major trading nations, and goes on to identify the key barriers that might be addressed in a SETA.

With respect to modes of services supply, the author focuses on investment restrictions that hinder 'commercial presence' (Mode 3) and impediments to the temporary movement of service providers that restrict the 'movement of natural persons' (Mode 4). Government procurement regulations have a significant impact on services related to sustainable energy, as the public sector is often the largest buyer. The author also identifies barriers affecting inputs, market structure, standards, transparency and administrative procedures.

By examining lessons from past and on-going negotiations, including the General Agreement on Trade in Service (GATS), preferential trade agreements (PTAs) and the Trade in Services Agreement (TiSA), the author suggests ways in which these key issues might be handled in a SETA. He also explores how the GATS Understanding on Financial Services, commitments made during the Doha Round and the ideas tabled in the TiSA talks, could feed into a SETA.

In Chapter 4, 'Governing Clean Energy Subsidies: What, Why, and How Legal?', by Arunabha Ghosh with Himani Gangania, after noting that clean energy subsidies are complex instruments that reflect the multiple motivations of governments, industry and NGOs, the authors examine the debate on subsidies beyond the realm of legal interpretation (principally WTO law). They provide a framework to classify supporters and opponents, depending on the rationales offered to defend the subsidies. They also develop a typology that distinguishes between financial transfers, taxes, regulations, infrastructure and trade rules. The authors argue that legal analysis alone is not sufficient to deliver policy clarity or to achieve the principal goal of subsidies, namely attracting more investment to the production of sustainable energy.

Five areas concerning the governance of clean energy subsidies are identified for improvement at the national and international levels. First, common metrics are needed to count subsidies and to increase the transparency of public policy. Second, the importance of rationalising fossil-fuel subsidy programmes as a precursor to promoting clean energy sources needs to be spelled out. Third, greater policy clarity is needed in setting out the purpose of government support. Fourth, independent assessments of the alleged adverse impacts of subsidy policies could help reduce the threat of unilateral trade penalties. And finally, fifth, international institutions with rules governing trade, energy flows and climate change need greater coordination.

Many governments have used procurement policies to promote domestic sustainable energy industries. In Chapter 5, 'Trade Law Implications of Procurement Practices in Sustainable Energy Goods and Services', by Alan Hervé and David Luff, the authors explore the distortions to trade in SEGS resulting from public procurement policies, and how to address such distortions within a SETA. They survey existing international regulations on public procurement and assess the compatibility of national practices with existing international instruments, particularly the WTO GPA.

The authors stress that a SETA would actively promote SEGS in public procurement, while ensuring that national policies will not simply give preferences to domestic suppliers. They argue that inclusion of the SETA within the WTO may be the best option to avoid 'forum shopping' among agreements and avoid any reluctance on the part of WTO panels and the Appellate Body to refer to the SETA. With regard to the procurement-related provisions of the SETA, the authors suggest that the provisions might vary in terms of enforceability. For instance, soft-law provisions might adequately address SEGS-related requirements and the exchange of information on best practices between SETA parties. Standards and labels related to SEGS in the technical specifications for public procurement awards could be gradually harmonised in the same way.

Standards can affect international trade in renewable energy equipment and consequently the development of the industry. They are among the key non-tariff measures that will need to be addressed as part of any sustainable energy trade initiative or agreement.

Given that solar PV equipment is the one of the most intensely traded items of sustainable energy equipment, and also accounts for the highest value,[21] Chapter 6, 'Selling the Sun Safely and Effectively: Solar Photovoltaic Standards and the Implications for Trade Policy', by Sunny Rai and Tetyana Payasova, focuses on solar PV and presents a comprehensive overview of the mandatory technical regulations and voluntary standards for the technology in the main world markets. The authors further seek to explore which WTO disciplines are applicable to these technical regulations and standards and whether they can accommodate industry concerns related to their harmonisation.[22] Finally, the authors offer recommendations from the industry perspective, which are further reflected in the suggested legal solutions within the framework of a future SETA. These include, among others, a strengthening of disciplines for technical regulations adopted by non-governmental bodies, enhancing mutual recognition and transparency obligations, providing new disciplines for solar PV-related services, establishing a special information system for sustainable energy technical regulations and standards, including solar PV technical requirements, and creating a special subcommittee to coordinate the notifications related to such requirements.

Governments use green industrial policies such as LCRs to gain local benefits from increased RE deployment. Through LCRs, they aim to

---

[21] The solar PV market alone generated some US$ 60 billion in world trade in both imports and exports (excluding intra-EU trade) in 2011. World trade fell sharply in 2012, largely due to falling prices. See, UN Environment Programme (UNEP), *South-South Trade in Renewable Energy: A Trade Flow Analysis of Selected Environmental Goods*, 2014, www.unep.org/greeneconomy/Portals/88/documents/Report/South-South%20Trade_LOW-%20RES_26june.pdf.

[22] While the first six sections of the chapter refer to the term 'standard' in a general manner from the industry perspective, the seventh section employs trade terminology where the technical requirements depending on their nature (mandatory or voluntary) are subdivided into two main categories – mandatory technical regulations and voluntary standards. This distinction and the difference between legal disciplines applicable to both categories are further explained in the final section.

achieve policy objectives, such as economic growth and employment, despite uncertainties about the long-term benefits and the legality of such measures under WTO law. Chapter 7, 'Addressing Local Content Requirements in a Sustainable Energy Trade Agreement', by Sherry M. Stephenson, argues that LCRs clearly run counter to WTO rules, as has been confirmed by the Appellate Body in the *Ontario Case*, in which Japan and the EU complained about LCRs for RE equipment in the Canadian province of Ontario. Chapter 7 goes beyond earlier research, which only monitored the frequency and effectiveness of LCRs, by offering more pointed policy recommendations in the context of a SETA.

After providing an overview of LCRs in the RE sector, the author assesses the rationale and effectiveness of such measures, illustrated by two case studies. She analyses local content requirements in the context of a SETA and argues for alternatives to LCRs, as well as showing that a SETA could provide an opportunity to change current approaches to LCRs for SEGS in a non-discriminatory manner.

The diffusion of sustainable energy technologies can make substantial contributions to climate change mitigation as well as sustainable development. A key objective of a SETA would therefore be enhancing the diffusion of such technologies. Chapter 8, 'International Technology Diffusion in a Sustainable Energy Trade Agreement', by Thomas L. Brewer, presents several levels of analysis, including products, industries and countries, as well as companies' modes of technology transfer in international business.

International investment licensing and international services transactions are integral to technology diffusion processes, and therefore a SETA should address all types of barriers to these modes of diffusion. The author argues that a SETA should take energy efficiency technologies, government procurement, and standards and testing into account in order to encourage technology diffusion.

Next, the author advocates that the SETA agenda should not only be about trade liberalisation; it should also be about finding a balance between the roles of governments and markets, and the role of governments in fixing failures in those markets. A balance should also be found with regard to the role of government and subsidies in facilitating efficient transitions to low-carbon economies.

The political economy of the patterns of interests and influence in international negotiations for a SETA are changing as more emerging and developing countries are rapidly becoming significant exporters, as well as importers, of sustainable energy technologies. In order to facilitate the introduction of new and evolving technologies, the author pleads for a flexible SETA agenda that can be expanded to include new technologies. He concludes that we need a new international institutional architecture that enables the diffusion of sustainable energy technologies.

In Chapter 9, 'Legal Options for a Sustainable Energy Trade Agreement', by Matthew Kennedy, the author evaluates the legal options available to any group of interested governments that wish to conclude an agreement covering trade in SEGS. He also considers several collateral issues, such as negotiating procedures, pathways for accession and a potential SETA's relationship to existing WTO rules and dispute settlement.

Since existing WTO rules do not cover the energy sector in a systematic way, the author argues that a SETA could be a way to address a variety of trade barriers to sustainable energy imports. Three possible legal forms are explored: (1) an optional WTO agreement similar to the ITA; (2) an optional WTO agreement similar to the GPA; and (3) a stand-alone agreement outside the WTO, which might be incorporated into the WTO framework at a later date. After examining the pros and cons of each option, he concludes that an agreement within the WTO would be the first-best option. The choice between an ITA-type agreement and a GPA-type agreement depends on what the new agreement would contain. An ITA-type agreement would operate through modifications of WTO Members' goods and services schedules so that it can include market access improvements, and even new sets of rules through the incorporation of a so-called 'reference paper'. However, it is limited in its subject matter to the scope of the General Agreement on Tariffs and Trade 1994 (GATT 1994) and the General Agreement on Trade in Services (GATS). And, most importantly, it may only yield rights, not diminish obligations. A GPA-type agreement would be added to Annex 4 of the WTO Agreement and would not be subject to the same substantive limitations as long as it was a trade agreement. However, it would face a procedural hurdle, as agreements can be added to Annex 4 only by a decision of the Ministerial Conference (or the General Council conducting its functions between sessions) taken 'exclusively by consensus'. This means that any Member present at a meeting where the decision is proposed could, in theory, block it.

Alternatively, a SETA could be implemented outside the WTO framework. Implementation outside the WTO would be a second-best option, because the potential agreement would not benefit from the WTO institutional framework, particularly its dispute settlement system.

## 6    The Way Forward

At a time when the international community is searching actively for new ways to boost economic growth and combat global warming, negotiation of a SETA presents a practical opportunity to make progress on both fronts. The SETA concept takes its inspiration from the ITA, which was designed to accelerate and enhance the uptake of another class of goods and services considered to generate positive externalities. But while the ITA addressed only tariff barriers, some of the most important governmental impediments to cross-border trade and investment in RE are of the non-tariff variety, notably trade-distorting or environmental perverse subsidies, including to fossil fuel, as well as technical and investment performance standards.

This argues for a wider approach than that taken by the ITA. Yet, each additional component or chapter of a SETA has the potential to complicate further the political task of reaching agreement, even in the context of a negotiation involving a limited number of states. Thus, a modular approach, in which participating countries assume a core set of obligations while committing to make future, step-by-step progress toward certain others, may be most feasible.

For example, SETA parties might seek to achieve agreement on a core agreement consisting of:
(a) elimination of tariffs on a relatively narrow set of tariffs lines;
(b) mutual recognition of a specified set of product technical standards;
(c) agreement on a narrow list of permissible 'green box' clean energy subsidies;
(d) agreement on a range of transitory measures which would allow for a gradual phase-out of prohibited LCRs while avoiding disputes; and
(e) commitment to measure and disclose all production- and consumption-related fossil-fuel subsidies according to a common methodology.

At the same time, they might agree on a built-in forward work programme of consultations with the potential to yield future commitments among some or all parties on one or more of the following topics:
(a) time-bound commitments on the reduction of certain fossil-fuel subsidies;
(b) restrictions on certain sustainable energy subsidies;
(c) an expanded list of zero-tariff products and services;
(d) an expanded list of technical standards subject to mutual recognition;
(e) market access commitments with respect to certain sustainable energy services; and
(f) provisions to actively promote SEGS in public procurement.

The choice of legal modality will inevitably influence the level of substantive ambition and the number of initial signatories. The most promising option in our view would be an agreement that, like the ITA, takes the form of an MFN set of commitments within the WTO. This presumably will only be possible if the overwhelming majority of production were covered by the agreement, a task that Hufbauer and Kim in Chapter 2 suggest might be possible but is by no means assured. However, the launch of the EGA negotiations in Davos and Geneva in 2014, confirmed as a plurilateral, critical-mass approach, with MFN outcomes, and 'within the WTO', implies that this might be within reach after all. The second-best option would be a non-MFN agreement that nevertheless is sanctioned by the WTO either through the passage of a waiver (which requires unanimous consent from the WTO's Membership) or through recognition of the agreement as a 'multilateral environmental agreement' (MEA).

The best approach for navigating the complex substantive and procedural landscape of an eventual SETA will become clearer only when interested parties begin to engage in direct informal discussions, beyond the initial tariffs approach of the EGA. These consultations need not take place within the WTO, notwithstanding its obvious competence, or any particular regional forum, notwithstanding the encouraging progress made recently on EGs tariffs in APEC. These two fora would be natural spaces for the process to take root, but there may be just as strong a logic to beginning the informal discussions among a limited number of a geographically diverse but substantively like-minded group of countries on an ad hoc basis, or as an offshoot of the G20 as suggested by the original WEF SEFTA proposal, or even as an adjunct to an existing multilateral environmental framework such as the UNFCCC or Rio + 20 follow-up process. The developments in the negotiation of the EGA can be seen as one example along these lines, which should be built on to encompass the key components of a SETA.

On their part, a range of regional trade agreements (RTAs) is including chapters on sustainable energy. One notable example is the RTA between the EU and Singapore, initialled in September 2013. The agreement includes one full chapter on addressing Non-Tariff Barriers to Trade and Investment in Renewable Energy Generation, including rules to facilitate trade and investment in equipment to generate renewable energy such as mutual acceptance of technical standards and conformity assessment reports. Others, such as the Canada–Peru and the Canada–Chile free trade agreement (FTA), provide access to environmental and associated services providers such as engineering well beyond WTO commitments. There are also indications that the mega-regionals like the Trans-Atlantic Trade and Investment Partnership (TTIP), or the Trans-Pacific Partnership (TPP), will do the same.

One possible way to start the diplomatic ball rolling on the multilateral front would be for countries in the EGA or others in the WTO to convene an informal, multi-stakeholder symposium on the SETA concept, perhaps in partnership with another relevant intergovernmental organisation (IGO) such as the UNEP, the IEA, the Energy Charter or the GGGI, now an IGO pursuant to a treaty signed during the UN Rio + 20 conference in 2012. The WTO, in conjunction with other IGOs or global think tanks, might then consider initiating a Joint Study on the topic, which in turn might be tabled for discussion in the G20, the WTO Committee on Trade and Environment (CTE), or an energy intergovernmental process.

Another avenue has just been opened. During the run-up to the Paris Climate Summit, the UNFCCC COP21, the world saw a hybrid agreement evolving through two parallel but connected tracks. The first was a top-down, rather basic and soft agreement, consisting of government pledges of contributions towards emissions abatement, and including adaptation enabling mechanisms. The second consisted of a set of PPP collaborations and sub-national initiatives organised on a platform and geared towards engaging a broader set of actors in action.[23] A brilliant political opportunity has thus opened up for governments to move boldly on the ideas presented in this book. Embracing the notion of more efficient, robust markets for clean energy technologies would underpin a more ambitious scale-up of further development, diffusion and deployment of these sources of energy, and make a critical contribution to the efforts to curb climate change. The ball is in the court of political leaders to seize the moment.

There are many possible pathways for progress on the SETA concept, as explored in the book you're about to read. Wider discussion based on more complete information is a prerequisite for a positive outcome, whichever path gets chosen. The chapters in this book have been contributed in that spirit.

**References**

McGlade, C. and Paul Ekins (2015) 'The geographical distribution of fossil fuels unused when limiting global warming to 2°C', *Nature*, 517, 8 January: 187–190

Samans, Richard, Klaus Schwab and Mark Malloch-Brown (eds.) (2010) *Global Redesign: Strong International Cooperation in a More Independent World*. World Economic Forum, Geneva

---

[23] *Lima–Paris Action Agenda Statement, Twentieth Conference of the Parties (COP20) to the UN Framework Convention on Climate Change* (UNFCCC), 14 January 2015, www.cop20.pe/en/18732/comunicado-sobre-la-agenda-de-accion-lima-paris/.

## 2 Issues and Considerations for Negotiating a Sustainable Energy Trade Agreement: May 2012

*Gary C. Hufbauer and Jisun Kim*

In 2001, recognising the importance of better market access for environmental goods and services (EGS), the original Doha Ministerial Declaration called for negotiations on 'the reduction or, as appropriate, elimination of tariff and non-tariff barriers to environmental goods and services'.[1] Over the past decade, Members of the World Trade Organization (WTO) have devoted considerable effort to the mandate but, like much else in the Doha Declaration, results have been elusive.

One major difficulty is the failure to agree on a single EGS list. Many WTO Members seek to maximise market access abroad for their EGS exports, while minimising market access at home for their EGS imports – in other words, the mercantilist motives that underlie most trade negotiations bloom when it comes to EGS talks.[2]

A second major difficulty (which interacts with the first) is that many WTO Members have little interest in the EGS agenda. Whatever the approach, they protect EGS products in their home markets and have nothing to offer that can compete on export markets.

That said, in the WTO EGS negotiations, better progress has been made on environmental services than on environmental good (EGs). Negotiations on environmental services have been conducted in the context of the General Agreement on Trade in Services (GATS). Environmental services identified by the WTO include sewage services, refuse disposal, sanitation and similar services, vehicle emission reduction servicess, noise abatement services, nature and landscape protection services and 'other' environmental services. According to the WTO, more than forty Members, at all levels of development, have undertaken specific commitments on these services. However, the engineering, maintenance and operational services connected

---

[1] See Doha Declaration, para 31 (iii).
[2] In grappling with the difficulty of agreeing on a single EGS list, the negotiators have explored four approaches: (1) the list approach; (2) the request and offer approach; (3) the project approach; and (4) the integrated approach. Each will be discussed in detail later.

with the production of renewable energy (RE) are not the object of specific commitments.

To minimise disagreement over the appropriate content of the EGS schedule, and to narrow the talks to those with a strong interest in liberalisation, some countries have explored the option of reaching an EGS agreement on a bilateral, plurilateral or regional basis. For example, the United States has been talking with Canada, the EU and Australia about eliminating tariffs on green technologies such as solar and wind power to spur their use. Similar discussions on improving EGS market access are underway within the Asia-Pacific Economic Cooperation (APEC) and the Group of 20 (G20).[3]

In this chapter, we look into three aspects of a possible future Sustainable Energy Trade Agreement (SETA). First, we focus on the tariff rates on sustainable energy products in selected countries and the trade flows between them. Second, we survey a number of non-tariff measures. In the final section, we examine the key issues relating to negotiating approaches and the institutional framework.

## 1 Tariff Barriers

While a SETA would cover both tariff barriers and NTBs, negotiations in the first instance would likely address tariffs. While tariffs are often not the highest barriers to trade in SETA products, they are easily quantified and less controversial than NTBs – especially LCRs, subsidies and standards. In our view, tariffs should be eliminated first, and other barriers should be phased out over a period of five or ten years.

### *Possible SETA List*

As with the EGS negotiations in the WTO, the prospect of a bilateral, plurilateral or regional pact that would liberalise trade in EGs alone has been thwarted by the difficult task of defining EGs.[4] Since there is no

---

[3] At the APEC summit held in Honolulu in November 2011, leaders agreed to develop a list of EGs in 2012 on which APEC economies would reduce applied tariffs to 5 per cent or less by 2015, and eliminate non-tariff barriers (NTBs), including local content requirements (LCRs) that distort EGS trade. Carrying out this promise, APEC leaders at the September 2012 Summit held in Vladivostok endorsed a list of fifty-four Environmental Goods (at the Harmonized System (HS) six-digit level) for liberalisation by the end of 2015.

[4] 'Environmental goods' are considered to be set of manufactured products, technologies and chemicals used in association with environmental services; that is, pollution and waste affecting water, soil and air. Discussions taking place in the EGS negotiations extended to encompass so-called 'environmentally preferable products' (EPPs). EPPs are not necessarily used for environmental purposes but have positive environmental characteristics

internationally agreed definition, countries have proposed their own lists. These lists were gathered in a WTO report for the Committee on Trade and Environment Special Session (CTESS), that lists the universe of EGs of interest to member countries (TN/TE/20, 21 April 2011). After matching EGs with tariff lines (based on the HS 2002 classification at the six-digit level), the WTO compilation covers 408 tariff lines, each of which is accompanied by a detailed description and related categories under six broad headings – air pollution control, RE, waste management and water treatment, environmental technologies, carbon capture and storage (CCS) and others.[5] So far, nothing has come of WTO negotiations over this ambitious EGs list.

In our view, a short list would better serve as a launching pad for plurilateral negotiations to establish a SETA. A longer list would create greater room for disagreement, for example over dual-use products, and would stimulate mercantilistic instincts to a larger extent. Accordingly, for the purposes of this chapter, we have drawn up our own short list that covers just thirty-nine EGs tariff lines at the HS six-digit level (see Table 2.1). Our list comprises thirty-two EGs found under the sole category of 'renewable energy' in the WTO compilation, plus six EGs based on our judgement that are described both as 'renewable energy' and another category such as 'environmental technologies', plus undenatured ethyl alcohol (HS 220710), commonly called 'ethanol', and widely used for fuel.

It is controversial whether ethanol and biofuels are good for the environment, but we included both ethanol and biodiesel in our short list since their production and use has grown rapidly with considerable political support and thus financial incentives. The United States and the EU, among others, provide substantial incentives to national ethanol and biofuel production, including subsidies, tax incentives and consumption mandates.[6] Some developing countries see ethanol and biofuel as

---

compared to substitute goods. The EPP concept draws on work undertaken by the UN Conference on Trade and Development (UNCTAD), which defines EPPs as products that cause significantly less environmental harm than alternative products serving the same purpose. 'Less environmental harm' is judged by four criteria: (a) use of natural resources and energy; (b) quantitative amount and hazardous quality of waste generated by the product along its life cycle; (c) impact on human and animal health; and (d) preservation of the environment (UNCTAD 2011; UNESCWA 2007).

[5] See Annex II.A, WTO (2011). Negotiations on EGs have taken place at the tariff line level (six-digit codes) rather than the product level (eight-digit or ten-digit codes). Tariffs are internationally consistent only at the HS six-digit level; 'over-inclusiveness' avoids contentious disagreements on product definitions (Hufbauer et al. 2010).

[6] According to the World Bank (2007a), in the United States, more than 200 support measures are provided to biofuel producers. With this push, the production of biofuels has expanded rapidly and the OECD has projected that production could again double in the next decade (OECD 2008).

Table 2.1 *List of SETA Products, Trade Values and Tariff Rates*[a]

|   | HS Code | Description | Imports from world (mill $, 2010) | MFN bound rates (%) | MFN applied rates (%) | Effective applied rates (%) | Binding coverage (%) |
|---|---|---|---|---|---|---|---|
| | | **Products categorized as 'renewable energy' only from the WTO (2011)** | | | | | |
| | 3907 | Polyacetals, other polyethers and epoxide resins, in primary forms; polycarbonates, alkyd resins, polyallyl esters and other polyesters, in primary forms | | | | | |
| 1 | 390799 | Other polyesters: Other | 5,858.4 | 11.12 | 5.07 | 4.09 | 78 |
| | 4016 | **Other articles of vulcanised rubber other than hard rubber** | | | | | |
| 2 | 401699 | Other: other | 8,329.1 | 11.57 | 5.38 | 4.25 | 77 |
| | 4707 | **Recovered (waste and scrap) paper or paperboard** | | | | | |
| 3 | 470710 | Recovered (waste & scrap) unbleached kraft paper/paperboard/corrugated paper/paperboard | 4,239.4 | 4.08 | 0.69 | 0.62 | 77 |
| 4 | 470720 | Other paper or paperboard made mainly of bleached chemical | 834.7 | 16.01 | 1.23 | 1.20 | 76 |
| 5 | 470730 | Paper or paperboard made mainly of mechanical pulp (for example, newspapers, journals and similar printed matter) | 2,410.0 | 4.73 | 0.59 | 0.55 | 76 |
| 6 | 470790 | Other, including unsorted waste and scrap | 2,075.8 | 9.87 | 1.47 | 1.19 | 76 |
| | 7308 | Structures (excluding prefabricated buildings of heading 94.06) and parts of structures (for example, bridges and bridge sections, lock gates, towers, lattice masts, roofs, roofing frameworks, doors and windows and their frames and thresholds for doors, shutters, balustrades, pillars and columns), of iron or steel; plates, rods, angles, shapes, sections, tubes and the like, prepared for use in structures, of iron or steel | | | | | |
| 7 | 730820 | Towers & lattice masts | 2,812.1 | 13.84 | 4.05 | 3.20 | 74 |

Table 2.1 (*cont.*)

| | HS Code | Description | Imports from world (mill $, 2010) | MFN bound rates (%) | MFN applied rates (%) | Effective applied rates (%) | Binding coverage (%) |
|---|---|---|---|---|---|---|---|
| | **8406** | **Steam turbines and other vapour turbines** | | | | | |
| 8 | 840682 | Other turbines of an output not exceeding 40MW | 486.9 | 17.84 | 4.30 | 3.66 | 79 |
| | **8418** | **Refrigerators, freezers and other refrigerating or freezing equipment, electric or other; heat pumps other than air conditioning machines of heading 84.15** | | | | | |
| 9 | 841861 | Other refrigerating or freezing equipment, heat pumps: Compression-type units whose condensers are heat exchangers | 1,490.9 | 11.00 | 4.41 | 2.85 | 82 |
| 10 | 841869 | Other refrigerating or freezing equipment heat pumps: Parts | 4,639.3 | 14.81 | 5.91 | 4.13 | 82 |
| | **8419** | **Machinery, plant or laboratory equipment** | | | | | |
| 11 | 841919 | Instantaneous or storage water heaters, non-electric: Other | 1,720.2 | 8.40 | 3.89 | 2.44 | 81 |
| | **8479** | **Machines and mechanical appliances having individual functions, not specified or included elsewhere in this Chapter** | | | | | |
| 12 | 847920 | Machinery for the extraction or preparation of animal or fixed vegetable fats or oils | 351.6 | 15.43 | 2.29 | 1.06 | 81 |
| | **8483** | **Transmission shafts (including cam shafts and crank shafts) and cranks; bearing housings and plain shaft bearings; gears and gearing; ball or roller screws; gear boxes and other speed changers, including torque converters; flywheels and pulleys, including pulley blocks; clutches and shaft couplings (including universal joints)** | | | | | |
| 13 | 848340 | Gears, gearing (excluding toothed wheels, chain sprockets and other transmission elements presented separately); ball or roller screws; gear boxes, other speed changers, including torque converters | 11,616.2 | 12.30 | 4.77 | 4.33 | 82 |
| 14 | 848360 | Clutches & shaft couplings (incl. universal joints) | 2,128.0 | 12.08 | 4.86 | 4.37 | 82 |
| | **8501** | **Electric motors and generators (excluding generating sets)** | | | | | |
| 15 | 850161 | AC generators (alternators), of an output not exceeding 75 Kva | 804.2 | 11.37 | 4.39 | 3.77 | 85 |

| | | | | | | | |
|---|---|---|---|---|---|---|---|
| 16 | 850162 | AC generators (alternator), of an output exceeding 75 kVA but not exceeding 375 kVA | 497.3 | 12.51 | 4.68 | 3.84 | 86 |
| 17 | 850163 | AC generators (alternator), of an output exceeding 375 kVA but not exceeding 750 kVA | 270.1 | 11.45 | 3.94 | 2.97 | 85 |
| 18 | 850164 | AC generators (alternator), of an output exceeding 750 kVA | 2,171.2 | 11.84 | 3.74 | 3.38 | 84 |
| | 8502 | Electric generating sets and rotary converters | | | | | |
| 19 | 850231 | Other generating sets: Wind-powered | 5,713.7 | 8.16 | 3.27 | 3.04 | 82 |
| 20 | 850239 | Other generating sets: other | 5,185.5 | 16.65 | 2.57 | 2.17 | 82 |
| | 8506 | Primary cells and primary batteries | | | | | |
| 21 | 850610 | Primary cells & primary batteries, manganese dioxide | 2,902.1 | 13.26 | 7.91 | 6.66 | 79 |
| 22 | 850630 | Primary cells & primary batteries, mercuric oxide | 19.8 | 14.31 | 9.99 | 9.43 | 79 |
| 23 | 850640 | Primary cells & primary batteries, silver oxide | 183.9 | 9.80 | 4.67 | 3.52 | 79 |
| 24 | 850650 | Primary cells & primary batteries, lithium | 1,817.5 | 10.95 | 5.00 | 3.96 | 79 |
| 25 | 850660 | Primary cells & primary batteries, air-zinc | 170.2 | 9.46 | 3.02 | 2.37 | 79 |
| 26 | 850690 | Parts | 246.3 | 11.59 | 2.67 | 2.22 | 79 |
| | 8507 | Electric accumulators, including separators therefor, whether or not rectangular (including square) | | | | | |
| 27 | 850720 | Other lead – Acid accumulators | 4,025.0 | 15.42 | 6.39 | 5.28 | 80 |
| 28 | 850740 | Nickel-iron | 39.1 | 15.87 | 7.02 | 4.15 | 80 |
| | 8537 | Boards, panels, consoles, desks, cabinets and other bases, equipped with two or more apparatus of heading 85.35 or 85.36, for electric control or the distribution of electricity, including those incorporating instruments or apparatus of Chapter 90, and numerical control apparatus, other than switching apparatus of heading 85.17 | | | | | |
| 29 | 853710 | For a voltage not exceeding 1,000 V | 25,271.4 | 10.29 | 4.85 | 3.94 | 81 |
| | 8541 | Diodes, transistors and similar semiconductor devices; photosensitive semiconductor devices, including photovoltaic cells whether or not assembled in modules or made up into panels; light emitting diodes; mounted piezo-electric crystals | | | | | |
| 30 | 854140 | Photosensitive semiconductor devices, including photovoltaic cells whether or not assembled in modules or made up into panels; light emitting diodes. | 64,126.0 | 0.55 | 0.04 | 0.04 | 84 |

Table 2.1 (cont.)

|  |  |  | | Tariff Rates (2010) | | |
|---|---|---|---|---|---|---|
|  | HS Code | Description | Imports from world (mill $, 2010) | MFN bound rates (%) | MFN applied rates (%) | Effective applied rates (%) | Binding coverage (%) |
|  | 9001 | **Optical fibres and optical fibre bundles; optical fibre cables other than those of heading 85.44; sheets and plates of polarising material; lenses (including contact lenses), prisms, mirrors and other optical elements, of any material, unmounted, other than such elements of glass not optically worked** | | | | | |
| 31 | 900190 | other | 6,238.9 | 7.35 | 5.53 | 5.14 | 75 |
|  | 9002 | **Lenses, prisms, mirrors and other optical elements, of any material, mounted, being parts of or fittings for instruments or apparatus, other than such elements of glass not optically worked** | | | | | |
| 32 | 900290 | other | 1,484.0 | 7.38 | 5.51 | 5.18 | 75 |
|  | **Other Products**[c] | | | | | | |
|  | 2207 | **Ethyl alcohol, undenat, n/un 80% alc, alcohol, denat**[d] | | | | | |
| 33 | 220710 | Undenatured ethyl alcohol of an alcoholic strength | 3,858.6 | 45.21 | 14.70 | 10.97 | 98 |
|  | 3824 | **Prepared binders for foundry moulds or cores; chemical products and preparations of the chemical or allied industries (including those consisting of mixtures of natural products), not elsewhere specified or included** | | | | | |
| 34 | 382490 | Other[e] | 30,788.5 | 9.67 | 5.02 | 3.95 | 77 |
|  | 8410 | **Hydraulic turbines, water wheels, and regulators therefor** | | | | | |
| 35 | 841011 | Hydraulic turbines and water wheels of a power not exceeding 1,000 kW. | 69.1 | 20.50 | 4.76 | 3.51 | 81 |
| 36 | 841090 | Parts, including regulators | 952.5 | 14.73 | 4.96 | 3.17 | 81 |
|  | 8411 | **Turbo-jets, turbo-propellers and other gas turbines** | | | | | |
| 37 | 841181 | Other gas turbines of a power not exceeding 5,000 kW | 1,385.7 | 8.33 | 1.73 | 1.37 | 81 |
| 38 | 841182 | Other gas turbines of a power exceeding 5,000 kW | 3,966.2 | 17.35 | 3.75 | 3.29 | 81 |

| 39 | 8503 | Parts suitable for use solely or principally with the machines of heading 85.01 or 85.02 | | | | |
|---|---|---|---|---|---|---|
| | 850300 | Parts suitable for use solely or principally with the machines of heading 85.01 or 85.02. | 13,549.2 | 13.30 | 3.88 | 3.10 | 82 |

**Memorandum:**

| | | | | | |
|---|---|---|---|---|---|
| All 39 SEFTA products | 224,728.5 | 8.26 | 3.31 | 2.73 | 80 |
| All products | 13,048,386.3 | 9.19 | 4.10 | 3.11 | 79 |
| Industrial Products | 11,570,139.7 | 8.15 | 3.65 | 2.74 | 76 |

*Notes:*

*a.* The calculations here use a list of 32 EGs found under the sole category of 'renewable energy' in the list found in the Annex II. A of the WTO (2011) plus 6 EGs based on our judgement that are described both as 'renewable energy' and another category such as 'environmental technologies', plus un-denatured ethyl alcohol (HS 220710). The WTO (2011) reviewed all proposals of EGs of interest, put forward by the members, and presented a comprehensive list which covers 408 EGs at the HS six digit level.

*b.* Tariff rates are weighted average; The effective applied rates take account of preferential trade agreements.

*c.* Other products are selected based on authors' judgment. Except for HS 220710, all other products are also found in the WTO (2011) but labeled not only 'renewable energy' but also something else (e.g. 'environmental technologies')

*d.* There is currently no specific customs classification for bioethanol for biofuel production. This product is traded under HS 2207, which covers un-denatured (HS220710) and denatured alcohol (HS 220720). While both denatured and un-denatured alcohol can be used for biofuel production, in this table, we only include un-denatured alcohol (HS 220710) since it seems more suitable for use as a fuel (denatured ethanol is often used as a solvent).

*e.* There is currently no specific customs classification for biodiesel for biofuel production. Biodiesel is classified under HS code 382490 which covers some other products.

*Source:* For a list of 39 product at the HS six digit level, World Trade Organization (2011); for trade values, UNCOMTRADE and for tariff rates, TRAINS (both accessed through the WITS on September 24, 2011)

promising exports; hence the commercial balance argues for their inclusion in a plurilateral negotiation.[7]

Most EGs in our list are lumped into codes that include unrelated products. For example, solar photovoltaic (PV) panels are included in part of a large sub-heading (HS 854140), which includes semiconductor devices and light emitting diodes (LEDs). While some countries want liberalisation to be confined to HS six-digit categories that have a single end-use, UNCTAD (2011) found that out of some 440 entries of EGs (at the HS six- digit level) compiled by the WTO, only a half a dozen were purely and singularly used for environmental purposes.[8] Multiple end-uses are a common feature of HS tariff codes.

Before we delve into the trade figures in our SETA list, it is worth noting that the APEC leaders at the 2012 Summit agreed on a list of fifty-four EGs for which applied tariff rates would be cut to 5 per cent or less by the end of 2015. While this is not a binding commitment among the twenty-one APEC economies, it is meaningful in that the major economies for the first time agreed on a list of EGs for tariff cuts. The list includes a number of RE and clean technologies such as solar panels, and gas and wind turbines, wastewater treatment technologies, air pollution control technologies and environmental monitoring and assessment equipment.[9] Table 2.7 at the end of the chapter shows the export values of APEC EGs to APEC countries and the import values of APEC EGs from the world. In 2010, APEC national exports of fifty-four EGs to the APEC region were about

---

[7] Ethanol and biodiesel are classified under broader HS six-digit categories that contain other products as well. Ethanol is traded under HS 2207, which covers undenatured (HS220710) and denatured alcohol (HS 220720). While both denatured and undenatured alcohol can be used for biofuel production, we only included undenatured alcohol (HS 220710) in our list since it is considered more suitable for use as a fuel (denatured ethanol is often used as a solvent). Biodiesel is classified under HS 382490 that includes products, preparations and residual products of the chemical or allied industries not elsewhere specified. Ethanol is classified as an agricultural product, while biodiesel is classified as an industrial product. Thanks to this difference, trade negotiations have so far been conducted within different WTO groups.

[8] UNCTAD (2011) has identified only two types of renewable energy equipment that can pass the single end-use test at the HS six-digit level: (1) hydraulic turbines (841011, 841012, 841013) and (2) wind-powered electricity generating sets (850231).

[9] According to a press release by the United States Trade Representative (USTR), the core products included in the APEC list are: RE and clean energy technologies such as solar panels and gas and wind turbines, on which tariffs in the region are currently as high as 35 per cent; wastewater treatment technologies, such filters and ultraviolet (UV) disinfection equipment, on which tariffs in the region are currently as high as 21 per cent; air pollution control technologies, such as soot removers and catalytic converters, on which tariffs in the region are as high as 20 per cent; solid and hazardous waste treatment technologies, such as waste incinerators; and crushing and sorting machinery, on which tariffs in the region are currently as high as 20 per cent; and environmental monitoring and assessment equipment, such as air and water quality monitors and manometers to measure pressure; and water delivery systems, on which tariffs in the region are currently as high as 20 per cent.

USD 115 billion, accounting for about 3 per cent of total merchandise exports of all products to the region. The top three exporters of the fifty-four APEC EGs are Japan, the United States and China. The APEC imports of fifty-four EGs from the world were about USD 195 billion, accounting for about 3 per cent of imports of APEC EGs from the world. APEC accounts for about 60 per cent of world imports of these products.

Interestingly, there is little overlap between our SETA list which contains thirty-nine products and the APEC list which contains fifty-four products – only seven products are in common between the two lists (see Table 2.8 at the end of the chapter).[10] Careful examination may be required to explain the differences between the lists, but our speculation is that it mainly reflects the fact that our list was mainly drawn from EGs under the category of 'renewable energy' in the WTO list, which originally covered 408 tariff lines categorised under six headings. By contrast, the APEC list covers products under broader categories than just RE technologies; they also include air pollution control technologies, environmental monitoring equipment, etc. A comparison between our SETA list and the World Bank list of forty-three EGs associated with clean energy technologies (such as solar, wind, energy efficient lighting and clean coal technologies) seems to confirm this speculation, since twenty products are found in common between the two (see Table 2.8).[11]

Returning to our own analysis, Table 2.1 shows world import values, the trade-weighted average world tariff rates and the binding coverage for each of the thirty-nine EGs in our short list. In 2010, the world imports of all thirty-nine EGs were about USD 225 billion, or roughly 1.7 per cent of total imports of all products. Among the thirty-nine EGs, PV cells and modules (HS 854140) were by far the largest, accounting for about 29 per cent of world imports of all thirty-nine EGs. With respect to tariff rates, while most favoured nation (MFN) bound, applied and effective applied rates range widely,[12] the trade-weighted average tariff rates for all thirty-nine EGs are in line with the average tariffs for all industrial products.

### *Core and Candidate Countries*

We have identified ten 'core countries' that seem highly likely to join any SETA negotiations if talks are launched: NAFTA, the EU, Chile,

---

[10] All products are specified at the HS six-digit level.
[11] For information, ten products are in common between the World Bank list and the APEC list, and five products are in common among the three lists.
[12] MFN bound rates range from almost zero per cent (HS 854140, which covers solar PV) to 45 per cent (HS 220710, which covers ethanol used for fuel); MFN applied rates range from almost zero per cent (HS 854140) to 15 per cent (HS 220710); and effective applied rates range from almost zero per cent (HS 854140) to 11 per cent (HS 220710).

Colombia, Peru, Australia, New Zealand, Singapore, Japan and Korea. We have also identified six 'candidate countries' that hold great commercial and environmental interest for the core countries: Brazil, India, China, Indonesia, Turkey and South Africa. However, several candidate countries protect the thirty-nine EGs products to a much higher extent than the core countries, and accordingly these candidates may have less interest in the SETA concept.[13] Average bound tariff rates for all thirty-nine EGs in Brazil, India, Indonesia, Turkey and South Africa are high, but China is a notable exception. China could possibly be the first candidate country to join the SETA.

These designations of core and candidate countries are, of course, merely illustrative. Once a SETA is launched, its sponsors would surely hope for wide participation by WTO Members. If obligations are phased in over time for developing countries, many might accede to the agreement.

*Tariff Rates*

Returning to the nuts and bolts, Table 2.2 presents trade-weighted averages for bound and applied tariff rates for each of the sixteen countries in our 'core' and 'candidate' groups. On average, most developing countries – whether core countries or candidate countries – maintain high bound and applied tariff rates. The rates range between 14 per cent (Turkey) and 37 per cent (Colombia) for MFN bound rates, and between almost zero per cent (Turkey) and 11 per cent (Brazil) for effective applied rates. By implication, there is also a good deal of 'water' in these tariff structures.[14] China is an exception as it maintains low bound and applied tariff rates. Likewise, tariff rates are already very low in most developed countries such as NAFTA, the EU-27, Australia, Singapore and Japan.

Several EGs in our list already have low tariffs. For example, applied tariff rates on solar PV (HS 854140) are very low in most of our selected countries – almost zero per cent in many cases. Applied tariff rates on wind turbines (HS 850231) are also below or around 5 per cent in most cases and around 10 per cent in a few countries such as Colombia, Korea, China and India. There is an upside and a downside to this situation. The upside is that, with respect to low-tariff products, negotiating the SETA should not be difficult. The downside is that the gains, in terms of

---

[13] Among the core countries, Chile, Colombia and Peru have high bound MFN tariff rates. However, their applied rates are low, so binding at zero tariff levels should not be difficult.
[14] The difference between bound and applied rates is often referred to as the 'binding overhang' or the 'water in the tariff'.

Table 2.2 Tariff Rates for SETA Products by Selected Countries, Tariff Year 2010[a]

| | HS Code | NAFTA MFN bound rates (%) | NAFTA MFN applied rates (%) | NAFTA Effective applied rates (%) | EU 27 MFN bound rates (%) | EU 27 MFN applied rates (%) | EU 27 Effective applied rates (%) | Chile MFN bound rates (%) | Chile MFN applied rates (%) | Chile Effective applied rates (%) | Colombia MFN bound rates (%) | Colombia MFN applied rates (%) | Colombia Effective applied rates (%) |
|---|---|---|---|---|---|---|---|---|---|---|---|---|---|
| | 3907 | Polyacetals, other polyethers and epoxide resins, in primary forms; polycarbonates, alkyd resins, polyallyl esters and other polyesters, in primary forms | | | | | | | | | | | |
| 1 | 390799 | 15.14 | 3.76 | 3.21 | 9.60 | 5.60 | 2.40 | 25.00 | 6.00 | 3.63 | 35.00 | 15.00 | 10.12 |
| | 4016 | Other articles of vulcanised rubber other than hard rubber | | | | | | | | | | | |
| 2 | 401699 | 11.79 | 2.37 | 1.61 | 10.80 | 6.25 | 2.53 | 25.00 | 6.00 | 5.63 | 35.00 | 14.29 | 13.07 |
| | 4707 | Recovered (waste and scrap) paper or paperboard | | | | | | | | | | | |
| 3 | 470710 | 23.17 | 0.00 | 0.00 | 9.88 | 0.13 | 0.07 | 25.00 | 6.00 | 5.77 | 35.00 | 5.00 | 4.67 |
| 4 | 470720 | 28.47 | 0.00 | 0.00 | 10.06 | 0.00 | 0.00 | 25.00 | 6.00 | 6.00 | 35.00 | 5.00 | 4.76 |
| 5 | 470730 | 3.94 | 0.00 | 0.00 | 13.15 | 0.03 | 0.00 | 25.00 | 6.00 | 4.73 | 35.00 | 5.00 | 3.60 |
| 6 | 470790 | 15.97 | 0.00 | 0.00 | 20.88 | 0.01 | 0.00 | 25.00 | 6.00 | 5.85 | 35.00 | 5.00 | 4.75 |
| | 7308 | Structures (excluding prefabricated buildings of heading 94.06) and parts of structures (for example, bridges and bridge sections, lock gates, towers, lattice masts, roofs, roofing frameworks, doors and windows and their frames and thresholds for doors, shutters, balustrades, pillars and columns), of iron or steel; plates, rods, angles, shapes, sections, tubes and the like, prepared for use in structures, of iron or steel | | | | | | | | | | | |
| 7 | 730820 | 0.13 | 0.02 | 0.02 | 12.24 | 0.91 | 0.45 | 25.00 | 6.00 | 1.17 | 35.00 | 15.00 | 14.13 |
| | 8406 | Steam turbines and other vapour turbines | | | | | | | | | | | |
| 8 | 840682 | 19.94 | 3.85 | 3.56 | 9.82 | 5.48 | 1.26 | 25.00 | 6.00 | 3.66 | 35.00 | 5.00 | 5.00 |
| | 8418 | Refrigerators, freezers and other refrigerating or freezing equipment, electric or other; heat pumps other than air conditioning machines of heading 84.15 | | | | | | | | | | | |
| 9 | 841861 | 5.84 | 1.95 | 0.99 | 9.35 | 5.16 | 1.50 | 25.00 | 6.00 | 5.10 | 35.00 | 15.00 | 10.72 |

Products categorized as 'renewable energy' only from the WTO (2011)

Table 2.2 (cont.)

|  | HS Code | NAFTA MFN bound rates (%) | NAFTA MFN applied rates (%) | NAFTA Effective applied rates (%) | EU 27 MFN bound rates (%) | EU 27 MFN applied rates (%) | EU 27 Effective applied rates (%) | Chile MFN bound rates (%) | Chile MFN applied rates (%) | Chile Effective applied rates (%) | Colombia MFN bound rates (%) | Colombia MFN applied rates (%) | Colombia Effective applied rates (%) |
|---|---|---|---|---|---|---|---|---|---|---|---|---|---|
| 10 | 841869 | 4.64 | 1.44 | 1.43 | 11.41 | 5.29 | 2.49 | 25.00 | 6.00 | 5.39 | 35.00 | 12.50 | 9.96 |
|  | 8419 | Machinery, plant or laboratory equipment |
| 11 | 841919 | 3.76 | 2.00 | 0.81 | 13.52 | 5.35 | 2.59 | 25.00 | 6.00 | 2.00 | 35.00 | 20.00 | 6.45 |
|  | 8479 | Machines and mechanical appliances having individual functions, not specified or included elsewhere in this Chapter |
| 12 | 847920 | 7.25 | 0.00 | 0.00 | 12.96 | 5.20 | 2.33 | 25.00 | 6.00 | 5.94 | 35.00 | 10.00 | 10.00 |
|  | 8483 | Transmission shafts (including cam shafts and crank shafts) and cranks; bearing housings and plain shaft bearings; gears and gearing; ball or roller screws; gear boxes and other speed changers, including torque converters; flywheels and pulleys, including pulley blocks; clutches and shaft couplings (including universal joints) |
| 13 | 848340 | 5.32 | 1.22 | 1.05 | 11.31 | 5.50 | 3.32 | 25.00 | 6.00 | 5.77 | 35.00 | 10.00 | 9.50 |
| 14 | 848360 | 7.82 | 3.10 | 2.73 | 10.00 | 5.31 | 3.34 | 25.00 | 6.00 | 5.68 | 35.00 | 7.50 | 7.30 |
|  | 8501 | Electric motors and generators (excluding generating sets) |
| 15 | 850161 | 3.43 | 2.24 | 1.67 | 10.67 | 6.19 | 4.56 | 25.00 | 6.00 | 5.74 | 35.00 | 8.33 | 7.05 |
| 16 | 850162 | 4.43 | 2.93 | 1.79 | 10.80 | 8.07 | 7.26 | 25.00 | 6.00 | 5.48 | 35.00 | 10.00 | 2.16 |
| 17 | 850163 | 3.53 | 2.61 | 1.04 | 15.97 | 6.98 | 4.95 | 25.00 | 6.00 | 5.13 | 35.00 | 10.00 | 2.58 |
| 18 | 850164 | 3.99 | 2.16 | 2.02 | 23.41 | 8.20 | 6.18 | 25.00 | 6.00 | 5.04 | 35.00 | 10.00 | 9.99 |
|  | 8502 | Electric generating sets and rotary converters |
| 19 | 850231 | 4.40 | 2.46 | 2.44 | 21.24 | 1.65 | 0.09 | 25.00 | 6.00 | 6.00 | 35.00 | 10.00 | 10.00 |
| 20 | 850239 | 6.30 | 2.60 | 1.84 | 17.04 | 6.14 | 4.19 | 25.00 | 6.00 | 5.82 | 35.00 | 12.50 | 7.59 |
|  | 8506 | Primary cells and primary batteries |
| 21 | 850610 | 8.61 | 2.52 | 1.26 | 16.09 | 7.61 | 3.27 | 25.00 | 6.00 | 5.96 | 35.00 | 10.71 | 10.70 |
| 22 | 850630 | 15.70 | 3.94 | 3.74 | 20.38 | 10.99 | 10.73 | 25.00 | 6.00 | 6.00 | 35.00 | 5.00 | 5.00 |
| 23 | 850640 | 5.37 | 3.44 | 2.35 | 13.08 | 6.86 | 5.06 | 25.00 | 6.00 | 6.00 | 35.00 | 5.00 | 5.00 |
| 24 | 850650 | 11.22 | 2.13 | 1.38 | 11.62 | 6.45 | 4.51 | 25.00 | 6.00 | 5.83 | 35.00 | 5.00 | 4.99 |
| 25 | 850660 | 5.51 | 4.15 | 2.88 | 11.12 | 7.11 | 3.50 | 25.00 | 6.00 | 5.99 | 35.00 | 5.00 | 5.00 |
| 26 | 850690 | 2.97 | 2.40 | 1.89 | 9.65 | 4.08 | 3.22 | 25.00 | 6.00 | 5.92 | 35.00 | 5.00 | 5.00 |

| | | | | | | | | | | |
|---|---|---|---|---|---|---|---|---|---|---|
| | 8507 | Electric accumulators, including separators therefor, whether or not rectangular (including square) | | | | | | | | |
| 27 | 850720 | 6.93 | 3.77 | 2.42 | 10.86 | 4.47 | 2.12 | 25.00 | 6.00 | 5.01 | 35.00 | 15.00 | 14.81 |
| 28 | 850740 | 9.94 | 2.76 | 2.24 | 8.45 | 4.48 | 4.05 | 25.00 | 6.00 | 5.93 | 35.00 | 5.00 | 4.92 |
| | 8537 | Boards, panels, consoles, desks, cabinets and other bases, equipped with two or more apparatus of heading 85.35 or 85.36, for electric control or the distribution of electricity, including those incorporating instruments or apparatus of Chapter 90, and numerical control apparatus, other than switching apparatus of heading 85.17 | | | | | | | | |
| 29 | 853710 | 5.84 | 2.51 | 1.34 | 12.97 | 5.00 | 2.85 | 25.00 | 6.00 | 5.10 | 35.00 | 15.00 | 13.10 |
| | 8541 | Diodes, transistors and similar semiconductor devices; photosensitive semiconductor devices, including photovoltaic cells whether or not assembled in modules or made up into panels; light emitting diodes; mounted piezo-electric crystals | | | | | | | | |
| 30 | 854140 | 5.57 | 0.00 | 0.00 | 5.45 | 1.15 | 1.15 | 25.00 | 6.00 | 5.58 | 35.00 | 5.00 | 4.90 |
| | 9001 | Optical fibres and optical fibre bundles; optical fibre cables other than those of heading 85.44; sheets and plates of polarising material; lenses (including contact lenses), prisms, mirrors and other optical elements, of any material, unmounted, other than such elements of glass not optically worked | | | | | | | | |
| 31 | 900190 | 10.54 | 2.28 | 1.76 | 9.81 | 3.82 | 2.15 | 25.00 | 6.00 | 5.97 | 35.00 | 5.00 | 4.96 |
| | 9002 | Lenses, prisms, mirrors and other optical elements, of any material, mounted, being parts of or fittings for instruments or apparatus, other than such elements of glass not optically worked | | | | | | | | |
| 32 | 900290 | 3.67 | 1.79 | 1.12 | 11.06 | 6.11 | 4.76 | 25.00 | 6.00 | 5.96 | 35.00 | 5.00 | 4.96 |
| Other Products | | | | | | | | | | | |
| | 2207 | Ethyl alcohol, undenat, n/un 80% alc, alcohol, denat | | | | | | | | |
| 33 | 220710 | 22.24[b] | 18.34[b] | 16.28[b] | 36.29 | 31.45 | 3.36 | 25.00 | 6.00 | 0.16 | 70.00 | 15.00 | 10.19 |
| | 3824 | Prepared binders for foundry moulds or cores; chemical products and preparations of the chemical or allied industries (including those consisting of mixtures of natural products), not elsewhere specified or included | | | | | | | | |
| 34 | 382490 | 11.80 | 2.65 | 1.98 | 12.24 | 5.20 | 1.65 | 25.00 | 6.00 | 4.96 | 35.00 | 7.25 | 6.65 |
| | 8410 | Hydraulic turbines, water wheels, and regulators therefor | | | | | | | | |
| 35 | 841011 | 8.05 | 3.31 | 2.58 | 11.52 | 4.54 | 1.83 | 24.00 | 6.00 | 5.96 | 35.00 | 15.00 | 12.98 |
| 36 | 841090 | 5.46 | 3.11 | 1.81 | 15.05 | 8.26 | 2.55 | 25.00 | 6.00 | 2.66 | 35.00 | 5.00 | 2.84 |
| | 8411 | Turbo-jets, turbo-propellers and other gas turbines | | | | | | | | |
| 37 | 841181 | 5.48 | 0.91 | 0.62 | 11.13 | 4.29 | 4.15 | 25.00 | 6.00 | 5.82 | 35.00 | 5.00 | 5.00 |
| 38 | 841182 | 10.57 | 1.21 | 0.34 | 11.56 | 5.14 | 3.49 | 25.00 | 6.00 | 6.00 | 35.00 | 5.00 | 5.00 |
| | 8503 | Parts suitable for use solely or principally with the machines of heading 85.01 or 85.02 | | | | | | | | |
| 39 | 850300 | 11.22 | 2.19 | 1.33 | 11.44 | 6.19 | 3.68 | 25.00 | 6.00 | 5.41 | 35.00 | 10.00 | 9.34 |
| All 39 SETA products | | 7.12 | 1.96 | 1.36 | 11.45 | 4.63 | 2.34 | 25.00 | 6.00 | 5.22 | 36.85 | 9.69 | 8.48 |

|  | HS Code | Peru MFN bound rates (%) | Peru MFN applied rates (%) | Peru Effective applied rates (%) | Australia MFN bound rates (%) | Australia MFN applied rates (%) | Australia Effective applied rates (%) | New Zealand MFN bound rates (%) | New Zealand MFN applied rates (%) | New Zealand Effective applied rates (%) | Singapore MFN bound rates (%) | Singapore MFN applied rates (%) | Singapore Effective applied rates (%) |
|---|---|---|---|---|---|---|---|---|---|---|---|---|---|
| \multicolumn{14}{l}{Products categorized as 'renewable energy' only from the WTO (2011)} |
|  | 3907 | \multicolumn{13}{l}{Polyacetals, other polyethers and epoxide resins, in primary forms; polycarbonates, alkyd resins, polyallyl esters and other polyesters, in primary forms} |
| 1 | 390799 | 30.00 | 0.00 | 0.00 | 10.00 | 5.00 | 2.97 | 13.00 | 2.50 | 1.34 | 6.50 | 0.00 | 0.00 |
|  | 4016 | \multicolumn{13}{l}{Other articles of vulcanised rubber other than hard rubber} |
| 2 | 401699 | 30.00 | 6.43 | 5.91 | 15.00 | 5.00 | 3.36 | 13.37 | 3.61 | 1.46 |  | 0.00 | 0.00 |
|  | 4707 | \multicolumn{13}{l}{Recovered (waste and scrap) paper or paperboard} |
| 3 | 470710 | 30.00 | 0.00 | 0.00 |  | 0.00 | 0.00 |  |  |  | 0.00 | 0.00 | 0.00 |
| 4 | 470720 | 30.00 | 0.00 | 0.00 | 0.00 | 0.00 | 0.00 |  |  |  | 0.00 | 0.00 | 0.00 |
| 5 | 470730 |  |  |  | 0.00 | 0.00 | 0.00 | 0.00 | 0.00 | 0.00 | 0.00 | 0.00 | 0.00 |
| 6 | 470790 | 30.00 | 0.00 | 0.00 | 0.00 | 0.00 | 0.00 | 0.00 | 0.00 | 0.00 | 0.00 | 0.00 | 0.00 |
|  | 7308 | \multicolumn{13}{l}{Structures (excluding prefabricated buildings of heading 94.06) and parts of structures (for example, bridges and bridge sections, lock gates, towers, lattice masts, roofs, roofing frameworks, doors and windows and their frames and thresholds for doors, shutters, balustrades, pillars and columns), of iron or steel; plates, rods, angles, shapes, sections, tubes and the like, prepared for use in structures, of iron or steel} |
| 7 | 730820 | 30.00 | 0.00 | 0.00 | 10.00 | 5.00 | 3.71 | 27.00 | 5.00 | 4.86 |  | 0.00 | 0.00 |
|  | 8406 | \multicolumn{13}{l}{Steam turbines and other vapour turbines} |
| 8 | 840682 | 30.00 | 0.00 | 0.00 | 21.00 | 0.00 | 0.00 | 0.00 | 0.00 |  |  |  |  |
|  | 8418 | \multicolumn{13}{l}{Refrigerators, freezers and other refrigerating or freezing equipment, electric or other; heat pumps other than air conditioning machines of heading 84.15} |
| 9 | 841861 | 30.00 | 0.00 | 0.00 | 25.00 | 5.00 | 4.29 | 25.00 | 5.00 | 4.49 | 10.00 | 0.00 | 0.00 |
| 10 | 841869 | 30.00 | 0.00 | 0.00 | 15.00 | 5.00 | 3.22 | 25.00 | 5.00 | 3.39 | 5.00 | 0.00 | 0.00 |
|  | 8419 | \multicolumn{13}{l}{Machinery, plant or laboratory equipment} |
| 11 | 841919 | 30.00 | 9.00 | 7.99 | 15.00 | 5.00 | 4.92 | 25.00 | 5.00 | 3.52 | 10.00 | 0.00 | 0.00 |
|  | 8479 | \multicolumn{13}{l}{Machines and mechanical appliances having individual functions, not specified or included elsewhere in this Chapter} |
| 12 | 847920 | 30.00 | 0.00 | 0.00 | 10.00 | 0.00 | 0.00 | 25.00 | 5.00 | 4.88 | 0.00 | 0.00 | 0.00 |

| | | | | | | | | | | |
|---|---|---|---|---|---|---|---|---|---|---|
| | 8483 | | Transmission shafts (including cam shafts and crank shafts) and cranks; bearing housings and plain shaft bearings; gears and gearing; ball or roller screws; gear boxes and other speed changers, including torque converters; flywheels and pulleys, including pulley blocks; clutches and shaft couplings (including universal joints) | | | | | | | |
| 13 | 848340 | 30.00 | 6.75 | 18.00 | 5.00 | 31.67 | 5.00 | 3.79 | 4.59 | 10.00 | 0.00 |
| 14 | 848360 | 30.00 | 0.00 | 13.00 | 5.00 | 25.00 | 5.00 | 3.71 | 4.24 | | 0.00 |
| | 8501 | | Electric motors and generators (excluding generating sets) | | | | | | | |
| 15 | 850161 | 30.00 | 0.00 | 10.00 | 5.00 | 16.50 | 5.00 | 2.76 | 4.35 | | 0.00 |
| 16 | 850162 | 30.00 | 0.00 | 10.00 | 5.00 | 16.50 | 5.00 | 3.17 | 4.66 | 10.00 | 0.00 |
| 17 | 850163 | 30.00 | 0.00 | 0.00 | 0.00 | 16.50 | 5.00 | 0.00 | 5.00 | 10.00 | 0.00 |
| 18 | 850164 | 30.00 | 0.00 | 0.00 | 0.00 | 16.50 | 5.00 | 0.00 | 4.25 | | 0.00 |
| | 8502 | | Electric generating sets and rotary converters | | | | | | | |
| 19 | 850231 | 30.00 | 0.00 | 5.00 | 2.50 | 16.50 | 2.50 | 2.21 | 2.50 | 10.00 | 0.00 |
| 20 | 850239 | 30.00 | 0.00 | 5.00 | 2.50 | 16.50 | 5.00 | 2.06 | 4.97 | 10.00 | 0.00 |
| | 8506 | | Primary cells and primary batteries | | | | | | | |
| 21 | 850610 | 30.00 | 9.00 | 8.68 | 15.00 | 18.25 | 0.00 | 0.00 | 0.00 | | 0.00 |
| 22 | 850630 | | | | 15.00 | 25.00 | 0.00 | 0.00 | 0.00 | | 0.00 |
| 23 | 850640 | 30.00 | 9.00 | 0.00 | 15.00 | 25.00 | 0.00 | 0.00 | 0.00 | | 0.00 |
| 24 | 850650 | 30.00 | 9.00 | 8.99 | 15.00 | 25.00 | 0.00 | 0.00 | 0.00 | | 0.00 |
| 25 | 850660 | 30.00 | 9.00 | 9.00 | 15.00 | 25.00 | 0.00 | 0.00 | 0.00 | | 0.00 |
| 26 | 850690 | 30.00 | 9.00 | 9.00 | 15.00 | 25.00 | 0.00 | 0.00 | 0.00 | 10.00 | 0.00 |
| | 8507 | | Electric accumulators, including separators therefor, whether or not rectangular (including square) | | | | | | | |
| 27 | 850720 | 30.00 | 9.00 | 8.92 | 18.00 | 22.00 | 5.00 | 3.97 | 3.95 | | 0.00 |
| 28 | 850740 | 30.00 | 9.00 | 8.98 | 21.00 | 22.00 | 5.00 | 2.00 | 3.13 | | 0.00 |
| | 8537 | | Boards, panels, consoles, desks, cabinets and other bases, equipped with two or more apparatus of heading 85.35 or 85.36, for electric control or the distribution of electricity, including those incorporating instruments or apparatus of Chapter 90, and numerical control apparatus, other than switching apparatus of heading 85.17 | | | | | | | |
| 29 | 853710 | 30.00 | 0.00 | 0.00 | 10.00 | 2.50 | 1.40 | 30.00 | 5.00 | 3.80 | 10.00 | 0.00 |
| | 8541 | | Diodes, transistors and similar semiconductor devices; photosensitive semiconductor devices, including photovoltaic cells whether or not assembled in modules or made up into panels; light emitting diodes; mounted piezo-electric crystals | | | | | | | |
| 30 | 854140 | 30.00 | 0.00 | 0.00 | 0.00 | 0.00 | 0.00 | 0.00 | 0.00 | 0.00 | 0.00 |

Table 2.2 (cont.)

| | HS Code | Peru MFN bound rates (%) | Peru MFN applied rates (%) | Peru Effective applied rates (%) | Australia MFN bound rates (%) | Australia MFN applied rates (%) | Australia Effective applied rates (%) | New Zealand MFN bound rates (%) | New Zealand MFN applied rates (%) | New Zealand Effective applied rates (%) | Singapore MFN bound rates (%) | Singapore MFN applied rates (%) | Singapore Effective applied rates (%) |
|---|---|---|---|---|---|---|---|---|---|---|---|---|---|
| | **9001** | \multicolumn{12}{l}{Optical fibres and optical fibre bundles; optical fibre cables other than those of heading 85.44; sheets and plates of polarising material; lenses (including contact lenses), prisms, mirrors and other optical elements, of any material, unmounted, other than such elements of glass not optically worked} | | | | | | | | | | | |
| 31 | 900190 | 30.00 | 9.00 | 9.00 | 3.50 | 2.50 | 1.49 | 7.50 | 2.50 | 2.12 | 0.00 | 0.00 | 0.00 |
| | **9002** | \multicolumn{12}{l}{Lenses, prisms, mirrors and other optical elements, of any material, mounted, being parts of or fittings for instruments or apparatus, other than such elements of glass not optically worked} | | | | | | | | | | | |
| 32 | 900290 | 30.00 | 9.00 | 9.00 | 1.00 | 0.00 | 0.00 | 14.00 | 5.00 | 2.95 | 0.00 | 0.00 | 0.00 |
| **Other Products** | | | | | | | | | | | | | | |
| | **2207** | \multicolumn{12}{l}{Ethyl alcohol, undenat, n/un 80% alc, alcohol, denat} | | | | | | | | | | | |
| 33 | 220710 | 30.00 | 9.00 | 2.92 | 10.00 | 5.00 | 2.61 | 2.43 | 0.00 | 0.00 | 0.00 | 0.00 | 0.00 |
| | **3824** | \multicolumn{12}{l}{Prepared binders for foundry moulds or cores; chemical products and preparations of the chemical or allied industries (including those consisting of mixtures of natural products), not elsewhere specified or included} | | | | | | | | | | | |
| 34 | 382490 | 30.00 | 3.44 | 3.03 | 10.00 | 1.00 | 0.58 | 3.42 | 0.29 | 0.18 | 6.50 | 0.00 | 0.00 |
| | **8410** | \multicolumn{12}{l}{Hydraulic turbines, water wheels, and regulators therefor} | | | | | | | | | | | |
| 35 | 841011 | 30.00 | 0.00 | 0.00 | 15.00 | 5.00 | 4.42 | 0.00 | 0.00 | 0.00 | | 0.00 | 0.00 |
| 36 | 841090 | 30.00 | 0.00 | 0.00 | 15.00 | 5.00 | 3.79 | 0.00 | 0.00 | 0.00 | 10.00 | 0.00 | 0.00 |
| | **8411** | \multicolumn{12}{l}{Turbo-jets, turbo-propellers and other gas turbines} | | | | | | | | | | | |
| 37 | 841181 | 30.00 | 0.00 | 0.00 | 0.00 | 0.00 | 0.00 | 0.00 | 0.00 | 0.00 | 10.00 | 0.00 | 0.00 |
| 38 | 841182 | 30.00 | 0.00 | 0.00 | 0.00 | 0.00 | 0.00 | 0.00 | 0.00 | 0.00 | 10.00 | 0.00 | 0.00 |
| | **8503** | \multicolumn{12}{l}{Parts suitable for use solely or principally with the machines of heading 85.01 or 85.02} | | | | | | | | | | | |
| 39 | 850300 | 30.00 | 0.00 | 0.00 | 15.00 | 5.00 | 4.65 | 16.83 | 5.00 | 4.87 | | 0.00 | 0.00 |
| **All 39 SETA products** | | 30.00 | 2.08 | 1.86 | 9.20 | 2.70 | 2.06 | 16.12 | 3.33 | 2.89 | 6.39 | 0.00 | 0.00 |

| | HS Code | Japan MFN bound rates (%) | Japan MFN applied rates (%) | Japan Effective applied rates (%) | Korea MFN bound rates (%) | Korea MFN applied rates (%) | Korea Effective applied rates (%) | Brazil MFN bound rates (%) | Brazil MFN applied rates (%) | Brazil Effective applied rates (%) | China MFN bound rates (%) | China MFN applied rates (%) | China Effective applied rates (%) |
|---|---|---|---|---|---|---|---|---|---|---|---|---|---|
| | | \multicolumn{13}{l}{Products categorized as 'renewable energy' only from the WTO (2011)} |
| | 3907 | \multicolumn{13}{l}{Polyacetals, other polyethers and epoxide resins, in primary forms; polycarbonates, alkyd resins, polyallyl esters and other polyesters, in primary forms} |
| 1 | 390799 | 0.00 | 3.10 | 2.19 | 6.50 | 6.50 | 5.76 | 20.00 | 8.00 | 7.76 | 6.50 | 6.50 | 5.79 |
| | 4016 | \multicolumn{13}{l}{Other articles of vulcanised rubber other than hard rubber} |
| 2 | 401699 | 0.00 | 0.00 | 0.00 | 10.40 | 6.40 | 5.72 | 35.00 | 9.00 | 8.87 | 9.00 | 9.00 | 8.29 |
| | 4707 | \multicolumn{13}{l}{Recovered (waste and scrap) paper or paperboard} |
| 3 | 470710 | 0.00 | 0.00 | 0.00 | 0.00 | 0.00 | 0.00 | 35.00 | 2.00 | 0.15 | 0.00 | 0.00 | 0.00 |
| 4 | 470720 | 0.00 | 0.00 | 0.00 | 0.00 | 0.00 | 0.00 | 35.00 | 2.00 | 2.00 | 1.00 | 0.00 | 0.00 |
| 5 | 470730 | 0.00 | 0.00 | 0.00 | 0.00 | 0.00 | 0.00 | 35.00 | 2.00 | 2.00 | 1.00 | 0.00 | 0.00 |
| 6 | 470790 | 0.00 | 0.00 | 0.00 | 0.00 | 0.00 | 0.00 | 35.00 | 2.00 | 0.25 | 1.00 | 0.00 | 0.00 |
| | 7308 | \multicolumn{13}{l}{Structures (excluding prefabricated buildings of heading 94.06) and parts of structures (for example, bridges and bridge sections, lock gates, towers, lattice masts, roofs, roofing frameworks, doors and windows and their frames and thresholds for doors, shutters, balustrades, pillars and columns), of iron or steel; plates, rods, angles, shapes, sections, tubes and the like, prepared for use in structures, of iron or steel} |
| 7 | 730820 | 0.00 | 0.00 | 0.00 | 13.00 | 8.00 | 8.00 | 35.00 | 14.00 | 13.79 | 8.40 | 8.40 | 8.40 |
| | 8406 | \multicolumn{13}{l}{Steam turbines and other vapour turbines} |
| 8 | 840682 | 0.00 | 0.00 | 0.00 | 5.00 | 5.00 | 5.00 | 32.50 | 14.00 | 14.00 | 5.00 | 5.00 | 5.00 |
| | 8418 | \multicolumn{13}{l}{Refrigerators, freezers and other refrigerating or freezing equipment, electric or other; heat pumps other than air conditioning machines of heading 84.15} |
| 9 | 841861 | 0.00 | 0.00 | 0.00 | 13.00 | 8.00 | 8.00 | 35.00 | 14.00 | 13.88 | 10.00 | 12.50 | 11.27 |
| 10 | 841869 | 0.00 | 0.00 | 0.00 | 13.00 | 8.00 | 7.98 | 35.00 | 13.71 | 13.60 | 15.00 | 10.00 | 2.97 |
| | 8419 | \multicolumn{13}{l}{Machinery, plant or laboratory equipment} |
| 11 | 841919 | 0.00 | 0.00 | 0.00 | 13.00 | 8.00 | 8.00 | 35.00 | 20.00 | 16.15 | 35.00 | 35.00 | 34.79 |
| | 8479 | \multicolumn{13}{l}{Machines and mechanical appliances having individual functions, not specified or included elsewhere in this Chapter} |

Table 2.2 (cont.)

| | HS Code | Japan MFN bound rates (%) | Japan MFN applied rates (%) | Japan Effective applied rates (%) | Korea MFN bound rates (%) | Korea MFN applied rates (%) | Korea Effective applied rates (%) | Brazil MFN bound rates (%) | Brazil MFN applied rates (%) | Brazil Effective applied rates (%) | China MFN bound rates (%) | China MFN applied rates (%) | China Effective applied rates (%) |
|---|---|---|---|---|---|---|---|---|---|---|---|---|---|
| 12 | 847920 | 0.00 | 0.00 | 0.00 | 13.00 | 8.00 | 8.00 | 35.00 | 14.00 | 5.07 | 10.00 | 10.00 | 8.99 |
| | 8483 | Transmission shafts (including cam shafts and crank shafts) and cranks; bearing housings and plain shaft bearings; gears and gearing; ball or roller screws; gear boxes and other speed changers, including torque converters; flywheels and pulleys, including pulley blocks; clutches and shaft couplings (including universal joints) | | | | | | | | | | | |
| 13 | 848340 | 0.00 | 0.00 | 0.00 | 12.00 | 6.75 | 6.75 | 30.71 | 14.00 | 13.71 | 8.00 | 8.00 | 7.80 |
| 14 | 848360 | 0.00 | 0.00 | 0.00 | 9.00 | 5.50 | 5.50 | 28.33 | 14.00 | 13.79 | 8.00 | 8.00 | 7.98 |
| | 8501 | Electric motors and generators (excluding generating sets) | | | | | | | | | | | |
| 15 | 850161 | 0.00 | 0.00 | 0.00 | 13.00 | 8.00 | 8.00 | 35.00 | 14.00 | 13.52 | 5.00 | 5.00 | 4.97 |
| 16 | 850162 | 0.00 | 0.00 | 0.00 | 13.00 | 8.00 | 8.00 | 35.00 | 14.00 | 9.15 | 12.00 | 12.00 | 10.05 |
| 17 | 850163 | 0.00 | 0.00 | 0.00 | 0.00 | 4.00 | 4.00 | 35.00 | 14.00 | 6.44 | 12.00 | 12.00 | 11.99 |
| 18 | 850164 | 0.00 | 0.00 | 0.00 | 0.00 | 0.00 | 0.00 | 35.00 | 14.00 | 13.89 | 7.27 | 7.27 | 7.13 |
| | 8502 | Electric generating sets and rotary converters | | | | | | | | | | | |
| 19 | 850231 | 0.00 | 0.00 | 0.00 | | 8.00 | 8.00 | 35.00 | 14.00 | 14.00 | 8.00 | 8.00 | 8.00 |
| 20 | 850239 | 0.00 | 0.00 | 0.00 | | 8.00 | 8.00 | 35.00 | 14.00 | 14.00 | 10.00 | 10.00 | 10.00 |
| | 8506 | Primary cells and primary batteries | | | | | | | | | | | |
| 21 | 850610 | 0.00 | 0.00 | 0.00 | | 11.33 | 10.86 | 35.00 | 16.00 | 16.00 | 20.00 | 20.00 | 17.67 |
| 22 | 850630 | | | | 13.00 | 8.00 | 8.00 | | | | 14.00 | 14.00 | 13.01 |
| 23 | 850640 | 0.00 | 0.00 | 0.00 | 13.00 | 8.00 | 8.00 | 35.00 | 8.00 | 8.00 | 14.00 | 14.00 | 13.97 |
| 24 | 850650 | 0.00 | 0.00 | 0.00 | 13.00 | 8.00 | 8.00 | 35.00 | 8.00 | 8.00 | 14.00 | 14.00 | 10.45 |
| 25 | 850660 | 0.00 | 0.00 | 0.00 | 13.00 | 8.00 | 8.00 | 35.00 | 8.00 | 8.00 | 14.00 | 14.00 | 13.20 |
| 26 | 850690 | 0.00 | 0.00 | 0.00 | 13.00 | 8.00 | 8.00 | 35.00 | 14.00 | 14.00 | 12.00 | 12.00 | 8.32 |
| | 8507 | Electric accumulators, including separators therefor, whether or not rectangular (including square) | | | | | | | | | | | |
| 27 | 850720 | 0.00 | 0.00 | 0.00 | | 8.00 | 6.34 | 25.00 | 18.00 | 16.54 | 10.00 | 10.00 | 8.04 |
| 28 | 850740 | 0.00 | 0.00 | 0.00 | | 8.00 | 8.00 | 35.00 | 18.00 | 18.00 | 12.00 | 12.00 | 12.00 |

| | | | | | | | | | | |
|---|---|---|---|---|---|---|---|---|---|---|
| 29 | 8537 | Boards, panels, consoles, desks, cabinets and other bases, equipped with two or more apparatus of heading 85.35 or 85.36, for electric control or the distribution of electricity, including those incorporating instruments or apparatus of Chapter 90, and numerical control apparatus, other than switching apparatus of heading 85.17 | | | | | | | | |
| | | 0.00 | 0.00 | 8.00 | 8.00 | 35.00 | 12.00 | 11.56 | 6.70 | 6.13 | 5.60 |
| | 853710 | Diodes, transistors and similar semiconductor devices; photosensitive semiconductor devices, including photovoltaic cells whether or not assembled in modules or made up into panels; light emitting diodes; mounted piezo-electric crystals | | | | | | | | |
| 30 | 8541 | 0.00 | 0.00 | 0.00 | 0.00 | 20.59 | 1.89 | 1.89 | 0.00 | 0.00 | 0.00 |
| | 854140 | Optical fibres and optical fibre bundles; optical fibre cables other than those of heading 85.44; sheets and plates of polarising material; lenses (including contact lenses), prisms, mirrors and other optical elements, of any material, unmounted, other than such elements of glass not optically worked | | | | | | | | |
| 31 | 9001 | 0.00 | 0.00 | 8.00 | 8.00 | 35.00 | 18.00 | 17.92 | 8.00 | 8.00 | 7.49 |
| | 900190 | Lenses, prisms, mirrors and other optical elements, of any material, mounted, being parts of or fittings for instruments or apparatus, other than such elements of glass not optically worked | | | | | | | | |
| 32 | 9002 | 0.00 | 0.00 | 13.00 | 6.33 | 35.00 | 16.00 | 16.00 | 15.00 | 15.00 | 14.61 |
| | 900290 | | | | | | | | | | |
| **Other Products** | | | | | | | | | | | |
| | 2207 | Ethyl alcohol, undenat, n/un 80% alc, alcohol, denat | | | | | | | | |
| 33 | 220710 | 27.20 | 1.67 | 130.00 | 103.33 | 103.33 | 35.00 | 20.00 | 16.32 | 40.00 | 38.55 |
| | 3824 | Prepared binders for foundry moulds or cores; chemical products and preparations of the chemical or allied industries (including those consisting of mixtures of natural products), not elsewhere specified or included | | | | | | | | |
| 34 | 382490 | 0.47 | 2.20 | 1.35 | 6.47 | 6.37 | 6.33 | 19.65 | 8.22 | 7.87 | 6.92 | 6.32 |
| | 8410 | Hydraulic turbines, water wheels, and regulators therefor | | | | | | | | |
| 35 | 841011 | 0.00 | 0.00 | 0.00 | 6.50 | 4.00 | 4.00 | 35.00 | 14.00 | 14.00 | 10.00 | 10.00 |
| 36 | 841090 | 0.00 | 0.00 | 0.00 | 6.50 | 4.00 | 4.00 | 30.00 | 14.00 | 13.91 | 6.00 | 6.00 |
| | 8411 | Turbo-jets, turbo-propellers and other gas turbines | | | | | | | | |
| 37 | 841181 | 0.00 | 0.00 | 0.00 | 9.00 | 6.33 | 6.33 | 10.00 | 0.00 | 0.00 | 15.00 | 15.00 |
| 38 | 841182 | 0.00 | 0.00 | 0.00 | 9.00 | 6.33 | 6.33 | 10.00 | 0.00 | 0.00 | 3.00 | 3.00 |
| | 8503 | Parts suitable for use solely or principally with the machines of heading 85.01 or 85.02 | | | | | | | | |
| 39 | 850300 | 0.00 | 0.00 | 0.00 | 13.00 | 8.00 | 7.80 | 30.00 | 14.00 | 13.95 | 6.50 | 5.51 |
| **All 39 SETA products** | | 1.51 | 0.48 | 0.33 | 6.51 | 5.63 | 5.58 | 29.32 | 11.30 | 11.02 | 4.78 | 4.54 | 4.08 |

|  | HS Code | India MFN bound rates (%) | India MFN applied rates (%) | India Effective applied rates (%) | Indonesia MFN bound rates (%) | Indonesia MFN applied rates (%) | Indonesia Effective applied rates (%) | Turkey MFN bound rates (%) | Turkey MFN applied rates (%) | Turkey Effective applied rates (%) | South Africa MFN bound rates (%) | South Africa MFN applied rates (%) | South Africa Effective applied rates (%) |
|---|---|---|---|---|---|---|---|---|---|---|---|---|---|
| Products categorized as 'renewable energy' only from the WTO (2011) | | | | | | | | | | | | | |
| 3907 | Polyacetals, other polyethers and epoxide resins, in primary forms; polycarbonates, alkyd resins, polyallyl esters and other polyesters, in primary forms | | | | | | | | | | | | |
| 1 | 390799 | 40.00 | 8.00 | 7.79 | 40.00 | 5.00 | 0.86 | 33.20 | 3.25 | 1.70 | 15.00 | 0.00 | 0.00 |
| 4016 | Other articles of vulcanised rubber other than hard rubber | | | | | | | | | | | | |
| 2 | 401699 | 10.00 | 10.00 | 9.98 | 40.91 | 6.67 | 2.25 | 14.44 | 2.32 | 0.41 | 21.82 | 5.77 | 3.88 |
| 4707 | Recovered (waste and scrap) paper or paperboard | | | | | | | | | | | | |
| 3 | 470710 | 40.00 | 10.00 | 9.63 | 40.00 | 5.00 | 3.81 | | 0.00 | 0.00 | 5.00 | 0.00 | 0.00 |
| 4 | 470720 | 40.00 | 10.00 | 9.89 | 40.00 | 5.00 | 4.26 | | 0.00 | 0.00 | | | |
| 5 | 470730 | 40.00 | 10.00 | 9.86 | 40.00 | 5.00 | 4.53 | | 0.00 | 0.00 | 5.00 | 0.00 | 0.00 |
| 6 | 470790 | 40.00 | 10.00 | 8.72 | 40.00 | 5.00 | 3.89 | | 0.00 | 0.00 | 5.00 | 0.00 | 0.00 |
| 7308 | Structures (excluding prefabricated buildings of heading 94.06) and parts of structures (for example, bridges and bridge sections, lock gates, towers, lattice masts, roofs, roofing frameworks, doors and windows and their frames and thresholds for doors, shutters, balustrades, pillars and columns), of iron or steel; plates, rods, angles, shapes, sections, tubes and the like, prepared for use in structures, of iron or steel | | | | | | | | | | | | |
| 7 | 730820 | 40.00 | 10.00 | 10.00 | 40.00 | 12.50 | 1.99 | | 0.00 | 0.00 | 15.00 | 7.50 | 6.20 |
| 8406 | Steam turbines and other vapour turbines | | | | | | | | | | | | |
| 8 | 840682 | 25.00 | 8.00 | 7.99 | 40.00 | 5.00 | 1.58 | | 2.70 | 0.01 | 0.00 | 0.00 | 0.00 |
| 8418 | Refrigerators, freezers and other refrigerating or freezing equipment, electric or other; heat pumps other than air conditioning machines of heading 84.15 | | | | | | | | | | | | |
| 9 | 841861 | 25.00 | 8.00 | 7.58 | 40.00 | 5.00 | 3.07 | | 2.20 | 0.16 | 30.00 | 12.50 | 9.03 |
| 10 | 841869 | 25.00 | 8.67 | 8.66 | 40.00 | 5.00 | 1.32 | 5.00 | 2.20 | 0.11 | 30.00 | 5.00 | 3.13 |
| 8419 | Machinery, plant or laboratory equipment | | | | | | | | | | | | |
| 11 | 841919 | 40.00 | 9.00 | 9.00 | 40.00 | 7.50 | 4.77 | | 2.60 | 0.01 | 12.50 | 7.50 | 6.50 |
| 8479 | Machines and mechanical appliances having individual functions, not specified or included elsewhere in this Chapter | | | | | | | | | | | | |
| 12 | 847920 | 25.00 | 8.00 | 7.98 | 30.00 | 5.00 | 0.16 | | 1.70 | 0.04 | 10.00 | 0.00 | 0.00 |
| 8483 | Transmission shafts (including cam shafts and crank shafts) and cranks; bearing housings and plain shaft bearings; gears and gearing; ball or roller screws; gear boxes and other speed changers, including torque converters; flywheels and pulleys, including pulley blocks; clutches and shaft couplings (including universal joints) | | | | | | | | | | | | |
| 13 | 848340 | 40.00 | 8.00 | 8.00 | 40.00 | 5.00 | 3.04 | 13.78 | 1.96 | 0.31 | 10.00 | 0.00 | 0.00 |
| 14 | 848360 | 40.00 | 8.00 | 7.99 | 40.00 | 5.00 | 2.34 | 13.33 | 1.35 | 0.25 | 12.50 | 0.00 | 0.00 |

| | | | | | | | | | | | |
|---|---|---|---|---|---|---|---|---|---|---|---|
| | 8501 | Electric motors and generators (excluding generating sets) | | | | | | | | | |
| 15 | 850161 | 25.00 | 8.00 | 40.00 | 10.00 | 1.12 | 13.33 | 1.35 | 0.14 | 20.00 | 7.50 | 5.56 |
| 16 | 850162 | 25.00 | 7.99 | 40.00 | 10.00 | 0.55 | 12.60 | 1.35 | 0.05 | 20.00 | 10.00 | 5.90 |
| 17 | 850163 | 25.00 | 8.00 | 40.00 | 10.00 | 1.23 | 12.00 | 2.70 | 0.00 | 20.00 | 0.00 | 0.00 |
| 18 | 850164 | 25.00 | 7.92 | 40.00 | 10.00 | 1.78 | 14.00 | 2.70 | 0.98 | 20.00 | 0.00 | 0.00 |
| | 8502 | Electric generating sets and rotary converters | | | | | | | | | |
| 19 | 850231 | 25.00 | 8.00 | 40.00 | 10.00 | 7.63 | 13.60 | 1.35 | 0.00 | 20.00 | 0.00 | 0.00 |
| 20 | 850239 | 40.00 | 7.95 | 40.00 | 10.00 | 2.91 | 12.93 | 1.80 | 0.01 | 15.00 | 11.25 | 4.77 |
| | 8506 | Primary cells and primary batteries | | | | | | | | | |
| 21 | 850610 | 40.00 | 9.53 | 40.00 | 10.00 | 0.75 | | 4.70 | 0.03 | 20.00 | 10.00 | 6.20 |
| 22 | 850630 | 40.00 | 10.00 | 40.00 | 10.00 | 10.00 | | 4.70 | 0.00 | 20.00 | 0.00 | 0.00 |
| 23 | 850640 | 40.00 | 9.81 | 40.00 | 10.00 | 5.76 | | 4.70 | 3.25 | 20.00 | 0.00 | 0.00 |
| 24 | 850650 | 40.00 | 8.89 | 40.00 | 10.00 | 0.57 | | 4.70 | 3.17 | 20.00 | 10.00 | 8.91 |
| 25 | 850660 | 40.00 | 10.00 | 40.00 | 10.00 | 1.49 | | 4.70 | 1.13 | 20.00 | 0.00 | 0.00 |
| 26 | 850690 | 40.00 | 9.76 | 40.00 | 0.00 | 0.00 | | 4.70 | 0.58 | 20.00 | 0.00 | 0.00 |
| | 8507 | Electric accumulators, including separators therefor, whether or not rectangular (including square) | | | | | | | | | |
| 27 | 850720 | 40.00 | 9.98 | 40.00 | 15.00 | 4.17 | | 3.70 | 0.17 | 15.00 | 0.00 | 0.00 |
| 28 | 850740 | 40.00 | 8.27 | 40.00 | 10.00 | 0.51 | | 2.70 | 0.00 | 15.00 | 0.00 | 0.00 |
| | 8537 | Boards, panels, consoles, desks, cabinets and other bases, equipped with two or more apparatus of heading 85.35 or 85.36, for electric control or the distribution of electricity, including those incorporating instruments or apparatus of Chapter 90, and numerical control apparatus, other than switching apparatus of heading 85.17 | | | | | | | | | |
| 29 | 853710 | 40.00 | 8.00 | 7.91 | 40.00 | 5.00 | 1.73 | 13.60 | 2.10 | 0.32 | 20.00 | 11.67 | 6.00 |
| | 8541 | Diodes, transistors and similar semiconductor devices; photosensitive semiconductor devices, including photovoltaic cells whether or not assembled in modules or made up into panels; light emitting diodes; mounted piezo-electric crystals | | | | | | | | | |
| 30 | 854140 | 0.00 | 0.00 | 0.00 | 0.00 | 0.00 | 0.00 | 0.00 | 0.00 | 10.00 | 0.00 | 0.00 |
| | 9001 | Optical fibres and optical fibre bundles; optical fibre cables other than those of heading 85.44; sheets and plates of polarising material; lenses (including contact lenses), prisms, mirrors and other optical elements, of any material, unmounted, other than such elements of glass not optically worked | | | | | | | | | |
| 31 | 900190 | 10.00 | 9.73 | 40.00 | 5.00 | 1.42 | 7.50 | 1.45 | 1.01 | 15.00 | 0.00 | 0.00 |
| | 9002 | Lenses, prisms, mirrors and other optical elements, of any material, mounted, being parts of or fittings for instruments or apparatus, other than such elements of glass not optically worked | | | | | | | | | |
| 32 | 900290 | 10.00 | 8.33 | 40.00 | 3.00 | 0.59 | 8.00 | 3.35 | 1.37 | 10.00 | 0.00 | 0.00 |

Table 2.2 (cont.)

|  | HS Code | India MFN bound rates (%) | India MFN applied rates (%) | India Effective applied rates (%) | Indonesia MFN bound rates (%) | Indonesia MFN applied rates (%) | Indonesia Effective applied rates (%) | Turkey MFN bound rates (%) | Turkey MFN applied rates (%) | Turkey Effective applied rates (%) | South Africa MFN bound rates (%) | South Africa MFN applied rates (%) | South Africa Effective applied rates (%) |
|---|---|---|---|---|---|---|---|---|---|---|---|---|---|
| **Other Products** | | | | | | | | | | | | | |
| | 2207 | Ethyl alcohol, undenat, n/un 80% alc, alcohol, denat | | | | | | | | | | | |
| 33 | 220710 | 150.00 | 150.00 | 150.00 | 70.00 | 30.00 | 13.57 | 102.00 | 27.50 | 26.60 | 597.00 | | 0.00 |
| | 3824 | Prepared binders for foundry moulds or cores; chemical products and preparations of the chemical or allied industries (including those consisting of mixtures of natural products), not elsewhere specified or included | | | | | | | | | | | |
| 34 | 382490 | 40.00 | 8.00 | 7.97 | 40.00 | 5.00 | 3.30 | 14.50 | 4.90 | 0.37 | 14.84 | 3.75 | 1.79 |
| | 8410 | Hydraulic turbines, water wheels, and regulators therefor | | | | | | | | | | | |
| 35 | 841011 | 25.00 | 8.00 | 8.00 | 30.00 | 5.00 | 0.42 | 4.50 | 4.50 | 0.00 | 0.00 | 0.00 | 0.00 |
| 36 | 841090 | 25.00 | 8.00 | 7.92 | 30.00 | 5.00 | 2.31 | 3.00 | 4.50 | 0.06 | 0.00 | 0.00 | 0.00 |
| | 8411 | Turbo-jets, turbo-propellers and other gas turbines | | | | | | | | | | | |
| 37 | 841181 | 25.00 | 8.00 | 8.00 | 33.33 | 5.00 | 1.46 | 4.00 | 2.05 | 1.46 | 0.00 | 0.00 | 0.00 |
| 38 | 841182 | 25.00 | 8.00 | 8.00 | 33.33 | 5.00 | 0.35 | 3.00 | 2.05 | 0.46 | 0.00 | 0.00 | 0.00 |
| | 8503 | Parts suitable for use solely or principally with the machines of heading 85.01 or 85.02 | | | | | | | | | | | |
| 39 | 850300 | 25.00 | 8.00 | 7.98 | 40.00 | 4.00 | 1.59 | 21.00 | 2.70 | 0.07 | 15.00 | 11.25 | 4.77 |
| **All 39 SETA products** | | 33.30 | 8.97 | 8.86 | 37.10 | 5.25 | 2.19 | 14.14 | 2.68 | 0.40 | 21.24 | 3.50 | 1.93 |

*Notes:*

a. Tariff rates are weighted average. Tariff years are 2010 except for India (2009).

b. Ethanol is classified for *trade* purposes as HS 220710, undenatured ethyl alcohol. The United States imposes a 2.5 percent ad valorem tariff and 54 cents per gallon specific duty on imported ethanol. Since tariff data from TRAINS does not reflect a specific duty, we calculated the ad valorem equivalent of the 54 cents/gallon, amounting to a roughly 15 %, and added it to the bound, applied and effective rates for NAFTA.

*Source:* TRAINS (accessed through the WITS on September 25, 2011)

increased trade, from eliminating the residual tariffs would be limited. 'No pain, no gain' is a familiar saying in trade policy. On the other hand, many low-tariff EGs face various NTBs, especially LCRs, licenses and standards that often have quite distortive impacts. Our hope is that, while tariff elimination would be the first deliverable, NTBs would be removed within a few years.

Looking at three basic clean energy technologies, namely solar, clean coal and efficient lighting, in the eighteen top greenhouse gas (GHG) emitting developing countries, the World Bank (2007b) estimated that the removal of tariffs for those technologies would result in trade gains of up to 7 per cent and the removal of both tariffs and NTBs could boost trade by as much as 13 per cent. These calculations suggest the promise of a SETA pact.

It is worth noting that many countries maintain high tariff rates for ethanol (HS 220710) – in some cases more than 100 per cent.[15] Developing countries generally favour trade liberalisation in ethanol and biofuels as they see the export potential. To make the SETA enticing to a wider group of members, we think that ethanol and biofuels should be on the list of covered products.

### Trade Flows

In trade negotiations, there is no getting away from calculations of commercial advantage where, to put it crudely, exports are good and imports are bad. For that reason, we have compiled Table 2.3, which shows values (imports and exports) for each of the thirty-nine EGs traded between three major economies – namely NAFTA, the EU and China – and the other core and candidate countries in our list.

In the North American Free Trade Agreement (NAFTA), the major partners for both imports and exports in all thirty-nine EGs are the EU, China and Japan. NAFTA is a net importer of all thirty-nine EGs with respect to its top three trading partners. The EU's top three import partners for those goods are China, NAFTA and Japan. The main import commodity is PV cells and modules (HS 854140), accounting for about 70 per cent of the total. Most of the imports come from China. For exports, the EU's top three partners are NAFTA, China and Turkey. Its main export commodities are control boards (HS 853710), gears (HS

---

[15] The United States imposes a 2.5 per cent ad valorem tariff and a 54 cents per gallon specific duty on imported ethanol. Since tariff data from the Trade Analysis Information system (TRAINS) does not reflect a specific duty, we calculated that the ad valorem equivalent of the 54 cents per gallon amounts roughly equals a 15 per cent tariff, and added this figure to the bound, applied and effective rates for NAFTA.

Table 2.3a *Trade with Other Core and Candidate Countries: NAFTA, 2010*

2.3.1 Imports of SETA Products from the Core and Candidate Countries

mill. US$

| | HS Code | EU 27 | Chile | Colombia | Peru | Australia | New Zealand | Singapore | Japan | Korea | Brazil | China | India | Indonesia | Turkey | South Africa |
|---|---|---|---|---|---|---|---|---|---|---|---|---|---|---|---|---|
| | | Core Countries | | | | | | | | | | | Candidate Countries | | | |

**Products categorized as 'renewable energy' only from the WTO (2011)**

**3907 Polyacetals, other polyethers and epoxide resins, in primary forms; polycarbonates, alkyd resins, polyallyl esters and other polyesters, in primary forms**

| 1 | 390799 | 187.3 | 0.0 | 0.0 | 0.0 | 0.0 | 0.0 | 0.4 | 53.2 | 12.2 | 2.6 | 26.2 | 1.0 | 0.0 | 1.4 | 0.0 |

**4016 Other articles of vulcanised rubber other than hard rubber**

| 2 | 401699 | 293.3 | 0.6 | 0.2 | 0.2 | 4.4 | 0.7 | 3.9 | 198.2 | 67.5 | 5.9 | 360.1 | 20.7 | 8.5 | 1.5 | 0.5 |

**4707 Recovered (waste and scrap) paper or paperboard**

| 3 | 470710 | 0.0 | 0.0 | 0.0 | 0.0 | 0.0 | 0.0 | 0.0 | 0.0 | 0.0 | 0.0 | 0.0 | 0.0 | 0.0 | 0.0 | 0.0 |
| 4 | 470720 | 0.6 | 0.0 | 0.0 | 0.0 | 0.0 | 0.0 | 0.0 | 0.0 | 0.0 | 0.0 | 0.0 | 0.0 | 0.0 | 0.0 | 0.0 |
| 5 | 470730 | 0.5 | 0.0 | 0.0 | 0.0 | 0.0 | 0.0 | 0.0 | 0.0 | 0.0 | 0.0 | 0.0 | 0.0 | 0.0 | 0.0 | 0.0 |
| 6 | 470790 | 0.3 | 0.0 | 0.0 | 0.0 | 0.0 | 0.0 | 0.0 | 0.0 | 0.0 | 0.0 | 0.0 | 0.0 | 0.0 | 0.0 | 0.0 |

**7308 Structures (excluding prefabricated buildings of heading 94.06) and parts of structures (for example, bridges and bridge sections, lock gates, towers, lattice masts, roofs, roofing frameworks, doors and windows and their frames and thresholds for doors, shutters, balustrades, pillars and columns), of iron or steel; plates, rods, angles, shapes, sections, tubes and the like, prepared for use in structures, of iron or steel**

| 7 | 730820 | 33.2 | 0.0 | 0.0 | 0.0 | 0.0 | 0.0 | 0.0 | 0.0 | 33.8 | 0.4 | 134.4 | 16.7 | 20.8 | 0.2 | 0.0 |

**8406 Steam turbines and other vapour turbines**

| 8 | 840682 | 1.0 | 0.2 | 0.0 | 0.0 | 0.0 | 0.0 | 0.0 | 6.7 | 0.0 | 3.9 | 4.4 | 0.0 | 0.0 | 0.0 | 0.0 |

**8418 Refrigerators, freezers and other refrigerating or freezing equipment, electric or other; heat pumps other than air conditioning machines of heading 84.15**

| 9 | 841861 | 13.4 | 0.0 | 0.0 | 0.0 | 0.0 | 0.0 | 0.2 | 4.2 | 0.6 | 0.0 | 25.8 | 0.2 | 0.0 | 0.0 | 0.0 |
| 10 | 841869 | 158.2 | 0.2 | 2.6 | 0.0 | 0.0 | 0.4 | 2.2 | 31.7 | 36.6 | 7.9 | 229.9 | 2.7 | 0.7 | 0.0 | 1.8 |

**8419 Machinery, plant or laboratory equipment**

| 11 | 841919 | 13.4 | 0.0 | 0.0 | 0.0 | 0.0 | 0.0 | 0.0 | 26.5 | 4.8 | 0.0 | 26.0 | 0.1 | 0.0 | 0.1 | 0.2 |

**8479 Machines and mechanical appliances having individual functions, not specified or included elsewhere in this Chapter**

| 12 | 847920 | 5.4 | 0.0 | 0.0 | 1.1 | 0.0 | 0.0 | 0.0 | 0.0 | 0.0 | 0.0 | 0.5 | 0.6 | 0.0 | 0.2 | 0.0 |

| | | | | | | | | | | | | |
|---|---|---|---|---|---|---|---|---|---|---|---|---|
| 13 | 848340 | 1,332.2 | 0.5 | 0.1 | 0.5 | 4.2 | 1.2 | 4.5 | 313.8 | 39.7 | 8.6 | 278.8 | 38.1 | 0.3 | 2.1 | 4.1 |
| 14 | 848360 | 180.7 | 0.0 | 0.0 | 0.0 | 0.6 | 0.1 | 0.2 | 27.6 | 11.4 | 1.9 | 28.0 | 4.6 | 0.0 | 9.3 | 0.3 |

8483 — Transmission shafts (including cam shafts and crank shafts) and cranks; bearing housings and plain shaft bearings; gears and gearing; ball or roller screws; gear boxes and other speed changers, including torque converters; flywheels and pulleys, including pulley blocks; clutches and shaft couplings (including universal joints)

8501 — Electric motors and generators (excluding generating sets)

| | | | | | | | | | | | | |
|---|---|---|---|---|---|---|---|---|---|---|---|---|
| 15 | 850161 | 65.2 | 0.0 | 0.0 | 0.0 | 0.2 | 0.0 | 0.2 | 5.4 | 0.5 | 2.5 | 35.6 | 3.6 | 0.0 | 0.0 | 0.1 |
| 16 | 850162 | 9.3 | 0.0 | 0.0 | 0.0 | 0.1 | 0.1 | 0.0 | 18.3 | 0.0 | 0.3 | 7.8 | 9.4 | 0.0 | 0.0 | 0.0 |
| 17 | 850163 | 8.3 | 0.0 | 0.0 | 0.0 | 0.0 | 0.0 | 0.0 | 0.3 | 0.0 | 1.7 | 0.5 | 0.0 | 0.0 | 0.0 | 0.0 |
| 18 | 850164 | 183.0 | 0.0 | 0.0 | 0.0 | 1.0 | 0.0 | 0.0 | 128.4 | 11.4 | 11.5 | 23.8 | 5.3 | 0.0 | 0.0 | 0.2 |

8502 — Electric generating sets and rotary converters

| | | | | | | | | | | | | |
|---|---|---|---|---|---|---|---|---|---|---|---|---|
| 19 | 850231 | 1,991.7 | 0.0 | 0.0 | 0.0 | 0.1 | 0.0 | 0.0 | 0.3 | 8.2 | 25.3 | 18.1 | 140.5 | 0.0 | 0.0 | 0.0 |
| 20 | 850239 | 84.8 | 0.0 | 0.0 | 0.0 | 0.0 | 0.0 | 0.0 | 333.7 | 0.4 | 14.7 | 15.4 | 0.2 | 0.0 | 0.0 | 0.0 |

8506 — Primary cells and primary batteries

| | | | | | | | | | | | | |
|---|---|---|---|---|---|---|---|---|---|---|---|---|
| 21 | 850610 | 30.7 | 0.0 | 0.6 | 0.0 | 0.0 | 0.0 | 4.9 | 4.6 | 2.1 | 0.0 | 205.2 | 0.1 | 53.8 | 0.0 | 2.1 |
| 22 | 850630 | 0.0 | 0.0 | 0.0 | 0.0 | 0.0 | 0.0 | 0.0 | 0.1 | 0.0 | 0.0 | 1.3 | 0.0 | 0.0 | 0.0 | 0.0 |
| 23 | 850640 | 2.2 | 0.0 | 0.0 | 0.0 | 0.0 | 0.0 | 0.0 | 8.4 | 0.0 | 0.0 | 2.4 | 0.0 | 0.0 | 0.0 | 0.0 |
| 24 | 850650 | 47.1 | 0.0 | 0.0 | 0.0 | 0.1 | 0.1 | 0.4 | 87.0 | 49.2 | 0.0 | 63.8 | 0.1 | 12.1 | 0.0 | 0.0 |
| 25 | 850660 | 25.5 | 0.0 | 0.0 | 0.0 | 0.0 | 0.0 | 0.0 | 0.7 | 0.1 | 0.0 | 1.1 | 0.0 | 0.1 | 0.0 | 0.0 |
| 26 | 850690 | 5.1 | 0.0 | 0.0 | 0.0 | 0.0 | 0.0 | 0.0 | 3.6 | 1.0 | 0.0 | 4.4 | 0.0 | 0.0 | 0.0 | 0.0 |

8507 — Electric accumulators, including separators therefor, whether or not rectangular (including square)

| | | | | | | | | | | | | |
|---|---|---|---|---|---|---|---|---|---|---|---|---|
| 27 | 850720 | 60.9 | 0.0 | 0.0 | 0.0 | 0.0 | 0.1 | 0.3 | 10.2 | 4.3 | 0.0 | 385.9 | 13.5 | 0.0 | 0.0 | 0.2 |
| 28 | 850740 | 1.0 | 0.0 | 0.0 | 0.0 | 0.0 | 0.2 | 0.1 | 2.6 | 0.0 | 0.0 | 2.2 | 0.0 | 0.0 | 0.0 | 0.0 |

8537 — Boards, panels, consoles, desks, cabinets and other bases, equipped with two or more apparatus of heading 85.35 or 85.36, for electric control or the distribution of electricity, including those incorporating instruments or apparatus of Chapter 90, and numerical control apparatus, other than switching apparatus of heading 85.17

| | | | | | | | | | | | | |
|---|---|---|---|---|---|---|---|---|---|---|---|---|
| 29 | 853710 | 1,087.1 | 0.2 | 0.5 | 0.0 | 7.6 | 2.3 | 277.8 | 453.3 | 127.3 | 19.0 | 985.2 | 64.8 | 22.2 | 0.7 | 1.5 |

8541 — Diodes, transistors and similar semiconductor devices; photosensitive semiconductor devices, including photovoltaic cells whether or not assembled in modules or made up into panels; light emitting diodes; mounted piezo-electric crystals

| | | | | | | | | | | | | |
|---|---|---|---|---|---|---|---|---|---|---|---|---|
| 30 | 854140 | 331.5 | 0.1 | 0.0 | 0.0 | 1.1 | 1.5 | 83.6 | 1,065.4 | 69.2 | 0.4 | 1,905.2 | 45.8 | 8.0 | 0.3 | 0.2 |

9001 — Optical fibres and optical fibre bundles; optical fibre cables other than those of heading 85.44; sheets and plates of polarising material; lenses (including contact lenses), prisms, mirrors and other optical elements, of any material, unmounted, other than such elements of glass not optically worked

Table 2.3a (cont.)

2.3.1 Imports of SETA Products from the Core and Candidate Countries

mill. US$

| | | HS Code | Core Countries | | | | | | | | | | Candidate Countries | | | | |
|---|---|---|---|---|---|---|---|---|---|---|---|---|---|---|---|---|---|
| | | | EU 27 | Chile | Colombia | Peru | Australia | New Zealand | Singapore | Japan | Korea | Brazil | China | India | Indonesia | Turkey | South Africa |
| 31 | | 900190 | 102.5 | 0.0 | 0.0 | 0.0 | 0.8 | 0.0 | 47.1 | 78.7 | 16.5 | 0.1 | 103.8 | 0.6 | 0.0 | 0.0 | 0.0 |
| | 9002 | Lenses, prisms, mirrors and other optical elements, of any material, mounted, being parts of or fittings for instruments or apparatus, other than such elements of glass not optically worked | | | | | | | | | | | | | | | |
| 32 | | 900290 | 74.6 | 0.0 | 0.0 | 0.0 | 0.2 | 0.1 | 16.6 | 43.1 | 1.1 | 0.0 | 30.9 | 1.3 | 0.0 | 0.0 | 0.0 |
| Other Products | | | | | | | | | | | | | | | | | |
| | 2207 | Ethyl alcohol, undenat, n/un 80% alc, alcohol, denat | | | | | | | | | | | | | | | |
| 33 | | 220710 | 2.5 | 0.2 | 0.0 | 0.0 | 5.1 | 0.0 | 0.0 | 0.0 | 0.0 | 281.6 | 0.1 | 0.0 | 0.0 | 0.0 | 26.3 |
| | 3824 | Prepared binders for foundry moulds or cores; chemical products and preparations of the chemical or allied industries (including those consisting of mixtures of natural products), not elsewhere specified or included | | | | | | | | | | | | | | | |
| 34 | | 382490 | 833.2 | 1.1 | 2.5 | 1.2 | 9.6 | 0.9 | 2.9 | 182.2 | 21.8 | 10.2 | 260.7 | 16.9 | 6.7 | 1.2 | 2.4 |
| | 8410 | Hydraulic turbines, water wheels, and regulators therefor | | | | | | | | | | | | | | | |
| 35 | | 841011 | 2.2 | 0.0 | 0.0 | 0.0 | 0.0 | 0.0 | 0.0 | 0.0 | 0.0 | 0.0 | 0.1 | 0.0 | 0.0 | 0.0 | 0.0 |
| 36 | | 841090 | 25.0 | 0.0 | 0.0 | 0.0 | 0.0 | 0.0 | 0.0 | 3.9 | 11.2 | 26.0 | 19.8 | 0.1 | 0.0 | 0.0 | 0.1 |
| | 8411 | Turbo-jets, turbo-propellers and other gas turbines | | | | | | | | | | | | | | | |
| 37 | | 841181 | 120.0 | 0.0 | 0.0 | 0.0 | 0.0 | 0.0 | 0.0 | 0.9 | 0.0 | 0.5 | 0.0 | 0.0 | 1.2 | 0.0 | 0.0 |
| 38 | | 841182 | 81.2 | 0.0 | 0.0 | 0.0 | 0.0 | 0.0 | 0.2 | 5.9 | 0.0 | 0.0 | 0.0 | 0.0 | 0.0 | 0.0 | 0.0 |
| | 8503 | Parts suitable for use solely or principally with the machines of heading 85.01 or 85.02 | | | | | | | | | | | | | | | |
| 39 | | 850300 | 524.4 | 0.0 | 0.0 | 0.0 | 4.9 | 0.0 | 39.1 | 151.0 | 19.8 | 94.9 | 343.0 | 66.9 | 10.0 | 0.5 | 4.0 |
| | Total Imports | | 7,918.2 | 3.1 | 6.5 | 2.9 | 41.2 | 7.2 | 484.4 | 3,249.7 | 550.6 | 519.9 | 5,530.4 | 453.8 | 144.4 | 17.7 | 43.9 |

2.3.2 Exports of SETA Products to the Core and Candidate Countries

mill. US$

| | | | Core Countries | | | | | | | | | Candidate Countries | | | |
|---|---|---|---|---|---|---|---|---|---|---|---|---|---|---|---|
| | HS Code | EU 27 | Chile | Colombia | Peru | Australia | New Zealand | Singapore | Japan | Korea | Brazil | China | India | Indonesia | Turkey | South Africa |

**Products categorized as 'renewable energy' only from the WTO (2011)**

**3907** Polyacetals, other polyethers and epoxide resins, in primary forms; polycarbonates, alkyd resins, polyallyl esters and other polyesters, in primary forms

| 1 | 390799 | 168.3 | 2.8 | 3.5 | 1.3 | 3.7 | 0.6 | 9.7 | 81.6 | 32.5 | 19.0 | 65.1 | 5.4 | 0.5 | 0.3 | 0.3 |

**4016** Other articles of vulcanised rubber other than hard rubber

| 2 | 401699 | 80.6 | 6.1 | 4.4 | 1.7 | 8.8 | 0.6 | 6.8 | 8.6 | 2.6 | 18.0 | 15.7 | 6.5 | 0.8 | 1.4 | 3.4 |

**4707** Recovered (waste and scrap) paper or paperboard

| 3 | 470710 | 19.1 | 14.5 | 4.9 | 2.1 | 0.2 | 0.0 | 0.0 | 3.7 | 53.5 | 1.6 | 768.6 | 128.0 | 21.6 | 3.0 | |
| 4 | 470720 | 22.2 | 5.3 | 5.5 | 1.2 | 0.8 | 0.0 | 0.1 | 5.2 | 9.4 | 0.2 | 256.3 | 42.0 | 2.7 | 0.4 | 0.0 |
| 5 | 470730 | 2.1 | 0.6 | 0.7 | 0.2 | 0.2 | 0.0 | 0.0 | 0.3 | 35.3 | 0.5 | 463.2 | 16.1 | 27.5 | 0.3 | 0.0 |
| 6 | 470790 | 26.2 | 4.1 | 10.7 | 3.0 | 0.0 | 0.0 | 0.1 | 8.1 | 93.7 | 0.0 | 442.7 | 62.7 | 17.6 | 0.6 | 0.3 |

**7308** Structures (excluding prefabricated buildings of heading 94.06) and parts of structures (for example, bridges and bridge sections, lock gates, towers, lattice masts, roofs, roofing frameworks, doors and windows and their frames and thresholds for doors, shutters, balustrades, pillars and columns), of iron or steel; plates, rods, angles, shapes, sections, tubes and the like, prepared for use in structures, of iron or steel

| 7 | 730820 | 2.4 | 0.1 | 0.1 | 0.5 | 2.7 | 0.5 | 0.7 | 0.1 | 0.1 | 0.4 | 0.3 | 0.2 | 0.4 | 0.0 | 0.2 |

**8406** Steam turbines and other vapour turbines

| 8 | 840682 | 9.3 | 0.1 | 3.0 | 0.0 | 1.5 | 1.2 | 1.2 | 4.9 | 7.3 | 0.1 | 6.7 | 2.3 | 1.0 | 0.3 | 0.3 |

**8418** Refrigerators, freezers and other refrigerating or freezing equipment, electric or other; heat pumps other than air conditioning machines of heading 84.15

| 9 | 841861 | 3.8 | 0.0 | 0.3 | 0.1 | 1.1 | 0.2 | 0.2 | 0.1 | 0.8 | 7.6 | 1.8 | 0.3 | 0.2 | 0.8 | 0.1 |
| 10 | 841869 | 75.0 | 4.9 | 13.9 | 5.3 | 38.8 | 3.1 | 15.0 | 16.7 | 22.8 | 43.2 | 39.4 | 15.3 | 6.3 | 13.1 | 5.9 |

**8419** Machinery, plant or laboratory equipment

| 11 | 841919 | 10.2 | 0.4 | 1.3 | 0.6 | 0.3 | 0.1 | 0.0 | 0.7 | 0.9 | 0.9 | 8.3 | 0.2 | 0.1 | 0.2 | 0.3 |

**8479** Machines and mechanical appliances having individual functions, not specified or included elsewhere in this Chapter

| 12 | 847920 | 2.1 | 0.0 | 1.0 | 0.3 | 0.6 | 0.0 | 0.3 | 0.3 | 0.3 | 3.2 | 4.9 | 0.1 | 0.0 | 0.0 | 0.0 |

**8483** Transmission shafts (including cam shafts and crank shafts) and cranks; bearing housings and plain shaft bearings; gears and gearing; ball or roller screws; gear boxes and other speed changers, including torque converters; flywheels and pulleys, including pulley blocks; clutches and shaft couplings (including universal joints)

| 13 | 848340 | 171.7 | 18.4 | 9.8 | 10.8 | 28.6 | 1.3 | 27.3 | 17.1 | 29.0 | 98.6 | 59.6 | 12.3 | 2.2 | 1.9 | 6.1 |
| 14 | 848360 | 45.2 | 4.6 | 2.2 | 2.5 | 4.8 | 0.2 | 6.4 | 7.8 | 1.9 | 10.2 | 11.8 | 4.6 | 0.6 | 1.1 | 1.9 |

Table 2.3a (cont.)

2.3.2 Exports of SETA Products to the Core and Candidate Countries

mill. US$

| | HS Code | Core Countries | | | | | | | | | | | Candidate Countries | | | | |
|---|---|---|---|---|---|---|---|---|---|---|---|---|---|---|---|---|---|
| | | EU 27 | Chile | Colombia | Peru | Australia | New Zealand | Singapore | Japan | Korea | Brazil | China | India | Indonesia | Turkey | South Africa |
| | **8501** | Electric motors and generators (excluding generating sets) | | | | | | | | | | | | | | | |
| 15 | 850161 | 26.4 | 0.3 | 1.1 | 0.8 | 1.1 | 0.0 | 1.5 | 0.7 | 0.1 | 2.3 | 3.9 | 0.4 | 0.1 | 0.7 | 0.1 |
| 16 | 850162 | 97.9 | 0.1 | 1.2 | 1.3 | 0.3 | 0.0 | 0.8 | 0.7 | 0.3 | 8.3 | 12.2 | 0.0 | 0.0 | 0.5 | 0.8 |
| 17 | 850163 | 2.0 | 0.4 | 1.3 | 1.4 | 0.4 | 0.0 | 0.3 | 0.0 | 0.0 | 5.8 | 0.4 | 0.0 | 0.0 | 0.1 | 0.0 |
| 18 | 850164 | 97.6 | 0.5 | 1.6 | 0.4 | 6.9 | 0.0 | 15.2 | 0.3 | 14.3 | 0.5 | 4.0 | 16.1 | 0.8 | 26.0 | 0.3 |
| | **8502** | Electric generating sets and rotary converters | | | | | | | | | | | | | | | |
| 19 | 850231 | 8.5 | 0.0 | 0.0 | 0.0 | 126.7 | 0.0 | 0.0 | 0.0 | 2.2 | 0.0 | 0.0 | 0.0 | 0.0 | 0.0 | 0.0 |
| 20 | 850239 | 90.7 | 0.2 | 5.2 | 8.4 | 0.0 | 0.0 | 0.5 | 1.4 | 78.7 | 31.3 | 90.0 | 7.2 | 21.4 | 7.4 | 0.1 |
| | **8506** | Primary cells and primary batteries | | | | | | | | | | | | | | | |
| 21 | 850610 | 24.7 | 2.4 | 2.1 | 0.4 | 4.5 | 0.4 | 19.6 | 2.3 | 0.4 | 3.9 | 11.0 | 0.1 | 0.0 | 0.0 | 0.3 |
| 22 | 850630 | 0.0 | 0.1 | 0.0 | 0.0 | 0.1 | 0.0 | 0.4 | 0.0 | 0.0 | 0.0 | 0.2 | 0.0 | 0.0 | 0.0 | 0.0 |
| 23 | 850640 | 5.6 | 0.0 | 0.1 | 0.0 | 0.3 | 0.0 | 0.5 | 0.0 | 0.4 | 0.5 | 0.0 | 0.0 | 0.0 | 0.0 | 0.2 |
| 24 | 850650 | 111.9 | 0.6 | 2.9 | 0.6 | 8.5 | 0.3 | 12.8 | 11.1 | 2.7 | 4.4 | 10.0 | 1.2 | 0.8 | 0.1 | 0.2 |
| 25 | 850660 | 4.6 | 0.1 | 0.0 | 0.0 | 0.5 | 0.1 | 0.2 | 0.1 | 0.1 | 0.2 | 0.0 | 0.0 | 0.0 | 0.0 | 0.0 |
| 26 | 850690 | 11.0 | 0.2 | 0.7 | 0.3 | 0.4 | 0.1 | 7.5 | 1.9 | 4.2 | 0.7 | 2.3 | 0.1 | 0.1 | 0.1 | 0.0 |
| | **8507** | Electric accumulators, including separators therefor, whether or not rectangular (including square) | | | | | | | | | | | | | | | |
| 27 | 850720 | 138.6 | 5.1 | 2.8 | 2.3 | 15.7 | 2.5 | 16.3 | 9.3 | 11.0 | 6.1 | 16.1 | 4.2 | 0.7 | 1.4 | 11.5 |
| 28 | 850740 | 0.2 | 0.0 | 0.1 | 0.0 | 0.3 | 0.0 | 0.0 | 0.3 | 0.0 | 0.0 | 0.0 | 0.0 | 0.0 | 0.0 | 0.0 |
| | **8537** | Boards, panels, consoles, desks, cabinets and other bases, equipped with two or more apparatus of heading 85.35 or 85.36, for electric control or the distribution of electricity, including those incorporating instruments or apparatus of Chapter 90, and numerical control apparatus, other than switching apparatus of heading 85.17 | | | | | | | | | | | | | | | |
| 29 | 853710 | 493.8 | 15.2 | 17.8 | 6.7 | 55.8 | 4.5 | 72.3 | 38.8 | 65.1 | 95.0 | 169.4 | 36.7 | 5.9 | 10.7 | 17.1 |
| | **8541** | Diodes, transistors and similar semiconductor devices; photosensitive semiconductor devices, including photovoltaic cells whether or not assembled in modules or made up into panels; light emitting diodes; mounted piezo-electric crystals | | | | | | | | | | | | | | | |
| | 854140 | 1,270.3 | 2.3 | 0.9 | 0.6 | 10.9 | 1.1 | 178.6 | 222.7 | 70.5 | 9.6 | 140.4 | 53.1 | 0.4 | 1.3 | 4.4 |
| 30 | **9001** | Optical fibres and optical fibre bundles; optical fibre cables other than those of heading 85.44; sheets and plates of polarising material; lenses (including contact lenses), prisms, mirrors and other optical elements, of any material, unmounted, other than such elements of glass not optically worked | | | | | | | | | | | | | | | |

| | | | | | | | | | | | |
|---|---|---|---|---|---|---|---|---|---|---|---|
| 31 | 900190 | 146.1 | 0.2 | 2.5 | 0.1 | 6.8 | 0.3 | 10.2 | 149.3 | 179.4 | 2.0 | 281.1 | 3.3 | 0.5 | 0.8 | 0.5 |
| | 9002 | Lenses, prisms, mirrors and other optical elements, of any material, mounted, being parts of or fittings for instruments or apparatus, other than such elements of glass not optically worked |
| 32 | 900290 | 59.8 | 0.3 | 0.1 | 0.1 | 0.6 | 0.1 | 5.1 | 11.8 | 7.6 | 0.9 | 4.9 | 0.9 | 0.2 | 0.2 | 0.1 |
| **Other Products** |
| | 2207 | Ethyl alcohol, undenat, n/un 80% alc, alcohol, denat |
| 33 | 220710 | 120.3 | 2.0 | 0.0 | 0.1 | 4.0 | 0.0 | 7.8 | 0.1 | 12.2 | 9.0 | 0.3 | 16.1 | 0.0 | 0.5 | 0.7 |
| | 3824 | Prepared binders for foundry moulds or cores; chemical products and preparations of the chemical or allied industries (including those consisting of mixtures of natural products), not elsewhere specified or included |
| 34 | 382490 | 753.7 | 39.5 | 31.2 | 41.1 | 50.8 | 4.9 | 141.8 | 155.6 | 87.8 | 109.3 | 309.0 | 80.5 | 27.3 | 12.0 | 15.7 |
| | 8410 | Hydraulic turbines, water wheels, and regulators therefor |
| 35 | 841011 | 2.8 | 0.4 | 0.1 | 0.0 | 0.0 | 0.0 | 0.0 | 0.0 | 0.1 | 0.3 | 0.0 | 0.2 | 0.1 | 0.0 | 0.0 |
| 36 | 841090 | 8.8 | 4.1 | 0.2 | 0.5 | 1.3 | 0.1 | 0.1 | 0.2 | 1.9 | 0.9 | 0.1 | 0.5 | 0.0 | 0.2 | 0.1 |
| | 8411 | Turbo-jets, turbo-propellers and other gas turbines |
| 37 | 841181 | 144.7 | 2.5 | 1.9 | 1.2 | 4.7 | 0.2 | 18.5 | 14.7 | 1.5 | 0.9 | 9.5 | 0.3 | 1.4 | 0.1 | 0.4 |
| 38 | 841182 | 699.7 | 18.4 | 15.4 | 0.0 | 83.5 | 13.7 | 45.3 | 171.3 | 35.3 | 34.2 | 177.9 | 275.4 | 31.5 | 24.0 | 0.0 |
| | 8503 | Parts suitable for use solely or principally with the machines of heading 85.01 or 85.02 |
| 39 | 850300 | 165.6 | 23.0 | 13.7 | 7.0 | 19.2 | 0.7 | 22.1 | 43.3 | 47.0 | 35.8 | 42.2 | 17.4 | 6.6 | 9.1 | 13.0 |
| | Total Exports | 5,123.3 | 179.6 | 164.1 | 103.1 | 495.0 | 36.8 | 645.0 | 990.6 | 912.4 | 564.9 | 3,429.2 | 809.8 | 179.2 | 118.6 | 84.3 |

*Source*: UN COMTRADE (accessed through the WITS on October 1, 2011)

Table 2.3b  *Trade with Other Core and Candidate Countries: EU-27, 2010*

mill. US$

### 3.1 Imports

| | | Core Countries | | | | | | | | | Candidate Countries | | | |
|---|---|---|---|---|---|---|---|---|---|---|---|---|---|---|
| | HS Code | NAFTA | Chile | Colombia | Peru | Australia | New Zealand | Singapore | Japan | Korea | Brazil | China | India | Indonesia | Turkey | South Africa |

Products categorized as 'renewable energy' only from the WTO (2011)

| | | | | | | | | | | | | | | | | |
|---|---|---|---|---|---|---|---|---|---|---|---|---|---|---|---|---|
| | 3907 | Polyacetals, other polyethers and epoxide resins, in primary forms; polycarbonates, alkyd resins, polyallyl esters and other polyesters, in primary forms |
| 1 | 390799 | 158.3 | 0.0 | 0.0 | 0.0 | 0.1 | 0.0 | 0.2 | 26.6 | 20.1 | 0.9 | 10.3 | 3.2 | 0.0 | 26.6 | 0.1 |
| | 4016 | Other articles of vulcanised rubber other than hard rubber |
| 2 | 401699 | 138.2 | 0.0 | 0.0 | 0.0 | 0.9 | 0.5 | 5.1 | 115.2 | 39.4 | 10.2 | 199.6 | 54.8 | 2.8 | 125.2 | 3.0 |
| | 4707 | Recovered (waste and scrap) paper or paperboard |
| 3 | 470710 | 6.3 | 0.0 | 0.0 | 0.0 | 0.0 | 0.0 | 0.0 | 0.0 | 0.0 | 0.0 | 0.1 | 0.0 | 0.0 | 0.1 | 0.0 |
| 4 | 470720 | 18.6 | 0.0 | 0.0 | 0.0 | 0.0 | 0.0 | 0.0 | 0.0 | 0.0 | 0.5 | 0.0 | 0.0 | 0.0 | 0.0 | 0.0 |
| 5 | 470730 | 22.7 | 0.0 | 0.0 | 0.0 | 0.0 | 0.0 | 0.0 | 0.0 | 0.0 | 0.0 | 0.0 | 0.0 | 0.0 | 0.3 | 0.0 |
| 6 | 470790 | 16.8 | 0.0 | 0.0 | 0.0 | 0.0 | 0.0 | 0.0 | 0.0 | 0.0 | 0.0 | 0.0 | 0.0 | 0.0 | 0.5 | 0.0 |
| | 7308 | Structures (excluding prefabricated buildings of heading 94.06) and parts of structures (for example, bridges and bridge sections, lock gates, towers, lattice masts, roofs, roofing frameworks, doors and windows and their frames and thresholds for doors, shutters, balustrades, pillars and columns), of iron or steel; plates, rods, angles, shapes, sections, tubes and the like, prepared for use in structures, of iron or steel |
| 7 | 730820 | 1.1 | 0.0 | 0.0 | 0.0 | 0.0 | 0.0 | 0.0 | 0.0 | 16.8 | 0.0 | 42.0 | 2.7 | 4.4 | 8.4 | 0.9 |
| | 8406 | Steam turbines and other vapour turbines |
| 8 | 840682 | 6.9 | 0.0 | 0.0 | 0.0 | 0.0 | 0.0 | 0.0 | 3.1 | 0.0 | 8.5 | 0.0 | 0.2 | 0.0 | 0.0 | 0.0 |
| | 8418 | Refrigerators, freezers and other refrigerating or freezing equipment, electric or other; heat pumps other than air conditioning machines of heading 84.15 |
| 9 | 841861 | 9.4 | 0.0 | 0.0 | 0.0 | 0.0 | 0.0 | 1.0 | 16.9 | 1.0 | 0.0 | 53.3 | 0.1 | 0.3 | 1.2 | 1.3 |
| 10 | 841869 | 69.3 | 0.0 | 0.0 | 0.0 | 0.0 | 0.0 | 0.8 | 18.3 | 6.6 | 0.0 | 116.3 | 2.1 | 0.0 | 3.3 | 0.0 |
| | 8419 | Machinery, plant or laboratory equipment |
| 11 | 841919 | 11.0 | 0.0 | 0.0 | 0.0 | 0.1 | 0.0 | 0.0 | 0.7 | 0.1 | 0.0 | 49.9 | 0.3 | 0.0 | 5.1 | 0.0 |

| # | Code | Description | | | | | | | | | | |
|---|------|-------------|---|---|---|---|---|---|---|---|---|---|
| 12 | 8479 | Machines and mechanical appliances having individual functions, not specified or included elsewhere in this Chapter | | | | | | | | | | |
| | 847920 | 0.8 | 0.0 | 0.0 | 0.0 | 0.0 | 0.4 | 1.2 | 1.9 | 0.0 | 7.3 | 0.0 |
| | 8483 | Transmission shafts (including cam shafts and crank shafts) and cranks; bearing housings and plain shaft bearings; gears and gearing; ball or roller screws; gear boxes and other speed changers, including torque converters; flywheels and pulleys, including pulley blocks; clutches and shaft couplings (including universal joints) | | | | | | | | | | |
| 13 | 848340 | 225.4 | 0.6 | 0.0 | 1.6 | 0.1 | 1.2 | 193.0 | 7.6 | 2.8 | 139.9 | 25.1 | 0.1 | 11.3 | 5.3 |
| 14 | 848360 | 63.0 | 0.0 | 0.0 | 0.6 | 0.1 | 1.2 | 39.2 | 6.1 | 3.6 | 34.7 | 9.3 | 0.0 | 2.7 | 3.7 |
| | 8501 | Electric motors and generators (excluding generating sets) | | | | | | | | | | |
| 15 | 850161 | 55.7 | 0.4 | 0.0 | 0.0 | 0.0 | 0.0 | 4.2 | 1.9 | 0.0 | 10.3 | 11.1 | 0.0 | 1.2 | 0.4 |
| 16 | 850162 | 30.8 | 0.0 | 0.1 | 0.2 | 0.5 | 0.1 | 0.2 | 0.1 | 0.0 | 9.0 | 1.3 | 0.0 | 0.2 | 0.1 |
| 17 | 850163 | 7.8 | 0.0 | 0.1 | 0.0 | 0.1 | 0.1 | 1.7 | 0.0 | 0.1 | 9.5 | 0.6 | 0.0 | 0.1 | 0.0 |
| 18 | 850164 | 68.3 | 0.0 | 0.0 | 0.0 | 0.0 | 0.0 | 0.5 | 14.2 | 2.1 | 56.3 | 6.9 | 0.0 | 0.5 | 0.0 |
| | 8502 | Electric generating sets and rotary converters | | | | | | | | | | |
| 19 | 850231 | 7.9 | 0.0 | 0.0 | 0.1 | 0.0 | 0.0 | 0.2 | 0.0 | 1.6 | 7.9 | 6.3 | 0.0 | 0.0 | 0.0 |
| 20 | 850239 | 88.8 | 0.0 | 0.0 | 1.2 | 0.0 | 0.0 | 0.5 | 4.1 | 0.4 | 11.4 | 0.1 | 0.0 | 0.4 | 0.0 |
| | 8506 | Primary cells and primary batteries | | | | | | | | | | |
| 21 | 850610 | 20.9 | 0.0 | 0.0 | 0.0 | 0.0 | 7.8 | 7.4 | 1.7 | 0.1 | 145.7 | 0.4 | 8.9 | 0.0 | 0.0 |
| 22 | 850630 | 0.0 | 0.0 | 0.0 | 0.0 | 0.0 | 0.0 | 0.2 | 0.0 | 0.0 | 1.6 | 0.0 | 0.0 | 0.0 | 0.0 |
| 23 | 850640 | 2.4 | 0.0 | 0.0 | 0.0 | 0.0 | 0.1 | 3.2 | 0.0 | 0.0 | 1.7 | 0.0 | 0.0 | 0.0 | 0.0 |
| 24 | 850650 | 130.9 | 0.0 | 0.0 | 1.2 | 0.2 | 3.8 | 55.1 | 16.6 | 0.0 | 46.5 | 0.1 | 11.3 | 0.6 | 0.0 |
| 25 | 850660 | 6.6 | 0.0 | 0.0 | 0.0 | 0.0 | 0.2 | 0.8 | 0.4 | 0.0 | 13.2 | 0.0 | 0.1 | 0.0 | 0.0 |
| 26 | 850690 | 9.2 | 0.0 | 0.0 | 0.0 | 0.0 | 0.1 | 1.4 | 0.2 | 0.0 | 1.9 | 0.2 | 0.1 | 0.0 | 0.1 |
| | 8507 | Electric accumulators, including separators therefor, whether or not rectangular (including square) | | | | | | | | | | |
| 27 | 850720 | 115.5 | 0.0 | 0.0 | 0.0 | 0.0 | 1.9 | 1.7 | 3.1 | 0.9 | 328.5 | 8.6 | 0.2 | 0.6 | 2.1 |
| 28 | 850740 | 0.5 | 0.0 | 0.0 | 0.0 | 0.0 | 0.0 | 0.4 | 0.0 | 0.0 | 3.0 | 0.0 | 0.0 | 0.0 | 0.0 |
| | 8537 | Boards, panels, consoles, desks, cabinets and other bases, equipped with two or more apparatus of heading 85.35 or 85.36, for electric control or the distribution of electricity, including those incorporating instruments or apparatus of Chapter 90, and numerical control apparatus, other than switching apparatus of heading 85.17 | | | | | | | | | | |
| 29 | 853710 | 593.0 | 0.3 | 0.0 | 0.0 | 17.8 | 332.3 | 104.5 | 16.1 | 342.5 | 12.1 | 12.0 | 13.1 | 5.1 |
| | 8541 | Diodes, transistors and similar semiconductor devices; photosensitive semiconductor devices, including photovoltaic cells whether or not assembled in modules or made up into panels; light emitting diodes; mounted piezo-electric crystals | | | | | | | | | | |

Table 2.3b (cont.)

3.1 Imports

mill. US$

| | HS Code | Core Countries | | | | | | | | | | Candidate Countries | | | | |
|---|---|---|---|---|---|---|---|---|---|---|---|---|---|---|---|---|
| | | NAFTA | Chile | Colombia | Peru | Australia | New Zealand | Singapore | Japan | Korea | Brazil | China | India | Indonesia | Turkey | South Africa |
| 30 | 854140 | 1,587.6 | 0.0 | 0.0 | 0.0 | 21.3 | 0.1 | 37.3 | 1,934.6 | 1,190.9 | 0.1 | 16,210.6 | 477.8 | 12.2 | 11.8 | 156.0 |
| | 9001 | Optical fibres and optical fibre bundles; optical fibre cables other than those of heading 85.44; sheets and plates of polarising material; lenses (including contact lenses), prisms, mirrors and other optical elements, of any material, unmounted, other than such elements of glass not optically worked | | | | | | | | | | | | | | |
| 31 | 900190 | 154.7 | 0.0 | 0.0 | 0.0 | 0.1 | 0.0 | 27.7 | 86.4 | 143.0 | 0.1 | 58.8 | 1.9 | 0.4 | 0.3 | 0.0 |
| | 9002 | Lenses, prisms, mirrors and other optical elements, of any material, mounted, being parts of or fittings for instruments or apparatus, other than such elements of glass not optically worked | | | | | | | | | | | | | | |
| 32 | 900290 | 41.8 | 0.1 | 0.0 | 0.0 | 0.2 | 0.0 | 6.7 | 31.5 | 4.1 | 0.0 | 22.0 | 0.6 | 0.1 | 0.2 | 0.2 |
| | Other Products | | | | | | | | | | | | | | | |
| | 2207 | Ethyl alcohol, undenat, n/un 80% alc, alcohol, denat | | | | | | | | | | | | | | |
| 33 | 220710 | 26.9 | 0.0 | 0.0 | 5.6 | 0.0 | 0.0 | 0.0 | 0.0 | 6.9 | 105.1 | 0.0 | 0.0 | 0.0 | 9.6 | 0.0 |
| | 3824 | Prepared binders for foundry moulds or cores; chemical products and preparations of the chemical or allied industries (including those consisting of mixtures of natural products), not elsewhere specified or included | | | | | | | | | | | | | | |
| 34 | 382490 | 725.3 | 0.1 | 10.6 | 0.0 | 4.2 | 0.2 | 13.7 | 145.2 | 13.3 | 127.5 | 153.5 | 41.7 | 268.4 | 16.3 | 12.2 |
| | 8410 | Hydraulic turbines, water wheels, and regulators therefor | | | | | | | | | | | | | | |
| 35 | 841011 | 0.3 | 0.0 | 0.0 | 0.0 | 0.0 | 0.0 | 0.0 | 0.0 | 0.0 | 0.1 | 1.0 | 0.1 | 0.6 | 0.0 | 0.0 |
| 36 | 841090 | 7.1 | 0.0 | 0.0 | 0.0 | 0.0 | 0.0 | 0.2 | 0.1 | 2.8 | 10.9 | 7.0 | 9.6 | 0.0 | 0.9 | 0.0 |
| | 8411 | Turbo-jets, turbo-propellers and other gas turbines | | | | | | | | | | | | | | |
| 37 | 841181 | 311.7 | 0.0 | 0.0 | 0.0 | 0.0 | 0.0 | 1.9 | 1.5 | 0.1 | 0.1 | 0.2 | 0.2 | 0.0 | 1.1 | 0.1 |
| 38 | 841182 | 576.3 | 0.0 | 1.6 | 0.6 | 5.2 | 0.0 | 2.5 | 18.0 | 0.0 | 16.0 | 23.7 | 6.8 | 0.0 | 4.9 | 0.0 |
| | 8503 | Parts suitable for use solely or principally with the machines of heading 85.01 or 85.02 | | | | | | | | | | | | | | |
| 39 | 850300 | 173.0 | 0.0 | 0.0 | 0.0 | 0.0 | 0.0 | 3.3 | 112.7 | 23.0 | 29.6 | 283.1 | 48.4 | 4.0 | 74.4 | 8.0 |
| | Total Imports | 5,490.7 | 1.5 | 12.3 | 6.2 | 85.6 | 11.3 | 134.6 | 3,152.5 | 1,628.3 | 337.5 | 18,395.9 | 734.2 | 325.8 | 327.8 | 198.7 |

## 3.2 Exports

mill. US$

| | HS Code | Core Countries ||||||||| Candidate Countries ||||||
|---|---|---|---|---|---|---|---|---|---|---|---|---|---|---|---|---|
| | | NAFTA | Chile | Colombia | Peru | Australia | New Zealand | Singapore | Japan | Korea | Brazil | China | India | Indonesia | Turkey | South Africa |

**Products categorized as 'renewable energy' only from the WTO (2011)**

**3907** Polyacetals, other polyethers and epoxide resins, in primary forms; polycarbonates, alkyd resins, polyallyl esters and other polyesters, in primary forms

1  390799  146.4  1.0  2.7  0.5  6.2  26.9  9.3  30.5  67.1  81.6  18.4  3.0  67.5  12.3

**4016** Other articles of vulcanised rubber other than hard rubber

2  401699  311.5  4.7  3.5  2.5  24.9  34.7  35.0  67.1  233.8  41.5  4.8  81.8  46.1

**4707** Recovered (waste and scrap) paper or paperboard

3  470710  0.8   0.0  0.1  0.1  1.7  0.4  2.2  0.1  583.3  38.2  63.8  7.1  0.0
4  470720  0.3   0.1  0.0  0.5  0.0  0.0  2.1  0.0  12.4  9.8  2.9  1.0  0.0
5  470730  1.1   0.4  0.0  0.0  0.7  0.0  8.5  0.0  97.3  38.9  46.0  2.4  0.3
6  470790  1.7   0.2  0.0  0.0  2.0  0.0  5.1  0.0  176.2  38.2  37.1  5.3  0.1

**7308** Structures (excluding prefabricated buildings of heading 94.06) and parts of structures (for example, bridges and bridge sections, lock gates, towers, lattice masts, roofs, roofing frameworks, doors and windows and their frames and thresholds for doors, shutters, balustrades, pillars and columns), of iron or steel; plates, rods, angles, shapes, sections, tubes and the like, prepared for use in structures, of iron or steel

7  730820  16.2  0.2  0.0  0.4  0.8  0.0  1.0  8.1  0.6  0.5  0.1  15.8  0.8

**8406** Steam turbines and other vapour turbines

8  840682  4.4  5.5  4.8  0.0  29.6  0.0  13.9  8.0  45.3  1.1  0.1  10.8  0.1

**8418** Refrigerators, freezers and other refrigerating or freezing equipment, electric or other; heat pumps other than air conditioning machines of heading 84.15

9   841861  25.2   0.6  0.0  12.1  1.4  4.2  1.6  0.8  14.3  3.3  0.4  21.3  18.8
10  841869  101.7  4.5  1.3  0.9  22.1  2.9  16.4  16.2  33.9  22.4  6.1  37.7  13.8

**8419** Machinery, plant or laboratory equipment

11  841919  30.5  1.6  0.0  0.0  2.9  1.1  1.0  0.4  7.4  2.6  2.0  7.4  1.5

**8479** Machines and mechanical appliances having individual functions, not specified or included elsewhere in this Chapter

12  847920  12.8  2.0  1.5  0.7  3.6  0.0  0.0  3.0  6.0  5.9  1.2  7.3  9.9

**8483** Transmission shafts (including cam shafts and crank shafts) and cranks; bearing housings and plain shaft bearings; gears and gearing; ball or roller screws; gear boxes and other speed changers, including torque converters; flywheels and pulleys, including pulley blocks; clutches and shaft couplings (including universal joints)

13  848340  1,066.9  23.5  9.2  8.2  92.5  62.3  198.0  179.6  807.7  244.2  18.9  108.0  67.2
14  848360  145.0   4.2   2.3  1.3  17.7  35.6  30.3  18.2  130.5  35.0  5.0  17.1  14.3

**8501** Electric motors and generators (excluding generating sets)

15  850161  51.7  0.5  0.1  0.1  3.7  0.5  6.1  0.9  1.3  0.4  10.3  5.1  0.3  8.1  2.1

Table 2.3b (cont.)

3.2 Exports

mill. US$

| | HS Code | Core Countries | | | | | | | | | | Candidate Countries | | | | | |
|---|---|---|---|---|---|---|---|---|---|---|---|---|---|---|---|---|---|
| | | NAFTA | Chile | Colombia | Peru | Australia | New Zealand | Singapore | Japan | Korea | Brazil | China | India | Indonesia | Turkey | South Africa |
| 16 | 850162 | 19.6 | 0.3 | 0.0 | 0.0 | 0.8 | 0.1 | 4.6 | 0.9 | 2.5 | 1.8 | 2.5 | 1.3 | 0.0 | 8.2 | 0.9 |
| 17 | 850163 | 34.1 | 0.1 | 0.0 | 0.0 | 0.7 | 0.3 | 2.2 | 0.1 | 2.7 | 0.1 | 1.2 | 2.4 | 0.1 | 6.1 | 1.8 |
| 18 | 850164 | 239.6 | 15.5 | 11.5 | 1.1 | 8.5 | 0.2 | 22.7 | 7.9 | 51.2 | 4.6 | 73.3 | 35.5 | 3.2 | 44.8 | 1.2 |
| | 8502 | Electric generating sets and rotary converters | | | | | | | | | | | | | | |
| 19 | 850231 | 1,206.4 | 0.0 | 0.0 | 0.0 | 26.0 | 32.5 | 0.0 | 25.1 | 0.0 | 51.5 | 5.7 | 0.1 | 0.0 | 187.9 | 1.2 |
| 20 | 850239 | 60.6 | 5.5 | 45.4 | 7.1 | 0.1 | 0.1 | 0.5 | 0.9 | 17.9 | 1.0 | 33.5 | 33.3 | 16.0 | 90.5 | 0.2 |
| | 8506 | Primary cells and primary batteries | | | | | | | | | | | | | | |
| 21 | 850610 | 23.3 | 0.0 | 1.5 | 0.0 | 4.8 | 0.2 | 0.2 | 0.2 | 0.0 | 6.3 | 1.0 | 1.1 | 0.2 | 10.5 | 4.3 |
| 22 | 850630 | 0.2 | 0.0 | 0.0 | 0.0 | 0.0 | 0.0 | 0.0 | 0.0 | 0.0 | 0.0 | 0.0 | 0.0 | 0.0 | 0.1 | 0.0 |
| 23 | 850640 | 2.6 | 0.0 | 0.0 | 1.3 | 0.0 | 0.0 | 0.2 | 0.0 | 0.0 | 0.0 | 0.1 | 0.0 | 0.0 | 0.0 | 0.1 |
| 24 | 850650 | 48.5 | 0.2 | 0.5 | 0.2 | 1.7 | 0.2 | 12.8 | 2.0 | 1.8 | 4.3 | 3.6 | 0.9 | 0.4 | 0.9 | 0.9 |
| 25 | 850660 | 22.6 | 0.0 | 0.0 | 0.0 | 1.8 | 0.5 | 2.8 | 3.5 | 1.2 | 0.8 | 0.5 | 0.1 | 0.0 | 1.3 | 0.4 |
| 26 | 850690 | 5.8 | 0.0 | 0.5 | 0.0 | 0.1 | 0.0 | 0.2 | 0.1 | 0.0 | 0.0 | 2.2 | 0.0 | 0.0 | 0.1 | 0.0 |
| | 8507 | Electric accumulators, including separators therefor, whether or not rectangular (including square) | | | | | | | | | | | | | | |
| 27 | 850720 | 48.0 | 5.1 | 1.6 | 5.1 | 13.4 | 0.8 | 10.7 | 6.8 | 2.6 | 8.1 | 21.8 | 1.8 | 2.1 | 23.6 | 9.8 |
| 28 | 850740 | 2.1 | 0.0 | 0.0 | 0.0 | 0.1 | 0.0 | 0.0 | 0.0 | 0.0 | 0.0 | 0.0 | 0.0 | 0.0 | 0.0 | 0.0 |
| | 8537 | Boards, panels, consoles, desks, cabinets and other bases, equipped with two or more apparatus of heading 85.35 or 85.36, for electric control or the distribution of electricity, including those incorporating instruments or apparatus of Chapter 90, and numerical control apparatus, other than switching apparatus of heading 85.17 | | | | | | | | | | | | | | |
| 29 | 853710 | 1,025.4 | 25.4 | 12.4 | 5.7 | 61.7 | 4.8 | 94.6 | 95.6 | 161.8 | 136.4 | 1,518.1 | 149.1 | 24.1 | 169.7 | 94.8 |
| | 8541 | Diodes, transistors and similar semiconductor devices; photosensitive semiconductor devices, including photovoltaic cells whether or not assembled in modules or made up into panels; light emitting diodes; mounted piezo-electric crystals | | | | | | | | | | | | | | |
| 30 | 854140 | 257.9 | 0.9 | 0.2 | 0.2 | 36.6 | 0.5 | 15.7 | 97.5 | 155.3 | 13.7 | 230.8 | 82.6 | 1.8 | 14.1 | 50.8 |
| | 9001 | Optical fibres and optical fibre bundles; optical fibre cables other than those of heading 85.44; sheets and plates of polarising material; lenses (including contact lenses), prisms, mirrors and other optical elements, of any material, unmounted, other than such elements of glass not optically worked | | | | | | | | | | | | | | |
| 31 | 900190 | 132.0 | 0.0 | 0.1 | 0.0 | 1.9 | 0.1 | 3.3 | 24.2 | 7.4 | 1.8 | 15.3 | 2.9 | 0.3 | 6.4 | 1.1 |
| | 9002 | Lenses, prisms, mirrors and other optical elements, of any material, mounted, being parts of or fittings for instruments or apparatus, other than such elements of glass not optically worked | | | | | | | | | | | | | | |

| | | | | | | | | | | | | |
|---|---|---|---|---|---|---|---|---|---|---|---|---|
| 32 | 900290 | 74.5 | 0.2 | 0.1 | 0.0 | 1.3 | 0.2 | 3.7 | 8.3 | 4.9 | 1.6 | 6.6 | 1.5 | 0.0 | 1.6 | 0.9 |
| Other Products | | | | | | | | | | | | |
| | 2207 | Ethyl alcohol, undenat, n/un 80% alc, alcohol, denat | | | | | | | | | | |
| 33 | 220710 | 7.2 | 0.4 | 0.1 | 0.1 | 0.1 | 0.1 | 0.9 | 0.4 | 0.4 | 0.4 | 2.0 | 1.4 | 0.5 | 0.6 | 0.1 |
| | 3824 | Prepared binders for foundry moulds or cores; chemical products and preparations of the chemical or allied industries (including those consisting of mixtures of natural products), not elsewhere specified or included |
| 34 | 382490 | 839.2 | 40.8 | 18.6 | 12.0 | 117.9 | 15.0 | 76.3 | 172.2 | 207.4 | 201.3 | 329.2 | 113.1 | 38.2 | 317.7 | 65.7 |
| | 8410 | Hydraulic turbines, water wheels, and regulators therefor |
| 35 | 841011 | 1.9 | 0.6 | 0.0 | 0.0 | 0.2 | 0.0 | 0.1 | 0.2 | 0.3 | 0.0 | 3.9 | 0.0 | 0.0 | 2.8 | 0.1 |
| 36 | 841090 | 21.9 | 14.5 | 0.0 | 1.3 | 6.8 | 0.8 | 2.1 | 0.0 | 9.1 | 19.6 | 21.0 | 33.5 | 9.0 | 49.2 | 0.1 |
| | 8411 | Turbo-jets, turbo-propellers and other gas turbines |
| 37 | 841181 | 167.9 | 0.0 | 6.5 | 0.0 | 22.0 | 0.5 | 17.1 | 3.5 | 5.8 | 6.5 | 14.4 | 3.0 | 3.9 | 3.3 | 7.9 |
| 38 | 841182 | 235.7 | 0.0 | 0.9 | 12.5 | 10.5 | 4.2 | 36.6 | 3.0 | 66.6 | 34.2 | 95.7 | 84.9 | 11.0 | 141.1 | 19.4 |
| | 8503 | Parts suitable for use solely or principally with the machines of heading 85.01 or 85.02 |
| 39 | 850300 | 746.1 | 18.9 | 2.7 | 17.8 | 55.1 | 10.3 | 14.8 | 39.2 | 56.9 | 149.7 | 368.7 | 116.0 | 13.1 | 135.0 | 29.9 |
| Total Exports | | 7,139.2 | 176.8 | 82.4 | 117.7 | 591.6 | 90.3 | 449.3 | 670.8 | 1,081.3 | 975.9 | 4,991.3 | 1,169.2 | 315.3 | 1,613.7 | 478.4 |

Source: UN COMTRADE (accessed through the WITS on October 1, 2011)

Table 2.3c  *Trade with Other Core and Candidate Countries: China, 2010*

2.3.1 Imports of SETA Products from the Core and Candidate Countries

mill. US$

| | | Core Countries | | | | | | | | | | Candidate Countries | | | | |
|---|---|---|---|---|---|---|---|---|---|---|---|---|---|---|---|---|
| | HS Code | NAFTA | EU27 | Chile | Colombia | Peru | Australia | New Zealand | Singapore | Japan | Korea | Brazil | India | Indonesia | Turkey | South Africa |

**Products categorized as 'renewable energy' only from the WTO (2011)**

| | 3907 | Polyacetals, other polyethers and epoxide resins, in primary forms; polycarbonates, alkyd resins, polyallyl esters and other polyesters, in primary forms |
|---|---|---|
| 1 | 390799 | 115.9  102.0  0.0  0.0  0.0  1.6  0.1  6.6  275.5  43.1  0.1  0.4  0.0  0.1  0.0 |
| | 4016 | Other articles of vulcanised rubber other than hard rubber |
| 2 | 401699 | 58.7  147.5  0.0  0.0  0.0  0.8  0.2  15.3  180.4  84.6  2.7  1.3  3.7  0.8  0.1 |
| | 4707 | Recovered (waste and scrap) paper or paperboard |
| 3 | 470710 | 1,473.3  1,073.6  0.0  0.0  0.0  78.7  11.5  0.1  408.1  8.9  0.0  0.0  0.0  2.0  0.1 |
| 4 | 470720 | 68.8  19.0  0.0  0.0  0.0  10.0  0.6  0.6  38.2  0.3  0.0  0.0  0.0  1.2  0.0 |
| 5 | 470730 | 808.1  173.2  0.0  0.0  0.0  40.6  1.9  0.0  111.3  7.0  0.0  0.0  0.0  0.3  0.0 |
| 6 | 470790 | 224.6  233.7  0.0  0.0  0.0  46.4  0.9  0.0  226.8  0.3  0.0  0.0  0.0  0.0  0.0 |
| | 7308 | Structures (excluding prefabricated buildings of heading 94.06) and parts of structures (for example, bridges and bridge sections, lock gates, towers, lattice masts, roofs, roofing frameworks, doors and windows and their frames and thresholds for doors, shutters, balustrades, pillars and columns), of iron or steel; plates, rods, angles, shapes, sections, tubes and the like, prepared for use in structures, of iron or steel |
| 7 | 730820 | 0.1  0.1  0.0  0.0  0.0  0.0  0.0  0.0  0.0  0.0  0.0  0.0  0.0  0.0  0.0 |
| | 8406 | Steam turbines and other vapour turbines |
| 8 | 840682 | 30.9  51.0  0.0  0.0  0.0  0.0  0.0  0.0  21.4  0.0  0.0  0.0  0.0  0.0  0.0 |
| | 8418 | Refrigerators, freezers and other refrigerating or freezing equipment, electric or other; heat pumps other than air conditioning machines of heading 84.15 |
| 9 | 841861 | 4.1  3.1  0.0  0.0  0.0  0.0  0.2  0.9  3.6  0.1  0.0  0.0  0.0  0.0  0.0 |
| 10 | 841869 | 55.5  64.9  0.0  0.0  0.0  0.1  0.0  417.4  36.7  12.7  0.0  0.0  0.2  0.0  0.0 |
| | 8419 | Machinery, plant or laboratory equipment |
| 11 | 841919 | 0.4  0.6  0.0  0.0  0.0  0.2  0.0  0.0  0.0  0.0  0.0  0.0  0.0  0.0  0.0 |
| | 8479 | Machines and mechanical appliances having individual functions, not specified or included elsewhere in this Chapter |

| | | | | | | | | | | |
|---|---|---|---|---|---|---|---|---|---|---|
| 12 | 847920 | 0.1 | 2.6 | 0.0 | 0.0 | 0.2 | 0.0 | 0.1 | 1.0 | 0.4 | 0.0 | 0.0 |
| | 8483 | Transmission shafts (including cam shafts and crank shafts) and cranks; bearing housings and plain shaft bearings; gears and gearing; ball or roller screws; gear boxes and other speed changers, including torque converters; flywheels and pulleys, including pulley blocks; clutches and shaft couplings (including universal joints) |
| 13 | 848340 | 132.7 | 1,087.4 | 0.0 | 4.8 | 0.3 | 8.7 | 96.5 | 2.0 | 9.2 | 0.2 | 0.3 |
| 14 | 848360 | 18.4 | 157.5 | 0.0 | 0.4 | 0.0 | 0.4 | 6.3 | 0.0 | 0.5 | 0.0 | 0.0 |
| | 8501 | Electric motors and generators (excluding generating sets) |
| 15 | 850161 | 5.7 | 4.2 | 0.0 | 0.3 | 0.0 | 0.7 | 0.1 | 0.0 | 0.0 | 0.0 | 0.0 |
| 16 | 850162 | 0.6 | 3.1 | 0.0 | 0.2 | 0.0 | 1.2 | 1.9 | 0.0 | 0.0 | 0.0 | 0.0 |
| 17 | 850163 | 1.5 | 2.7 | 0.0 | 0.0 | 0.0 | 0.0 | 1.4 | 0.0 | 0.0 | 0.0 | 0.0 |
| 18 | 850164 | 7.6 | 83.4 | 0.0 | 0.0 | 0.0 | 2.4 | 17.0 | 0.0 | 7.0 | 0.0 | 0.0 |
| | 8502 | Electric generating sets and rotary converters |
| 19 | 850231 | 0.0 | 9.6 | 0.0 | 0.0 | 0.0 | 0.0 | 0.7 | 0.0 | 0.0 | 0.0 | 0.0 |
| 20 | 850239 | 66.0 | 32.8 | 0.0 | 0.0 | 0.0 | 3.2 | 0.0 | 0.0 | 0.0 | 0.0 | 0.0 |
| | 8506 | Primary cells and primary batteries |
| 21 | 850610 | 17.7 | 1.8 | 0.0 | 0.0 | 0.0 | 22.2 | 0.7 | 0.0 | 0.0 | 6.9 | 0.0 |
| 22 | 850630 | 0.0 | 0.0 | 0.0 | 0.0 | 0.0 | 0.0 | 0.0 | 0.0 | 0.0 | 0.0 | 0.0 |
| 23 | 850640 | 0.0 | 0.1 | 0.0 | 0.0 | 0.0 | 19.2 | 0.0 | 0.0 | 0.0 | 0.0 | 0.0 |
| 24 | 850650 | 10.3 | 5.9 | 0.0 | 0.6 | 0.0 | 66.4 | 4.9 | 0.0 | 0.0 | 38.4 | 0.0 |
| 25 | 850660 | 0.4 | 1.2 | 0.0 | 0.0 | 0.0 | 0.0 | 0.0 | 0.0 | 0.0 | 0.2 | 0.0 |
| 26 | 850690 | 2.0 | 1.3 | 0.0 | 0.0 | 0.0 | 1.1 | 0.0 | 0.0 | 0.0 | 0.1 | 0.0 |
| | 8507 | Electric accumulators, including separators therefor, whether or not rectangular (including square) |
| 27 | 850720 | 24.7 | 23.2 | 0.0 | 0.0 | 0.0 | 1.8 | 1.1 | 0.0 | 0.8 | 0.0 | 0.0 |
| 28 | 850740 | 0.0 | 0.0 | 0.0 | 0.0 | 0.0 | 0.1 | 0.0 | 0.0 | 0.0 | 0.0 | 0.0 |
| | 8537 | Boards, panels, consoles, desks, cabinets and other bases, equipped with two or more apparatus of heading 85.35 or 85.36, for electric control or the distribution of electricity, including those incorporating instruments or apparatus of Chapter 90, and numerical control apparatus, other than switching apparatus of heading 85.17 |
| 29 | 853710 | 250.3 | 1,501.7 | 0.0 | 9.7 | 1.0 | 23.1 | 959.1 | 149.5 | 2.6 | 7.0 | 3.2 |
| | 8541 | Diodes, transistors and similar semiconductor devices; photosensitive semiconductor devices, including photovoltaic cells whether or not assembled in modules or made up into panels; light emitting diodes; mounted piezo-electric crystals |
| 30 | 854140 | 170.6 | 267.8 | 0.0 | 0.0 | 0.0 | 39.9 | 1,695.4 | 728.9 | 0.0 | 47.2 | 97.2 | 0.0 |

Table 2.3c (cont.)

### 2.3.1 Imports of SETA Products from the Core and Candidate Countries

mill. US$

| | HS Code | NAFTA | EU27 | Chile | Colombia | Peru | Australia | New Zealand | Singapore | Japan | Korea | Brazil | India | Indonesia | Turkey | South Africa |
|---|---|---|---|---|---|---|---|---|---|---|---|---|---|---|---|---|
| | 9001 | Optical fibres and optical fibre bundles; optical fibre cables other than those of heading 85.44; sheets and plates of polarising material; lenses (including contact lenses), prisms, mirrors and other optical elements, of any material, unmounted, other than such elements of glass not optically worked | | | | | | | | | | | | | | |
| 31 | 900190 | 99.2 | 62.4 | 0.0 | 0.0 | 2.1 | 0.0 | 20.4 | 1,068.9 | 934.3 | 0.0 | 0.2 | 0.7 | 0.0 | 0.0 |
| | 9002 | Lenses, prisms, mirrors and other optical elements, of any material, mounted, being parts of or fittings for instruments or apparatus, other than such elements of glass not optically worked | | | | | | | | | | | | | | |
| 32 | 900290 | 13.8 | 16.7 | 0.0 | 0.0 | 0.0 | 0.0 | 0.9 | 50.0 | 52.5 | 0.0 | 0.0 | 0.0 | 0.0 | 0.0 |
| **Other Products** | | | | | | | | | | | | | | | | |
| | 2207 | Ethyl alcohol, undenat, n/un 80% alc, alcohol, denat | | | | | | | | | | | | | | |
| 33 | 220710 | 0.1 | 0.3 | 0.0 | 0.0 | 0.0 | 0.0 | 0.0 | 0.2 | 0.0 | 0.0 | 0.0 | 0.0 | 0.0 | 0.0 |
| | 3824 | Prepared binders for foundry moulds or cores; chemical products and preparations of the chemical or allied industries (including those consisting of mixtures of natural products), not elsewhere specified or included | | | | | | | | | | | | | | |
| 34 | 382490 | 1,174.2 | 623.7 | 0.2 | 0.0 | 0.0 | 7.3 | 0.7 | 87.4 | 1,486.9 | 379.7 | 4.0 | 11.3 | 35.7 | 11.3 | 19.0 |
| | 8410 | Hydraulic turbines, water wheels, and regulators therefor | | | | | | | | | | | | | | |
| 35 | 841011 | 0.0 | 4.5 | 0.0 | 0.0 | 0.0 | 0.0 | 0.0 | 0.0 | 0.0 | 0.0 | 0.0 | 0.0 | 0.0 | 0.0 |
| 36 | 841090 | 0.0 | 12.9 | 0.0 | 0.0 | 0.0 | 0.0 | 0.0 | 1.1 | 6.9 | 9.7 | 0.0 | 0.0 | 0.0 | 0.0 |
| | 8411 | Turbo-jets, turbo-propellers and other gas turbines | | | | | | | | | | | | | | |
| 37 | 841181 | 2.2 | 3.3 | 0.0 | 0.0 | 0.0 | 0.0 | 0.0 | 0.4 | 0.0 | 0.0 | 0.0 | 0.0 | 0.0 | 0.0 |
| 38 | 841182 | 66.2 | 28.2 | 0.0 | 0.0 | 0.0 | 0.0 | 0.0 | 0.0 | 0.0 | 0.0 | 0.0 | 0.0 | 0.0 | 0.0 |
| | 8503 | Parts suitable for use solely or principally with the machines of heading 85.01 or 85.02 | | | | | | | | | | | | | | |
| 39 | 850300 | 57.3 | 316.1 | 0.0 | 0.0 | 0.0 | 0.3 | 0.0 | 6.4 | 370.3 | 80.8 | 3.2 | 23.1 | 1.6 | 0.3 | 0.0 |
| | Total Imports | 4,961.7 | 6,123.0 | 0.2 | 0.0 | 0.0 | 204.5 | 17.4 | 652.4 | 7,545.9 | 2,620.2 | 23.9 | 103.9 | 191.8 | 16.5 | 22.6 |

2.3.2 Exports of SETA Products to the Core and Candidate Countries

mill. US$

| | | | | Core Countries | | | | | | | | | | | |
|---|---|---|---|---|---|---|---|---|---|---|---|---|---|---|---|
| | HS Code | NAFTA | EU27 | Chile | Colombia | Peru | Australia | New Zealand | Singapore | Japan | Korea | Brazil | India | Indonesia | Turkey | South Africa |

Products categorized as 'renewable energy' only from the WTO (2011)

**3907** Polyacetals, other polyethers and epoxide resins, in primary forms; polycarbonates, alkyd resins, polyallyl esters and other polyesters, in primary forms
1 | 390799 | 16.3 | 17.0 | 0.2 | 0.7 | 0.0 | 3.3 | 0.1 | 5.2 | 16.6 | 65.8 | 3.5 | 19.5 | 3.6 | 3.5 | 1.1

**4016** Other articles of vulcanised rubber other than hard rubber
2 | 401699 | 140.2 | 77.7 | 1.8 | 0.6 | 0.4 | 8.7 | 0.8 | 5.4 | 72.1 | 21.9 | 4.1 | 8.6 | 6.9 | 3.6 | 2.3

**4707** Recovered (waste and scrap) paper or paperboard
3 | 470710 | 0.0 | 0.0 | 0.0 | 0.0 | 0.0 | 0.0 | 0.0 | 0.0 | 0.0 | 0.0 | 0.0 | 0.0 | 0.0 | 0.0 | 0.0
4 | 470720 | 0.0 | 0.0 | 0.0 | 0.0 | 0.0 | 0.0 | 0.0 | 0.0 | 0.0 | 0.0 | 0.0 | 0.0 | 0.0 | 0.0 | 0.0
5 | 470730 | 0.0 | 0.0 | 0.0 | 0.0 | 0.0 | 0.0 | 0.0 | 0.0 | 0.0 | 0.0 | 0.0 | 0.0 | 0.0 | 0.0 | 0.0
6 | 470790 | 0.0 | 0.0 | 0.0 | 0.0 | 0.0 | 0.0 | 0.0 | 0.0 | 0.0 | 0.0 | 0.0 | 0.0 | 0.0 | 0.0 | 0.0

**7308** Structures (excluding prefabricated buildings of heading 94.06) and parts of structures (for example, bridges and bridge sections, lock gates, towers, lattice masts, roofs, roofing frameworks, doors and windows and their frames and thresholds for doors, shutters, balustrades, pillars and columns), of iron or steel; plates, rods, angles, shapes, sections, tubes and the like, prepared for use in structures, of iron or steel
7 | 730820 | 59.2 | 14.0 | 0.8 | 0.0 | 3.1 | 13.5 | 0.6 | 1.2 | 16.1 | 0.2 | 0.0 | 6.2 | 15.2 | 0.0 | 4.3

**8406** Steam turbines and other vapour turbines
8 | 840682 | 0.0 | 0.0 | 0.0 | 0.0 | 0.0 | 0.0 | 0.0 | 3.4 | 0.0 | 0.2 | 0.2 | 30.2 | 19.4 | 8.8 | 1.4

**8418** Refrigerators, freezers and other refrigerating or freezing equipment, electric or other; heat pumps other than air conditioning machines of heading 84.15
9 | 841861 | 20.0 | 71.7 | 2.3 | 0.2 | 0.2 | 6.2 | 1.0 | 3.2 | 6.7 | 2.5 | 9.5 | 6.0 | 0.6 | 0.8 | 5.3
10 | 841869 | 74.3 | 205.2 | 6.3 | 1.2 | 3.0 | 26.4 | 2.5 | 29.5 | 17.0 | 22.7 | 10.6 | 77.0 | 26.7 | 11.4 | 10.6

**8419** Machinery, plant or laboratory equipment
11 | 841919 | 19.4 | 30.5 | 2.0 | 0.1 | 0.2 | 2.6 | 0.3 | 0.3 | 0.2 | 3.1 | 0.9 | 4.5 | 0.4 | 7.7 | 13.7

**8479** Machines and mechanical appliances having individual functions, not specified or included elsewhere in this Chapter
12 | 847920 | 2.1 | 0.6 | 0.0 | 0.0 | 0.3 | 0.1 | 0.0 | 0.0 | 0.0 | 0.0 | 0.1 | 0.6 | 7.6 | 0.0 | 2.7

Table 2.3c (*cont.*)

2.3.2 Exports of SETA Products to the Core and Candidate Countries

mill. US$

|    | HS Code | NAFTA | EU27 | Chile | Colombia | Peru | Australia | New Zealand | Singapore | Japan | Korea | Brazil | India | Indonesia | Turkey | South Africa |
|----|---------|-------|------|-------|----------|------|-----------|-------------|-----------|-------|-------|--------|-------|-----------|--------|--------------|
|    | 8483 | Transmission shafts (including cam shafts and crank shafts) and cranks; bearing housings and plain shaft bearings; gears and gearing; ball or roller screws; gear boxes and other speed changers, including torque converters; flywheels and pulleys, including pulley blocks; clutches and shaft couplings (including universal joints) | | | | | | | | | | | | | | | |
| 13 | 848340 | 270.5 | 155.8 | 2.1 | 1.2 | 2.2 | 8.0 | 0.5 | 20.2 | 75.8 | 25.1 | 20.5 | 83.9 | 22.2 | 7.7 | 5.7 |
| 14 | 848360 | 29.3 | 30.8 | 0.6 | 0.8 | 0.2 | 1.8 | 0.1 | 1.7 | 16.3 | 3.8 | 8.7 | 7.3 | 9.4 | 1.0 | 2.7 |
|    | 8501 | Electric motors and generators (excluding generating sets) | | | | | | | | | | | | | | | |
| 15 | 850161 | 24.4 | 3.9 | 0.2 | 0.3 | 1.1 | 2.1 | 0.0 | 1.8 | 3.9 | 0.5 | 3.1 | 1.2 | 15.6 | 3.9 | 0.3 |
| 16 | 850162 | 11.7 | 8.0 | 0.0 | 0.1 | 0.0 | 0.6 | 0.0 | 4.1 | 3.1 | 2.9 | 6.0 | 0.9 | 5.4 | 7.0 | 0.1 |
| 17 | 850163 | 0.2 | 11.4 | 0.0 | 0.0 | 0.2 | 0.2 | 0.0 | 4.7 | 1.0 | 0.8 | 0.5 | 4.3 | 2.1 | 3.1 | 0.2 |
| 18 | 850164 | 23.1 | 44.2 | 0.0 | 0.7 | 0.5 | 0.2 | 0.2 | 16.3 | 18.1 | 0.5 | 0.9 | 52.3 | 10.9 | 29.6 | 0.9 |
|    | 8502 | Electric generating sets and rotary converters | | | | | | | | | | | | | | | |
| 19 | 850231 | 10.7 | 7.2 | 2.9 | 0.0 | 0.0 | 19.4 | 0.0 | 0.1 | 0.0 | 0.1 | 0.1 | 10.1 | 0.1 | 0.2 | 0.2 |
| 20 | 850239 | 10.4 | 4.3 | 0.0 | 0.1 | 0.2 | 0.1 | 0.3 | 2.1 | 0.0 | 0.0 | 0.5 | 3.3 | 3.5 | 18.7 | 0.0 |
|    | 8506 | Primary cells and primary batteries | | | | | | | | | | | | | | | |
| 21 | 850610 | 186.8 | 234.6 | 2.6 | 3.6 | 0.8 | 35.0 | 2.5 | 13.1 | 41.3 | 11.7 | 20.3 | 16.6 | 4.1 | 11.6 | 2.9 |
| 22 | 850630 | 0.0 | 0.0 | 0.0 | 0.0 | 0.0 | 0.0 | 0.0 | 0.0 | 0.0 | 0.0 | 0.0 | 0.0 | 0.0 | 0.0 | 0.0 |
| 23 | 850640 | 0.9 | 0.1 | 0.0 | 0.0 | 0.0 | 0.0 | 0.0 | 0.0 | 0.1 | 0.0 | 0.0 | 0.0 | 0.0 | 0.0 | 0.0 |
| 24 | 850650 | 12.8 | 11.5 | 0.0 | 0.0 | 0.0 | 2.8 | 0.1 | 0.6 | 4.0 | 0.4 | 2.1 | 2.3 | 0.0 | 0.6 | 0.2 |
| 25 | 850660 | 0.0 | 0.9 | 0.0 | 0.0 | 0.0 | 0.0 | 0.0 | 0.0 | 0.0 | 0.0 | 0.0 | 0.0 | 0.0 | 0.0 | 0.0 |

| #  | HS Code | Description / Values |
|----|---------|----------------------|
| 26 | 850690  | 2.5  0.1  0.0  0.0  0.0  0.0  0.0  0.5  0.0  0.7  1.0  0.0 |
|    | **8507** | Electric accumulators, including separators therefor, whether or not rectangular (including square) |
| 27 | 850720  | 295.6  390.1  6.6  10.1  2.3  40.9  6.2  54.6  14.8  22.5  45.0  129.4  21.7  40.1  19.0 |
| 28 | 850740  | 0.1  0.0  0.0  0.0  0.0  0.0  0.0  0.0  0.0  0.0  0.0  0.0  0.0 |
|    | **8537** | Boards, panels, consoles, desks, cabinets and other bases, equipped with two or more apparatus of heading 85.35 or 85.36, for electric control or the distribution of electricity, including those incorporating instruments or apparatus of Chapter 90, and numerical control apparatus, other than switching apparatus of heading 85.17 |
| 29 | 853710  | 502.8  490.3  4.5  1.1  1.1  26.4  1.6  61.6  458.6  113.9  36.6  85.4  33.2  20.1  5.5 |
|    | **8541** | Diodes, transistors and similar semiconductor devices; photosensitive semiconductor devices, including photovoltaic cells whether or not assembled in modules or made up into panels; light emitting diodes; mounted piezo-electric crystals |
| 30 | 854140  | 1,459.2  19,579.0  2.0  1.2  2.4  785.4  1.9  66.9  467.2  382.9  8.6  100.4  13.6  9.2  37.9 |
|    | **9001** | Optical fibres and optical fibre bundles; optical fibre cables other than those of heading 85.44; sheets and plates of polarising material; lenses (including contact lenses), prisms, mirrors and other optical elements, of any material, unmounted, other than such elements of glass not optically worked |
| 31 | 900190  | 103.6  55.1  0.0  0.0  3.2  0.2  46.1  219.1  70.6  1.8  1.2  3.4  4.0  0.1 |
|    | **9002** | Lenses, prisms, mirrors and other optical elements, of any material, mounted, being parts of or fittings for instruments or apparatus, other than such elements of glass not optically worked |
| 32 | 900290  | 29.6  18.0  0.0  0.0  0.0  0.3  0.0  4.0  71.5  14.7  0.0  0.3  0.1  0.0 |
|    | **Other Products** | |
|    | **2207** | Ethyl alcohol, undenat, n/un 80% alc, alcohol, denat |
| 33 | 220710  | 0.0  0.0  0.0  0.0  2.2  0.0  1.4  9.9  49.0  0.0  1.4  0.0  0.0  0.0 |
|    | **3824** | Prepared binders for foundry moulds or cores; chemical products and preparations of the chemical or allied industries (including those consisting of mixtures of natural products), not elsewhere specified or included |
| 34 | 382490  | 176.5  245.1  6.6  2.9  2.1  25.6  4.4  26.6  272.0  175.1  29.5  40.6  33.9  18.9  15.1 |
|    | **8410** | Hydraulic turbines, water wheels, and regulators therefor |
| 35 | 841011  | 0.1  0.9  0.0  0.0  0.0  0.0  0.0  0.0  0.0  0.0  0.0  0.9  0.7  0.0 |
| 36 | 841090  | 28.1  8.9  0.6  0.0  0.1  0.0  0.0  0.5  6.0  1.1  30.3  6.0  20.5  0.0 |
|    | **8411** | Turbo-jets, turbo-propellers and other gas turbines |
| 37 | 841181  | 0.0  0.0  0.0  0.0  0.0  0.0  0.0  0.0  0.0  0.0  0.0  0.0  0.0  0.0 |

Table 2.3c (cont.)

2.3.2 Exports of SETA Products to the Core and Candidate Countries

mill. US$

| | HS Code | NAFTA | EU27 | Chile | Colombia | Peru | Australia | New Zealand | Singapore | Japan | Korea | Brazil | India | Indonesia | Turkey | South Africa |
|---|---|---|---|---|---|---|---|---|---|---|---|---|---|---|---|---|
| | | | | | | | Core Countries | | | | | | | | | |
| 38 | 841182 | 2.5 | **Parts suitable for use solely or principally with the machines of heading 85.01 or 85.02** | | | | | | | | | | | | | |
| | 8503 | 567.4 | 410.4 | 1.2 | 0.7 | 0.5 | 13.5 | 0.2 | 14.3 | 177.3 | 93.3 | 38.6 | 324.8 | 88.8 | 13.4 | 20.7 |
| 39 | 850300 | | | | | | | | | | | | | | | |
| | Total Exports | 4,080.0 | 22,127.1 | 43.0 | 25.4 | 20.6 | 1,028.4 | 23.4 | 388.6 | 1,988.6 | 1,085.6 | 281.6 | 1,024.5 | 355.8 | 245.7 | 152.7 |

*Source:* COMTRADE (accessed through the WITS on October 1, 2011)

848340) and biodiesel (HS 382490). For China, the top three import partners are Japan, the EU and NAFTA, and the main import commodities are biodiesel (HS 382490), recovered paper (HS 470710) and PV cells and modules (HS 854140). China's top three export destinations are the EU, NAFTA and Japan, and its main export commodity is PV cells and modules (HS 854140). China is also the top exporter of wind turbine towers, static converters, solar batteries for energy storage in off-grid PV systems and other items.

While trade negotiators may enjoy delving into details like these, the key conclusion we derive is that – from the perspective of eliminating tariff barriers – China should have a strong commercial interest in joining the SETA talks. However, since the SETA would eventually point to the elimination of LCRs and harmonisation of standards, China might take a wait-and-see attitude.

Table 2.4 summarises total trade – imports and exports – in all thirty-nine EGs between each country and the other core and candidate countries. The total imports of all thirty-nine EGs from all the core and candidate countries were USD 102 billion in 2010, without counting intra-NAFTA imports and intra-EU imports. Taking into account imports within the NAFTA pact and within the EU, the total imports of all thirty-nine EGs between all the core and candidate countries were about USD 176 billion.[16] Calculated in this way, SETA trade accounts for about 78 per cent of world imports of all thirty-nine EGs. In other words, if both the core and candidate countries sign up, the SETA coverage would amount to a 'critical mass'; however, if only the core countries are present at the launch, the trade coverage would be 'substantial', but not up to the 'critical mass' standard.

The top three importing countries for all thirty-nine EGs are the EU 27, China and NAFTA, followed by Korea and Japan. The top three exporting countries are China, the EU 27 and Japan, followed by NAFTA and Korea. Among all the countries, only China and Japan are net exporters of all thirty-nine EGs; all others are net importers. The total exports of all thirty-nine EGs between the core countries themselves (not counting internal NAFTA and EU trade) were USD 36 billion; and the total exports from the core countries to the candidate countries were USD 30 billion. Candidate exports to the core countries were USD 34 billion, while candidate exports to other candidate countries only amounted to USD 3 billion.

From this perspective, the candidate countries have a commercial incentive to join a SETA, merely to ensure access to the core markets.

---

[16] In 2010, the total imports of all thirty-nine EGs within the NAFTA pact and within the EU 27 were USD 14 billion and USD 60 billion, respectively.

Table 2.4 Trade of All 39 SETA Products between Selected Countries, 2010[a]

mill US $

|  | Core Countries ||||||||||| Candidate Countries |||||| |
|---|---|---|---|---|---|---|---|---|---|---|---|---|---|---|---|---|---|---|
|  | NAFTA | EU 27 | Chile | Colombia | Peru | Australia | New Zealand | Singapore | Japan | Korea | Imports from core | Brazil | China | India | Indonesia | Turkey | South Africa | Imports from candidates | Imports (core + candidates) |
| NAFTA |  | 7,918 | 3 | — | 3 | 41 | 7 | 484 | 3,250 | 551 | 12,264 | 520 | 5,530 | 454 | 144 | 18 | 44 | 6,710 | 18,974 |
| EU 27 | 5,491 |  | 1 | 12 | 6 | 86 | 11 | 135 | 3,153 | 1,628 | 10,523 | 337 | 18,396 | 734 | 326 | 328 | 199 | 20,320 | 30,843 |
| Chile[b] | 264 | 279 |  | 1 | 1 | 8 | 0 | 3 | 86 | 49 | 690 | 64 | 59 | 1 | 7 | 0 | 2 | 134 | 824 |
| Colombia | 189 | 80 | 1 |  | 13 | 0 | 0 | 3 | 4 | 3 | 293 | 32 | 35 | 28 | 3 | 0 | 0 | 98 | 391 |
| Peru | 167 | 75 | 13 | 23 |  | 1 | 0 | 2 | 4 | 2 | 289 | 19 | 31 | 11 | 2 | 4 | 1 | 69 | 358 |
| Australia | 656 | 569 | 1 | 0 | 0 |  | 27 | 59 | 239 | 46 | 1,596 | 13 | 1,046 | 79 | 21 | 2 | 10 | 1,170 | 2,766 |
| New Zealand | 66 | 73 | 0 | 0 | 0 | 33 |  | 4 | 23 | 10 | 211 | 3 | 36 | 1 | 3 | 0 | 2 | 45 | 256 |
| Singapore | 876 | 673 | 0 | 0 | 0 | 13 | 6 |  | 366 | 51 | 1,986 | 20 | 457 | 27 | 219 | 0 | 9 | 733 | 2,719 |
| Japan | 1,220 | 843 | 0 | 0 | 0 | 3 | 1 | 41 |  | 501 | 2,608 | 288 | 2,457 | 12 | 156 | 6 | 3 | 2,923 | 5,531 |
| Korea[b] | 870 | 1,529 | 0 | 0 | 0 | 8 | 1 | 49 | 2,793 |  | 5,250 | 53 | 1,243 | 16 | 43 | 0 | 1 | 1,357 | 6,607 |
| Exports to core | 9,799 | 12,040 | 19 | 44 | 23 | 194 | 53 | 780 | 9,917 | 2,842 | 35,710 | 1,349 | 29,290 | 1,364 | 924 | 358 | 272 | 33,558 | 69,268 |
| Brazil | 621 | 977 | 2 | 0 | 0 | 5 | 1 | 20 | 183 | 68 | 1,878 |  | 390 | 156 | 7 | 5 | 2 | 559 | 2,437 |
| China | 4,962 | 6,123 | 0 | 0 | 0 | 204 | 17 | 652 | 7,546 | 2,620 | 22,125 | 24 |  | 104 | 192 | 17 | 23 | 359 | 22,484 |
| India[b] | 523 | 1,000 | 0 | 0 | 0 | 15 | 4 | 65 | 133 | 61 | 1,800 | 32 | 498 |  | 6 | 2 | 2 | 540 | 2,340 |
| Indonesia | 193 | 437 | 0 | 0 | 0 | 53 | 14 | 286 | 444 | 78 | 1,505 | 2 | 437 | 23 |  | 6 | 3 | 471 | 1,976 |
| Turkey | 170 | 2,095 | 0 | 0 | 0 | 1 | 0 | 4 | 42 | 58 | 2,370 | 9 | 635 | 64 | 10 |  | 1 | 719 | 3,089 |
| South Africa | 138 | 393 | 0 | 0 | 0 | 7 | 0 | 4 | 22 | 14 | 579 | 20 | 122 | 32 | 1 | 1 |  | 176 | 755 |
| Exports to candidates | 6,607 | 11,026 | 3 | 0 | 0 | 285 | 36 | 1,031 | 8,370 | 2,899 | 30,257 | 87 | 2,082 | 379 | 216 | 30 | 30 | 2,824 | 33,081 |
| Exports (core + candidates) | 16,406 | 23,065 | 22 | 44 | 23 | 478 | 90 | 1,810 | 18,288 | 5,740 | 65,967 | 1,436 | 31,373 | 1,743 | 1,140 | 388 | 302 | 36,382 | 102,349 |

Notes:

a. Rows are imports and columns are exports.

b. Data for year 2009

Source: UN COMTRADE (accessed through the WITS on October 3, 2011)

While trade in sustainable energy products between the candidate countries will certainly expand, the commercial promise of these flows is the future, not the present.

## 2    Non-Tariff Barriers

Despite commitments by world leaders to resist protectionism, the use of trade restrictive measures continues to flourish.[17] A wide range of government supports has favoured renewable energy, giving credence to the idea of 'green protectionism' and a 'green race'. Trade frictions between the United States and China are particularly worrisome. In his 2012 State of the Union address, President Obama praised his Administration for bringing trade cases against China at nearly twice the rate of the Bush Administration, and went on to announce the creation of a Trade Enforcement Unit to investigate unfair practices (with attention to China foremost).[18] Solar and wind energy are at the forefront of bilateral trade friction, and a public backlash against US measures is already gathering steam in China.

In recent years, countries have extended various forms of public support to promote RE. According to REN21 (2011), at least 118 countries had some type of renewable support policy at the national level in early 2011; in addition, many sub-national governments (cities, states and provinces) are pursuing their own renewable policies. From an environmental perspective, these support measures are welcome. They create new industries and jobs, facilitate technological innovation and reduce carbon emissions. In fact, global demand for RE heavily depends on governmental support in various forms, including LCRs, subsidies and standards. The big downside is that such programmes inevitably raise the cost of deploying a given amount of RE as they deviate from the national treatment principles that undergird the world system of trade and investment. In this section, we discuss non-tariff measures that play a key role in boosting RE but also act as barriers to efficient trade and investment.

*Local Content Requirements*

Among the most troublesome barriers are LCRs. They mandate the use of locally produced components or services in government-sponsored

---

[17] See Global Trade Alert (2011).
[18] For the full text of President Obama's remarks in the State of the Union Address, see www.whitehouse.gov/the-press-office/2012/01/24/remarks-president-state-union-address.

projects, and became rather popular in the wake of the global financial crisis (2008–2009). The common goal is to foster local jobs.

LCRs are especially prevalent in large-scale RE projects, at both national and sub-national levels of government. China is one country that has used LCRs as an important policy tool; LCRs for wind farm projects have been in place for many years. During the Ninth Five-Year Plan (1996–2000), a policy requiring a minimum of 40 per cent local content for wind turbine equipment was approved by the National Development and Reform Commission (NDRC). This figure was increased to 70 per cent in 2005. The NDRC revoked the 70 per cent mandate LCR in 2009, but domestic content remains a factor in awarding wind farm concessions. Kirkegaard et al. (2009) argue that the 70 per cent LCR policy, coupled with China's large market, led many non-Chinese turbine firms to establish production facilities in China.

Despite the irreversible outcome of Chinese industrial policy, the United States raised a dispute over Chinese wind power subsidies contingent on LCRs. That case was settled through consultations in 2011,[19] but Chinese subsidies for other renewable energy remain on the agenda.

Another example is Ontario's feed-in tariff (FIT) programme. Canada's largest province enacted its Green Energy and Green Economy Act in May 2009. Among other features, the act enabled a FIT programme that included a domestic content requirement. In September 2010, Japan requested consultations with Canada, claiming that the measures were inconsistent with WTO rules.[20] In August 2011, the EU requested consultations with Canada for the same FIT program; Japan and the United States joined those consultations as well.[21]

In still another LCR case, Renault received a €100 million state loan for electric vehicles in February 2010. The *quid pro quo* was Renault's pledge of best efforts to raise the local content of electric car development up to 70 per cent, and to maintain its employment within France.[22]

Table 2.5 lists several LCRs currently in place. Many countries are conditioning their support for FITs on LCRs. While their effectiveness in boosting jobs can be questioned, LCRs clearly reduce competition by excluding foreign suppliers, thereby raising the cost of renewable energy.

---

[19] See the WTO website, *China – Measures Concerning Wind Power Equipment*, WT/DS419.
[20] See the WTO website, *Canada – Certain Measures Affecting the Renewable Energy Generation Sector*, WT/DS412.
[21] See the WTO website, *Canada – Measures Relating to the Feed-in Tariff Program*, WT/DS426.
[22] See 'France: state loan for Renault in return of a pledge to increase local content and to maintain employment', *Global Trade Alert*, March 24, 2010.

Table 2.5 *Some Cases of LCRs in Place, as of January 2012*

| Country/Province | Project/Program | Description |
| --- | --- | --- |
| United States/ Ohio | Renewable energy in general | Mandating half of its renewable energy to be supplied through in-state production |
| United States/ Colorado, Missouri | Renewable energy in general | Applying a 1.25 multiplier to renewable energy certificates produced from in-state resources |
| Canada/ Ontario | Solar and wind projects (Green Energy and Green Economy Act 2009) | Requiring, from 2009–2011, 25% of costs for wind projects to be local and 50% of solar projects. From 2012 onward, these restrictions increase to 50% and 60% respectively to benefit from FIT program |
| Canada/ Quebec | Calls for tenders for the creation of wind farms | Requiring wind farms be located in the Gaspesie region of the province with regional content requirement of 30 % minimum |
| Italy | Solar ContoEnergia 4 (solar subsidy law, approved in May2011) | Offering PV projects that contain at least 60 % of local content with an additional 5–10 percent premium over normal FIT rates |
| India | Solar Jawaharlal Nehru National Solar Mission (JNNSM, launched in 2010) | Mandating cells and modules for solar PV projects based on crystalline silicon to be manufactured in India and for solar thermal, mandated 30% project to have domestic content to benefit from FIT program |
| Portugal | Wind tender | Awarding contracts only to bidders engaged in research collaborations with local universities |
| Ukraine | Renewable energy in general (Law of Ukraine on the Electric Power Industry, passed in June 2011) | Requiring producers of electricity from renewable energy sources to meet a certain share of materials, works and/or services of Ukrainian origin that is used to construct a power plant in order to benefit from FIT program |

*Sources*: Global Trade Alert website and various sources.

*Subsidies*

Responding to petitions filed by US solar companies, which have lost a large share of their market to Chinese solar panels, the International Trade Administration (ITA) in the Department of Commerce, and the independent International Trade Commission (ITC), initiated an investigation to determine whether China had been subsidised its solar panel industry, thereby injuring US solar firms.[23] In early 2012, preliminary rulings issued by the ITA and ITC favoured the US petitioners. This case may well be a harbinger of trade disputes ahead in the RE space.

To be sure, the overall level of public support for RE is coming under short-term pressure on account of budget woes in Europe and the United States. But the International Energy Agency (IEA) projected that the share of non-hydro renewables in power generation would increase from 3 per cent in 2009 to 15 per cent in 2035, and that this gain, if it happened, would be underpinned by public subsidies. According to the IEA (2011), global RE subsidies were about USD 66 billion in 2010, and will reach almost USD 250 billion in 2035.[24] Going forward, these are large numbers.

Yet, while huge RE subsidies linked to LCRs clearly distort markets, they may be critical for RE to compete with coal, oil and natural gas. Many RE technologies are not expected to achieve 'grid parity' – the point at which alternative means of generating electricity becomes not more expensive than existing means of generating energy – until 2020. While the subsidy cost per unit of output is expected to decline, the IEA (2011) forecasts that most RE sources will need continued support for several decades to compete in electricity markets. Besides, governments continue to subsidise fossil-fuel use on a vast scale, contradicting sensible economic and environmental goals. In 2010, world fossil-fuel subsidies amounted to USD 409 billion. Without reform, spending on fossil-fuel consumption subsidies is projected to reach USD 660 billion in 2020, some 0.7 per cent of global gross domestic product (GDP) (IEA 2011).

Since fossil-fuel subsidies are about five to six times the subsidies going to RE, and since few RE sources are yet able to compete with fossil fuels, it seems unreasonable to limit RE subsidies as part of a SETA accord. As a

---

[23] Solar World Industries America, a solar panel manufacturer, and six other solar companies filed petitions in October 2011 alleging that China was flooding the US market with under-priced solar panels and subsidising its solar industry in violation of WTO rules. See 'International Trade Commission: Chinese solar imports might threaten US companies', *The Hill*, December 2, 2011.

[24] The economic case for renewables is strongest when environmental externalities are taken into account or when potential customers are dispersed and do not have access to an electricity grid (Steenblik et al. 2006).

political matter, it is probably impractical to insist that national subsidies be extended to foreign firms based in the SETA area. Rather, SETA Members should seek agreement on phasing out LCRs that are linked to RE subsidies.

That said, some reduction of RE subsidies may be in the stars, independently of any future SETA. At a symposium held in Washington in November 2011, Amy Porges, an international trade lawyer, pointed out that Brazil was now a net importer of ethanol from the United States on account of high sugarcane prices. Meanwhile, many US politicians are fed up with ethanol subsidies. The next five years could be an auspicious period to forge a SETA deal that covers ethanol tariffs, phases out ethanol subsidies and brings Brazil on board. In the same spirit, some countries, including France and Germany, have curtailed their renewable subsidies in response to budget realities and price comparisons. As for solar energy, a main reason is that the solar panels produced by China and Taiwan are so cheap that domestic firms cannot compete, whatever the subsidy. These examples suggest that some SETA Members are, of their own accord, travelling a downward path with respect to RE subsidies.

Yet we think that SETA disciplines on RE subsidies are not a promising topic in the near term (ethanol may be an exception). From both environmental and economic standpoints, curtailing fossil-fuel subsidies is more important and probably more practical. The leaders of the major economies have repeatedly made commitments to reform fossil-fuel subsidies, but so far they have little to show for all their words. Perhaps a SETA pact could begin the process of meaningful discipline.

*Standards*

Mandatory technical regulations and voluntary standards achieve both public and commercial objectives. Regulations and standards in the energy space include efficiency standards, labelling requirements and private systems for identifying green products. These measures are instituted at the Federal, state, city or corporate level and play a key role in promoting economic integration, averting negative externalities and fostering innovation. Yet at the same time regulations and standards can discriminate against foreign firms, limit competition and act as trade barriers.

Recognising the right of WTO Members to protect human health, safety and the environment, the Agreement on Technical Barriers to Trade (TBT Agreement) asks WTO Members to ensure that regulations and standards serve legitimate objectives and do not create unnecessary obstacles to trade. The TBT encourages governments to apply international standards where they exist. In this regard, RE industries operate under international standards promulgated by the International Electrotechnical Commission (IEC) and the International Organization for Standardization (ISO).

The number of measures notified as 'specific trade concerns' to the TBT Committee has increased, and many of them relate to RE. For example, in 2010, the United States raised its concern to the TBT Committee over a Korean standard for thin-film solar panels (KS 61646:2007 thin-film terrestrial PV modules), arguing that the Korean standard locked US solar panel producers out of the Korean market. While the Korean standard incorporates much of the IEC 61646 (the international standard that applies to all types of thin-film solar panels), the Korean standard only applies to one type of thin-film solar panel – amorphous silicon (A-Si) type. The Korean standard does not require mandatory compliance, but solar panels must be certified by the Korea Energy Management Corporation (KEMCO) in order to be sold in the Korean market, and the KEMCO only certifies one type of thin-film solar panels based on that standard.[25] The net result is to exclude other thin-film solar panels that would meet the specifications of IEC 61646.

Since standards are often voluntary as a legal matter, but mandatory as a practical matter, they are hard to harmonise between countries. In one paper discussing this problem in the context of liberalising trade in environmental goods, Vikhlywaev (2009) suggested that members of a trade group could consider a 'smorgasbord' approach along the lines of a current trend in the ISO: declare certain national, regional or international standards as equivalent rather than trying to negotiate a single standard. The SETA could adopt this approach. Also, SETA Members may consider adopting the features of a mutual recognition agreement (MRA) for assessing national conformity to agreed standards. Finally, the SETA might require its Members to consult with each other in confidence before publishing the pre-announcement of a new standard.

*Investment*

According to Bloomberg, despite the sluggish global economy, total global investment in clean energy has increased rapidly over the past few years, reaching a new record of USD 260 billion in 2011 – 5 per cent above 2010 levels and almost five times the total of USD 54 billion in

---

[25] The Korean standard does not apply to cadmium telluride, copper indium selenide and gallium arsenide solar panels. Korea is the only country in the world that specifically restricts application of the IEC standard to only one of the three leading types of thin film panels. See WTO (2011b) and USTR (2011)

2004.[26] In 2011, the United States ranked top, with about USD 60 billion in clean energy investment, and China was second with about USD 47 billion.

Despite this rapid increase, non-tariff measures such as subsidies confined to national firms and local content requirements have delayed new investment flows. Some countries have adopted measures that directly discriminate against foreign-controlled firms, such as capital or ownership requirements. For example, the registered capital requirement for a foreign or Sino-foreign wind power project is a minimum of 33 per cent of total assets, while the requirement for Chinese companies is 10 per cent (NFTC 2010). Moreover, foreign companies operating in China are required to enter into a joint venture in which Chinese partners control 51 per cent ownership.

In light of this background, the SETA should include a Chapter that phases out capital and ownership requirements, as well as other measures that directly discriminate against foreign investors.

*Services*

A Peterson Institute policy brief, *Framework for the International Services Agreement*, explores the details of a possible post-Doha Agreement designed to achieve significant liberalisation of services trade and investment by self-selected WTO Members.[27] Here we need only sketch the outlines as they apply to RE. In general, trade in services occurs through four modes of supply (enumerated in the GATS: (1) cross-border trade; (2) consumption abroad; (3) commercial presence; and (4) temporary movement of natural persons.[28] Modes 1 and 3 are most relevant for a SETA.

If the SETA is concluded before the International Services Agreement (ISA), a major issue will be dual-use energy services, similar to the problem of dual-use energy goods (goods with both renewable and non-renewable

---

[26] See 'Solar surge drives record clean energy investment in 2011', press release, *Bloomberg New Energy Finance*, January 12, 2012.

[27] See Hufbauer and Schott (2012).

[28] Mode 1: cross-border trade is defined to cover services flows from the territory of one WTO Member into the territory of another Member (e.g. banking or architectural services transmitted via telecommunications or mail); Mode 2: consumption abroad refers to situations where a service consumer (e.g. tourist or patient) moves into another Member's territory to obtain a service; Mode 3: commercial presence implies that a service supplier of one Member establishes a territorial presence, including through ownership or lease of premises, in another Member's territory to provide a service (e.g. domestic subsidiaries of foreign insurance companies or hotel chains); and Mode 4: temporary movement of natural persons, which consists of persons of one Member entering the territory of another Member to supply a service (e.g. accountants, doctors or teachers).

applications). The UN Central Product Classification (CPC) list, which serves as the basis for the WTO Services Classification list, is not specific enough to identify either environmental or energy services. To address this concern, the EU, Canada and Colombia have put forward their own proposals (ICTSD 2011).

Our recommendation is that, within the SETA arrangement, the Members should focus on a limited list of services principally applied to RE. To this end, they could adopt a project approach, focusing on services directly connected with the installation, operation and maintenance of RE projects. For these services, the SETA should call for free trade and investment (Modes 1 and 3), phased in over a reasonable period of time.

## 3   Negotiating Approaches

A related paper authored by the Peterson Institute, *Will the WTO Have a Bright Future?*, pictures a post-Doha world in which WTO negotiations centre on a series of plurilateral agreements, negotiated by self-selected WTO Members and applied on a conditional MFN basis.[29] In this policy brief, we sketch the application of that post-Doha scenario to a SETA accord, potentially one of the plurilateral agreements.

### *ITA or GPA Model?*

The Ministerial Declaration on trade in information technology (IT) products, commonly called the Information Technology Agreement (ITA), is often cited as a model for a SETA. The ITA calls for free trade in covered products. As the first post-Uruguay Round agreement, the ITA was signed at the WTO Ministerial Conference held in Singapore in December 1996, and took effect in March 1997.[30] Its membership steadily increased from the twenty-nine original signatories in 1996 to seventy-three members in 2010, representing about 97 per cent of world trade in IT products.[31] From the get-go, the ITA was applied on an

---

[29] See Hufbauer and Schott (2012).
[30] At that time, twenty-nine countries or separate customs territories signed the Declaration, but it was unclear whether the ITA could go into effect since the original twenty-nine signatories did not reach the 90 per cent trade coverage criterion stipulated by the Declaration. However, during the period leading up to the April 1997 deadline for notification, a number of other countries were added to the signatories. As a result the 90 percent criterion was met and the ITA went into effect on 13 March 1997.
[31] There are three principles that one must abide by to become an ITA member. (1) all products listed in the Declaration must be covered, (2) all must be reduced to a zero tariff level and (3) all other duties and charges must be bound at zero. There are no exceptions to product coverage, but for an extended implementation period is allowed for sensitive

unconditional MFN basis – whether or not signatories, all WTO countries could benefit from free access to the signatory markets.

While highly successful, it is questionable whether the ITA could serve as a model for a SETA. Fliess (2009) spelled out the differences between the context in which the ITA was negotiated and the context of an agreement on EGS. Favourable conditions for the ITA – such as strong industry support, high market maturity level, a critical role in the broader economy, and developing countries as major players – are weak for EGS. Fliess makes a strong argument, even though China and India are already key players in renewable markets and leading global companies have backed the SETA concept. Realistically, it is highly unlikely that a SETA can be launched with a coverage extending to 90 per cent of global trade. Without coverage near that level, the 'free-rider' problem becomes an insuperable political obstacle to a SETA that permits unconditional MFN access to its market opening provisions.

The main alternative to the ITA model for SETA is the model established by the Government Procurement Agreement (GPA). The GPA was part of the Uruguay Round accords that established the WTO, signed at Marrakesh in April 1994, and entered into force in January 1995. The GPA was concluded as one of the plurilateral agreements covered by Annex 4 of the Marrakesh Agreement establishing the WTO. The GPA establishes a framework of rights and obligations among the signatories with respect to their national laws, regulations, procedures and practices in the area of government procurement. The GPA, unlike the ITA, is based on the conditional MFN principal; its basic thrust is to open selected portions of government procurement markets only to GPA signatories.

Business firms with export interests in the SETA space will strongly prefer the GPA model to the ITA model, since benefits are confined to Member Countries. Otherwise, the SETA will have little or no attraction for the founders. However, to be an agreement within the WTO, on account of its conditional MFN application, the SETA will require a waiver under the Marrakesh Agreement, Article IX (3) (akin to waiver for the GPA).[32] As Hufbauer and Schott (2012) argue, similar waivers

---

items. The commitments undertaken under the ITA in the WTO are applied on an unconditional MFN basis.

[32] Article IX (3) reads as follows: 'In exceptional circumstances, the Ministerial Conference may decide to waive an obligation imposed on a Member by this Agreement or any of the Multilateral Trade Agreements, provided that any such decision shall be taken by three fourths (4) of the Members unless otherwise provided for in this paragraph. (a) A request for a waiver concerning this Agreement shall be submitted to the Ministerial Conference for consideration pursuant to the practice of decision-making by consensus. The Ministerial Conference shall establish a time-period, which shall not exceed 90 days, to

would be required for other plurilateral agreements in the post-Doha Round trading world.

If a waiver for a SETA proves impossible, then like-minded governments could negotiate a SETA outside the WTO framework. But in that event, the agreement would not enjoy access to the WTO Dispute Settlement Mechanism (DSM). That is an important shortcoming, since enforcement of the Agreement between its Members will be critical for any SETA's long-term viability. The case brought against Canada by Japan illustrates the sort of questions that will inevitably arise. In September 2010, Japan requested consultations with Canada with respect to the domestic content requirements in Canada's FIT programme for RE, claiming that the measures discriminated against foreign producers. The consultations were inconclusive and, upon Japan's request, a WTO panel was established in October 2011.[33] Eventually the panel and the Appellate Body will resolve the dispute. Without this mechanism, the disagreement could go on for years, and the same indeterminate application could await provisions in if the pact is established as a free-standing agreement outside the WTO framework.

*Goods and Services Coverage*

In grappling with the definition of EGs, the EGS talks in the Doha Round considered four approaches: (1) the list approach; (2) the request-and-offer approach; (3) the project approach; and (4) the integrated approach.

The list approach[34] seeks an agreed list of EGS for special treatment, such as tariff reductions.[35] Recognising the difficulty of agreeing on a single list, some countries suggested a request-and-offer approach that aims to provide special treatment to specific goods and services in response to requests and offers between individual WTO Members. Another idea is the project approach – originally proposed by India – which gives special

---

consider the request. If consensus is not reached during the time-period, any decision to grant a waiver shall be taken by three fourths of the Members. (b) A request for a waiver concerning the Multilateral Trade Agreements in Annexes 1A or 1B or 1 C and their annexes shall be submitted initially to the Council for Trade in Goods, the Council for Trade in Services or the Council for TRIPS, respectively, for consideration during a time-period which shall not exceed 90 days. At the end of the time-period, the relevant Council shall submit a report to the Ministerial Conference.'

[33] See the WTO website for a summary of the dispute, *Canada – Certain Measures Affecting the Renewable Energy Generation Sector*, DS412.

[34] In the absence of an internationally agreed definition, several lists of EGs were proposed and circulated by international organisations, including the OECD and APEC, in search of a starting point for the negotiations. Referring to the lists proposed by members of WTO, the WTO Special Session of the Committee on Trade and Environment (CTE) has tried to nail down a single list.

[35] See Krenicki (2010).

treatment to goods and services associated with a specific environmental project approved by a designated national authority. This approach was driven by concerns over the dual use of many environmental products enumerated in the HS codes.

Due to the limitations inherent in each approach, some countries have proposed hybrid approaches. For example, Argentina proposed an integrated approach that resembles the project approach with multilaterally agreed pre-identified categories of goods used in the approved projects. Some countries suggested a link to clean development mechanism (CDM) projects. Also, some countries, notably Singapore, Australia, Hong Kong and China, have proposed a core list of single-use EGs supplemented by self-selected lists and request-and offer-procedures (UNCTAD 2011).

It may be less difficult to draw up a list of SEGS covered by a SETA than to agree on EGS coverage, due to its relatively narrow scope. However, the issues raised in the EGS negotiations would still need to be addressed in the SETA talks. For example, the current classification systems are not specific enough to isolate SEGS; dual-use thus remains an issue.

To avoid a significant dual-use problem, SETA Member Countries could adopt the integrated approach that evolved from the EGS negotiations. Following this model, the SETA negotiations could start with an agreed short list of single-use sustainable energy products and services. Then, the Members could add products to be liberalised, based both on the request-and-offer approach and the project approach.

*Country Coverage*

For the success of a SETA, the major players in RE markets should participate. In the earlier section on tariff barriers, we listed ten core countries – NAFTA, the EU, Chile, Colombia, Peru, Australia, New Zealand, Singapore, Japan and Korea – that seem highly likely to join any SETA negotiations. We also listed six candidate countries – Brazil, India, China, Indonesia, Turkey and South Africa – that hold great commercial and environmental interest. In practical terms, a SETA might start out with a few members, perhaps the United States, the EU, Japan and a few other advanced countries, as well as one or two big developing countries such as Brazil and China.

UNCTAD (2010) noted that the markets for renewable energies and other EGs either do not exist or are weak in many developing countries, but a few developing countries have become key players. China's participation is critical. As a leading exporter of RE goods, China has strong commercial reasons to join.

One big question is whether unanimity or super-majorities will be required to accept new Members into the SETA, and to enlarge the product and subject matter coverage. Our thinking is that unanimous consent (the consensus rule) would freeze the SETA at its starting position. Rather, in our view, it should be possible to expand country, product and subject matter coverage upon an affirmative vote of 75 per cent of the existing Members, calculated in both of two ways: (1) one country, one vote; and (2) weighted according to their combined imports and exports of SETA products. The dual test would ensure against a bulldozer expansion, either by small or by large countries. Moreover, as a last resort, a dissenting SETA Member could give notice and leave the Agreement.

*Business Engagement*

As witnessed in other sectoral negotiations, support from business is critical for success. The ITA was strongly backed by leading companies such as IBM and Toshiba; the Basic Telecommunications Agreement (BTA) was backed by a large number of telecom firms.

RE markets are dominated by a handful of companies, concentrated in a few countries. As Table 2.6 shows, the top 10 manufacturers for

Table 2.6 *Top 10 Solar PV Cell/Wind Turbine Manufacturers, 2010 (market share %)*

| | Solar PV Cell | | | Wind Turbine | |
|---|---|---|---|---|---|
| | Company (Country) | Share | | Company (Country) | Share |
| 1 | Suntech Power (China) | 7% | 1 | Vestas (Denmark) | 14% |
| 2 | JA Solar (China) | 6% | 2 | Sinovel (China) | 11% |
| 3 | First Solar (USA) | 6% | 3 | GE Wind (USA) | 9% |
| 4 | Yingli Green Energy (China) | 5% | 4 | Goldwind (China) | 9% |
| 5 | Trina Solar (China) | 5% | 5 | Enercon (Germany) | 7% |
| 6 | Q-Cells (Germany) | 4% | 6 | Suzlon Group (India) | 7% |
| 7 | Kyocera (Japan) | 3% | 7 | Dongfang (China) | 7% |
| 8 | Motech (Taiwan) | 3% | 8 | Gamesa (Spain) | 6% |
| 9 | Sharp (Japan) | 3% | 9 | Siemens Wind Power (Denmark) | 6% |
| 10 | Gintech (Taiwan) | 3% | 10 | United Power (China) | 4% |
| Total | | 42% | Total | | 80% |

*Source:* REN 21 (2011).

solar PV cells and wind turbines accounted for about 42 per cent and 80 per cent of the world market share in 2010, respectively. Among leading global companies, some have already expressed their support for a SETA. For example, John Krenicki, Vice President of GE and CEO and president of GE Energy, called for an International Green Free Trade Agreement, aimed at removing trade barriers to renewable energy.[36] Vestas is another enthusiastic supporter. In the run up to the G20 summit held in Cannes in November 2011, the B20 working group, led by Ditlev Engel, President and CEO of Vestas, called on the G20 leaders to create a safe haven for the free trade in EGS by agreeing first to eliminate trade barriers among the G20 countries.[37]

Along with support from leading firms, the SETA could also enlist the backing of business organisations such as the World Business Council for Sustainable Development (WBCSD) which numbers more than 200 Members.

## 4    Conclusion

Over the past decade, the demand for clean technologies and products has soared. The demand will only get stronger as the consequences of climate change become more apparent. The SETA talks provide a modest but practical way to make progress on one aspect of the overarching climate challenge.

In this chapter, we have examined trade and tariff data for thirty-nine narrowly defined EGs, both for the world as a whole, and for selected 'core' and 'candidate' countries. In 2010, world imports of the thirty-nine EGs accounted for about 1.7 per cent of the total imports of all products. While this does not seem like much, our starting point is modest, only thirty-nine tariff lines out of the more than 5,000 tariff lines identified at the HS six-digit level. In our view, a short list will better serve as a launching pad, but once the SETA is off the ground, its product coverage should be enlarged.

In general, developing countries are more protective than developed countries, maintaining both higher bound and applied tariff rates. The mercantilist approach that dominates trade negotiations will very likely colour trade negotiations in EGs, making it harder to bring developing countries within a SETA. However, the spectre of climate change has

---

[36] G20 Business Summit press release, 9 November 2011.
[37] Many developing countries have identified tariffs as one of the impediments to technology transfer (UNFCCC 2009).

Table 2.7 *Trade of APEC 50 EGs, 2010*[a]

|  | Exports of APEC EGs to APEC | | Imports of APEC EGs from World | |
|---|---|---|---|---|
|  | USD (million) | % of Total Exports to APEC | USD (million) | % of Total Imports from World |
| Australia | 916.7 | 0.6 | 5,530.5 | 2.9 |
| Brunei Darussalam[b] | n.a | n.a | n.a | n.a |
| Canada | 4,073.2 | 1.3 | 9,175.1 | 2.4 |
| Chile | 40.5 | 0.1 | 1,020.1 | 1.7 |
| China | 15,831.5 | 1.7 | 49,249.7 | 3.8 |
| Hong Kong | 109.8 | 1.2 | 9,077.2 | 2.1 |
| Indonesia | 223.1 | 0.2 | 3,206.4 | 2.4 |
| Japan | 37,065.0 | 6.9 | 13,353.3 | 1.9 |
| Korea, Rep | 9,455.5 | 3.1 | 24,432.6 | 5.7 |
| Malaysia | 2,968.8 | 2.2 | 5,040.3 | 3.1 |
| Mexico | 5,632.4 | 2.2 | 8,304.4 | 2.8 |
| New Zealand | 94.4 | 0.5 | 500.3 | 1.7 |
| Papua New Guinea[b] | n.a | n.a | n.a | n.a |
| Peru | 4.1 | 0.0 | 553.3 | 1.8 |
| Philippines | 273.4 | 0.7 | 1,275.2 | 2.2 |
| Russia | 185.7 | 0.3 | 7,182.7 | 2.9 |
| Singapore | 9,846.6 | 3.9 | 10,755.8 | 3.5 |
| Taipei[b] | n.a | n.a | n.a | n.a |
| Thailand | 1,667.5 | 1.3 | 4,662.9 | 2.6 |
| United States | 26,634.5 | 4.1 | 39,881.6 | 2.0 |
| Viet Nam | 195.7 | 0.4 | 1,993.4 | 2.3 |
| Total | 1,15,218.4 | 2.9 | 1,95,194.8 | 2.8 |

*Notes*:
a. While the original APEC EG list contains fifty-four products at the HS six-digit levels, we only include fifty products (forty-eight from the HS 2002 and two from HS 2007) here since four products identified in the APEC list are from the HS 2012 for which trade data is not available in UN COMTRADE.
b. Trade data from HS 2002 not available.
*Source*: UN COMTRADE (accessed through the WITS on 20 September 2012).

brought a greater appreciation of the benefits of importing EGs and acquiring associated technology.[34] Moreover, a few developing countries, notably China and India, have surged as big exporters of certain EGs to the world market. These events raise the potential for both developed and developing countries to join the SETA talks.

While tariff rates range widely among our core and candidate countries, tariffs are not necessarily the biggest impediment to trade in SETA

Table 2.8 Comparison of SETA, World Bank and APEC Lists

| | HS Code | Description | World Bank (43 Products, 2007) | APEC (54 Products, 2012) |
|---|---|---|---|---|
| **Products categorised as 'renewable energy' only from WTO (2011)** | | | | |
| | 3907 | **Polyacetals, other polyethers and epoxide resins, in primary forms; polycarbonates, alkyd resins, polyallyl esters and other polyesters, in primary forms** | | |
| 1 | 390799 | Other polyesters: Other | | |
| | 4016 | **Other articles of vulcanised rubber other than hard rubber** | | |
| 2 | 401699 | Other: Other | | |
| | 4707 | **Recovered (waste and scrap) paper or paperboard** | | |
| 3 | 470710 | Recovered (waste & scrap) unbleached kraft paper/paperboard/corrugated paper/paperboard | | |
| 4 | 470720 | Other paper or paperboard made mainly of bleached chemical | | |
| 5 | 470730 | Paper or paperboard made mainly of mechanical pulp (for example, newspapers, journals and similar printed matter) | | |
| 6 | 470790 | Other, including unsorted waste and scrap | | |
| | 7308 | **Structures (excluding prefabricated buildings of heading 94.06) and parts of structures (for example, bridges and bridge sections, lock gates, towers, lattice masts, roofs, roofing frameworks, doors and windows and their frames and thresholds for doors, shutters, balustrades, pillars and columns), of iron or steel; plates, rods, angles, shapes, sections, tubes and the like, prepared for use in structures, of iron or steel** | | |
| 7 | 730820 | Towers and lattice masts | o | |
| | 8406 | **Steam turbines and other vapour turbines** | | |
| 8 | 840682 | Other turbines of an output not exceeding 40MW | | |
| | 8418 | **Refrigerators, freezers and other refrigerating or freezing equipment, electric or other; heat pumps other than air conditioning machines of heading 84.15** | | |
| 9 | 841861 | Other refrigerating or freezing equipment, heat pumps: Compression-type units whose condensers are heat exchangers | o | |

Table 2.8 (cont.)

| | HS Code | Description | World Bank (43 Products, 2007) | APEC (54 Products, 2012) |
|---|---|---|---|---|
| 10 | 841869 | Other refrigerating or freezing equipment heat pumps: Parts | o | |
| | **8419** | **Machinery, plant or laboratory equipment** | | |
| 11 | 841919 | Instantaneous or storage water heaters, non-electric: Other | o | o |
| | **8479** | **Machines and mechanical appliances having individual functions, not specified or included elsewhere in this Chapter** | | |
| 12 | 847920 | Machinery for the extraction or preparation of animal or fixed vegetable fats or oils | | |
| | **8483** | **Transmission shafts (including cam shafts and crank shafts) and cranks; bearing housings and plain shaft bearings; gears and gearing; ball or roller screws; gear boxes and other speed changers, including torque converters; flywheels and pulleys, including pulley blocks; clutches and shaft couplings (including universal joints)** | | |
| 13 | 848340 | Gears, gearing (excluding toothed wheels, chain sprockets and other transmission elements presented separately); ball or roller screws; gear boxes, other speed changers, including torque converters | o | |
| 14 | 848360 | Clutches and shaft couplings (incl. universal joints) | o | |
| | **8501** | **Electric motors and generators (excluding generating sets)** | | |
| 15 | 850161 | AC generators (alternators), of an output not exceeding 75 kVA | o | |
| 16 | 850162 | AC generators (alternator), of an output exceeding 75 kVA but not exceeding 375 kVA | o | |
| 17 | 850163 | AC generators (alternator), of an output exceeding 375 kVA but not exceeding 750 kVA | o | |
| 18 | 850164 | AC generators (alternator), of an output exceeding 750 kVA | | o |
| | **8502** | **Electric generating sets and rotary converters** | | |
| 19 | 850231 | Other generating sets: Wind-powered | o | o |
| 20 | 850239 | Other generating sets: other | | o |
| | **8506** | **Primary cells and primary batteries** | | |
| 21 | 850610 | Primary cells and primary batteries, manganese dioxide | | |

| | | | |
|---|---|---|---|
| 22 | | 850630 | Primary cells and primary batteries, mercuric oxide |
| 23 | | 850640 | Primary cells and primary batteries, silver oxide |
| 24 | | 850650 | Primary cells and primary batteries, lithium |
| 25 | | 850660 | Primary cells and primary batteries, air-zinc |
| 26 | | 850690 | Parts |
| | 8507 | | Electric accumulators, including separators therefor, whether or not rectangular (including square) |
| 27 | | 850720 | Other lead – Acid accumulators o |
| 28 | | 850740 | Nickel-iron |
| | 8537 | | Boards, panels, consoles, desks, cabinets and other bases, equipped with two or more apparatus of heading 85.35 or 85.36, for electric control or the distribution of electricity, including those incorporating instruments or apparatus of Chapter 90, and numerical control apparatus, other than switching apparatus of heading 85.17 |
| 29 | | 853710 | For a voltage not exceeding 1,000 V o |
| | 8541 | | Diodes, transistors and similar semiconductor devices; photosensitive semiconductor devices, including photovoltaic cells whether or not assembled in modules or made up into panels; light emitting diodes; mounted piezo-electric crystals |
| 30 | | 854140 | Photosensitive semiconductor devices, including photovoltaic cells whether or o not assembled in modules or made up into panels; light emitting diodes |
| | 9001 | | Optical fibres and optical fibre bundles; optical fibre cables other than those of heading 85.44; sheets and plates of polarising material; lenses (including contact lenses), prisms, mirrors and other optical elements, of any material, unmounted, other than such elements of glass not optically worked |
| 31 | | 900190 | Other o |
| | 9002 | | Lenses, prisms, mirrors and other optical elements, of any material, mounted, being parts of or fittings for instruments or apparatus, other than such elements of glass not optically worked |
| 32 | | 900290 | Other o |
| Other products[a] | | | |
| | 2207 | | Ethyl alcohol, undenat, n/un 80 per cent alc, alcohol, denat[b] |
| 33 | | 220710 | Undenatured ethyl alcohol of an alcoholic strength |
| | 3824 | | Prepared binders for foundry moulds or cores; chemical products and preparations of the chemical or allied industries (including those consisting of mixtures of natural products), not elsewhere specified or included |

Table 2.8 (cont.)

| | HS Code | Description | World Bank (43 Products, 2007) | APEC (54 Products, 2012) |
|---|---|---|---|---|
| 34 | 382490 | Other[c] | | |
| | 8410 | Hydraulic turbines, water wheels, and regulators therefor | | |
| 35 | 841011 | Hydraulic turbines and water wheels of a power not exceeding 1,000 kW | o | |
| 36 | 841090 | Parts, including regulators | o | |
| | 8411 | **Turbo-jets, turbo-propellers and other gas turbines** | | |
| 37 | 841181 | Other gas turbines of a power not exceeding 5,000 kW | o | |
| 38 | 841182 | Other gas turbines of a power exceeding 5,000 kW | o | o |
| | 8503 | **Parts suitable for use solely or principally with the machines of heading 85.01 or 85.02** | | |
| 39 | 850300 | Parts suitable for use solely or principally with the machines of heading 85.01 or 85.02 | o | o |

*Notes:*

*a.* Other products are selected based on authors' judgement. Except for HS 220710, all other products are also found in the WTO (2011) but labelled not only 'renewable energy' but also something else (e.g. 'environmental technologies').

*b.* There is currently no specific customs classification for bioethanol for biofuel production. This product is traded under HS 2207, which covers undenatured (HS220710) and denatured alcohol (HS 220720). While both denatured and undenatured alcohol can be used for biofuel production, in this table, we only include u-denatured alcohol (HS 220710) since it seems more suitable for use as a fuel (denatured ethanol is often used as a solvent).

*c.* There is currently no specific customs classification for biodiesel for biofuel production. Biodiesel is classified under HS code 382490 which covers some other products.

*Sources*: For the World Bank list, see World Bank (2007). The APEC list is available at www.apec.org/Press/News-Releases/2012/~/link.aspx?_id=357A0FFAFA184C278FC840F9F6E9DF36&_z=z.

products. In fact, many EGs that already enjoy low tariff rates still face severe NTBs, such as restrictive standards and LCRs (usually associated with government procurement or state-owned enterprises, SOEs), subsidies limited to national firms and restrictive standards. In our view, while tariff elimination for EGs should be the first priority in any SETA talks, NTBs must also be addressed, and several of them should be phased out. Our preliminary suggestion as to the priority order of NTBs is: LCRs, services, investment, subsidies and standards.

The Peterson Institute has examined negotiating approaches in companion papers.[38] Their application to the SETA can be summarised in four precepts:

- The plausible model is the GPA, not the ITA – in other words, benefits extended on a conditional MFN basis, but the agreement should be open to all WTO Members who subscribe to the obligations.
- If acceptable to three fourths of WTO Members, the SETA should be part of the WTO framework. The benefit to SETA Members would be access to the WTO DSM; the benefit to non-Members is the possibility of accession at a later date.
- It should be possible to extend SETA coverage as to products and countries by a super-majority of Members; unanimity should not be required.
- Support from the business community is vital. Leading firms and associations should be deeply involved at every stage of SETA negotiations.

## References

Fliess, Barbara (2009) 'The WTO negotiations on environmental goods and services: need for a change in mindset away from a free-standing sectoral deal'. *Trade and Environment Review*. UNCTAD, New York and Geneva

Global Trade Alert (2011) *Trade Tensions Mount: The 10th GTA Report*. CEPR/Global Trade Alert, London

Hufbauer, Gary et al. (2010) *Figuring Out the Doha Round*. Peterson Institute for International Economics, Washington, DC

(2012) *Framework for the International Services Agreement*. Peterson Institute for International Economics, Washington, DC

Hufbauer, Gary and Jeffrey Schott (2012) *Will the WTO Have a Bright Future?* Report to the ICC Foundation. Peterson Institute for International Economics, Washington, DC

ICTSD (2011) *Fostering Low Carbon Growth: The Case for a Sustainable Energy Trade Agreement*. ICTSD, Geneva

---

[38] See Hufbauer et al. (2012) and Hufbauer and Schott (2012).

IEA (2011) *World Energy Outlook 2011*. International Energy Agency, Paris
Kirkegaard, Jacob et al. (2009) *It Should Be a Breeze: Harnessing the Potential of Open Trade and Investment Flows in the Wind Energy Industry*. Peterson Institute for International Economics, Washington, DC
Krenicki, John (2010) *Tearing Down Barriers to Green Trade*. 2009 Citizenship Report, www.ge.com/citizenship
NFTC (2010) *China's Promotion of the Renewable Electric Power Equipment Industry: Hydro, Wind, Solar and Biomass*. National Foreign Trade Council, Washington, DC
OECD (2008) *Biofuel Support Policies: An Economic Assessment*. Organisation for Economic Cooperation and Development, Paris
REN21 (2011) *Renewables 2011 Global Status Report*. REN21, Paris
Steenblik, Ronald et al. (2006) *Synergies between Trade in Environmental Services and Trade in Environmental Goods*. OECD, Paris
UNCTAD (2010) *Trade and Environment Review 2009/2010*. United Nations Conference on Trade and Development, New York and Geneva
   (2011) *WTO Negotiations on Environmental Goods: Selected Technical Issues*. United Nations Conference on Trade and Development, New York and Geneva
United Nations Economic and Social Commission for Western Asia (2007) *The Liberalization of Trade in Environmental Goods and Services in the ESCWA and Arab Regions*. United Nations, New York
UNFCCC (2009) *Second Synthesis Report on Technology Needs Identified by Parties Not Included in Annex I to the Convention*. United Nations Framework Convention on Climate Change, Bonn
USTR (2011) *Report on Technical Barriers to Trade*. USTR, Washington, DC
Vikhlywaev, A. (2009) 'WTO negotiations on environmental goods and services: the case of renewables', *Trade and Environment Review 2009/2010*. UNCTAD, New York and Geneva
World Bank (2007a) *World Development Report 2008: Agriculture for Development*. World Bank, Washington DC
   (2007b) *International Trade and Climate Change*. World Bank, Washington, DC
World Economic Forum (2010) *Global Agenda Council Reports 2010: Summaries of Global Agenda Council Discussions from the Summit on the Global Agenda 2009*. WEF, Geneva
WTO (2011a) *Committee on Trade and Environment in Special Session*. World Trade Organization, Geneva
   (2011b) *Specific Trade Concerns Raised in the TBT Committee*. World Trade Organization, Geneva

# 3   Trade in Sustainable Energy Services: October 2013

*Joachim Monkelbaan*

Although services related to sustainable energy are largely neglected in international negotiations, they should be a key component of the Sustainable Energy Trade Agreement (SETA) currently under consideration.

Including services in a SETA, however, poses a number of challenges. The first of these lies in identifying a reasonable set of sustainable energy-related services that could be subject to trade liberalisation negotiations. Given that these services are spread across multiple sectors, identifying them could be a daunting task. Another challenge arises from the current disconnect between negotiations on environmental goods (EGs) and negotiations on environmental services at the World Trade Organization (WTO). One incentive for including trade in sustainable energy services in a SETA is that this could both facilitate the diffusion of associated sustainable energy technologies and enable countries to obtain access to such services and the related knowledge transfers.

This chapter attempts to respond to these challenges by identifying services that are directly linked to the diffusion of sustainable energy technologies and analysing specific commitments made by the major trading countries of these services.

Services regulation has connections to many other issues likely to be addressed in a SETA. Facilitating trade in 'services complementary to sustainable energy technologies' goes beyond the boundaries of the General Agreement on Trade in Services (GATS), as it is not limited to the issues of market access and national treatment. Domestic legislation, regulatory measures and administrative rules could also affect trade in these services. Addressing the issue of trade liberalisation in complementary services of sustainable energy technologies in tandem with government procurement issues is vital for the development of any SETA.

Energy plays a crucial role in realising the wider promises embodied by a commitment to sustainable development. These promises include climate change mitigation, prevention of water and air pollution and related health benefits, global access to modern forms of energy, transfer of knowledge and technology, and increased employment opportunities due to the labour intensity of renewable forms of energy.

It will require both domestic and international support frameworks to realise such win–win outcomes. A SETA could provide a framework for the massive scale-up of both goods and services related to sustainable energy, focusing initially on services related to the construction and information communication and technology sectors.

## 1 Services and a Sustainable Energy Trade Agreement

The need to scale-up the use of renewable energy (RE) sources is becoming increasingly urgent in light of rapidly growing global demand for energy- and greenhouse gas (GHG)- reducing technologies. It is thus essential to make RE[1] goods and services readily accessible. Given that cost remains one of the most prohibitive factors to large-scale adoption of RE technologies, cost reduction measures should be given high priority. One logical approach is to liberalise markets and investment regimes in activities related to trade in RE services, thereby insuring that consumers gain access to a wider selection of services at competitive prices. That may lead to further economic benefits, such as the growth of domestic RE goods and services firms that may eventually export to the world market. Trade can also provide new incentives for innovation and investment in related climate-friendly technologies.

Considerable potential lies in the dissemination of sustainable energy goods and services (SEGS) through the liberalisation of trade. The World Bank, for example, has calculated that removing tariff barriers and non-tariff barriers (NTBs) to clean energy technologies could result in a nearly 14 per cent increase in their trade volume For some energy efficient products, the removal of trade barriers could increase trade by up to 60 per cent (World Bank 2007a). Trade should thus be seen only as a means to growth, green jobs and mitigation of GHG emissions. Carefully crafted trade policies should contribute to the massive and rapid deployment of more efficient and cleaner technologies that promote clean growth and energy security.

Local content requirements (LCRs) are a typical example of an NTB to trade in sustainable energy goods. These are banned under the General Agreement on Tariffs and Trade (GATT), Article III:4, the Agreement on Trade-Related Investment Measures (TRIMs), Article 2, and of the Subsidies and Countervailing Measures (SCM) Agreement, Articles 3

---

[1] RE sources include solar energy, wind power, geothermal energy, hydropower, biomass and ocean energy.

and 5. Some governments have also adopted LCRs for sustainable energy services. These may include requirements to transfer technology and provide services using local facilities or infrastructure. While some governments see LCRs as a desirable policy tool, services exporters may perceive them as protectionist. If a trade initiative were to tackle them, services exporters are likely to lobby their governments to participate.

Many political and technical difficulties remain with respect to the liberalisation of environmental goods and services within the Doha Round. In addition, renewable energy policy measures are increasingly challenged in trade disputes.

Considering these factors, the time appears right to create global enabling frameworks for trade in SEGS through sustainable energy trade initiatives, which could eventually lead to the conclusion of a SETA.[2] In addition to reducing barriers to trade, a SETA could create clarity on what types of support governments can give their sustainable energy industries, transparency and predictability for businesses and investors, and an overall regulatory framework in support of a rapid scale-up of sustainable energy.

A SETA could be conceived in many different forms, including, for example, as a plurilateral agreement, such as the Information Technology Agreement (ITA) or the Government Procurement Agreement (GPA).[3]

The objective of this chapter is to promote innovative approaches to negotiating sustainable energy services. To lay a stable basis for such negotiations, it will be necessary to (1) identify the coverage and classification of services that are directly related and complementary to the diffusion of sustainable energy technologies, (2) analyse commitments made unilaterally and through international negotiations (including in preferential trade agreements or PTAs) so as to assess the degree of openness of the services market and (3) propose ways to address these findings in sustainable energy trade initiatives and in a SETA. The wider implications for sustainable development will be considered as well.

To realise a successful diffusion of SEGS, understanding the synergies between trade in goods and services is crucial, as certain sustainable energy goods are indispensable for delivering associated services, and vice versa. For instance, an analysis of environmental goods (EGs) associated with service contracts carried out by the Organisation for Economic Cooperation and Development (OECD) demonstrates that many of these EGs are used in the performance of environmental services

---

[2] For more information on SETA, see International Centre for Trade and Sustainable Development (2011).
[3] See, Kennedy (2012).

(OECD 2005).[4] In addition, an empirical study shows that trade in sustainable energy technologies is often impeded by trade restrictions on associated services (Steenblik and Kim 2008). Furthermore, several empirical studies reveal that some of the key services required for sustainable energy options, ranging from energy efficiency projects to utility-scale wind power projects, are often unavailable in the host countries (Steenblik and Geloso Grosso 2011; Sterk et al. 2007). A wide variety of services is required for the creation of wind and solar power projects. Given the complexity and degree of specialisation, local enterprises do not always have the competence to provide the full range of such services.

## 2    Scope and Outline of This Chapter

Because many different services could relate to sustainable energy, this study narrows the scope of services subject to an analysis of the major trading countries and their specific commitments. It also aims to capture lessons that could be learned for a SETA.

From the onset, this study reviews negotiations related to (sustainable energy) services and how these negotiations have been challenged by classification issues. Second, it provides an overview of sustainable energy technologies and practices in the energy supply sector (the key focus of the Fourth Assessment Report of the Intergovernmental Panel on Climate Change, see IPCC 2007). The chapter focuses on services that could directly influence the diffusion of sustainable energy technologies, known as 'complementary services of sustainable energy technologies'. Given the inseparable links between SEGS, it is important to highlight those services that are directly linked to sustainable energy technologies.

Third, based on the 'complementary services of sustainable energy technologies' discussed in the energy-supply sector, the chapter identifies corresponding services categories in terms of the United Nations Central Product Classification (CPC, version 2).[5] The most important sustainable energy services – in particular, those related to solar and wind power projects – were compiled by the International Centre for Trade and Sustainable Development (ICTSD) based on consultations and surveys with solar and wind energy companies and industry associations.

---

[4] OECD (2005).
[5] Although neither the GATS sectoral classification list (W/120) nor the CPC are a compulsory instruments in the WTO, due to data limitations this chapter uses trade data based on both systems. There is no common understanding as to whether and to what extent CPC, version 2 might offer guidance in addressing the inadequacies of W/120, although CPC, version 2 presents an updated and more detailed classification of environmental services.

Trade in Sustainable Energy Services 89

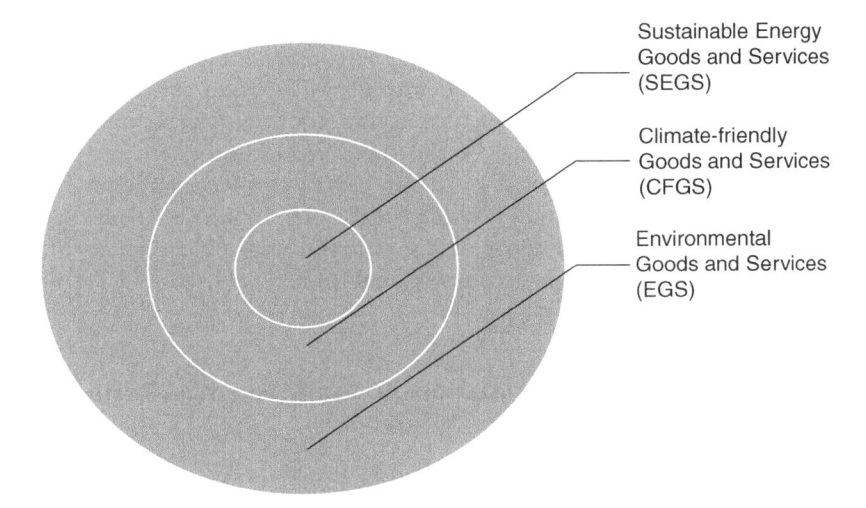

Figure 3.1: The Relation Between EGS, CFGS and SEGS

Sustainable Energy Goods and Services (SEGS)

Climate-friendly Goods and Services (CFGS)

Environmental Goods and Services (EGS)

Fourth, in analysing the major trading countries' specific commitments under the GATS on the services concerned, this chapter further narrows the scope of services by focusing on the services groups that most frequently appear across multiple economic sectors. Given the data limitations on trade in these services, the major trading countries of these services are identified at the CPC group level. Their specific commitments are analysed at sectoral or sub-sectoral levels within each CPC group.

Finally, this chapter looks at the sustainable development aspects of trade in sustainable energy services, including how the lessons learned could be used for crafting a SETA. Building on earlier work by ICTSD on environmental goods and services (EGS) and on climate-friendly goods and services (CFGS)[6] the present study has a more specific focus on a limited number of categories[7] of goods and services, as shown in Figure 3.1.

The different terms EGS, CFGS and SEGS will be used throughout the chapter (see Figure 3.1 for a depiction of how these acronyms relate to one another). Whilst the terms EGS and CFGS are used, the main focus is specifically on SEGS.

---

[6] See Kim (2011).
[7] While EGS would typically include sewage, sanitation and noise technologies, and climate-friendly goods related to the agriculture, forestry and waste sectors, the category of SEGS does not include them.

One limitation of this chapter is that it is based mostly on the GATS Schedules of Services Commitments. These Schedules date from two decades ago (they were finalised before the WTO even existed and have not been touched since except for telecom and financial services). This limits the representativeness of the data. In addition, environmental services were not a high priority in the 1990s, and were largely ignored during the Uruguay Round of trade negotiations. In sum, there are severe limitations to analysing the existing commitments. To obviate these, commitments in the more recent regional trade agreements (RTAs) will be examined.

## 3  International Negotiations on Trade and the Classification of Sustainable Energy Services

As trade in goods and services related to sustainable energy remains under discussion in the Doha Round – for example, under EGS[8] – it is important to understand the dynamics of these negotiations. The negotiations on EGS are taking place simultaneously in two different WTO fora: negotiations on EGs at the Special Session of the Committee on Trade and Environment (CTE SS) and negotiations on environmental services at the Special Session of the Committee on Trade in Services (CTS SS).

According to the chair of the CTE SS,[9] one option to reconcile the two sets of negotiations would be to draft textual elements cross-referencing the work in the CTS SS related to enhanced commitments on environmental services. Another possibility would be to associate enhanced commitments on environmental services with the EGs or an agreed set of EGs.[10]

In any case, progress in both fora has been slow. The WTO adopted the request-and-offer approach for the negotiation of specific market access commitments in services. By 2008, seventy initial offers and thirty revised offers across all services sectors had been submitted to the WTO, but since then few offers have been received, and they are considered to be out of date by now.

According to Mirodout and van der Marel, there are several reasons for the lack of domestic export interest and appetite for multilateral services liberalisation in general:

---

[8] Doha Ministerial Declaration, para 31 (iii), calls for the 'reduction or, as appropriate, elimination of tariffs and non-tariff barriers on environmental goods and services'.
[9] TN/TE/20, 21 April 2011.
[10] JOB(07)193/Rev.1 (European Communities, United States, 6 December 2007), para 3.

(1) Unilateral services reform before and after the Uruguay Round in most countries had reduced the incentive to lobby for further commitments. Service exporters perceive the current climate as relatively open. Barriers are still higher, though, for a number of countries across sectors (transport services) and modes (temporary movement of labour), both for developing and developed economies.
(2) It is difficult to reform services on a discriminatory basis. The nature of a services barrier makes it harder to distinguish between partner countries.
(3) Increased mutual interdependence over the last several decades has made business interests think that a reversal of the current openness is unlikely.
(4) Services liberalisation will be dealt with in the Doha Round only once the agriculture and non-agricultural modalities are resolved. This may motivate the business community to wait and see what happens before starting to lobby actively.
(5) Developing countries' commitments are not as extensive, since most of these countries share small markets. They are, therefore, not of great interest to high-income countries, which lowers the incentive for developing countries to negotiate greater market access in GATS.

This underlines the need for innovative approaches to free up the flow in sustainable energy services. Such novel approaches are needed all the more as the negotiating session on EGs has been struggling to identify a list of goods that are of interest to the majority of WTO Members. Meanwhile, the negotiations on environmental services are facing the challenge of updating the current GATS classification, as it does not reflect the rapidly evolving structure of the environmental services industry.

The biggest challenge of the GATS itself is that it provides for only positive list commitments based on the out-of-date nomenclature of W/120.[11] The current classification (W/120) of environmental services, for example, focuses largely on infrastructure services despite the fact that 'non-infrastructure' services – such as air pollution control[12] or environmental consulting – have emerged as important activities in recent years, primarily owing to increasingly demanding environmental regulations (Cossy 2011; Cottier and Baracol-Pinhão 2009; Nartova 2009).[13] As noted previously, during the Uruguay Round, there was no organised concept of either environmental services or energy services. In the last ten

---

[11] See n. 3 above.  [12] Air pollution control is covered in W/120 under 'other services'.
[13] The W/120 classification list used for the negotiations dates from 1991. No WTO Member has yet scheduled commitments under W/120/Rev 1.

or fifteen years, the wider energy services business community has lobbied WTO Members to obtain commitments on energy services as part of WTO accession packages, but those cover only a few markets and are focused on fossil-fuel-based energy. The fact that these commitments exist demonstrates that it is possible to get WTO-compatible commitments that are not tied to W/120.

Several proposals on an updated classification are under examination by WTO Members. Some have based their proposals on the classification developed by the OECD/the Statistical Office of the European Community (EuroStat), which includes three categories of environmental services: pollution management, cleaner technologies and resource management.[14] The EU has proposed seven sub-sectors based on the environmental media (air, water, soil, waste, noise, etc.) to supplement the basic classification scheme in order to preserve the mutually exclusive character of the W/120 list.

Despite several proposals on the development of a more comprehensive classification system for environmental services, a 'dual-use' problem – the overlap between certain environmental services and services classified within other sectors – presents a serious challenge. The fact that services related to sustainable energy are spread across multiple sectors classified in W/120 only heightens the issue of dual use. Several proposals have been put forward to address this issue. The EU, for instance, has proposed a 'cluster' approach, in which services used for environmental as well as other purposes (dual-use services) would be classified separately and be subject to a 'checklist' during other sectoral negotiations. Canada supports the EU's approach, encouraging liberalisation in all modes of delivery. However, Canada differentiates between the present list of environmental services (core services) and other related services (non-core or dual-use services) and stresses the importance of liberalising both services at the sub-sectoral level. Proposals by the United States and Switzerland are largely in line with the classification of 'core' versus 'non-core' services (Nartova 2009).

Opinions are divided as to whether an appropriate classification is a pre-requisite for scheduling meaningful commitments in supporting the development of sustainable energy use and production. For instance, Cossy (2011) argues that the absence of an appropriate classification does not prevent Members from negotiating on climate change-related

---

[14] 'Pollution management' comprises activities that produce technology or services to treat or remove adverse environmental effects; 'cleaner technologies' comprise any activity that continuously improves, reduces, or eliminates the environmental impact of technologies, processes or products; and 'resource management' includes activities that prevent environmental damage to air, water and/or soil.

services. More important, she stresses, is to ensure that each schedule is internally coherent by avoiding overlap among sectors and defining the scope of the commitments clearly and precisely.[15]

In a note[16] to WTO Members, the WTO Secretariat suggested several ways in which clean energy services could be classified. The Secretariat started by confirming that neither W/120 nor the UN's CPC contains an explicit reference to services related to RE or energy efficiency. Under both systems, the classification of energy-related services is neutral with respect to the energy source (sustainable energy services cannot be distinguished from services related to fossil fuels). The only explicit reference made to renewable energy is found in 'engineering services for power projects' (CPC2 83324).[17]

According to the note, Members may want to give further consideration to the classification of services associated with emerging technologies. CCS, for example, involves various services, such as the identification of a suitable geological formation or carbon dioxide ($CO_2$) capture at the point of emission, transport to the reservoir and storage on a long-term basis.[18] On the one hand, it could be considered that CCS involves several services classified in different sectors and sub-sectors of W/120, particularly in business and transport services. On the other hand, some CCS-related services – or CCS altogether – might constitute 'new' services, in which case they would at a minimum fall under a residual 'other' category, because CPC and W/120 are deemed to be exhaustive (i.e. they include all services).

'Smart grids' is another emerging technology that may deserve further consideration from a classification point of view. The International Energy Agency (IEA) defines a smart grid as 'an electricity network that uses digital and other advanced technologies to monitor and manage the transport of electricity from all generation sources to meet the varying

---

[15] One issue that is important in relation to the classification of environmental services is how to classify 'new' activities, particularly in a sector undergoing significant technological development. The field of carbon capture and storage (CCS) may be a case in point (Cossy 2011).
[16] JOB/SERV/94.
[17] The explanatory note describes this sector as covering 'the application of physical laws and principles of engineering in the design, development, and utilization of machines, materials, instruments, structures, processes, and systems for electricity generation, transmission, and distribution projects. This sub-class includes: (i) engineering services related to facilities that generate electrical power from coal and other fossil-fuel energy, such as oil and gas; nuclear energy; the energy in falling water; other energy, such as solar power, wind power, geothermal power including cogeneration facilities; (ii) engineering services related to overhead or underground electrical power transmission and distribution lines.'
[18] *Environmental Services – Overview of Classification Issues.* Informal Note by the WTO Secretariat, JOB/SERV/84, 31 August 2011, Section F.

electricity demands of end-users'.[19] Smart grids are expected to make an important contribution to the promotion of energy efficiency and RE sources by allowing consumers to make informed choices and directly control and manage their individual electricity consumption. From a classification point of view, smart grids services are likely to cut across several W/120 sectors, including telecom and computer services, and perhaps also services involving energy distribution.

The GATS offers significant flexibility for specifying the scope of commitments in Members' schedules.[20] All WTO Members are subject to general obligations under the agreement, including most-favoured nation (MFN) treatment and transparency, which apply to all service sectors. However, the main GATS obligations, namely market access and national treatment (NT), apply only in sectors where members undertake the 'specific commitments' listed in their national schedules. Members can select the sectors and modes of supply for which they are ready to undertake specific commitments, with various types of limitations for the purpose of meeting national policy objectives.

In sum, Members are free to specify their commitments on the related services across different sectors in their schedules within the current structure of classification once they agree on the scope of services that support sustainable energy. For instance, they could in their schedules under 'engineering services' specify 'engineering services for power projects or industrial projects aimed at mitigating climate change through energy efficiency improvement'; 'building projects that aim at improving energy performance'; or 'transportation projects that are based on modal shifts from road transport to public transport only'.

## 4  Services Relevant to the Diffusion of Sustainable Energy Supply Technologies

Sustainable energy cuts across almost all economic sectors, ranging from energy and transport to buildings and industry. As seen in the previous section, a variety of services across multiple sectors classified in W/120 appear to be related to sustainable energy activities. For instance, telecom services are relevant to saving energy and improving energy efficiency in sectors such

---

[19] IEA (2010).
[20] The GATS is also flexible enough to accommodate sector-specific intentions. In the case of telecom services, for instance, Members specified 'additional commitments' in their schedules, to reflect elements of the reference paper in the telecom sector with regard to transparency requirements, competition disciplines and institutional obligations concerning on issues such as the creation of an independent regulator in the sector (Adlung 2009).

as utilities, transport and buildings. Smart information and communication technology (ICT) applications are emerging as useful cornerstones for 'smart' buildings, grids, transportation and industrial processes. Research and development (R&D) services are related to a variety of sustainable energy technologies across almost all sectors since technological innovation is an integral part of accelerating low-carbon development.

In an effort to narrow the scope of services in this study, this section discusses key sustainable energy technologies, along with the associated services that could be complementary to the diffusion of such technologies. For a further discussion of how a SETA could contribute to the diffusion of sustainable energy technologies, see Brewer (2012), who emphasises the link between technology diffusion and trade in services.

Making the energy supply sector more sustainable depends on the application of a wide range of available low- and zero-carbon technologies, including the widespread use of hydropower, bioenergy and other renewables. Several different services can be linked to these sustainable energy options. For instance, pre-construction power plant services include 'technical testing and analysis services' for a feasibility studies, as well as services related to site selection.

Improving the efficiency of power plants through technologies, such as combined heat and power (CHP) would require both 'construction services for facilities' and 'engineering services for power projects' that optimise the environmental performance of energy facilities.[21] Engineering services for power projects would also be needed not only to build facilities that generate electrical power from various energy sources (e.g. nuclear energy, solar power, wind power, and geothermal power), but also to build so-called capture-ready new power plants (Gibbins et al. 2006).[22]

Most renewable energy power plants are likely to require monitoring services once they are constructed, which will eventually reduce operation and maintenance costs. The General Electric Company (GE), for instance, provides remote wind-turbine monitoring services to increase the reliability and capacity of wind farms (Steenblik and Geloso Grosso 2011).

In order to promote the use of renewables, governments could require producers or distribution companies and retail suppliers to buy 'renewable energy credits or certificates' (RECs), which prove that a minimum share of the electricity generated or supplied to the retail consumer has come from RE sources (Delimatsis and Mavromati 2009). RECs are considered intangible financial assets, which could be traded in order to

---

[21] The EU and the United States proposed this as 'energy-related services'.
[22] According to IPCC's 4th Assessment Report (2007), detailed reports on CCS-ready plant-design studies are not yet in the public domain.

> **Box 3.1. Services Involved in Solar and Wind Power Projects**
>
> - Assessment of solar and wind resources (i.e. potential for producing electricity)
> - Site analysis
> - Project development
> - Real estate services
> - Project financing
> - Due diligences (technical, regulatory, financial, legal)
> - Project licensing and legal services
> - Project engineering and design
> - Environmental impact analysis
> - Construction of solar and wind power facilities
> - Solar field quality and performance testing
> - Retail sale of solar panels, mirrors and wind turbines
> - Installation of equipment
> - Maintenance of equipment
> - Operation of solar and wind power facilities
> - Transmission, distribution and sale of electricity generated by solar and wind power.

comply with the minimum obligatory quota related to renewables.[23] Trading in RECs, therefore, involves various intermediary financial services, such as brokerage, banking and insurance services.

Smart grids could improve energy efficiency by integrating both electricity and thermal storage technologies and reducing transmission and distribution losses (IEA 2010). Successful application of smart grids, however, requires modifications to the design, operation and deployment of electricity networks – a process that involves engineering services as well as services related to energy distribution.

Services involved in the technical testing and analysis of air quality are useful both for assessing the carbon-offset resulting from improved energy efficiency of power plants and for transporting $CO_2$ for storage. In the latter case, testing ensures that a possible rupture or leaking of pipelines will not lead to the accumulation of a dangerous level of $CO_2$ in the air.

The importance and diversity of service transactions in the solar and wind power industry are suggested by the list of services in Box 3.1, which

---

[23] There is an ongoing debate over whether RECs should be classified as 'goods' or 'services'. Many argue that they should be considered as 'financial services', as international trade applies to certificates, not the energy (Kim 2011; Cottier and Baracol-Pinhão 2011; and Delimatsis and Mavromati 2009).

are associated with solar and wind power projects. This list was developed by ICTSD, based on consultations and surveys with solar and wind energy companies and industry associations. Negotiators may want to examine where these different services should be classified in W/120. Table 3.6 at the end of the chapter shows the trade commitments key countries made under the GATS on some of these specific services related to wind energy.

Providers of other services include large energy and engineering companies that supply a wide range of vertically integrated products and services, including solar and wind farm developers, and a large number of small firms specialising in the provision of niche solar and wind energy services such as small-scale solar photovoltaic (PV) installations. Turbine manufacturers frequently participate in the wind power market through providing services related to the post-sale operation and maintenance of their turbines, or through developing wind power facilities at which their turbines are installed. The German firm Siemens, for example, provides training, repair and monitoring services in conjunction with the sale of its turbines, while the Japanese firm Mitsubishi supplies design, construction and installation services to its customers.

The UN Committee for Progress and Coordination (CPC) system (see Table 3.1) lists five broad services categories related to energy supply: construction services; financial and related services; other professional and business services; telecom, broadcasting and information supply; sewage and waste collection, treatment and disposal and other environmental services. The services categories associated with the energy supply sector that appear the most frequently are other professional technical and business services and construction services.

## 5 Construction and Professional Services Are Key to the Diffusion of Sustainable Energy Technologies

This section focuses on two services categories (at the CPC group level) that most frequently cut across multiple mitigation sectors: construction services and other professional services (see also Table 3.1). The commitments of major trading countries in these categories will be illustrated.

Due to data limitations on trade in services at the sub-sectoral level, this chapter uses data from major importers and exporters of services at the CPC group level and reviews their specific commitments on such services.[24] Some trade barriers at the group level are also discussed at the end of the chapter.

---

[24] Overall, data on trade in services is very limited. Readily available data are largely at the aggregate level (the sectoral level based on the W/120 classification or the CPC group level).

Table 3.1 *Key Sectoral Mitigation Technologies, and Corresponding Services Categories*[25]

| Sector | Key Mitigation Technologies and Practices Currently Commercially Available* | Corresponding Division in the CPC (version 2) | Related Services at UN CPC (version 2) Class and Sub-Class Levels |
|---|---|---|---|
| Energy supply | Improved supply and distribution efficiency; fuel switching from coal to gas; nuclear power; renewable heat and power; early applications of $CO_2$ capture and storage (CCS) (e.g. storage of removed $CO_2$ from natural gas) | Construction services [54] | – General construction services of power plants [54262]<br>– Site preparation services [543]<br>– Installation services [546] |
| | | Financial and related services [71] | – Financial services, expert investment banking, insurance services and pension services [711]<br>– Services auxiliary to financial services other then to insurance and pensions [715]<br>– Services auxiliary to insurance and pensions [716]<br>– Services of holding financial assets [717] |
| | | Other professional, technical and business services [83] | – Management consulting and management services; IT services [831]<br>– Engineering services for power projects [83324]<br>– Surface surveying services [83421]<br>– Composition and purity testing and analysis services [83441] |

[25] The classification of these services is based on the UN CPC version 2.

| | |
|---|---|
| Telecom, broadcasting and information supply services [84] | – Other technical testing and analysis services; radiological inspection of welds [83449]<br>– Other professional, technical and business services n.e.c. [839]<br>– Private network services [8414]<br>– Data transmission services [8415]<br>– Internet communication services [842]<br>– On-line content [843] |
| Sewage and waste collection, treatment and disposal and other environmental protection services [94] | – Hazardous waste treatment and disposal services [9432] |

### Other Professional, Technical and Business Services

Engineering services are key among the category of 'other professional, technical and business services' in effective electricity generation, transmission and distribution. Engineering services, which predominantly entail advisory, design, consulting and project management functions, complement construction services. Many firms provide integrated packages of engineering and construction services. As new channels of electronic supply continue to create business opportunities and international sourcing of engineering services becomes increasingly common, developing country exports of engineering services are on the rise (Cattaneo et al. 2010).

While developed countries have historically dominated the market in many sustainable energy services, emerging markets are expected to provide 14–20 per cent of the industry's estimated USD 1 trillion turnover in 2020 (Booz Allen Hamilton 2006). Although trade data on sustainable energy services at the national level is hard to come by, some existing data reveal that countries such as Brazil, the Republic of Korea, the Russian Federation and Singapore are already large exporters of 'other professional, technical and business services'. As an importer, Kazakhstan appears to be a big player in this area (see Table 3.2).

Most of the services discussed in this section are provided in all four modes of supply. The predominant modes are commercial presence (Mode 3), movement of natural persons (Mode 4) and cross-border trade (Mode 1) (see Box 3.2).

Table 3.2 *Major Exporters and Importers of Architectural, Engineering and Other Technical Services (USD million)*

| Exporters | Value | Importers | Value |
| --- | --- | --- | --- |
| EU (27) | 39.212 | EU (27) | 25.169 |
| Extra-EU (27) | 22.657 | Extra-EU (27) | 10.331 |
| India | 7.360 | India | 2.746 |
| United States | 5.020 | Canada | 2.560 |
| Canada | 4.066 | Brazil | 1.708 |
| Brazil | 3.033 | Russian Federation | 1.616 |
| Norway | 2.144 | Kazakhstan | 1.289 |
| Russian Federation | 1.571 | Singapore | 977 |
| Singapore | 1.398 | Norway | 579 |
| Australia | 955 | Korea, Rep. | 531 |
| Korea, Rep. | 253 | Australia | 370 |

*Source:* Cattaneo et al. (2010) derived from WTO (2007).

> **Box 3.2. The Four Modes of Services Supply**
>
> According to the GATS, service suppliers are either natural or legal persons. The modes of supply differ depending on the location of the service provider and the location of the service consumer. The GATS defines the four modes of supply in trade in services as follows:
>
> **Cross-border supply (Mode 1)**: Non-resident suppliers deliver services cross-border into a client's territory. For example, a Norwegian engineer sends a design sketch of a 'capture-ready' power plant to a client in Brazil via the Internet.
>
> **Consumption abroad (Mode 2)**: A service is supplied under Mode 2 when consumers from one country make use of a service in another country. For example, nationals of country A travel to country B as tourists, students or patients to consume the service.
>
> **Commercial presence (Mode 3)**: Foreign suppliers of services establish, operate, or expand their commercial presence in a client's territory, such as a branch, agency or wholly-owned subsidiary. For instance, a French architectural consulting firm opens an office in China to provide advisory services on building an energy efficient exhibition centre.
>
> **Movement of natural persons (Mode 4)**: This involves the entry and temporary stay in a client's territory of foreign individuals to supply a service.
>
> Source: Derived from Cattaneo et al. (2010).

A review of the sectoral commitments made by nine key exporters and importers in this area during the Uruguay Round (see Table 3.3) during the Uruguay Round reveals that although all nine countries scheduled commitments in this sector,[26] only Australia made commitments across all sub-sectors. Seven of the nine countries excluded 'services incidental to energy distribution' from their schedules.[27] Brazil, the EU, India and Singapore also excluded 'related scientific and technical consulting

---

[26] The EU's national schedule submitted during the Uruguay Round covered only the twelve original Member States. Fifteen new Member States (Austria, Cyprus, Czech Republic, Estonia, Finland, Hungary, Latvia, Lithuania, Malta, Poland, Slovak Republic, Slovenia and Sweden) submitted their individual schedules separately. In 2006, the EU certified a new schedule of commitments that covered all twenty-seven Member States. Among these, Cyprus and Malta have not made any commitments on this group of services.
[27] In its revised offer, the EU made a new limited commitment in this sub-sector.

Table 3.3 Sectoral Commitments on Other Professional, Technical and Business Services

| Major Exporters/ Importers | Architectural Services | Engineering Services | Integrated engineering Services | Management Consulting Services | Technical Testing and Analysis services | Services Incidental to Energy Distribution | Related Scientific and Technical Consulting Services |
|---|---|---|---|---|---|---|---|
| Australia (E/I)* | √ | √ | √ | √ | √ | √ | √ |
| Brazil (E/I) | o | o | x | √ | x | x | x |
| Canada (E/I) | o | o | o | o | √ | x | o |
| EU **(E/I) | o | o | o | o | o | x [o] | x |
| India (E/I) | x | o | x | x | o | x | x |
| Korea, Rep. (E/I) | o | √ | √ | √ | o | x | √ |
| Norway (E/I) | √ | √ | √ | √ | √ | x | √ |
| Singapore (E/I) | √ | o | x | √ | x | x | x |
| United States (E/I) | o | o | o | √ | x | √ | √ |

*Note:* √ = Unrestricted commitment, x = No commitment, o = Limited commitment.
[] = A new commitment included in the EU's 'revised offer' during the Doha Round.
*E/I = Major exporters as well as major importers.
**Among the then EC Member States, Cyprus and Malta have not made any commitment on the 'other professional, technical and business services' group.

*Source:* Derived from the WTO Services Database on Members' Commitments Schedule and Initial Offers as well as Revised Offers (TN/S/O and TN/S/Orev.1).

services', while several other countries made no commitments on integrated engineering services or technical testing and analysis services.[28]

Table 3.4 shows GATS commitments more specifically linked to the wind energy sector. It should again be noted that most of these schedules are out of date and should be placed in the context of 1993. Commitments conditionally made in revised offers under the Doha Round of services negotiations were not bound.

Most of the major exporting and importing countries have scheduled commitments in all four modes, except for Brazil and India, both of which have left Modes 1 and 2 largely unbound. Several EU Member States have also left Mode 1 unbound across all sub-sectors. That number has increased, according to the revised offer the EU submitted to the WTO (Table 3.5). However, the importance of cross-border supply in this area is growing thanks to vastly improved ICT systems. For instance, the Internet (or other means of telecommunication) are increasingly used for the transmission of architectural and engineering specifications, design plans for environmental projects, reports of specialist environmental consultants, environmental quality testing and analysis results and computer modelling simulations. Among the key developed countries in this sector, Canada has made limited Mode 1 commitments in almost all sub-sectors by requiring a commercial presence and residency for accreditation from certain service providers.[29]

While most of the countries surveyed made commitments in Mode 3, these were often subject to conditions on market access. For instance, six of the nine countries scheduled their commitments on engineering services with market access limitations in Mode 3. The majority of market access limitations in architectural services were also in Mode 3. Brazil and Canada restrict foreign architectural services suppliers from forming legal entities. Brazil stipulates that foreign firms or individuals wishing to join domestic suppliers must enter into in a 'specific type of legal entity', which must 'take the form of a sole proprietorship or partnership'. India allows market access for foreign engineering services suppliers only through 'incorporation with a foreign equity ceiling of 51 per cent'. Korea requires an 'economic needs test' for the establishment of a commercial presence. Specific limitations on Mode 3 that restrict market access are summarised in Table 3.6.

With regard to national treatment restrictions, Canada has made a specific limitation requiring non-resident firms to have a 'higher percentage

---

[28] The classification of sub-sectors in all the tables is based on W/120.
[29] Canada's revised offer removed the limitations on Mode 1 in architectural services and on Modes 1 and 2 in engineering and integrated engineering.

Table 3.4 *Snapshot of GATS Commitments in Services Related to Wind Energy*

| | United States | EU | China | India | Canada | Japan | Australia | Turkey | Brazil | Egypt |
|---|---|---|---|---|---|---|---|---|---|---|
| Certain related scientific and technical consulting services | ◉ | ○ | ◉ | ○ | ◉ | ◉ | ● | ○ | ○ | ○ |
| Services incidental to energy distribution | ● | ○ | ○ | ○ | ○ | ○ | ● | ○ | ○ | ○ |
| Certain professional services, including engineering and integrated engineering services | ● | ◉ | ◉ | ◉ | ◉ | ● | ● | ● | ◉ | ◉ |
| Distribution services, including commission agents, wholesale trade, and retail trade services that apply to fuels, related products, and brokerage of electricity | ◉ | ◉ | ◉ | ○ | ● | ● | ◉ | ○ | ◉ | ○ |
| Maintenance and repair of equipment, except transport-related equipment | ● | ● | ○ | ○ | ● | ◉ | ○ | ○ | ○ | ○ |
| Management consulting and related services | ● | ● | ◉ | ○ | ◉ | ● | ◉ | ● | ◉ | ○ |
| Construction and related engineering services | ◉ | ◉ | ◉ | ◉ | ◉ | ◉ | ◉ | ◉ | ◉ | ◉ |
| Technical testing and analysis services | ○ | ◉ | ◉ | ◉ | ● | ○ | ○ | ○ | ○ | ○ |

*Notes:*
- ● – Full commitment.
- ◉ – Partial commitment.
- ○ – No commitment.

Most measures regarding the supply of services through the presence of natural persons (Mode 4) are addressed in a Member Country's horizontal commitments. For the purposes of this table, a 'full commitment' means any commitment that grants full market access or NT to foreign individuals or firms that provide RE services through cross-border supply (Mode 1), consumption abroad (Mode 2) and commercial presence (Mode 3).

This table is not a comprehensive assessment of GATS commitments. In many cases, commitments apply to only part of a sector, and specific limitations may be in place. For full details regarding commitments, see the GATS schedules of individual countries.

*Source:* Compiled by the US International Trade Commission from individual countries' GATS Schedules of Specific Commitments.

Table 3.5 *Market Access and National Treatment Limitations on Mode 1: Other Professional, Technical and Business Services*

| Major Exporters/ Importers | Sub-Sectors | Limitation |
|---|---|---|
| **Market access** | | |
| Canada | Architectural services | [Citizenship requirement for accreditation (architects)]. |
| | Engineering/Integrated engineering services | Requirement of a commercial presence for accreditation (engineers): [Requirement of a commercial presence for accreditation (consulting engineers)]. Requirement of permanent residency for accreditation (engineers). ([Citizenship requirement for accreditation (engineers)]). |
| | Other business services: Management consulting services | Permanent residency requirement for accreditation (agrologists): [Citizenship requirement for accreditation (Professional administrators and certified management consultants or professional incorporation of administrators). Citizenship requirement for use of title (industrial relations counselors)]. |
| | Other business services: Related scientific and technical consulting services | Requirement of permanent residency and citizenship (free miner). Requirement for a commercial presence, permanent residency and citizenship for accreditation (Canadian corporation or a partnership of the foregoing land surveyors). Citizenship requirement for accreditation (subsurface surveying services, professional technologist, chemists). |
| EU | Architectural services | BE, [GR]CY, EL, IT, MT, PT, PL, SI: Unbound. |
| | Engineering services | [GR]CY, EL, IT, MT, PT: Unbound. |
| | Integrated engineering services | CY, EL, IT, MT, PT, PL[GR]: Unbound. |
| | Other business services: Technical testing and analysis services | IT: Unbound for the profession of biologist and chemical analyst. CY, CZ, MT, PL, SK, SE: Unbound. |
| | Other business services: Services incidental to energy distribution | All Member States except HU, LV, LT, SI: Unbound (HU, LV, LT, SI; Unbound). |
| Korea | Architectural services | Requirement of a commercial presence. Acquisition of Korean architectural license by passing an examination. |

Table 3.5 (*cont.*)

| Major Exporters/ Importers | Sub-Sectors | Limitation |
|---|---|---|
| | | Supply of services by foreign architects through joint contracts with architects licensed in Korea. |
| **National treatment** | | |
| Canada | Architectural services | Residency requirement for accreditation (architects; landscape architects). |
| | Engineering services | Engineers: residency requirement for accreditation (engineers). |
| | Other business services: Related scientific and technical consulting services | Differential tax measures (Federal and sub-national treatment for expenditures of services performed in Canada related to the exploration and development of a mineral resource, petroleum or natural gas (mineral and petroleum exploration and development). Residency requirement for accreditation (applied science technologist/technical). Residency requirement for accreditation (Cadastral surveying). Residency requirement for accreditation (geoscientists, land surveyors. Requirement of training for accreditation (land surveyors)). |
| EU | Architectural services | DE (application of the national rules on fees and emoluments for all services which are performed from abroad). BE, [GR], CY, EL, IT, MT, PT, PL: Unbound. |
| | Engineering services | AT, SI for planning services. [GR], CY, EL, IT, MT, PT: Unbound. |
| | Integrated engineering services | AT, SI for planning services. CY, EL, IT, MT, PT, PL [GR]: Unbound. |
| | Other business services: Technical testing and analysis services | T: Unbound for the profession of biologist and chemical analyst. CY, CZ, MT, PL, SK, SE: Unbound. |
| | Other business services: Services incidental to energy distribution | All Member States except HU, LV, LT, SI: Unbound. |

*Source:* Derived from the WTO Services Database on Members' Commitments Schedule and Initial Offers as well as Revised Offers (TN/S/O and TN/S/O rev. 1). EU Member States: AT (Austria), BE (Belgium), CY (Cyprus), CZ (Czech Republic), DE (Denmark), EE (Estonia), EL (Greece), ES (Spain), FI (Finland), FR (France), HU (Hungary), IE (Ireland), IT (Italy), LT (Latvia), LT (Lithuania), LU (Luxembourg), MT (Malta), NL (Netherlands), PL (Poland), PT (Portugal), SE (Sweden), SI (Slovenia), SK (Slovak Republic), UK (United Kingdom).

Table 3.6 *Market Access and National Treatment Limitations on Mode 3: Other Professional, Technical and Business Services*

| Major Exporters/ importers | Sub-Sectors | Limitation |
|---|---|---|
| **Market access** | | |
| Brazil (E/I) | Architectural services | Foreign service suppliers must join Brazilian service suppliers in a specific type of legal entity (*consorcio*); the Brazilian partner shall maintain the leadership. The contract establishing the *consorcio* must clearly define its objective. |
| | Engineering services | Same conditions as a architectural services. |
| Canada (E/I) | Architectural services | Commercial presence of architects must take the form of a sole proprietorship or partnership. |
| | Other business services: Related scientific and technical consulting services | Permanent residency and citizenship requirement for a commercial presence for accreditation of sub-surface surveying services, professional technologist, chemists. |
| EU (E/I) | Architectural services | Restrictions on access of certain natural persons (ES: access is restricted to natural persons). FR: provision through SEL (*anonyme, à responsabilité limitée ou en commandite par actions*) or SCP only. IT, PT: access is restricted to natural persons. Professional associations (no incorporation) among natural persons permitted. LV: practice of three years in Latvia in the field of projecting and university degree required to receive the licence enabling engagement in business activity with full range of legal responsibility and rights to sign a project. |
| | Engineering services | ES: access is restricted to natural persons. IT, PT: access is restricted to natural persons. Professional association (no incorporation) among natural persons permitted. |
| | Integrated engineering services | ES: Access is restricted to natural persons. IT, PT: access is restricted to natural persons. Professional association (no incorporation) among natural persons permitted. |

Table 3.6 (cont.)

| Major Exporters/ importers | Sub-Sectors | Limitation |
|---|---|---|
|  | Other business services: Technical testing and analysis services | ES: access for chemical analysis through natural persons only. IT: access for the profession of biologist and chemical analyst through natural persons only. Professional association (no incorporation) among natural persons is permitted. PT: access for the profession of biologist and chemical analyst through natural persons only. CY, CZ, MT, PL, SK, SE: Unbound. |
|  | Other business services: Services incidental to energy distribution | All Member States except HU, LV, LT, SI: Unbound (HU, LV, LT, SI: None) |
|  | Other business services: Related scientific and technical consulting services | All Member States except ES, FR, IT, PT: None. ES: Access to profession of surveyors and geologists through natural persons only. FR: 'Surveying': access through a SEL (*anonyme, à responsabilité limitée ou en commandite par actions*), SCP, SA and SARL only. IT: for certain exploration services activities related to mining (minerals, oil, gas, etc.), exclusive rights may exist. IT: access to profession of surveyors and geologists through natural persons only. Professional association (no incorporation) among natural persons permitted. PT: access restricted to natural persons. |
| India (E/ I) | Engineering services | Only through incorporation with a foreign equity ceiling of 51 per cent |
|  | Other business services: Technical testing and analysis services | Only through incorporation with a foreign equity ceiling of 51 per cent |
| Korea, Rep. (E/ I) | Other business services: (Composition and purity testing and analysis services) | Requirement of economic needs test for the establishment of a commercial presence Main criteria: the number of and impact of existing domestic suppliers, protection of public health, safety and environment.* |
| Singapore (E/ I) | Engineering services | Limited corporations: only registered professional engineers or allied professionals (registered architects or |

Table 3.6 (cont.)

| Major Exporters/ importers | Sub-Sectors | Limitation |
|---|---|---|
| United States (E/ I) | Architectural services | land surveyors) shall be director of the corporations. Two-thirds of the officers, partners, and/or directors of an architectural firm in Michigan must be licensed in Michigan as architects, professional engineers and/ or land surveyors. |
| **National treatment** | | |
| Canada | Architectural services | Non-resident firms are required to maintain a higher percentage of practitioners in a partnership. |
| | Other business services: Related scientific and technical consulting services | Residency and training requirement for accreditation of cadastral surveying, geoscientists, land surveyors. |
| EU | Architectural services | Non-resident firms are required to maintain a higher percentage of practitioners in a partnership. |
| | Other business services: Related scientific and technical consulting services | Residency and training requirement for accreditation of cadastral surveying, geoscientists, land surveyors. |
| | Architectural services | EE: none except that at least one responsible person (project manager or consultant) must be a resident of Estonia. |
| | Engineering services | EE: none except that at least one responsible person (project manager or consultant) must be a resident of Estonia. |
| | Integrated engineering services | EE: none except that at least one responsible person (project manager or consultant) must be a resident of Estonia. |
| | Other business services: Technical testing and analysis services | All Member States except CY, CZ, MT, PL, SK, SE: none (CY, CZ, MT, PL, SK, SE: Unbound). |

*Source:* Derived from the WTO Services Database on Members' Commitments Schedule and Initial Offers as well as Revised Offers (TN/S/O and TN/S/O rev. 1).
*Note:* E/I = Major exporter as well as importer.

of practitioners in a partnership' in architectural services and 'residency and training for accreditation of certain service providers' in related scientific and technical consulting services such as cadastral surveying, geoscientists and land surveyors. Some EU Member States' revised offers – in particular, that of Estonia – have added limitations on national treatment in most sub-sectors by requiring residency of 'at least one responsible person'.[30] Most EU Member States have left 'services incidental to energy distribution' unbound, while a few of them have done so with regard to technical testing and analysis services.

Very few countries have bound commitments in Mode 4. Many have indicated horizontally applicable limitations predominantly concerning restrictions on the entry and temporary stay of service providers, including intra-corporate transferees, contractual service suppliers, business visitors, salespersons and independent professionals. Other restrictions include limited recognition of foreign diplomas (required to practice regulated professional services); restrictions on foreign nationals' acquisition of land and real estate; restrictions of foreign service providers on public monopolies, and limited eligibility of foreign nationals for subsidies, including tax benefits.[31]

A few countries have put specific limitations in Mode 4. Canada, for instance, restricts market access by requiring 'permanent residency and citizenship for accreditation of certain types of services suppliers'.[32] Several EU Member States require that certain foreign service providers fulfil 'academic and professional qualification requirements and membership requirements of the relevant professional body in the home country'.[33]

Canada has also placed limitations on national treatment, requiring residency for accreditation of landscape architects and other service providers related to scientific and technical consulting services. A few EU Member States have limitations on national treatment as well, such as a residency requirement for certain service providers in architectural and other business services. Almost all Member States require residency for providers of engineering and integrated engineering services.

In the case of engineering services, trade barriers are not limited to just market access and national treatment (Cattaneo et al. 2010). Trade opportunities for engineering firms hinge largely on a variety of laws, regulations and administrative rules at home and abroad that can have a substantial impact on their financial options and operations. For instance,

---

[30] The EU's initial and revised offers include the commitment schedules of its twenty-seven member states.

[31] A review of the revised offers by nine major trading countries in this sector reveals that the nature of horizontal limitations on modes of supply remains largely the same, although some countries clarified the criteria applied to relevant limitations or loosened the language to a limited extent. For instance, Canada has increased the minimum value of Canadian business that could be subject to foreign acquisition from 'no less than CAD 153 million' to CAD 250 million: TN/S/O; TN/S/O rev. 1.

[32] Canada removed the residency requirement for the accreditation of landscape architects in its revised offer submitted to the WTO: TN/S/O; TN/S/O rev. 1.

[33] Overall, the number of EU Member States that put specific limitations in Mode 4 has decreased in its revised offer: TN/S/O rev. 1.

national or sub-federal rules in foreign countries may limit engineering firms' legal entity or joint venture structure, including arbitrary equity limitations.

An engineering industry with a 3–4 per cent profit margin typically needs access to competitive financing through credit extension (Tulacz 2008). In many developing countries, limited access to finance often puts engineering firms at a competitive disadvantage. Tunisia, for instance, limits credit to 5 per cent of the engineering firm's output, while 10 per cent of the output is allowed for credit in other sectors, such as tourism and manufacturing (World Bank 2007b).

Rules concerning public procurement also affect trade in engineering services. For instance, distorted administrative practices – such as a lack of publicly available information about project requirements and the bidding process – impede the integrity and transparency of procurement, thereby negatively affecting performance. The GPA prohibits the use of measures that discriminate against foreign providers and addresses various aspects of procurement procedures, including 'criteria for the qualification of suppliers and technical specifications of products and services; tendering procedures; and the provisions for transparency'.[34]

Rules governing nationality and residency requirements for service providers, as well as their qualification and recognition procedures, can also influence trade. While professional qualification requirements are fundamental drivers in the industry, arduous qualification requirements and licensing procedures can hamper the delivery of services (Cattaneo et al. 2010). Excessively restrictive visa fees or unpredictable and time-consuming work permit procedures can also create trade barriers for services.

*Construction Services*

Construction services are involved in implementing various climate change mitigation options across multiple sectors, including energy supply, transport, buildings, industry and waste. Construction is one of the major service sectors in most economies in terms of employment and value added. In 2005, global spending on construction exceeded USD 4 trillion, representing 9–10 per cent of world gross domestic product (GDP) (Tulacz 2005). The most important driver for the development of services in this sector, particularly in the developed world, is increased spending on infrastructure and non-residential development (Butkeviciene 2005).

---

[34] Under the GPA, market access has been committed only for those entities specifically listed in the GPA schedules.

Table 3.7 *Major Exporters and Importers of Construction Services (USD million)*

| Exporters | Value | Importers | Value |
| --- | --- | --- | --- |
| EU (25) | 26.142 | EU (25) | 18.743 |
| Extra-EU (25) | 14.171 | Extra-EU (25) | 7.957 |
| Japan | 7.224 | Japan | 4.765 |
| United States | 4.139 | Russian Federation | 4.034 |
| China | 2.593 | Kazakhstan | 1.941 |
| Russian Federation | 2.209 | China | 1.619 |
| Turkey | 882 | Azerbaijan | 1.499 |
| India (estimated) | 828 | Angola | 1.323 |
| Malaysia | 811 | Malaysia | 1.087 |
| Singapore | 566 | United States | 1.039 |
| Egypt, Arab Rep. | 503 | India (estimated) | 774 |

*Source:* Engman (2010) derived from WTO (2008).

Transportation (25.6 per cent), building (23.8 per cent) and petroleum extraction (25.8 per cent) made up three fourths of the global construction market as of 2008 (ENR 2008). The public sector is clearly the largest client segment for the construction sector. In many developing countries, government procurement covers almost 50 per cent of construction expenditures (Mburu 2008).[35]

While many developing countries largely remain importers of construction services, several emerging economies as well as economies in transition are quickly becoming successful exporters of these services. The EU, Japan and the United States appear to be major exporters in this sector, followed by countries such as China, India, Malaysia and Egypt (Table 3.7).

The construction sector is characterised by a limited number of large international companies and a big number of local small- and medium-sized businesses. In 2007, exports by the top 225 international contractors amounted to USD 310 billion, accounting for more than one third of world total. Fifty-one of the 225 companies were Chinese and twenty-three were Turkish. Fourteen other countries, ranging from Brazil to Serbia, had at least one and, in some cases, three companies on the top 225 list (Engman 2010).

---

[35] The market share of construction services by relevant sectors is as follows: power (5.5 per cent); industrial (4.9 per cent); water/sewer/waste (4.4 per cent); and manufacturing (2.3 per cent).

A review of the sectoral commitments made by nine major exporting and importing countries of construction services during the Uruguay Round[36] reveals that all have scheduled commitments in this sector,[37] but none has made a full commitment. Most WTO Members have also placed limitations across all sub-sectors.

Construction projects require local production, because they are highly intensive in both labour and materials. The local characteristics of the construction business imply that commercial presence (Mode 3) is the preferred mode of supply, complemented by temporary movement of natural persons (Mode 4).

In general, restrictions on commercial presence are the most common barriers to trade in the construction services sector (Table 3.8). Market access may be limited on foreign investment; the type of legal entity for a foreign company (mandatory local incorporation); the number of suppliers; and the value of transactions or assets. While Egypt, India and Malaysia restrict the formation of a legal entity and foreign capital equity, China restricts the types of construction projects that foreign-owned enterprises can carry out. The EU excludes rights for construction, maintenance and management of highways and airports in certain Member States. Restrictions on national treatment in Mode 3 include registration and authorisation requirements; performance and technology transfer requirements; licensing, standards and qualification; and nationality and residency requirements (WTO 1998). China has also made a specific limitation in national treatment, lowering the registered capital requirements for joint venture construction enterprises.

Limitations on the temporary movement of natural persons, which are often included in labour market regulations, can impede trade in construction services, given the sector's labour intensive nature. These restrictions take different forms, ranging from bans and quotas to economic needs tests and residency requirements. Such restrictions can result in increased operating costs, project delays and unpredictability of project execution. The significance of Mode 4 in the construction sector depends largely on the entry strategy used, as the entry strategy varies depending on the duration of the project (Gelosso Grosso et al. 2008). The pattern that has been evolved over the past two decades seems to be one of market establishment aimed at a more permanent presence. Empirical evidence shows, however, that contractors facing high entry restrictions in the host market tend to

---

[36] Data on the commitment schedules of Azerbaijan and Kazakhstan are unavailable since they are not WTO members.
[37] Among the new EU Member States, Cyprus, Hungary and Malta have not submitted their commitment schedules on this group of services. Finland has made a partial commitment.

Table 3.8 *Market Access and National Treatment Limitations on Mode 3: Construction Services*

| Major Exporters/ Importers | Limitation |
|---|---|
| **Market access** | |
| China | Restrictions are in place on several types of construction projects by foreign-owned enterprises: (i) projects wholly financed by foreign investment and/or grants, (ii) projects financed by loans from international financial institutions (IFIs) and awarded through an international tendering process, (iii) joint Chinese–foreign projects with at least 50 per cent foreign investment, and (iv) Chinese-invested construction projects that are difficult to implement by domestic enterprises alone can be jointly undertaken by Chinese and foreign companies with the approval of the provincial government. |
| EU | Certain Member States do not grant exclusive rights for the construction, maintenance and management of highways and airports. A nationality condition for managers or the board of directors of construction companies supplying the public sector. |
| Egypt | Restrictions on the formation of a legal entity (only through a joint venture) and foreign capital equity (a ceiling of 49 per cent of the total capital requirement for the project). |
| India | Restrictions on the formation of a legal entity (only through incorporation) and foreign equity (a ceiling of 51 per cent). |
| Malaysia | Restrictions on the formation of a legal entity (only through a representative office, regional office or locally incorporated joint venture with Malaysian individuals or Malaysian-controlled corporations or both). Foreign shareholding in the joint venture may not exceed 30 per cent. |
| Turkey | Approval required from the Under Secretariat of Foreign Trade for the establishment of an ordinary partnership under the Civil Code, excluding ordinary partnerships formed for international tenders in Turkey by non-residents. |
| **National treatment** | |
| China | Differential treatment of registered capital requirements for joint venture construction enterprises, which have the obligation to undertake foreign-invested projects. |

*Source:* Derived from the WTO Services Database on Members' Commitments Schedule and Initial Offers as well as Revised Offers (TN/S/O and TN/S/Orev. 1).

resort to short-term rather than permanent entry (Chen 2008). The major importing and exporting countries in this sector have chosen to keep Mode 4 unbound and rely on their horizontal commitments to provide access.[38]

[38] Restrictions in Mode 4 may arise from a country's overall immigration policy, or specific labour market conditions. Consequently, specific commitments under the GATS tend to

Differential treatment of subsidies, along with other incentive schemes often provided for export promotions, can also have a discriminatory effect on trade in construction services. Restrictions on the movement of capital equipment and building materials can negatively affect trade, as they may give rise to unnecessary costs for imports of construction machinery (Geloso Grosso et al. 2008).

A review of specific commitments and limitations by the major exporting and importing countries in this area reveals that, with the exception of Singapore and Turkey, all countries have left Mode 1 unbound. While Singapore has made full commitments in Mode 1, Turkey has placed extensive limitations in Mode 1 on both market access and national treatment.[39]

In terms of national treatment limitations, foreign nationals frequently have limited eligibility for subsidies, including tax benefits. They also face limited recognition of qualifications from third countries, as well as restrictions on the acquisition of land and real estate. Restrictions on land and real estate use or ownership, along with other restrictions, can have a big impact on the provision of construction services, as these restrictions prevent property developers from acquiring real estate under construction until the completion of the project (Geloso Grosso 2008).

Many types of domestic regulatory measures can affect trade in construction services if these measures are discriminatory or unnecessarily burdensome. For instance, building regulations and associated technical requirements, as well as regular inspection requirements for safety, are related to the provision of construction services. Rules on the temporary admission of construction equipment can also hinder the market entry of foreign companies (Geloso Grosso et al. 2008).

---

be made at the 'horizontal' level (i.e. applied to all service sectors), with sector-specific qualifications. A review of horizontal commitments shows that almost all these countries have put horizontal market access limitations in Mode 4 by restricting the entry or temporary stay of service providers. Common examples of specific conditions for approval of the entry of service providers include: labour market testing; residency requirements for intra-corporate transferees and a requirement that the foreign company employ specific numbers of local staff; authorisation subject to non-availability of locals; authorisation subject to performance requirements (employment creation, transfer of technology or on-going level of investment). There is a perceived need for special fast-track visa procedures regardless of whether providers are based in developed or developing countries.

[39] The market access limitations in Mode 1 concern 'approval requirements by the government of foreign specialists involved with engineering or architecture-related works and qualification requirements of foreign engineers or architects (e.g. temporary member of the related Union of Chambers)'. Turkey removed this limitation in its revised initial offer.

Regulations concerning the administration of construction permits can also affect trade in construction services. According to Engman (2010), the administrative performance of construction permits tends to vary across countries in terms of the constructions' duration, procedure and cost. Opaque, expensive and overly bureaucratic administrative processes increase transaction costs and business risks, resulting in lower investment in new infrastructure and buildings.

China and the Russian Federation appear to have relatively expensive and bureaucratic processes. Getting a construction permit in Russia takes, on average, twenty-three months. There are fifty-four procedures to go through – the worst record among all the major exporting markets. In China, constructors must wait eleven months and go through thirty-seven procedures. The Indian administration is more rapid (224 days), but more expensive.

Government procurement practices are also crucial to trade in construction services given that its largest client segment is the public sector. In Germany and the United Kingdom, for instance, government procurement accounts for about 35 per cent of the share of construction activity, while it accounts for almost 50 per cent in the United States (Kim 2011).

Preferential treatment or minimum requirements for financial support to local companies often hinder market entry for foreign providers, creating trade barriers. Excessively strict standards applied to government procurement also tend to exclude many small- and medium-sized enterprises (SMEs) from developing countries (Tulacz 2005). Some foreign companies have experienced difficulty in entering the US market, as they are required to register and be licensed in each individual state, often with strict liability implications on equipment failure (Teljeur and Stern 2002).

*Commitments on Sustainable Energy Services in Preferential Trade Agreements*

The fact that services liberalisation at the multilateral level has been slow does not mean that there has been a similar standstill in other configurations (bilateral trade agreements and RTAs). In fact, countries generally take on more services commitments in RTAs than at the multilateral level. This is commonly known as the 'commitments gap'. Based on economic modelling, Mirodout and van der Marel (2012) found that both economic and non-economic factors play a role in determining the magnitude of the commitments gap. Factors that affect the negotiations and level of GATS-plus commitments are geography, market size and the role of mid-skilled labour endowments.

Asymmetries between countries and the quality of governance are also strong political economy factors for the commitments gap. While these conclusions hold on average for all services sectors, construction and financial services (both crucial for sustainable energy) are characterised by different patterns compared to other services sectors.

The commitments gap is higher in North–North agreements, which involve countries that are engaged in intra-industry trade and have a high GDP. Miroudot and van der Marel (2012) think that plurilateral negotiations, such as those currently underway on the Trade in Services Agreement (TiSA), are more likely to succeed than the Doha Round at the WTO.[40] One way to overcome opposition to North–South commitments could be the identification of specific concerns, as well as changing the views of governments and negotiators before considering deepening multilateral services commitments.

There are some notable examples of liberalisation and cooperation on sustainable energy services in specific preferential trade agreements. A few free-trade agreements (FTAs) address environmental and climate change-related services specifically:

- the EU–Colombia–Peru FTA, Article 275, para 5 (1), recognises the effects of climate change and commits the parties to consider mitigation actions 'facilitating the removal of trade and investment barriers to access to innovation, development, and deployment of goods, services, and technologies that can contribute to mitigation or adaptation, taking into account the circumstances of developing countries'.
- The EU–Mexico FTA, Article 34, notes the need to preserve the environment. Para 2 of the Article talks about cooperation between the parties in order to achieve this objective, including the promotion of training in human resources, launching joint research projects, and education.
- The EU–South Korea FTA, Article 13.6, reconfirms that the parties recognise sustainable development in all of its dimensions. Para 2 of the Article states in part: 'The Parties shall strive to facilitate and promote trade and foreign direct investment in environmental goods and services, including environmental technologies, sustainable renewable energy, energy efficient products and services, and ecolabelled goods, including through addressing related non-tariff barriers.'
- Article 8 of the India–Japan Agreement states that '[e]ach Party shall endeavour to ... encourage trade and dissemination of environmentally sound goods and services'.

---

[40] See also Hufbauer et al. (2012) for a discussion of the International Trade in Services Agreement (TiSA).

- In Article 9 of the Japan–Switzerland Agreement, the Parties agreed to 'encourage trade and dissemination of environmental products and environment-related services in order to facilitate access to technologies and products that support environmental protection and development goals, such as improved sanitation, pollution prevention, sustainable promotion of renewable energy and climate change-related goals'.

Significant commitments have also been made in several PTAs with regard to sectors linked to sustainable energy services (e.g. construction, financial and other services). In the Australia–ASEAN–New Zealand FTA, for example, the financial services schedule shows more commitments under the 'banking and other financial services' sub-sector than those present in Australia's GATS schedule – especially with regard to market access. All the differences affect Mode 3 (commercial presence) while some also concern Mode 1 (cross-border supply). Other FTAs that have progressed on construction and financial services include EU–Caribbean Forum (CARIFORUM), EU–South Korea, India–Japan and US–South Korea. These commitments are key since access to financing is often a critical factor in facilitating investments and establishing energy projects.

## 6 Sustainable Energy Services in the Wider Context of Sustainable Development

*Introduction*

Energy is central to achieving the economic, social and environmental aims of sustainable development. Energy services play a vital role in providing efficient access to energy in support of this endeavour. This section considers how sustainable energy can lead to win–win outcomes for both socioeconomic development and the environment. It also explores the importance of sustainable energy for jobs creation.

Developing countries are faced with the twin of challenges of achieving more reliable, affordable,[41] clean and efficient access to energy, as well as obtaining a greater share of the energy 'business'. The pursuit of these goals requires access to financial knowledge, expertise, technology and managerial know-how that will allow developing countries to improve their energy delivery using renewable energy resources.

Developing country energy producers are major importers of traditional energy services, such as those related to oil and gas exploration, wells and

---

[41] According to the International Labour Organization (ILO), in much of Asia, Africa and Latin America, the proportion of expenditure on energy by poor households is three times – and can be as much as twenty times – that of richer households.

pipelines, and drilling services. However, they have made few commitments in the energy sector in their GATS schedules, which – despite the alleged lack of definite certainty it engenders – would nonetheless allow them the flexibility to liberalise where this is deemed most consistent with domestic energy policy objectives and to seek important reciprocal concessions.

Only a limited number of developing countries have experience in structural reform in the energy sector. As a consequence, they have not developed the services that usually flow from the development of integrated market-based energy systems and the introduction of competition. The design of effective domestic energy policies would be better promoted by gaining understanding on the experiences of those countries that have implemented reforms in their energy sectors and permitted the emergence of competitive energy markets.

*Emerging Energy Services and Emissions Trading: Opportunities for Developing Countries*

Table 3.9 outlines the clean development mechanism to reduce emissions.

Table 3.9 *The Clean Development Mechanism and Services*

The Clean Development Mechanism (CDM) developed under the Kyoto Protocol is the only vehicle for trading emissions rights with developing countries. In fact, CDM projects are increasingly hosted in least-developed countries (LDCs). In the EU Emissions Trading System (ETS), for instance, credits from CDM projects registered after 2012 can be used only if they originate from a defined list of LDCs, excluding large emerging economies such as China and India. The exclusion of such countries is intended to motivate them to reduce emissions in ways other than the generation of carbon credits, and thus step up their regulatory approaches to mitigating the effects of climate change.

Emissions trading is expected to become one of the largest commodity markets in the world. This may allow a lucrative service sector to develop, bringing important financial flows to developing countries. It is estimated that the potential size of the CDM market will be in the range of USD 5 –10 billion per annum.

Services under the CDM facility would mainly consist of project-specific activities related to the design and implementation of projects. Most of the services involved require substantial expertise, which currently falls beyond the capacity of many developing countries. There is a risk that developing countries will be passive recipients of financial flows rather than proactive architects in the design of an emissions market. This has some important implications in relation to the achievement of CDM objectives, namely providing cost-effective compliance options for developed countries and helping them achieve sustainable development. The equitable achievement of these dual objectives, however, is likely to depend heavily on how individual transactions are actually shaped. The services component becomes crucial for this purpose.

### Securing Win–Win Outcomes

WTO liberalisation in the area of environmental services in general has been widely advocated for more than a decade as a means of enhancing developing countries' access to private capital, technology and management expertise, as well as improving market access for exports of environmental services (Hoekman et al. 2002). Many have argued that, by improving access to environmental know-how and technology, liberalisation will lead to greater environmental protection, thereby providing a win–win outcome for the economy and the environment (Andrew 2000; OECD 2000). Proposals for the liberalisation of environmental services under the GATS framework have also stimulated considerable public debate (Bisset et al. 2003).

The argument that trade liberalisation in environmental services (and goods) will result in a win–win outcome is open to a number of different interpretations, and the conclusions that can be drawn from theoretical and empirical studies can vary according to how we define 'win–win'. In this chapter, it is assumed that win–win outcomes occur when trade liberalisation and/or changes in trade rules have positive economic, environmental and social impacts. A combination of classical trade and welfare theory can be used to deduce that, under ideal market conditions, trade liberalisation will lead to increased economic welfare and optimal environmental quality. However, in imperfect market conditions, win–win outcomes are not guaranteed. In real world situations, both losers and winners should be expected. Win–win outcomes may be potentially realisable, but whether this is achieved in practice may depend on the nature and extent of the flanking and other supporting measures that are taken.

Flanking measures could come in the form of international support to build domestic regulatory capacity in developing countries. There are several reasons why such support would be particularly effective for regulatory capacity related to the infrastructure sector. First, many developing countries around the world lack even the most basic infrastructure. Infrastructure is important, not just the economy or the provision of basic services. It also allows the poorest communities in the world to gain access to modern forms of energy and greater livelihood possibilities. Second, the infrastructure sector depends heavily on government actions and policies. And, third, opening up the infrastructural sustainable energy services sector in developing countries could result in significant potential benefits in terms of investment, technology and management expertise.

To realise these potential benefits requires an effective regulatory framework, which can control anti-competitive behaviour, safeguard the public interest and contribute to social objectives in terms of poverty alleviation and equity. Where these regulatory frameworks are absent or ineffective, the gains will be less, the outcome for sustainable development more uncertain and public opposition more intense. The GATS acknowledges the right of WTO Members to enact regulations, and Members have the discretion to impose limitations on national treatment and market access. This means that sustained international support to build domestic regulatory capacity in developing countries is critical. While such support is needed to make progress in reaching agreement on international rules for the liberalisation of trade in sustainable energy services in general, it is particularly important for infrastructure-related services.

### Sustainable Energy Services and Green Jobs

Employment is a key element of the social pillar of sustainable development. According to the ILO, the transition to a greener economy could generate 15–60 million additional jobs globally over the next two decades and lift tens of millions of workers out of poverty.[42] At least half of the global workforce – the equivalent of 1.5 billion people – will be affected by this transition, which will be felt throughout the economy. The energy sector is expected to play a central role in this.

Tens of millions of jobs have already been created by the transition to a greener economy. For example, the RE sector employed close to 5 million workers in 2012, more than doubling the number of jobs from 2006–2010. Conversely, only 8–10 per cent of the workforce in the industrialised nations is employed in the industries that generate 70–80 per cent of the world's $CO_2$ emissions.[43] Energy efficiency is another important source of green jobs, particularly in the construction industry, the sector hardest hit by the economic crisis. Figure 3.2 shows the expected increase in employment in the EU's wind sector alone.

Net gains in green employment between 0.5 to 2 per cent of total employment are possible. In emerging economies and developing countries, the gains are likely to be higher than in industrialised countries, as the former can in some cases 'leapfrog' to green technology in the course of replacing obsolete and resource intensive infrastructure.

Wei et al. (2010) show that all RE technologies have higher labour intensity than fossil energy technologies (jobs per MWh). They also

[42] ILO (2012a).  [43] *Ibid.*

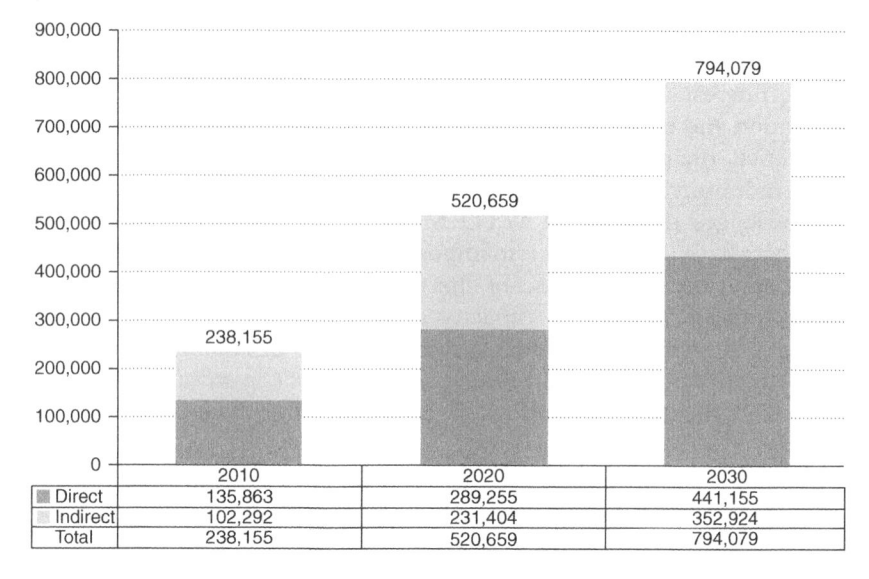

|  | 2010 | 2020 | 2030 |
|---|---|---|---|
| Direct | 135,863 | 289,255 | 441,155 |
| Indirect | 102,292 | 231,404 | 352,924 |
| Total | 238,155 | 520,659 | 794,079 |

Figure 3.2: Contribution of the Wind Energy Sector in the EU to the Direct and Indirect Employment Forecast, 2020 and 2030

generate more jobs than fossil fuels do. Projected investments of USD 630 billion by 2030 would translate into at least 20 million additional jobs in the renewable energy sector (ILO 2008)[44]

The quality of the jobs in the RE sector is another aspect of social and economic sustainability (Figure 3.3). Many essential jobs in this sector require a skilled workforce. Industry surveys in Germany have suggested that, on average, RE jobs are relatively high-skilled across both fuel-free and fuel-based technologies: 82 per cent of employees in the industry have vocational qualifications, and almost 40 per cent of these have a university degree, compared with an average for the whole industrial sector of 70 per cent and 10 per cent, respectively (IRENA 2011).[45] Evidence suggests that jobs in the RE industry are equivalent to or better quality than those in the fossil-fuel industry (ILO 2012b).[46]

Finally, specialist service providers typically have access to the latest know-how and technology for protecting the environment. This access is good not only for job creation in the communities in the vicinity of the service providers' operations; it also provides a conduit to importing developing countries for knowledge about sustainable energy. The most

[44] ILO (2008).  [45] IRENA (2011).  [46] See n. 44.

important method of knowledge transfer is often through trade and investment, and this effect is strengthened when the service provider employs local people.

*Distribution of Jobs in the PV Sector*

Table 3.10 shows the distribution of jobs in the PV sector.

Table 3.10 *Distribution of Jobs in the PV Sector*

---

While government policy in many countries is focused on job creation in the manufacturing part of the value chain for sustainable energy (e.g. on the production of solar panels and wind turbines), in reality far more jobs are created in other parts of the value chain (e.g. in the installation and maintenance of solar panels).

For every solar panel installed in Europe – even if produced in China – no less than 70 per cent of the value creation remains local. The EU solar industry provides employment to about 300,000 personnel, of whom 80 per cent or more are employed in upstream (e.g. equipment manufacturers and raw material suppliers) and downstream (e.g. importers, distributors, engineers, system integrators and installers). In fact, only 18 per cent of jobs related to solar PV derive from manufacturing, as compared to 62 per cent from installation.

This means that the majority of jobs in the solar industry are generated in the country where the solar power plant is sold, installed and serviced. As the European side is delivering a large proportion of the supply chain before and after the manufacture of cells, a high number of jobs in the EU depend on Chinese manufacturers.

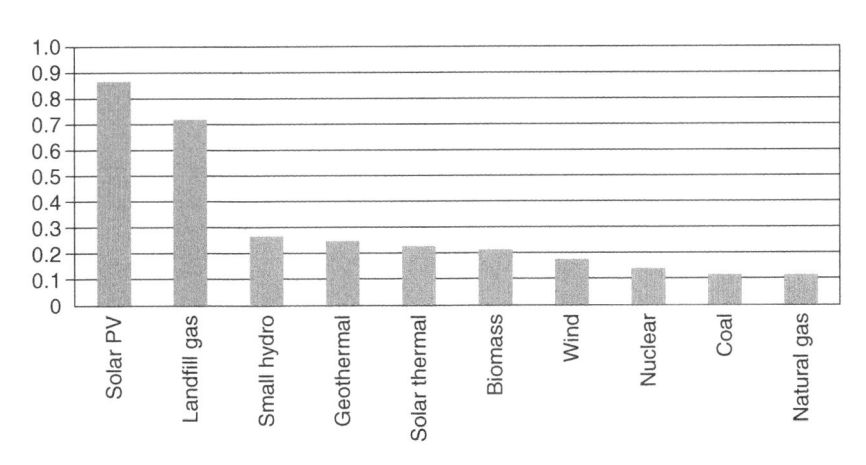

---

Figure 3.3: Comparison of Job-Years Across Technologies (Job-Years/GWh)
Source: Wei et al. (2010).

## 7 The Way Forward for Services in a SETA

*Background*

The UN Sustainable Energy for All (SE4All) initiative underlines the importance of sustainable energy services to human well-being. Liberalisation of such services can contribute to the advancement of international development goals, but to do so effectively requires policy coherence across a spectrum of trade liberalisation areas within the WTO negotiations framework and between the various international bodies with policy responsibilities in this area. Trade reform in the area of sustainable energy services must be designed in a way that is consistent with, and contributes to, the wider goals of poverty reduction and sustainable development.

The realisation of potential sustainable development benefits from liberalising sustainable energy services requires countries to give careful consideration to the potential economic, social and environmental impacts of liberalisation. This will help governments to identify the sectors and modes of supply where liberalisation is conducive to the fulfilment of national development goals. Effective mitigation measures, which may include a regulatory institutional framework that can safeguard the public interest, are an important precondition for ensuring an outcome that contributes to sustainable development. As current institutional frameworks seem unable to foster a massive scale-up of sustainable energy, it will be necessary devise innovative policy approaches.

A SETA would be one way to craft trade policies that contribute to the rapid deployment of more efficient and cleaner technologies that promote climate change mitigation, clean growth and energy security.[47] By spurring trade, a SETA could provide new incentives for innovation and investment in sustainable energy technologies.

*Specific Services to Focus on in a SETA*

Three highly concentrated sectors have a critical mass of countries that together make up 90 per cent of trade in these services. Two of these, construction and ICT, are directly related to sustainable energy services. Emerging economies such as China and India have high export competitiveness in these sectors. In addition, big emerging countries, and China in particular, are shifting manufacturing toward higher value-added products, emphasising the tertiary sector and searching for new market opportunities abroad. This means that – despite the general reluctance of emerging economies to fully open their domestic service markets – they

---

[47] See n. 2.

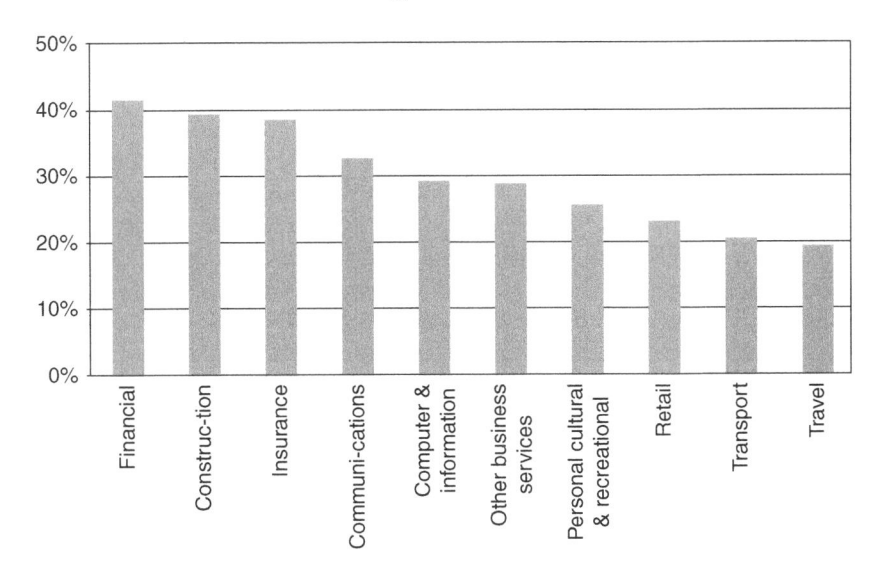

Figure 3.4: Trade Restrictiveness in Services Trade (Tariff ad valorem Equivalents) by Sector Weighted by Trade Volumes
Source: European Centre for international Political Economy (2012).

could be willing to give concessions in the construction and ICT sectors. Their inclusion in the SETA should be the highest priority.[48]

As the costs of the existing trade barriers in these sectors are significant, reform would lead to important gains. Figure 3.4 shows that construction follows the financial industry as the most protected sector.

*The Importance of Legal and Strategic Aspects of Services in a SETA*

Trade policy experts have identified several possible scenarios with regard to the creation of a SETA from a legal standpoint.[49] These include the possibility of either integrating the agreement into the WTO legal framework or leaving it outside of the WTO.

Some lessons could be learned from the on-going plurilateral negotiations on the TiSA. The WTO allows countries to negotiate stand-alone agreements, such as the TiSA, if they represent a 'critical mass'. Based on this, the twenty-seven countries representing 90 per cent of services trade could unite to withdraw all or some of the current barriers constraining the services market.[50]

[48] European Centre for International Political Economy (2012).   [49] Kennedy (2012).
[50] See n. 48.

In January 2013, WTO Members involved in the TiSA negotiations adopted a joint proposal by the EU and Australia that follows the GATS in terms of scheduling commitments, which would make it easier to attract countries that are not yet part of the plurilateral effort.

The proposal would allow countries to open only those sectors that they specifically schedule on a so-called 'positive list' to foreign competition. According to the GATS, improvements in market access for services apply on an MFN basis, subject to any relevant exemptions that participating members may have entered under the GATS, Article II. For example, certain Members maintained broad MFN exemptions based on reciprocity, or limited MFN exemptions, with regard to financial services after the conclusion of the Second GATS Protocol (1995) and, to a lesser extent, after the Fifth GATS Protocol (1997).

In contrast, the proposal would oblige countries to adopt a 'negative list' approach to national treatment obligations, which means they would have to treat foreign competitors no less favourably than domestic ones, even in sectors not scheduled for market access in the plurilateral agreement. This combination of positive and negative lists to schedule commitments (labelled a 'hybrid' approach) was adopted in late 2013 by the participants in the TiSA initiative.

One key attraction of the negative list approach is that it automatically liberalises all services unless the agreement provides otherwise. Many of the most dynamic and future-oriented services did not exist in the 1980s – such as sustainable energy services, cloud computing or e-payments. If the TiSA is moving forward on anything other than a pure positive list approach, it would arguably be easier to attract interest among the participants to agree on specific Sustainable Energy Trade Initiative (SETI) arrangements that have also adopted a negative list approach.

In terms of national treatment obligations, the EU–Australia approach would oblige countries to extend their unilateral concessions to all other TiSA signatories. It would also prohibit them from going back on the NT commitments that exist under the so-called 'standstill commitment'. Many see the EU–Australia proposal as an attempt to find a balance between getting ambitious commitments and staying within the GATS. TiSA participants ultimately backed the EU–Australia approach because they believed that it was easier to accept when applied only to NT rather than market access commitments.

Given that less than one-third of WTO Members have made services offers since 2006, negotiations on sustainable energy services could begin plurilaterally. The possibility of a plurilateral SETA that includes services might be best suited to achieving a 'critical mass', MFN-based approach. If the SETA were an optional, plurilateral MFN-based agreement within the WTO framework (an ITA-type agreement), it could be implemented through modifications to participating Members' services schedules.

Admittedly, it is quite unlikely that agreement on liberalising services trade can be reached between the countries representing 90 per cent of market share in the sector, particularly as China and India – which together account for more than 12 per cent of world trade – seem reluctant to enter into a trade-liberalising agreement on services.[51] However, the WTO legal framework sometimes allows Members to bargain outside of the formal negotiating rules. This flexibility could create an opportunity for a smaller group of countries with a large share of the services market, such as the Really Good Friends of Services (RGF), to negotiate a new WTO Agreement on services on its own under a non-MFN basis.

If the SETA were developed as a PTA, it would be excluded from MFN obligations in GATT 1994 and the GATS, i.e. the benefits it confers need not be extended to all non-participating Members. This could increase the incentive to participate. However, if the SETA were to be conceived as a GPA-type agreement, it would require the consensus of all WTO Members.

The GATS, Article V, allows parties to enter into commercial agreements outside of the WTO framework. In this context, another initiative for the liberalisation of services trade could be built following the example of the Trans-Pacific Partnership (TPP).[52] The GATS, Article V, applies to economic integration agreements. Among other things, these agreements must have 'substantial sectoral coverage', which is explained in the GATS as follows: 'This condition is understood in terms of number of sectors, volume of trade affected and modes of supply. In order to meet this condition, agreements should not provide for the a priori exclusion of any mode of supply.'[53] A SETA would not qualify if it were limited to sustainable energy services, even if those services were spread among different sectors, due to the volume of trade affected (Kennedy 2012).

However, according to Kennedy (2012), it could qualify of the GATT 1994, Article XXIV, and the GATS, Article V, if it were negotiated as part of a comprehensive FTA covering trade in goods and services. In that case, the enormous synergies between trade in goods and services related to sustainable energy would be captured. Such an agreement could be negotiated either as an amendment to an existing FTA or customs union, or as part of a new FTA or customs union. The modalities for such negotiations would necessarily require substantial coverage of trade in goods or substantial sectoral coverage in trade in services.

[51] *Ibid.*  [52] *Ibid.*  [53] GATS, Article V:1(a) and n. 1.

*Political Feasibility of Inclusion of Sustainable Energy Services in a SETA*

The EU, the world's leading services exporter and largest services economy, can play a very important role in shaping the inclusion of sustainable energy services in a SETA. The EU represents one-quarter of world trade in services and more than one-third of trade within the RGF group.[54]

Two main options seem possible for the liberalisation of services: to craft an agreement outside of the WTO or to focus on sectoral agreements, such as the SETA.

Some arguments against the value of a plurilateral SETA are based on the lack of an incentive, especially for big emerging countries, to reform and join the Agreement. This, however, is only partially correct given that most of the commitments made in trade agreements in Mode 1, Mode 2 and Mode 3 are open to all parties. Even though countries engage in bilateral agreements, national law is commonly drafted so that preferential agreements are not conferred on a specific state. Mode 4 commitments that are closely linked to domestic regulations may result in liberalising the foreign service providers' market in certain countries that meet specific conditions. Thus, incentives exist for parties outside a plurilateral SETA, depending on whether the offensive and defensive interests of emerging countries are in Mode 3 or domestic regulations and Mode 4.[55]

However, when assessing the feasibility of including sustainable energy services in the SETA, other points need to be prioritised and clarified, including the opportunity costs of setting services free from the Doha Round and the probability of success. In addition, a plurilateral agreement on services could be quite fragile, since if just one major player remains in standby mode, the prospect of an agreement under the WTO would be shattered. Finally, the trade-offs between the participants and non-participants must be addressed in such an Agreement. These negotiations need to take place before drawing any final conclusions about the feasibility of this initiative.

For regulatory issues in Mode 3, significant unilateral, bilateral and regional liberalisation already exists, so countries could simply bind the status quo. Mode 4 in general and services standards (e.g. construction codes) would be more difficult to liberalise from the beginning of a SETA.

The feasibility and size of the gains from sustainable energy services liberalisation will depend to a significant extent on domestic political institutions and reforms, which strengthen the economic environment for private investment and support market competition. Regulation is

---

[54] See n. 48.   [55] *Ibid.*

required, particularly in monopolistic markets, to ensure that potential gains from services liberalisation are maximised. Appropriate institutional and policy frameworks that take into account the potential economic, environmental and social impacts of liberalisation are necessary precursors to good policies, but capacity-building is often needed to support the establishment of such institutions.

## 8 Key Findings

*The Sustainable Services Landscape: Modes of Supply and Major Trading Countries*

This chapter demonstrates that services related to sustainable energy technologies are spread over a number of CPC groups, such as professional, technical, and business services; construction services; and other environmental protection services.

Due to the insufficiency of concrete references and commitments in WTO Members' services schedules with regard to energy services as a specific sector, this chapter is constrained to looking primary at services related to sustainable energy in a context-setting platform for analysis. The predominant modes of supply for sustainable energy services are Mode 3 and Mode 4, since providing services to construct and engineer power production projects, energy efficient buildings, or industrial plants and wastewater treatment plants requires the establishment of a commercial presence. The provision of such services also must be complemented by a range of relevant professional, technical and business services, supplied by a temporary movement of qualified service providers.

The provision of services through Mode 1 is increasing due to new channels of electronic supply, particularly in the (1) other professional, technical and business services, and (2) environmental services sectors. For this reason, ICT-related services will play an ever-larger role. Therefore, WTO Members' commitments in Mode 1 across all three CPC groups are becoming increasingly important in facilitating trade in sustainable energy services.

The EU and the United States, followed by Japan and Canada, are the biggest exporters of the three groups of services outlined in the paragraph above. A few emerging economies, as well as economies in transition, are also becoming major exporters, including India (other professional, technical and business services), China (construction services and energy goods and services), the Russian Federation (construction services) and Chinese Taipei (energy goods and services).

### Commitments on Trade in Services Related to Sustainable Energy and Remaining Barriers

An analysis of major trading countries' specific WTO commitments on these services groups reveals that only a handful of the countries have made a full commitment. Australia, for instance, has made a full commitment across the selected sub-sectors of professional, technical and business services. Canada and Chinese Taipei have done so with regard to environmental services. None of the major trading countries has made a full commitment on relevant construction services.

It appears that the principal modes of supply (Modes 3 and 4) for the complementary services of sustainable energy technologies are largely limited. Most of the trading countries concerned have put specific, as well as horizontal, limitations on both modes across the three groups of services.[56] Among others, common specific limitations in Mode 3 take the form of:
- restrictions on the formation of foreign companies' legal entity;
- requirement of an economic needs test for the establishment of a commercial presence;
- restriction on foreign investment (e.g. foreign capital equity); and
- nationality or residency requirements for accreditation of certain types of service providers (in terms of national treatment limitations).

Commonly seen forms of horizontal limitations in Mode 3 include:
- restrictions on the acquisition of land and real estate; and
- limited eligibility for subsidies, including tax benefits.

While a majority of the trading countries concerned have left Mode 4 unbound, most of them have put horizontal limitations on this category. The most frequent forms of such limitations are restrictions on entry and temporary stay of various services providers, including intra-corporate transferees, contractual service suppliers, business visitors, services salespersons and independent professionals.

In terms of national treatment in Mode 4, the following limitations appear frequently:
- limited recognition of third-country diplomas required to practice regulated professional services;
- restrictions on foreign nationals' or companies' acquisition of land and real estate;
- restrictions of foreign service providers on public monopolies; and
- limited eligibility of foreign nationals for subsidies, including tax benefits.

The degree of commitments in Mode 1 appears to vary across the three groups of services. While the majority of trading countries considered

---

[56] Many of the specific limitations in Mode 3 for environmental services were removed in the revised offers by countries such as Japan and Korea.

Mode 1 inapplicable to construction services, and hence left it unbound, a few countries (Brazil, India and some EU Member States) have left Mode 1 unbound for professional, technical and business services, as well as environmental services, with the exclusion of sanitation and similar services in the case of EU Member States.

Few countries appear to have offered new commitments across the three groups of services in their initial or revised offers during the Doha Round. The only new commitments made in these are the EU's limited commitments on services incidental to energy distribution and Australia's new commitments on other environmental services.

No discernible progress seems to have been made on horizontal limitations in the initial or revised offers. A review of these shows that the nature of horizontal limitations, and modes of supply limitations, remain largely the same. Notably, however, many countries that initially left Mode 1 unbound in the environmental services sector have since put limited commitments in their offers. Given the increasing importance of Mode 1 in providing complementary services of sustainable energy technologies, improved commitments – particularly in the category of other professional, technical and business services – could help facilitate trade in these services.

Market access and national treatment in Mode 3 (investment restrictions), qualification and licensing requirements, and services standards (e.g. construction codes) would be 'low hanging fruit' for a SETA. Investment restrictions in the form of foreign equity limits, legal form and economic needs tests are some of the elements that need to be revisited by policy makers in this sector.

Examples of impediments to the temporary movement of service providers that need to be similarly re-examined include quotas, labour market tests and limitations on the duration of stay for foreign providers. Discriminatory subsidies and taxes might also be important.

These are largely domestic regulatory issues, where more progress has been achieved at the regional level (e.g. in the EU and the North American Free Trade Agreement, NAFTA). For regulatory issues in Mode 3, significant unilateral, bilateral and regional liberalisation already exists, so countries could simply bind the status quo. Mode 4 in general and services standards (construction codes) would be more difficult to liberalise from the beginning of a SETA.

Since public procurement is an important driver for demand in a range of services related to sustainable energy, procurement regulation can have a significant impact on trade in these services. In addition to traditional government procurement, public–private partnerships (PPPs) – such as concessions and build-operate-transfer (BOT) contracts – have also emerged to facilitate private participation in infrastructure and service development. Related practices may affect trade in these services as well.

The construction sector and related architecture and engineering services are characterised by the importance of building regulations and technical requirements. In addition, contractors for projects related to sustainable energy depend on bringing in technologically sophisticated equipment to the project site from other countries. Therefore, standards affecting the mobility of goods and technologies may be important, and the harmonisation of standards may benefit trade and development in areas such as energy efficiency.[57]

While multilateral negotiations are inly inching forward in the Doha Round, there has been considerably more movement on services related to sustainable energy in FTAs and the TiSA.

*Ways Forward*

The complementarity between trade in sustainable energy technologies and trade in services cannot be emphasised enough. A wide range of products and technologies are connected with the provision of services related to sustainable energy. Energy efficiency programmes, for example, often utilise new electronic controls, energy efficient boilers and HVAC equipment. Across the spectrum of examples discussed in this chapter, for projects in most developing countries, a great deal of technologically sophisticated equipment must be imported – e.g. turbines for power projects, centrifugal blowers for methane capture projects, electricity sub-meters for energy-efficiency projects and electronic control equipment for many types of projects – while many construction materials are procured locally. The general implication of studies on developing economies is that the potential benefits of simultaneously liberalising trade in environmental services and environmental goods are likely to be much greater than liberalising trade in only one or the other.[58]

Sustainable energy services should be viewed in the wider context of sustainable development. Effective domestic and international frameworks that support sustainable energy will be crucial for realising the development benefits of access to sustainable energy for all. In particular, job creation can benefit from open trade in goods and services related to sustainable energy.

Against this background, there is a need to continue to pursue a variety of options for liberalising trade in environmental services more broadly and specifically in sustainable energy services.

First of all, in the context of the Doha Round, negotiations will continue on updating the GATS. While some see the agreement as limited to market

---

[57] Janssen (2010).   [58] Ravnborg et al. (2007).

access and national treatment, others consider it a flexible instrument with scope for good outcomes for environmental services. Classification instruments do not determine the scope of the GATS. Nothing prevents WTO Members from making broad commitments or taking a 'clustering approach' for commitments in energy-related services sectors. Also, there is opportunity in the GATS to address regulatory issues.

The GATS framework provides examples of options that could be explored to make progress in the liberalising of sustainable energy services. For instance, an Understanding was developed in financial services,[59] which takes a negative list approach to commitments in the sector, thus allowing Members who chose to adopt the Understanding to schedule limitations on specific commitments or sub-sectors. Arguably, sustainable energy services could be a good candidate for a similar approach.

Another option could be an approach similar to the WTO telecom reference paper,[60] which provides guidelines for a regulatory framework that Members should follow to support the transition of the telecom sector to a competitive marketplace. Members who wish to adopt the reference paper could make an additional commitment in their existing GATS schedules. Perhaps a similar approach could be developed for environmental services to address regulations related to sustainable energy services.

The limitation with these approaches is that they do not provide a framework for the massive scale-up of both goods and services related to sustainable energy – enough to have a significant impact on climate change – envisaged in a SETA. Crafting a SETA that includes services would require deeper consideration among stakeholders on the technical, legal and political aspects of trade in services related to sustainable energy.

Useful lessons can be drawn here from the TiSA negotiations. This raises the question of 'what was necessary as a catalyst for action on TiSA?' First, there needed to be a critical mass of like-minded countries. This group needed strong and commonly agreed objectives and identified benefits (supported by quantitative work). As is the case for any trade initiative, there needed to be a strong push from the business community (also see Hufbauer et al. 2012).

However, the TiSA is mostly about market access and will probably not bring any new initiatives concerning regulation. In addition, the agreement will most probably be outside of the WTO (at least in the first instance). Finally, following the TiSA approach means excluding Brazil, Russia, India, China and other developing countries.

---

[59] See the WTO website for the text of the Understanding on Commitments in Financial Services, www.wto.org/english/docs_e/legal_e/54-ufins_e.htm.
[60] See the WTO website on the Post-Uruguay Round negotiations on basic telecommunications, www.wto.org/English/tratop_e/serv_e/telecom_e/tel23_.htm.

The relationship between the TiSA (covering all services) and the proposed climate-related services within a SETA is difficult to predict at this point. Overlaps could exist, as they do between PTAs. Other avenues for sustainable energy trade initiatives could be the inclusion of protocols or annexes to FTAs, such as those on sustainable development attached to some FTAs between the EU and members of the Association of Southeast Asian Nations (ASEAN).

Based on these and other considerations in this chapter, stakeholders in trade in sustainable energy services may want to discuss the following:
- Should the classification of RE services and energy efficiency services be made more visible? What would be the best approach?
- Should it be left to each member wishing to undertake specific commitments on CCS, smart grids and other rapidly evolving technologies to decide how to classify and define relevant services?
- Is it appropriate to apply CPC definitions elaborated in 1991 to novel technologies that emerged several years later? If not, what would be the options?
- More generally, have Members encountered difficulties in classifying services associated with renewable energy or energy efficiency? If so, what are these and how could they be solved?
- What is the best way to appreciate the links between sustainable energy goods and services, and how can these linkages be taken into consideration when crafting trade policies?
- Can sustainable energy services be progressively liberalised through sustainable energy trade initiatives and a SETA and, if so, how? How can linkages with other agreements and initiatives be established?

## 9   Concluding Remarks

The potential benefits of liberalising trade in low-carbon technologies and services in tandem have been widely touted. To include sustainable energy services in a SETA, it is crucial to identify the services that could be complementary to the diffusion of sustainable energy technologies and to understand the current level of market access for such services.

Specific commitments made under the GATS may have a stronger impact on regulatory competence than tariff bindings do in goods trade, creating favourable conditions for investment and access to technology when an adequate regulatory framework is provided (Cossy 2011). The same would likely hold for commitments made in sustainable energy services in a SETA. Analysis of the major trading countries' specific commitments on the complementary services of sustainable energy technologies reveals that the principal modes of supply for these groups of services are limited, and little progress has been made on WTO Members' initial or revised offers so far. Reviewing bilateral, regional and unilateral liberalisation commitments could be useful when considering the inclusion of services related to sustainable energy in a SETA.

In addition, several empirical studies have shown that some of the key services required for producing and using energy more sustainably – ranging from energy efficiency projects to utility-scale wind power projects – are often unavailable in host countries (Steenblik and Geloso Grosso 2011; Sterk et al. 2007). Liberalising trade in these services might not only facilitate the diffusion of associated sustainable energy technologies, but also give countries easy access to such services. Although concerns have been raised that the 'complementary services of sustainable energy technologies' discussed in this chapter might exacerbate the persistent problem of dual use as the services cut across multiple sectors, a SETA should allow ample flexibility for specifying the scope of commitments in Members' schedules. If they wish to increase the market access for sustainable energy services through plurilateral trade negotiations, they could specify their commitments on such services in their schedules within the current GATS structure of classification.

It should be kept in mind that facilitating trade in complementary services of sustainable energy technology goes beyond the GATS, since trade barriers to these services are not limited to the issues of market access and NT. For instance, given that the public sector appears to be the largest client across all three groups of services, regulations concerning government procurement could have a significant impact on trade in these services. An empirical study shows how some existing practices, as well as limited transparency in this area, could create barriers to trade in EGs and associated services.[61] It goes without saying that regulations on government procurement, standards and qualifications for services providers play an important role in the EGS market. Given the close links between the two, liberalising trade in services associated with sustainable energy technologies should be addressed in conjunction with discussions on the plurilateral agreement on government procurement in the WTO.

Furthermore, domestic laws, regulatory measures and administrative rules all have the potential to affect trade in these services. Examples include domestic regulatory measures, such as financial thresholds, building regulations and associated technical requirements, or regular inspection requirements for safety. In facilitating trade in complementary services of sustainable energy technologies, relevant regulatory measures and administrative rules should be addressed in tandem.

---

[61] For detailed examples, see Fliess and Kim (2008).

Trade in services related to sustainable energy should respond to the demands of clients in developing countries. Those demands are being driven in some cases by tighter regulations and in others by corporate policy and corporate social responsibility in particular.

Benefits to businesses that engage outside experts to carry out sustainable energy services are manifold. Outsourcing allows them to concentrate on core activities while shifting some of the liability of meeting environmental regulations to other companies. Often, it allows the facilities involved to be built to an optimal scale, which may be larger than that required for a single client. The resulting economies of scale allow costs of sustainable energy to be reduced, and, because several clients may be served, introduce greater flexibility into contractual arrangements. Keeping the door open to imports of sustainable energy services and goods also helps ensure vigorous competition, which keeps down the price of goods and helps make their supply more reliable.

The main point of this chapter, however, is that the potential benefits of simultaneously liberalising trade in sustainable energy goods and services are likely to be much greater than liberalising trade in either one or the other. These benefits include better environmental performance of local industries (which increases a country's attractiveness to foreign direct investment or FDI); increased availability of services that benefit the environment and the health of the population; as well as reduced costs of spurring innovation. They also include increased local capacity to produce goods and provide sustainable energy services – capacities that can be translated into increased export opportunities.[62]

## References

Adlung, Robert (2009) 'GATS commitments on environmental services: "hover through the Fog and Filthy Air"', in Thomas Cottier, Olga Nartova and Sadeq Z. Bigdeli (eds.), *International Trade Regulation and the Mitigation of Climate Change*. Cambridge University Press

Andrew, Dale (2000) *Services Trade Liberalisation: Assessment of the Environmental Effects*. Paper presented at the Conference on Trade, Poverty and the Environment: Methodologies for Sustainability Impact Assessment of Trade Policy. University of Manchester

Bisset, Ron et al. (2003) *Sustainability Impact Assessment of Proposed WTO Negotiations: Environmental Services*. Report prepared for European Commission, DG Trade, Brussels

Brewer, Thomas (2012) *International Technology Diffusion in a Sustainable Energy Trade Agreement (SETA): Issues and Options for Institutional Architecture*.ICTSD, Geneva

---

[62] Steenblik et al. (2005).

Butkeviciene, Jolita (2005) *Managing Request-Offer Negotiations under GATS: The Case of Construction and Related Engineering Services*. UNCTAD, Geneva

Cattaneo, Olivier, Linda Schmid and Michael Engman (2010) 'Engineering services: how to compete in the most global of the professions', in Olivier Cattaneo, Michael Engman, Sebastién Sáez and Robert M. Stern (eds.), *International Trade in Services: New Trends and Opportunities for Developing Countries*. World Bank, Washington, DC

Chen, Chuan (2008) 'Entry mode selection for international construction markets: the influence of host country related factors', *Construction Management and Economics*, 26(3): 303–314

Cossy, Mireille (2011) 'Environmental services and the General Agreement on Trade in Services (GATS): legal issues and negotiating stakes at the WTO', in Christoff Harmann and Jörg P. Terhechite (eds.), *European Yearbook of International Economic Law*. Springer, New York

Cottier, Thomas and Donah Baracol-Pinhão (2009) 'Environmental goods and services: the environmental area initiative approach and climate change', in Sadeq Z. Bigdeli, Thomas Cottier and Olga Nartova (eds.), *International Trade Regulation and Climate Change*. Cambridge University Press

Delimatsis, Panagiotis and Despina Mavromati (2009) 'GATS, financial services and trade in RECs', in Sadeq Z. Bigdeli, Thomas Cottier and Olga Nartova (eds.), *International Trade Regulation and Climate Change*. Cambridge University Press

Engman, Michael (2010) 'Building empires overseas: internationalization in the construction services sector', in Olivier Cattaneo, Michael Engman, Sebastién Sáez and Robert Stern (eds.), *International Trade in Services: New Trends and Opportunities for Developing Countries*. World Bank, Washington, DC

European Centre for International Political Economy (2012) *The International Services Agreement (ISA) from the European Vantage Point*. ECIPE, Brussels

Fliess, Barbara and Joy Kim (2008) 'Non-tariff barriers (NTBs) facing trade in environmental goods and associated services', *Journal of World Trade*, 42 (3): 535–562

Geloso Grosso, Massimo, Anna Jankowska and Frédéric Gonzales (2008) *Trade and Regulation: The Case of Construction Services*. OECD, Paris

Gibbins, Jon et al. (2006) 'Scope for future $CO_2$ emission reductions from electricity generation through the deployment of carbon capture and storage technologies', in Hans Schellnhuber et al. (eds.), *Avoiding Dangerous Climate Change*. Cambridge University Press

Booz Allen Hamilton. (2006) *Globalization of Engineering Services: The Next Frontier for India*. NASSCOM, New Delhi

Hoekman, Bernard, Aaditya Mattoo and Philip English (2002) *Development, Trade and the WTO: A Handbook*. World Bank, Washington, DC

Hufbauer, Gary, Bradford Jensen and Sherry Stephenson (2012) *Framework for the International Services Agreement*. Policy Brief, No. PB12-10. Peterson Institute for International Economics, Washington, DC

Intergovernmental Panel on Climate Change (IPCC) (2007) *Fourth Assessment Report: Climate Change*: Synthesis Report. IMO/IPCC, Geneva

International Centre for Trade and Sustainable Development (ICTSD) (2011) *Fostering Low Carbon Growth: The Case for a Sustainable Energy Trade Agreement.* ICTSD, Geneva

International Energy Agency (IEA) (2010) *Energy Technology Perspectives: Scenarios & Strategies to 2050.* OECD/IEA, Paris

International Labour Organization (ILO) (2008) *Green Jobs: Facts and Figures.* ILO, Geneva

(2012a) *Sustainable Development, Green Growth and Quality Employment.* ILO, Geneva

(2012b) *Working Towards Sustainable Development: Opportunities for Decent Work and Social Inclusion in a Green Economy.* ILO, Geneva

International Renewable Energy Agency (IRENA) (2011) *Renewable Energy Jobs: Status, Prospects and Policies.* IRENA, Masdar City

Janssen, Rod (2010) *Harmonising Energy Efficiency Requirements: Building Foundations for Co-Operative Action.* ICTSD, Geneva

Kennedy, Matthew (2012) *Legal Options for a Sustainable Energy Trade Agreement.* ICTSD, Geneva

Kim, Joy Aeree (2011) *Facilitating Trade in Services Complementary to Climate-Friendly Technologies.* ICTSD, Geneva

Mburu, E. (2008) *Construction Services: Contribution to Sustainable Development and Issues on Trade in Services.* Paper presented at ICTSD, the African Economic Research Consortium and the Economic and Social Research Foundation, Tanzania, on 'Regional Dialogue for Eastern and Southern Africa in Trade in Services and Sustainable Development', Dar es Salaam, 30–31 October

Miroudot, Sébastien and Erik van der Marel (2012) *The Economics and Political Economy of Going Beyond the GATS.* GEM Working Paper, Sciences-Po, Paris, February

Nartova, Olga (2009) 'Assessment of GATS' impact on climate change mitigation', in Sadeq Z. Bigdeli, Thomas Cottier and Olga Nartova (eds.), *International Trade Regulation and the Mitigation of Climate Change.* Cambridge University Press

Organisation for Economic Co-operation and Development (OECD)(2000) *Environmental Goods and Services: An Assessment of the Environmental, Economic and Development Benefits of Further Global Trade Liberalisation.* COM/TD/ENV(2000)86/FINAL. OECD, Paris

(2005) *Trade that Benefits the Environment and Development: Opening Markets for Environmental Goods and Services.* OECD, Paris

Ravnborg, Helle, Mette Damsgaard and Kim Raben (2007) *Payments for Ecosystem Services: Issues and Pro-Poor Opportunities for Development Assistance.* DIIS Report 2007:6, Copenhagen

Steenblik, Ronald, Dominique Drouet and George Stubbs (2005) *Synergies between Trade in Environmental Services and Trade in Environmental Goods.* OECD, Paris

Steenblik, Ronald and Massimo Geloso Grosso (2011) *Trade in Services Related to Climate Change: An Exploratory Analysis.* OECD, Paris

Steenblik, Ronald and Joy Aeree Kim (2008) *Facilitating Trade in Selected Climate Change Mitigation Technologies in the Energy Supply, Buildings and Industry Services.* OECD, Paris

Sterk, Wolfgang et al. (2007) *Renewable Energy and the Clean Development Mechanism – Potential, Barriers and Ways Forward: A Guide for Policy Makers*. Federal Ministry for the Environment, Nature Conservation and Nuclear Safety, Berlin

Teljeur, E. and Stern, M. (2002) *Understanding the South African Construction Services Industry: Towards a GATS Negotiating Strategy*. Paper presented at the Trade and International Policy Strategies Forum 2002: Global Integration, Sustainable Development and the Southern African Economy, Muldersdrift

Tulacz, Gary (2005) 'World construction spending nears $4 trillion for 2004', *Engineering News-Record*, 254(1)

  (2008) 'The top 500 design firms: watching for signs of a market slowdown', *Engineering News-Record*, 260(13)

Wei, M., S. Patadia and D. Kammen. (2010) 'Putting renewables and energy efficiency to work: how many jobs can the clean energy industry generate in the US?', *Energy Policy*, 38: 919–931

World Bank (2007a) *International Trade and Climate Change: Economic, Legal and Institutional Perspectives*. World Bank, Washington DC

  (2007b) *Morocco's Backbone Services Sectors: Reforms for Higher Productivity and Deeper Integration with Europe*. Report 39755-MA. World Bank, Washington, DC

World Trade Organization (WTO) (1998) *Construction and Related Engineering Services*, Background Note by the Secretariat. Document S/C/W/38, Council for Trade in Services. WTO, Geneva

  (2011) Document TN/TE/20. WTO, Geneva

## Further Reading

CIEL and WWF (2003) International Discussion Paper. WWF, Gland

Geloso Grosso, Massimo (2005) 'Managing request-offer negotiations under the GATS: the case of environmental services', in *Trade that Benefits the Environment and Development: Opening Markets for Environmental Goods and Services*. OECD, Paris

Higgott, Richard and Heloise Weber (2005) 'GATS in context: development, an evolving *lex mercatoria* and the Doha Agenda', *Review of International Political Economy*, 12 (3): 434–455

UNCTAD (2000) *Regulation and Liberalization: Examples in the Construction Services Sector and its Contribution to the Development of Developing Countries*, TD/B/COM.1/32. Trade and Development Board, UNCTAD, Geneva

  (2001) *Energy Services in International Trade: Development Implications*. United Nations, Geneva

UNCTAD, UNDP and UNIDO (2007) *The Clean Development Mechanism – Building International Public–Private Partnerships under the Kyoto Protocol: Technical, Financial and Institutional Issues*. United Nations, Geneva

US International Trade Commission (USITC) (2005) *Renewable Energy Services: An Examination of US and Foreign Markets*. Publication No. 3805, Washington, DC

Vrolijk, Christiaan (2009) *The Potential Size of the Clean Development Mechanism*. Royal Institute of International Affairs, London

# 4 Governing Clean Energy Subsidies: What, Why, and How Legal?: August 2012

*Arunabha Ghosh with Himani Gangania*

Nearly 2 billion people have no access to modern sources of energy. Increasing energy access is one of the key ingredients for human development. At the same time, energy-related carbon dioxide ($CO_2$) emissions are expected to increase over the next two decades, especially in developing countries. Clean energy subsidies are, therefore, being used to support two simultaneous transitions: from no energy to energy access; and from fossil-fuel-based energy to a low-carbon energy pathway.

At least four sets of policy tensions are driving a growing international debate on the governance of clean energy subsidies. The first of these relates to the environment. Renewable electricity technologies and renewable transport fuels are one set of responses to climate change, but they have incremental costs over and above fossil fuels. In the absence of adequate international funding to transfer to cleaner technologies, many governments use subsidies to support these technologies and reduce their costs.

Second, tensions arise on economic grounds thanks to decisions to invest in clean energy sectors. The role of subsidies to smooth the fluctuations in clean energy sectors and increase investor confidence has become an international concern for sustaining investments in the face of the climate challenge.

The third source of tension is technology. Recent years have witnessed significant growth in manufacturing capacity and deployment of clean energy generation capacity in a number of countries, but many technologies still remain at the research and development (R&D) stage or have not been deployed at a scale that would make them commercially viable just yet and require different forms of financial support.

Economic drivers also manifest through a fourth tension: trade policy. Governments may design subsidy programmes to concentrate spending on domestic firms and discriminate against foreign firms operating within the country or against imports of clean energy products. Subsidies can be granted to promote clean energy product exports, making domestic firms more competitive in the international market.

There are also four arguments for using subsidies to promote clean energy. The public good argument stems from a desire to increase energy access and recognition of market failures. Another rationale offered for clean energy support is industrial policy. Firms demand government support to help secure access to patents and new technologies; to spread risk burdens; to get easy lines of credit; to secure land, water and other resources and to gain entry into the power market or access to the grid. A related rationale is job creation (and economic stimulus) in pursuit of 'green jobs' in the renewable energy (RE) sector. In recession-hit economies, the jobs argument has particular resonance to justify stimulus spending. And, finally, the tit-for-tat argument suggests that if one country supports its domestic industry and labour force, other governments, fearing unfair trade competition and loss of potential market opportunities, will seek to level the playing field.

## 1   Why Subsidise Clean Energy?

RE resources are naturally replenishing and virtually inexhaustible, but the amount of energy available per unit of time is limited. These resources include: biomass, geothermal, hydro, ocean thermal, wave action, tidal action, solar power and heating and wind. For the purposes of this chapter, we consider clean energy to be that supplied by RE resources for electricity generation or made available by energy efficiency measures. Definitions vary across data sources. In some instances where large hydropower is excluded from the estimates, the chapter will state it explicitly.

Globally, there was a surge in RE investments from 2004 to 2007. Countries like Brazil, China and India experienced compound annual growth rates in RE investments of 171 per cent, 104 per cent and 52 per cent, respectively, during 2004–2008.[1] But the onset of the global economic crisis halted the upward trend. Although the crisis did not fully manifest itself until late 2008 and through 2009, quarterly investment data show that the slowdown in sustainable energy investments began early in 2008.

In recent years, RE investments have picked up again, with new investments rising 32 per cent in 2010 to a record USD 211 billion (Figure 4.1). The gains have mostly come from sharp increases in small-scale distributed projects, in government-sponsored R&D and in asset finance for utility-scale projects. Also in 2010, new investment (asset finance, venture capital,

---

[1] Ghosh (2010a).

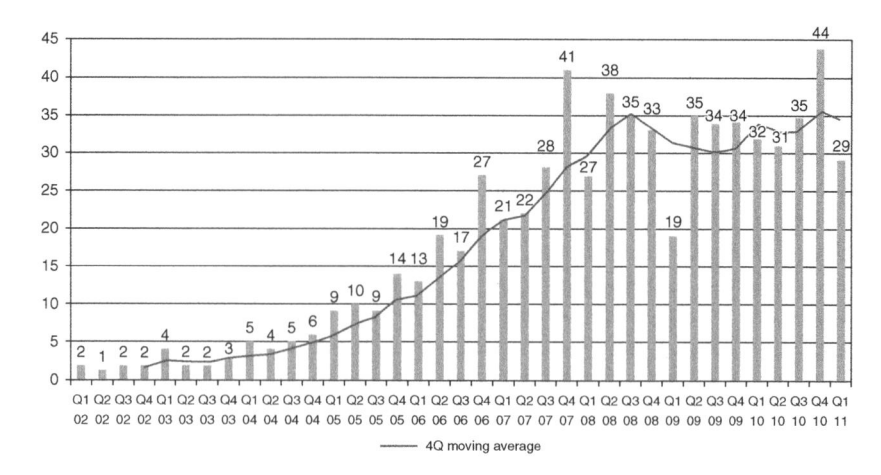

Figure 4.1: Global Financial New Investment in RE, Quarterly Trend, 2002–2011
Source: McCrone et al. (2011).

private equity and public markets) in developing economies (USD 72 billion) exceeded the developed countries' mark of USD 70 billion (investments in China and India were up 28 and 25 per cent, respectively).[2]

However, the share of RE in energy systems remains small. In electricity generation, for example, RE (excluding large hydro) remains small at only 8 per cent of power capacity and 5.4 per cent of electricity generation. While rising investments have ensured that RE accounts for 34 per cent of the net addition to global power capacity and 30 per cent of the net increase in power generation, these encouraging trends are tempered when taking into account the significant upstream investments in fossil-fuel energy (exploration, mining, unconventional sources, etc.). Of the combined investment in energy of USD 1.2 trillion in 2010, RE's share was only one-sixth.[3]

Clean energy clearly has a long way to go before it can credibly threaten the dominance of fossil fuel in electricity capacity and generation. The question is whether market trends would be sufficient to push investments into the sector at a pace fast enough to meet the twin goals of energy access and climate change mitigation. If not, why and under what circumstances would clean energy subsidies be needed? This section will explore the different arguments offered for subsidising clean energy – and, as discussed earlier, underscore many of the policy tensions surrounding

[2] McCrone et al. (2011): 11.   [3] McCrone et al. (2011): 26.

such support measures. Each set of arguments suggests costs and benefits for different constituencies, at the national and international levels.

### *The Public Good Argument*

One argument for subsidising clean energy stems from a desire to increase energy access and recognition of market failures. For example, increasing energy access to dispersed population settlements becomes harder the further they are from the electricity grid. This is particularly problematic for rural households. Even in densely populated regions, low electricity demand from rural households can make the installation of secondary and tertiary transmission lines and distribution systems uneconomical. Such households are unlikely to enjoy energy access unless part of the capital cost is subsidised.

Subsidies to increase energy access are not necessarily net costs for governments. The use of traditional biomass fuels for cooking and heating has severe health implications, especially for women. Access to modern energy sources means improved health outcomes, benefits that are often not included in economic cost-benefit analyses (CBAs). Such omissions result in a market failure, whereby energy utilities have no incentive to extend transmission lines when the social benefits of better health outcomes are not internalised in their balance sheets. In these cases, subsidising off-grid electricity systems might make even more sense.

Market failures also emerge for clean energy technologies. Left to their own devices, private technology and project developers under-invest in RE sectors, because the wider social benefits are disregarded. Not counting the positive environmental and health externalities of switching to cleaner sources of electricity – lower greenhouse gas (GHG) emissions and cleaner air – means that a calculation based on private returns would preclude clean energy investments. But, when the benefits of shifting away from traditional fuels are added along with the avoided fuel costs of using diesel or kerosene, subsidised off-grid RE applications have yielded significant returns. Solar photovoltaic (PV) systems, for instance, have offered economic returns and consumer surpluses of 27 to 94 per cent for projects in Bolivia, China, Indonesia, the Philippines and Sri Lanka.[4]

*Lowering RE Costs* Many RE sources account for only a small share of electricity generation in most countries (Figure 4.2) because the average costs of additional capacity installation remain much higher than for the dominant fossil-fuel sources, like coal or gas. The costs of new

---

[4] World Bank (2008a): 4.

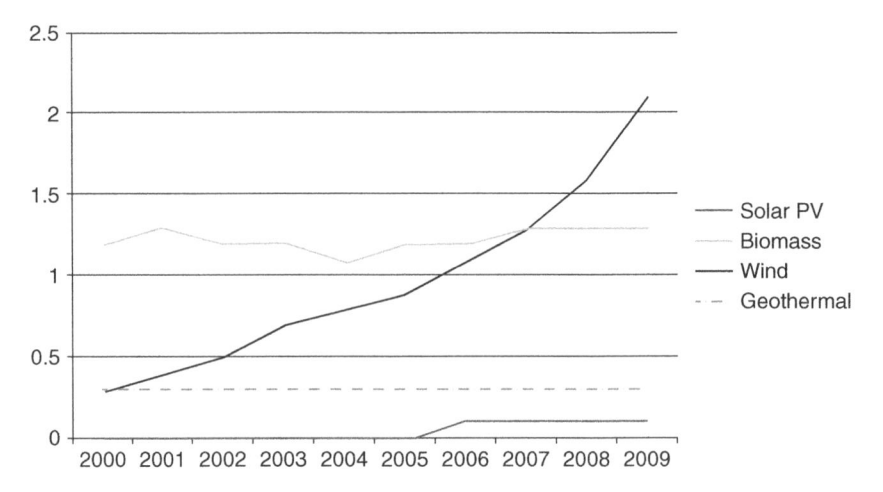

Figure 4.2: Worldwide Share of Many RE Sources in Total Electricity Generation Remains Small (per cent)
Source: Gelman (2010): 53.

technologies fall as the cumulative installed capacity increases, at a proportion known as the 'learning rate'. Subsidies that increase deployment of new technologies increase learning and help to bring costs for RE closer to those of non-renewable sources.[5]

It should be noted, however, that subsidies need not be the first-best option to correct market failures and provide public goods. A tax on fossil-fuel consumption in electricity generation, or the removal of fossil-fuel subsidies, are often better options.[6] There has been a long history of using subsidies to support energy infrastructure – often wastefully. In the United States, the biggest energy consumer historically, land was granted for timber and coal infrastructure in the 1800s. In the twentieth century, subsidies were offered to the oil and gas and nuclear energy industries. Offshore drilling, for instance, benefited from royalty waivers and favourable leases from the Federal government.[7] During the oil price spikes of the late 1970s and early 1980s, tax incentives for fossil fuels exceeded USD 12 billion.[8]

REs, however, have not been treated the same way. Supporters of clean energy, like former California Governor Arnold Schwarzenegger, argue

[5] Bridle et al. (2011).
[6] Thanks to an anonymous peer reviewer for highlighting this point.
[7] Arnold Schwarzenegger, 'An unfair fight for renewable energies', *Washington Post*, December 5, 2011.
[8] Robert B. Semple, Jr., 'Oil and gas had help. why not renewables?', *New York Times*, October 16, 2011: SR10.

that government policy must 'level the playing field for renewable energies' in order to develop a more competitive energy market, ensure energy and national security and create jobs at home.[9] This is not an argument, *per se*, for subsidising RE if better alternatives for removing distortions in energy markets exist.

However, the public good argument – accounting for the wider social and environmental benefits – is not sufficient to use clean energy subsidies if the subsidy programme is badly designed. Where government subsidies promote an industry or a sub-sector, participants are often wary of the longevity and credibility of commitments. Changes in government or in the economic condition can result in policy reversals and rapidly wipe out any expected long-term private and social gains from investing in clean energy. The sudden decision in October 2011 to almost halve the feed-in-tariff (FIT) for roof-top solar installations in the United Kingdom, which has given rise to legal disputes, is a case in point. Such a 'stop and go' approach gives rise to concerns for investor decisions and undermines the credibility of policy pronouncements in the future as well.[10]

### The Industrial Policy Argument

Another rationale offered for clean energy support is industrial policy. New energy technologies represent an opportunity for countries to demonstrate technological leadership and create an ecosystem that could support the development of new sectors of the economy. Should such technologies gain widespread demand, within and outside the country, domestic firms could gain a competitive advantage. In the early stages of a sector's development, firms demand support from the government for multiple reasons: to help secure access to patents and new technologies; to spread the burden of risk; to get easy lines of credit, especially for the high capital investments for renewable energies compared to fossil-fuel sources; to secure land, water and other resources; to gain entry into the power market or access to the grid, and so forth.

China, for instance, has elevated alternative energy and environmentally friendly and energy efficient technologies to the level of 'strategic emerging industries'.[11] For its 12th Five-Year Plan it planned USD 300 billion of investment each year to be divided among seven strategic industries. The objective is to take advantage of market trends and close

---

[9] Schwarzenegger, 'An unfair fight for renewable energies'.
[10] Sarah Murray, 'Carrot and stick approach supports but can distort', *Financial Times*, 31 October, 2011: SR1-2.
[11] *People's Daily Online*, 'Strategic emerging industries likely to contribute 8% of China's GDP by 2015', 19 October 2010.

the relatively small gap between emerging and developed economies in these new sectors, according to its then-Vice Premier Li Keqiang.[12]

This aggressive industrial policy has yielded private and collective dividends. China now accounts for three-fifths of the world's solar panel production, 95 per cent of which is exported.[13] Chinese firms, of course, benefit, but the global impact has been to drive panel prices down. In 2008, 1 watt (W) of solar capacity cost USD 3.30/W; by the end of 2011, the cost was closer to USD 1–1.20/W.[14]

*Picking winners...* The trouble is that picking winners is seldom easy. The forms of financial and other types of commercial support required at different stages of research, development, demonstration and widespread deployment of new technologies vary. At very early stages, government investment in basic R&D can create a body of knowledge and expertise necessary to stimulate further innovation. During 1961–2008, the US Federal government spent USD 172 billion on basic R&D for advanced energy technologies.[15] The question is whether basic R&D can easily translate into more risky commercial investments in emerging industries and firms. The Federal loan guarantee scheme, worth USD 10 billion and strongly supported by then-US Energy Secretary Steven Chu, has become the centre of a controversy following the collapse of Solyndra, a Californian solar company that had received USD 535 million in loan guarantees. Critics argue that governments have a long history of failed commercial investments in the energy sector. Lawrence Summers, former presidential economic adviser, described the government as 'a crappy VC [venture capitalist]'![16] VC investments routinely fail, but are justified as long as the upside from investment in one innovative technology makes up for the losses on other bets. The problem is that private VC investments often leave the riskiest projects for governments to cover, thus lowering the prospects for success.

*Risking policy capture* There are other dangers in using subsidies to support industrial policy, namely perversely distorting markets and credit flows, encouraging rent-seeking and other anti-competitive practices, locking in existing technologies at the cost of future innovation, or

---

[12] Melanie Hart, 'China eyes competitive edge in renewable energy', Center for American Progress, August 24, 2011.
[13] Keith Bradsher, 'Trade war in solar takes shape', *New York Times*, November 10, 2011: B1.
[14] Bradsher 'Trade war in solar takes shape': B1.   [15] Dooley (2008): 1.
[16] Steven Mufson, 'Before Solyndra, a long history of failed government energy projects', *Washington Post*, November 12, 2011.

simply adding excess capacity relative to demand. China's National Reform and Development Commission (NDRC), its power planning body, observed that its wind energy sector was already suffering from over-capacity, thereby questioning the need for and ability to absorb such large-scale investments.[17] In the United States, Federal subsidies for clean energy increased nearly threefold in 2007–2010 to reach USD 14.7 billion.[18] Thanks to a set of policies that amounted to a 'gold rush of subsidies', RE firms were able to benefit from support in the form of construction loan guarantees, cash grants, property tax breaks, long-term power purchase agreements (PPAs) and FITs, and wanted to do as much of this business as they could get their hands on.[19] While the policies go some way to level the playing field with fossil fuels and nuclear energy, the danger is whether too much government subsidy reduces the incentives for investors to bear commercial risks if they have too 'little skin in the game'.[20] The difficulty of making appropriate policy decisions is that the line between necessary government support and excessive largesse is not always clearly defined, especially if, as Michael Graetz of Columbia Law School warns, bets are made based on political motivations rather than scientific metrics.[21] In other words, policy design matters – such as degression in FITs based on differentiated technologies – in mitigating the potential emergence of perverse distortions from.[22]

### Green Jobs and the Economic Stimulus Argument

Another rationale is job creation, with some arguing that promoting clean energy industries could create millions of 'green jobs'.[23] Many economies hope to tap into the large potential for employment in the RE sector: the United Kingdom is aiming at creating 400,000 jobs over eight years; India sees 900,000 jobs in biomass gasification by 2025; and Nigeria's biofuel industry could lead to 200,000 jobs.[24] The German RE industry already employs 380,000 people with 108,000 in the solar PV industry alone.[25] Spain recorded nearly 116,000 jobs (mostly in wind) by 2010.[26] Thanks

---

[17] *China Daily*, 'China mulls $1.5 t boost for strategic industries', 3 December 2010.
[18] US Energy Information Administration (2011).
[19] Eric Lipton and Clifford Krauss, 'Rich subsidies powering solar and wind projects', *New York Times*, November 12, 2011: A1.
[20] See n. 19.   [21] Cited in Mufson, 'Before Solyndra'.
[22] Thanks to an anonymous peer reviewer for clarifying the argument here.
[23] See International Renewable Energy Agency (2011).
[24] United Nations Environment Programme (2009): 21.
[25] Adam Vaughan and Fiona Harvey, 'Solar subsidies to be cut by more than half', *Guardian*, 28 October 2011.
[26] Spain, Ministry of Industry, Tourism and Trade (2010): 157.

to the 'green' sector growing ten times faster than other industries since 2005, the state of California hosts a quarter of all US jobs in the solar sector and attracts a third of clean tech VC.[27]

The argument for green jobs has particular resonance in recession-hit economies, justifying billions of dollars of stimulus spending in pursuit of reducing unemployment rates.[28] When the economic crisis began, major economies announced up to USD 194 billion in 'green stimulus' spending (including RE, energy smart technologies, carbon capture and storage and transport).[29] Comparisons between countries are difficult, partly because the investments are expected to flow over several years, but estimates suggest that the United States led the field with USD 65.1 billion, followed by China (USD 46.1 billion) and South Korea (USD 32.2 billion).[30] Of the total commitments, about half was spent in 2009 (20.3 billion) and 2010 (USD 74.5 billion).[31] The expectation was that spending would amount to USD 68 billion in 2011, USD 21.4 billion in 2012 and USD 9.7 billion in 2013.[32]

Despite large commitments, stimulus spending has varied by country. Germany spent over half its allocation by 2010 and China spent 69 per cent, compared with only 36 per cent in the United States and 7 per cent in Brazil.[33] Nevertheless, the results have been striking: China now employs 1 million people in its RE sector (60 per cent in solar)[34] and California witnessed a 3 per cent increase in green jobs between January 2008 and January 2009 (in the middle of the global slowdown).[35]

While such efforts could increase employment in clean energy sectors, they need not create additional employment if job losses in fossil-fuel-based energy sectors are taken into account. This is because subsidised clean energy has the impact of lowering energy prices overall. If the price elasticity of demand for energy is not high, i.e. lower prices do not proportionately increase energy demand, the substitution of clean energy for dirty energy would necessarily lead to job losses in fossil-fuel sectors. The net effect would depend on the balance of job creation in clean energy versus job losses in dirty energy sectors. Moreover, the overall contribution could be small in advanced economies where energy production might not account for a large share of the economy.[36]

In developing countries, however, the price elasticity of energy demand would be high given the large share of income that energy (modern and

---

[27] See n. 9.  [28] Robins et al. (2010).  [29] McCrone et al. (2011): 30.
[30] World Economic Forum (2011): 23.  [31] See McCrone et al. (2011): 40.
[32] See World Economic Forum (2011).  [33] See World Economic Forum (2011): 24.
[34] United Nations Environment Programme (2009): 21.
[35] David Baker, 'Green jobs up 3% in 2008, Next 10 report finds', *San Francisco Chronicle*, January 19, 2011.
[36] Michael Levi, Michael (2011) 'New energy jobs won't solve the US unemployment problem', *Foreign Affairs Snapshot*, October 18, 2011.

traditional sources) costs poor households. In these economies, government subsidies would drive down RE prices and could encourage job creation across the supply chain (manufacturing, installation, financing, servicing, etc.). If these policies drive overall energy prices down, the demand for fossil-fuel energy would also increase (since many households hitherto unconnected to the grid would have an opportunity to access modern sources of energy). Therefore, job losses in the fossil-fuel sectors need not occur.

That said, in the long run, subsidised energy need not have a large positive impact on jobs if clean energy sectors face over-capacity. The clean energy market, valued at USD 240 billion per year, already faced a glut in 2011. Supplies of wind turbines and solar panels have exceeded demand, threatening jobs in the sector.[37] In other words, a 'green' energy policy could alter the mix of jobs in the energy sector but need not increase total jobs in the long run.

### The Tit-for-Tat Argument

The use of clean energy subsidies to promote industrial policy or create jobs has mercantilist outcomes. That is, if one country supports its domestic industry and labour force, other governments may fear unfair trade competition and loss of potential market opportunities. Domestic industry lobbies in the latter countries could protest: 'If others are doing it, so should we!' The main rationale for such measures – to level the playing field – need not result in efficient outcomes. If all major clean energy producers and exporters engaged in competitive subsidising, public funds would be diverted to support firms that need not be competitive otherwise. There might be a justification for such a policy only if the revenues from additional market share in other jurisdictions exceeded the costs to the public. But with other governments following a similar policy, the probability of greater market share is diminished. Unless tit-for-tat is used as a temporary strategy to 'punish' an errant country for violating an international agreement on energy subsidies, adopting a strategy of subsidising one's domestic industry solely because others are doing it is unlikely to present net positive outcomes.

### Who Wins and Who Loses

Behind each rationale for clean energy subsidies is a group of potential supporters and opponents (Table 4.1). The strongest rationale – the

---

[37] Leslie Hook, 'China to probe US clean energy subsidies', *Financial Times*, 25 November 2011.

Table 4.1 *Clean Energy Subsidies Have Supporters and Opponents*

|  | Rationale | Potential Supporters | Potential Opponents |
| --- | --- | --- | --- |
| Public good | Energy access; market failures owing to externalities | Energy consumers; environment; public health agencies | Fossil-fuel-based utilities; taxpayers or electricity rate payers (depending on length of subsidy) |
| Industrial policy | New technologies; competitive advantage | Clean energy product manufacturers; exporters; consumers if lowers costs passed through | Other industries (depending on source of subsidy); taxpayers if mostly for exports rather than lower energy consumption costs |
| Employment | Stimulus during recession; politically attractive | Skilled workers (if focus on manufacturing) | Lower-skilled or unskilled workers (if little attention to labour intensive activities in supply chain) |
| Tit-for-tat | Level the playing field | Domestic clean energy firms | Importers of clean energy components; multinational companies (MNCs) |

*Source:* Author.

provision of a public good in the presence of market failures – benefits energy consumers by driving down the prices of clean energy. It also benefits the environment and public health agencies, since externalities are reduced. But fossil fuel-based utilities would be expected to oppose these measures, since their market shares would be adversely affected by a growing role for RE in the energy mix of the economy. However, if the subsidies last a long time, taxpayers might also oppose the measures, especially if the benefits of energy access accrue to one section of society while the revenues to fund clean energy are drawn from another (say, richer consumers).

Any shift to a low-carbon energy sector would require changing the skill profile of the energy sector workforce. Industrial policy clearly benefits clean technology firms and exporters of clean energy products and services. Consumers would benefit only if lower costs were passed on to them; if most products are exported, domestic consumers would not necessarily benefit from the lower costs of the subsidised energy products. Moreover, if the subsidies are financed by withdrawing resources promised to other industries, those sectors would oppose the measures. On the jobs front – notwithstanding the long-term marginal impact on job creation – in the short run skilled workers are expected to benefit from policies that support high capital intensive RE sectors, especially if the policies the focus on manufacturing. Unskilled and lower-skilled workers might oppose these measures if little attention is paid to those parts of the value chain in RE electricity that utilise their services, such as installation and maintenance services.

Finally, tit-for-tat strategies benefit domestic clean energy firms but hurt more internationally integrated firms. The latter import raw materials and components of clean energy products. If tit-for-tat strategies undermine competitive markets, the prices of components would rise and adversely impact such firms. Multinational clean energy companies with operations in several countries are also expected to lose, since they are affected by declining profitability in foreign countries where they have business interests.

In other words, although clean energy subsidies have several rationales, the political economy of domestic and international support for them would depend on the balance of winners and losers affected by an evolving industry, changing technologies and competitiveness, and shifting domestic regulation and international rules governing trade, energy and climate change.

## 2  What are Clean Energy Subsidies?

The International Energy Agency (IEA) defines an energy subsidy as any government action primarily related to the energy sector that lowers the cost of energy production, raises the price received by energy producers or lowers the price paid by energy consumers.[38] The US Energy Information Administration (EIA) further defines energy subsidies as any government action whose purpose is to influence energy market outcomes, whether these actions take the form of financial incentives, regulation, R&D or public enterprise.

### *A Typology of Clean Energy Subsidies*

There are various kinds of clean energy subsidies; Table 4.2 offers a typology. One way to classify them is by the form they take. Subsidies for RE could be delivered directly as financial transfers, or indirectly by virtue of preferential tax treatment. Government support might also be regulatory in nature, whereby changes in laws create the incentives to invest in clean energy infrastructure or offer disincentives to firms and consumers that continue using fossil-fuel-based energy. Another type of support is related to physical infrastructure or access to natural resources,

---

[38] International Energy Agency, *Looking at Energy Subsidies: Getting the Price Right*, IEA, 1999.

Table 4.2 *A Typology of Clean Energy Subsidies*

| | Direct Financial transfers | Preferential Tax Treatments | Regulation | Infrastructure Support | Trade/Investment Restrictions |
|---|---|---|---|---|---|
| **Clean energy access/consumption** | Consumer subsidy | Tax credits for consumers | Grid connection | Grid access for consumers; Net metering | |
| **Clean energy generation and capacity** | FIT (financial transfer compo-nent); Long-term power purchase agreements (PPAs); Prefere-ntial credit | Accelerated depreci-ation; Investment tax credits; Production tax credits | Mandatory grid connection for RE firms FIT (regulatory component-such as compulsory purchase/ off-take of electricity generated) Demand guarantees (RPOs); Trading of renewable energy certificates (RECs); Government procurement | Grid access for RE firms; Land (below market price); Access to water; Energy-related services from government | Investment restrictions on foreign power firms |
| **Domestic clean energy equipment and services manufacturing/Production** | Equipment production subsidy | Excise duty rebate; Accelerated depreciation | Government procurement; Compulsory licensing of intellectual property (IP) | Land (below market price); Access to water resources | Market access restrictions on imported equipment and services, e.g. tariffs, quotas; technical standards; local content requirements (LCRs) |
| **Clean energy goods and services exports** | Export subsidy | Export tax rebate | Special Economic Zones (SEZs) | Land (below market price); Energy-related services from government | |

*Source*: Author.

which allows RE developers to lower the costs of production or supply electricity to consumers more easily. Finally, trade restrictions against foreign competitors can offer a competitive edge to domestic manufacturers or project developers.

A second way to classify clean energy subsidies is by understanding their purposes. As discussed in the previous section, government support is required if energy access has to be expanded to a larger share of the population. Market failure in clean energy development (thanks to unaccounted positive and negative externalities) adds to the case for subsidies to either extend access to clean energy or stimulate greater consumption of energy from clean sources. Subsidies may also support the creation of more clean energy generation capacity. This is only a second-best solution, since generation capacity does not necessarily translate into electricity feeding into the grid (if access to the grid is problematic or incentives such as accelerated depreciation are linked only to investment and not actual power generation) or to higher power consumption (if electricity costs remain prohibitively high). Nevertheless, generation capacity is a necessary step towards greater clean energy consumption. Government support is also offered with the primary purpose of building up domestic industrial capacity, for instance to manufacture solar panels or wind turbines. A related purpose, which also supports domestic industry, is when subsidies are geared toward boosting exports of clean energy products and services.

Using the framework described in Table 4.2, several clean energy support mechanisms may be identified. It should be noted that several types of subsidies could be used for multiple purposes and subsidy programmes seldom have only singular goals. Yet, this framework is useful because it establishes an explicit link between types of subsidies and their applicability for different purposes. This is relevant when interpreting WTO rules (see the final section on conclusions and policy recommendations).

*Direct Financial Transfers*

**Consumer Subsidies**
These subsidies are given directly to consumers to encourage them to substitute RE for fossil fuel. In Argentina and China, consumer subsidies have been offered as rebates in electricity bills. For off-grid electricity, the subsidy can take the form of a grant to cover initial capital costs (as in the cases of Benin, Bolivia and Togo). Off-grid RE products, such as biogas digesters, cleaner cook stoves, solar water heaters and solar home systems, are different. Bolivia, Laos, Nepal, Papua New Guinea, the Philippines, Tanzania and Zambia are among the countries that have used subsidies to

expand rural electrification. The World Bank has provided subsidies for solar home systems in various countries to cover between 12 and 90 per cent of the costs.[39] Government subsidies for off-grid projects have expanded capacity on a large scale: solar home systems (500,000 in Bangladesh, 400,000 in China and 600,000 in India); biogas digesters in India (4.2 million); and solar lanterns in India (800,000) by early 2010.[40] One-off subsidies via multilateral funding sources have also been used to offset capital costs for off-grid connections, such as the Bolivia Decentralised Electricity for Universal Access Project, which received USD 5.2 million from the Global Partnership for Output Based Aid (GPOBA) to install 7,000 PV systems for rural households, schools, clinics and small businesses.[41]

**Feed-In Tariffs and Power Purchase Agreements**
FITs offer investors in RE a preferential price (higher than fossil-fuel tariff rates) and are often guaranteed by long-term PPAs and grid access. They have a regulatory component, for instance when they require utilities or intermediaries to purchase electricity from renewable generators at the higher tariff price.[42] As explained further below, if the state bears the burden of this higher price (that is usually above the prevailing market price for electricity) there is a financial transfer component involved. The long horizon for a FIT policy is important, because it allows investors (whether large project developers or individual households) to calculate the period within which they would be able to recoup their investments. Since Germany adopted this policy in 2000, more than sixty-three countries have started offering FITs.[43]

Two policy questions are relevant to the governance of clean energy subsidies: what should the FIT rate be, and who would pay for it? In some cases, the FIT rate is gradually reduced to encourage greater efficiency and lower overall costs. How transparently rates are set matters in order to ensure that a few RE investors do not capture windfall profits from a high FIT. One option is to regulate the tariffs and compensate the energy service provider with a subsidy to cover costs (this was done for solar PV micro-grids in China). Another strategy is to use a reverse auctioning mechanism, and grant the concession to the bidder requiring the lowest level of subsidy. This policy was used for the Renewable Energy for Rural Markets Projects (PERMER) for off-grid concessions in Argentina in

---
[39] World Bank (2008a).
[40] World Bank (2010): 10. Climate Investment Funds Meeting of the SREP Sub-Committee, SREP/SC.4/Inf.2, October 28.
[41] World Bank (2010): 11. Climate Investment Funds Meeting of the SREP Sub-Committee, SREP/SC.4/Inf.2, October 28.
[42] Thanks to an anonymous reviewer for clarifying the distinction.   [43] REN21 (2009).

1999[44] and, more recently, for bidding under the first phase of the National Solar Mission in India.[45]

The other question – who pays – is also important, because the extra cost could be passed on to distribution companies, to final consumers or assumed by the government. FIT programmes could be considered financial support under three scenarios: first, when a public body uses public funds to execute the programme and provides the necessary funding; second, when the government asks a private body to execute the programme but pays for it; and third, when the government directs a private body to both execute the programme and pay for it through a reallocation of costs or other means.[46] The first model is used in Ontario, the second in the United Kingdom for small-scale projects and the third in Germany.

In developing countries, like India, distribution companies suffer from poor financial health thanks to the burden of other subsidy programmes, inefficient infrastructure, transmission and distribution (T&D) losses and theft. Burdening consumers in poor countries with higher rates is also challenging, unless the rates are progressive (with a higher share of the FIT cost falling on richer consumers). As argued earlier, if the government assumes the cost, it is a financial transfer. But, if the tariff is available only to domestic firms, foreign investors could argue that these are unfair trade practices.

**Preferential Credit**

For many commercial or development banks, investments in RE projects entail technology and policy risks. Such institutions could be offered low-cost credit lines and partial risk guarantees by the government to encourage them to offer preferential lending to RE project developers. The Indian Renewable Energy Development Agency (IREDA) has used financial support from the World Bank, the Asian Development Bank (ADB) and other bilateral donor agencies to offer concessional credit.

**Production Subsidies for Equipment Manufacturing**

Capital grants or low-interest loans to RE producers lower the cost of equipment production or help expand manufacturing capacity. Once again, although the type of subsidy might be similar, it is important to distinguish between subsidies directed toward expanding electricity generation capacity from those supporting local manufacturers. In India, the capital costs for installing biomass projects were subsidised to lower the eventual prices charged to consumers. Although clean energy transportation

---

[44] World Bank (2008a): 110.  [45] Council on Energy (2012): 8.
[46] Wilke (2011): 3–6.

fuels is not explicitly within the scope of this study, it is worth noting that to stimulate ethanol production, the US Congress passed legislation in 2005 and 2007 to set rising shares of ethanol in gasoline and offered the industry a 45 cent/gallon subsidy (USD 6 billion per year). By 2011, two-fifths of US corn production was being used for ethanol rather than food (this usage exceeded consumption of corn for livestock feed).[47]

## Export Subsidies

These subsidies are provided, again in the form of direct grants or concessional loans, to encourage local manufacturers and other firms offering clean energy products and services. Even when these subsidies are not explicitly intended for exports, their use can be distorted. One such perverse outcome of production subsidies driving exports is illustrated by corn ethanol in the United States. In addition to the production subsidy (see above), an import tariff of USD 54 cent/gallon was imposed. These policies not only restricted imports of cheaper Brazilian cane-based ethanol, but also made possible the export of 397 million gallons of ethanol.[48]

*Preferential Tax Credits*

### Tax Credits for Consumption

Tax exemptions, for example from personal taxes, can be used to encourage consumers to adopt RE.

### Accelerated Depreciation

This form of support allows project developers to use higher depreciation rates on their RE assets and thus receive related tax breaks. Accelerated depreciation also promotes investments in manufacturing and production capacity by domestic firms.

### Investment Tax Credits

These credits have been used in India, Sri Lanka and the United States to attract more foreign capital in RE. Income tax breaks (in India and the Philippines) offer investors the attraction of higher profits, thereby encouraging them to enter an otherwise riskier RE sector. In the United States, annual Federal tax support now amounts to USD 20 billion, three-quarters of which accrues to the wind, solar, ethanol and energy efficiency sectors, the first two sectors garnering USD 7 billion, corn ethanol getting another USD 6 billion and energy efficiency securing another USD 2 billion in 2010.[49]

---

[47] Melissa C. Lott, 'The US now uses more corn for fuel than for feed', *Scientific American*, October 7, 2011.
[48] Steven Rattner, 'The great corn con', *New York Times*, June 25, 2011: A19.
[49] Semple, Jr., 'Oil and gas had help. Why not renewables?': SR10.

**Production tax credits**
The credits, or generation-based incentives, are paid per KWh of electricity produced over and above the guaranteed power tariff. The amount paid depends on the actual amount of power produced, so there is an incentive for producers to generate more (unless the payments are capped at a certain level of electricity production). If they are not guaranteed for a certain period, however, they could be subject to shifts in policy.

**Excise Duty Rebates**
Rebates on sales, royalties and other levies are targeted at increasing RE production or manufacturing capacity. Governments in Kenya and Tanzania have provided duty exemptions for solar PV systems.

**Export Tax Rebates**
Like export subsidies, tax concessions could be used to encourage the export of RE products and services.

*Regulation*

**Grid Connections**
One of the most difficult challenges for scaling-up the infrastructure for RE is access to the grid. Without grid connections, projects dispersed across locations that are not close to population centres will remain unviable. Regulatory measures could be introduced that force utilities to extend transmission lines to RE project sites or to build electricity substations when projects are announced. These provisions help project developers secure loans, because financial institutions become more convinced about the viability of the proposed project to sell electricity to the grid. In China, for instance, until early 2010, a third of wind turbines (mostly located in the sparsely populated western provinces) were not connected to the national grid. In December 2009, China passed a law requiring grid operators to pay an RE project at twice the rate for the electricity that could be distributed.[50] Further grid-connection regulations could also ensure that consumers in rural areas or dispersed locations benefit from extensions of the grid.

**Demand Guarantees (RPOs)**
The guarantee of ever-increasing demand for RE helps bring prices down, owing to economies of scale. RPOs require power utilities to acquire a certain share of their electricity from RE producers. Over time, the share could increase, thereby creating and expanding the RE market.

---

[50] Keith Bradsher, 'China leading global race to make clean energy', *New York Times*, January 31, 2010.

### Trading via Renewable Energy Certificates

Imposing RPOs entails a cost on utilities, which seek a transition period to increase their portfolio of RE sources of power. Renewable energy certificates (RECs) are instruments signifying the shares of RE used by a utility. Those that do not meet their obligations are able to purchase surplus RECs from those that surpass their targets. This regulatory system creates a market mechanism to encourage RE adoption while reducing the costs of meeting targets. An REC mechanism also allows entities or provinces located in regions not conducive to RE generation to purchase clean energy from producers in faraway jurisdictions. A successful REC programme, therefore, depends on an integrated and efficient national or even cross-border electricity grid.

### Government Procurement

A major source of leverage is the purchasing power of governments. Regulatory changes that require governments to purchase more energy-efficient products or consume greater shares of RE could significantly influence market signals. The US General Services Administration (GSA) has a dual ambition to reduce the Federal government's environmental impact as well as stimulate innovation in clean technologies. It is able to do this because it has 9,800 buildings and 360 million square feet under its management, offering a captive market on a large scale.[51] If a country is a signatory to the plurilateral WTO Agreement on Government Procurement (GPA), and its policies favour domestic firms (that is, have an adverse impact on foreign firms), the measure can be challenged. But, if the country is not a party to the GPA, procurement policies favouring local project developers or equipment manufacturers will be seen as a clean energy support measure.

### Compulsory Licensing of Intellectual Property

RPOs and RECs are primarily geared to increasing RE generation capacity by expanding the market. If the objective is to promote domestic manufacturing capacity, compulsory licensing of IP could be one means to ensure that firms have access to the best technologies. Compulsory licensing, which has been used to increase supplies of pharmaceuticals, is potentially controversial and could invite trade disputes. Nevertheless, at the UN Framework Convention on Climate Change (UNFCCC) technology mechanism negotiations, developing countries have insisted that the option should be available to them. The case for compulsory licensing might yet have to be established, especially if production is not geared toward increasing energy access but for, say, export of clean energy equipment.

---

[51] Sarah Murray, 'Carrot and stick approach supports but can distort': SR1-2.

*Infrastructure Support*

**Grid Access**

While regulation to force grid connections to RE projects is one route (see above), direct government investment in grid infrastructure is another way to lower costs. In 2009, the Chinese government spent USD 45 billion to upgrade the electricity grid with state-owned banks providing the financing.[52] Similarly, net metering allows customers to earn revenue from selling surplus RE to the grid, which also helps utilities by helping to meet peak load demand. Net metering has been adopted in South Africa, Sri Lanka and the United States, among other countries. In rural and remote areas, extension of grid access could enable populations to benefit from larger renewable energy projects that are usually grid connected.

**Land Acquisition and Access to Other Natural Resources**

Infrastructure support can also be offered to build RE generation capacity and local manufacturing bases or to promote exports. Land acquisition is a critical factor in the RE ecosystem. In Gujarat, for instance, the government has established a solar park by acquiring land before selling it to project developers. Heavily subsidised land has been one factor in making China internationally competitive in wind and solar energy.[53] Similarly, governments can provide access to other natural resources, like water, for RE plants. These are potentially controversial policies, owing to competition for land and water resources for other agricultural and industrial purposes.

*Investment and Trade Restrictions*

**Market Access Restrictions and Investment Measures**

Support for the domestic RE sector also draws on trade measures. Regulations could prohibit foreign firms from participating as project developers in setting up RE generation capacity. Tariff barriers could also favour domestic equipment suppliers. Such measures might not lower costs for consumers. But, by restricting the number of players in the market, they offer indirect support to indigenous project developers (if investment measures were used) or domestic manufacturers (if tariffs were used).

**Other Non-Tariff Measures that Act as Incentives to Local Firms**

**Quotas:** Another form of trade restriction to promote domestic production is the use of quotas to restrict imports of foreign RE products. Even

---

[52] See n. 50.
[53] Keith Bradsher, 'On clean energy, China skirts rules', *New York Times*, September 8, 2010.

though a quota does not constitute a strict subsidy within the meaning of the WTO Agreement on Subsidies and Countervailing Measures (SCM), its economic impact is to shift the financial burden away from a single domestic firm to tax payers.

Quotas may also be applied on exports of raw materials that are used in RE products. In 2010, China curbed the export of rare earths, elements that are used in high-tech equipment, including clean energy products. In the first half of 2011, China cut export quotas again by 35 per cent. Speculation over the reasons for these curbs varies. However, given that China accounts for about 95 per cent of the global supply, the export restrictions triggered panic in industrialised countries (global prices for rare earths rose fourfold in 2010 and doubled again by April 2011). These measures prompted questions about whether existing trade rules were sufficient to adequately regulate or prohibit measures that threaten RE industries in other countries.[54]

**Technical Standards:** Countries can also restrict imports by promoting technical standards adopted by domestic manufacturing companies. For instance, if foreign firms produce wind turbines of a certain capacity that is different from domestic firms, a tendering process for procuring new wind turbines could specify the domestic standard. If standards vary across countries, technology R&D costs rise as the products and components have to be modified for each potential market. This is a disadvantage for foreign firms, but it benefits domestic firms that are already using the specified standard.

**Local Content Requirements (LCRs):** Finally, procurement policies could require a minimum share of local content in final RE products. The purpose of LCRs is to gradually promote the domestic manufacturing sector – allowing local firms to familiarise themselves with better technology – or to promote local jobs. If the market for RE is expected to grow and governments are worried about the security and quality of the RE infrastructure and continued supply of spare parts, there may be a case for promoting a domestic manufacturing base. The question is how to do so in the least discriminatory way. Governments could require that the products be manufactured within a given jurisdiction, even if the firm were foreign. Indeed, India's LCR rules for its National Solar Mission (JNNSM) do not discriminate between foreign and domestic firms, so long as a percentage of the final solar PV module is made at home.[55]

The problem with LCRs, however, is usually not discrimination against foreign firms that may be free to invest in domestic manufacturing

---

[54] James Bacchus, 'A rare-earths showdown looms', *Wall Street Journal*, May 20, 2011.
[55] Council on Energy (2012): 19–20.

facilities but rather against foreign imports. By imposing an LCR, imports are restricted and domestic (more expensive) production is favoured. This imposes a financial burden both on domestic consumers or taxpayers as well as on foreign producers of the equipment. Thus, both the motivation for the requirement, as well as the extent of harm caused to others, would matter in governing the use of LCRs. Moreover, LCR provisions can distort markets in unanticipated ways. For instance, the JNNSM imposed LCRs only on silicon technology, not on thin-film technology. One consequence has been that thin-film technology has been adopted on a much larger scale (half of the installations) compared with its global share in silicon technology of approximately 14 per cent.[56] Although the choice of technology is partly dependent on its appropriateness in different climatic conditions, this example still highlights the importance of using support measures that bear in mind potential market distortions.

*Potential Risks with Subsidy Schemes*

The range of government support for clean energy suggests that governing these measures would be complex, especially as this requires specific methods to identify and evaluate the motivations behind the support measures in a transparent manner. Governments and other stakeholders need to be cognisant of at least four potential risks. The first is that the support mechanisms would be non-transparent. This problem has particularly plagued subsidies for fossil-fuel sectors, where tax breaks on all sorts of indirect costs (such as labour costs, repair work, etc.) have been used to abuse concessions for oil and gas drilling activities. In the RE sectors, reverse auctioning has been used to ensure that subsidies accrue to the most efficient project developers, depending on their bid prices. However, reverse auctioning, if pursued too aggressively, could also result in an adverse selection of firms that have little experience or capacity to deliver projects at the low prices they promise. In those circumstances, subsidy schemes have to be supplemented by strict penalty clauses and enforcement measures.

A second potential demerit relates to the policy risks associated with government programmes. Project developers and their investors are unlikely to assume the long-term technology risks associated with RE if they are also worried about the credibility and longevity of government subsidies. For instance, in October 2011, the UK government suddenly decided to cut the FIT for households installing solar panels by more

[56] See n. 55: 21.

than half (from GBP0.433/kWh to GBP0.21/kWh) with only a few weeks' notice. The rationale for this change was that progressively reducing the burden on the taxpayer or energy consumer would also reduce the risk (as in China or Spain) that continuing premiums on tariffs would create a bubble and over-capacity, resulting in more painful industrial restructuring and job losses in the future.[57] This sudden move extended the period for households to recoup their investments from ten to eighteen years, thus reducing the incentive to install panels. Although the courts have questioned the legality of this policy shift, continuing legal appeals and challenges have prolonged the uncertainty in the market.[58] At the same time, completely rigid policies are not helpful if the financial commitments make the programme financially unviable in the long term. While policy stability is desirable, policies should also be adaptable to changing circumstances and the evidence of their impact and the maturing of the market.

Subsidies could lock in existing technologies or support incumbent firms. As the previous section argued, government intervention is justified when market failures either preclude investments in RE capacity or result in under-investment in R&D for future technologies. When the purpose of subsidies is considered – ranging from increasing energy access and generation capacity to manufacturing and export promotion – the public authorities have to balance their efforts to increase access and capacity today with supporting newer technologies in the future. Excessive support to the former tends to create market-entry barriers for newer, more innovative firms. As a result, energy subsidies do not necessarily benefit the intended population or stimulate the RE market as originally planned.[59]

Another challenge is that stakeholder participation and consultation may get ignored in the rush to implement subsidy programmes for clean energy. This is particularly important when RE projects are intended to provide energy access to far-flung communities. In the past, policies that have promoted off-grid systems have seldom factored in issues of maintenance, local capacity development, the need to ensure quality of service, etc. As a result, communities have tended to lose faith not only in the projects, but also in RE as a whole. Successful projects, by contrast, have given households a stake in managing and maintaining the systems. (In the UK example of household solar panels, consultations on FIT rates were set to conclude nearly two weeks after the revised FITs came

---

[57] Vaughan and Harvey, 'Solar subsidies to be cut by more than half'.
[58] Damian Carrington, 'Solar subsidies cuts: UK government loses court appeal', *Guardian*, 25 January 2012. See also John Vidal, 'Solar subsidy cuts legally flawed, high court rules', *Guardian*, 22 December 2011.
[59] Pokharel (2002).

into effect!) In other words, financial subsidies for RE are only one part of the ecosystem of government support needed to ensure successful uptake of new technologies at scale.

## 3  How Large Are the Subsidies and Where Do They Go?

This section will present empirical evidence from six countries to highlight variations in the scale, schedule, sector focus and rationale listed for clean energy subsidies. The purpose is not to offer a comprehensive listing of all subsidy programmes, but to illustrate the diversity of subsidies and their underlying policy premises. The six selected countries are Brazil, China, Germany, India, Spain and the United States. The last five of these are among the top five countries in the world in terms of installed RE capacity, if hydropower were not included in RE estimates.[60] If hydropower were included, Brazil would be among the leading five nations; it has also recorded high annual rates of growth in RE investment and was among the top five for new RE investment in 2010.[61]

### *Measurement Difficulties*

There is no widely accepted methodology for calculating energy subsidies or a harmonised reporting mechanism (even in such a unified market as the EU). This makes it difficult to estimate the actual level of subsidies directed to the energy sector, especially fossil-fuel industries. For instance, tax breaks and direct financial support may be calculated. But should the estimates also include the cost of environmental pollution (externalities not covered by industry), subsidised access to natural resources via state-owned property (i.e. mountains, aquifers, etc.) or military expenditures to protect oil shipping lanes?[62] It is unclear *a priori* whether some of these kinds of support measures, such as subsidised access to natural resources, would be considered subsidies under the WTO definition. But if their monetary value could be computed, the financial contribution of such support would become both more apparent and relevant to the discussion.

Likewise, precise information on global subsidies for RE is not readily available. The first problem is the lack of a common definition. As shown

---

[60] The accounting of hydropower capacity in RE estimates varies in different datasets. Since large hydro is a significant source of electricity, some estimates exclude it from RE estimates, so as not to offer a distorted impression about the development of newer renewable sources of energy (such as wind, solar, biomass, oceans and geothermal).
[61] REN21 (2011): 15.
[62] Mike Casey, 'Why we still don't know how much money goes to fossil energy', renewableenergywourl.com, 2011.

above, a broad understanding of clean energy support measures would include regulation, access to infrastructure and trade policies in addition to financial transfers and tax exemptions. Even classifying a single measure, such as a FIT, could be complicated depending on how it is designed, who bears the cost and who benefits.

The second, related, problem is that some of these additional measures are hard to compute. If transmission lines were extended only to connect a RE project, one could compute the cost of the extension and include it under infrastructure support. But if with the expansion of the grid both RE and fossil fuel-based projects benefited, how would the support for clean energy be distinguished? Similarly, an import restriction on components for RE projects might be considered a trade policy measure in favour of the domestic industry. But, if the components had multiple uses (in other industries or for fossil-fuel power as well), then the extent of support conferred only for clean energy purposes would have to be determined not by the increased tariffs, but by the proportion of the imported product whose end-use is solely in the RE sector.

Third, there are no common reporting requirements or procedures whereby information from different countries may be collected and collated. Most available statistics are estimates by reputable agencies such as the IEA and the EIA. For this study, information on clean energy investments as presented in annual reports by the United Nations Environment Programme (UNEP) and Bloomberg New Energy Finance (BNEF) is used. However, these reports do not cover all forms of government support apart from a small component of government-funded R&D and stimulus spending during the recession.

The absence of common reporting guidelines leads to the fourth challenge, namely the difficulty of comparing data across countries. Where subsidy programmes run for multiple years, the overall budget of the programme may be available, but it is difficult to find information on annual spending. Also, while concessional loans may be calculated, loan guarantees are harder to compute, unless the project fails and the guarantee is invoked by the commercial lenders. Access to infrastructure is particularly difficult to estimate. There are many claims about Chinese firms getting subsidised access to land for RE projects, but there are no reliable market rates for the land. Comparisons also matter when considering the purposes of the support measures. In the case of fossil fuels, for instance, energy subsidies as a whole in industrialised countries tend to be geared more toward production while developing countries use the subsidies to support energy consumption.[63]

[63] World Bank (2006).

Noting these challenges, the following sub-sections describe government support policies in different countries but do not make an attempt to compare the numbers across countries or over time. The country studies outline the types of subsidies, their sector focus, their rationale (whether for energy access and capacity installation or for manufacturing and exports), and their scale and timeline (where information is available). It should be noted, however, that these are not comprehensive listings of all past and current subsidies. This section lists some major policies that show the range of subsidies that have been applied at different times across major economies with significant RE capacity. Furthermore, these cases may also include some examples of subsidies for cleaner transport fuels that fall outside the scope of this chapter.

*Brazil*

Financial transfers and regulatory support in the past decade have been responsible for increasing Brazil's RE capacity and diversifying its sources of electricity. The *Luz para Todos* electrification programme was announced in 2003 with the goal of extending electricity access to 12 million people (10 million in rural areas).[64] While not solely dependent on RE, the programme did rely on distributed energy systems and isolated networks using RE sources (including 130,000 PV systems and mini-grid-based biomass projects). Funding for the programme drew mostly (more than 70 per cent) on the *Reserve Global de Reversão* (RGR) for loans and the *Conta de Deasenvolvimento Energético* (CDE) for subsidies; the remainder came from Federal states, municipalities and power supply companies, in equal measure. National funds were also used to subsidise up to 90 per cent of initial investment for regions with low electrification rates; consumers were spared the cost of network expansion.

Another major scheme was the Programme of Incentives for Alternative Electricity Sources (PROINFA), which began in 2002. In its first phase, its aim was to develop 3.3 gigawatts (GW) of RE (wind, biomass and small hydroelectricity) before 2007 with subsidies from the Energy Development Account.[65] Eletrobrás, the largest power utility, purchases electricity at pre-set preferential prices with a guarantee for a minimum 70 per cent of the contracted energy for the long-term (twenty years) PPAs and full coverage to exposure risks to short-term markets. In addition, the Brazilian National Development Bank (BNDES) provides special financing (up to 70 per cent of capital costs,

---

[64] Details about this programme are drawn from the International Energy Agency (2010c).
[65] World Resources Institute (2011).

low interest rates, amortisation for ten years, etc.) for PROINFA projects. In its second phase, the programme introduced RECs to promote further investment and source 10 per cent of electricity from renewable sources within twenty years.[66]

In addition to energy access, other objectives also underlie PROINFA. These include: job creation (150,000 jobs); attracting private sector investment in the RE sector (expected USD 2.6 billion); and support to domestic industry (with a minimum of 60 per cent of construction costs drawn on national companies).[67]

*China*

In February 2005, the 10th National People's Congress enacted a law to promote RE supply and improve the country's energy infrastructure. Additional supporting laws were enacted for the development of the RE industry in the medium and long term. These included measures for managing special capital for RE development, administrative regulations and price and cost-sharing arrangements for RE power generation.

China's RE industry has picked up since 2007, when it developed a blueprint to develop the wind, solar and biofuel sectors. The Golden Sun programme, introduced in 2009, was designed to offer national- and provincial-level subsidies to grid and off-grid solar PV projects, aiming for 500 megawatts (MW) of PV installed capacity by 2012.[68] At the national level, grid-connected projects (with minimum peak capacity of 300 kW) and off-grid projects were eligible for subsidies to cover installation costs of 50 per cent and 70 per cent, respectively (a capacity limitation of 20 MW per province was imposed). Developers had to satisfy quality standards set by the grid company, which would be certified by other agencies. In June 2011, the 50 per cent installation subsidy was replaced by a fixed tariff.[69] Eventually, 640 MW of projects were proposed, with a total investment of RMB 20 billion.[70]

The Wind Power Concession programme (2003–2007) invited domestic and international companies to bid for large-scale projects (100–200 MW).[71] Projects would be selected based on both the price/kWh as well as the share of domestic components used. The bid price was guaranteed as the FIT for the first 30,000 full load hours achieved; subsequently, the applicable return was the average local FIT on the power market at a

---

[66] International Energy Agency (2010c).   [67] International Energy Agency (2010b).
[68] Ying (2009): 8.   [69] International Energy Agency (2011c).
[70] Eric Martinot and Li Junfeng, 'Renewable energy policy update For China', Renewableenergyworld.com, July 21, 2010.
[71] REEEP (2009–2010): 19.

given time. Concessions were offered mostly for twenty-five years (shorter periods in some cases). The programme added 3.35 GW via annual competitive bidding.[72]

One of the ostensible aims of China's strategy is to increase the share of RE capacity in its energy mix: from 8 per cent currently to 15 per cent by 2020 (this is equivalent to shifting an economy like Italy to renewables). As a result, in 2010 alone, USD 54.4 billion was invested in clean energy, putting China in the lead globally for investments.[73] The National Development and Reform Commission (NDRC) expects to award about twenty projects by 2012 to reach 20 GW of installed wind capacity in 2020. In biofuels, for instance, the Agricultural Biological Mass Energy Industrial Development Programme (2007–2015), operating under the Department of Agriculture, aims to produce 23.3 billion m$^3$ of marsh gas annually and build 8,000 large- or medium-sized marsh gas projects.[74]

FITs have been used in both the wind and solar energy sectors, with visible impact. China is now the world's biggest wind farm operator. It also seeks to increase solar power capacity tenfold in five years. In 2011 it announced a FIT rate of 1 RMB (16 US cents) per KWh of power fed into the grid.[75]

The other aim, as indicated previously, is to develop a domestic manufacturing base as the foundation for RE investments. Concessions in the wind sector, for instance, have used LCRs effectively. Three projects (100–200 MW) in Jiangsu, Inner Mongolia, and Jilin provinces had 70 per cent LCRs. Two Chinese firms, Sinovel and Xinjiang Goldwind, are now among the top three wind turbine makers in the world. Wind turbine costs have fallen by two-thirds since 2007. In solar, seven of the top ten PV module manufacturers are Chinese, compared with only two in 2007. Solar panel prices fell 40 per cent in 2010–2011, again benefiting consumers throughout the world.[76]

The danger is, of course, over-capacity, threatening not only Chinese firms but also the support policies that have had significant impacts in a short period. Both solar and wind manufacturers complained of rising inventories with, in the latter case, about 27 per cent of turbines lying idle. If such trends continue, the government could either reduce subsidies

---

[72] Moerenhout, Liebert and Beaton (2012): 4. See also Martinot and Junfeng, 'Renewable energy policy update for China'.
[73] Leslie Hook and Ed Crooks, 'China's rush into renewables: the way the world turns', *Financial Times*, 29 November 2011: 10.
[74] Zhang Peidong, Yang Yanli, Shi Jin, Zheng Yonghong, Wang Lisheng and Li Xinrong (2009), 'Opportunities and challenges for renewable energy policy in China', *Renewable and Sustainable Energy Reviews*, 13(2), 2009: 442.
[75] Leslie Hook, 'Electricity: solar power feed-in tariff', *Financial Times*, 2 August 2011.
[76] Hook and Crooks, 'China's rush into renewables': 10.

for manufacturing or promote more installations at home rather than exports abroad.

In addition to FITs, LCRs and capital subsidies, regulations have helped. In December 2009, the 2005 RE law was updated to emphasise better coordination between the RE sector and the electricity sector at local and national levels. Utilities were required to purchase all RE generated and could be penalised for failing to do so. In addition, the RE fund was strengthened, allowing the Ministry of Finance to supplement the fund from general revenues.[77]

Building on the success of the RE sector in China, its 12th Five-Year Plan now aims for significant additions to generation as well as manufacturing capacity.[78] Solar installations are expected to grow tenfold in the next five years to reach the 10 GWp (gigawatt peak) target. The aim is to have one or two firms with a polysilicon production capacity of 50,000 tonnes and another two to four firms with 10,000 tonnes capacity, as well as one to two firms with 5 GW cell production capacity and eight to ten firms with 1 GW capacity.[79] A target of 90 GWp of installed capacity has been set for the wind sector, including six large onshore and two offshore bases (with a 70 GW capacity addition).[80] There are also plans to support the growing RE capacity with smart grids and regional power transmission channels. The planned investment in grid capacity extension alone is USD 400 billion during the plan period.

*Germany*

The share of RE in total electricity consumption in Germany has risen consistently over the past decade (Figure 4.3). This was, in part, because the country was the first to introduce FITs to encourage RE investments under the Renewable Energy Sources Act (EEG) in 2000. The EEG's core principles were priority access for RE to the grid, giving priority to transmission and distribution; equalisation of additional costs between all RE suppliers and grid operators; and financial support via the FIT but subject to ratcheting down the subsidy periodically (degression).[81] The EEG was revised for the period 2004–2008 to increase targets for the share of RE in total electricity supply (12.5 per cent by 2010 and 20 per cent by 2020).[82] The EEG

---

[77] Martinot and Junfeng, 'Renewable energy policy update For China'.
[78] China Direct (2011): 13.
[79] Chihheng Liang, 'The role of PV in China's energy policy through 2015', *Digitimes Research*, 11 October 2011.
[80] Seligsohn and Hsu (2011). [81] European Renewable Energy Council (2009a): 5.
[82] Mez (2005): 7.

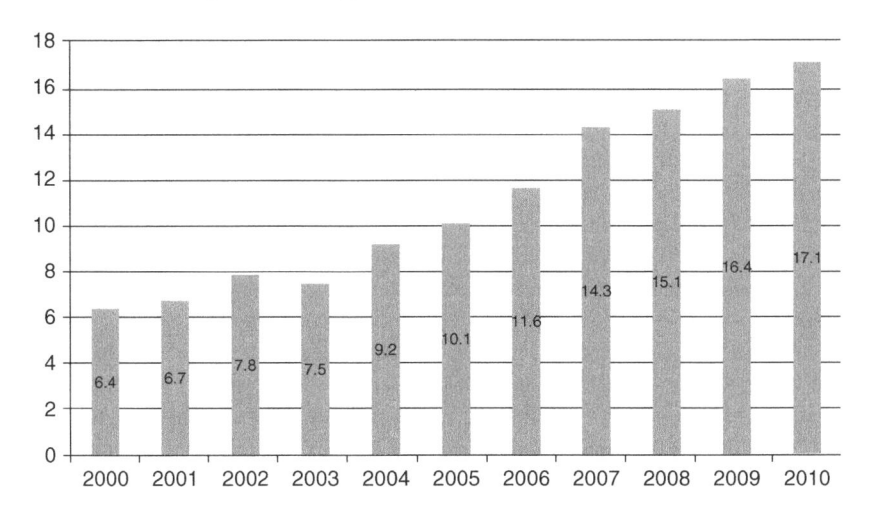

Figure 4.3: RE's Share in Germany's Electricity Consumption has Risen Steadily (per cent)
Source: Federal Ministry for the Environment, Nature Conservation and Nuclear Safety, *Development of Renewable Energy Sources in Germany 2010*, 2011.

also obliged grid operators to purchase and transmit all electricity generated from renewable sources but split the costs: project developers would bear the cost of connecting to the grid while grid operators would pay to upgrade the infrastructure. In addition, the Act sought nationwide equalisation of electricity volumes and FITs so as to avoid arbitrage in purchase contracts. The details had to be made publicly available.

Further revisions to the EEG came into effect in 2009, 2010 and again in 2012. The aim was to offer attractive FITs to encourage both onshore and offshore investments. RE is now expected to account for 35 per cent of electricity generation by 2020, 50 per cent by 2030, and 80 per cent by 2050. Onshore wind FITs are currently at EUR 8.93 cents/kWh (with tariff reduction at 1.5 per cent per year). Offshore wind projects can expect EUR 15 cents/kWh until 2018 (previously it was only up to 2015). In the hydropower, solar, geothermal and biomass sectors, FITs vary by size of project, reducing for larger-sized projects (tariff structures have been simplified since 2009).[83]

---

[83] International Energy Agency (2012).

The costs for the FITs were passed on to consumers, a cumulative burden of EUR 85.4 billion during 2000–2010.[84] With economies of scale and technological improvements, the need to maintain high FITs has also diminished. Thus, the grid regulator plans to reduce solar FITs by 15 per cent from early 2012.[85] Even then, for projects commissioned during 2004–2008, the FIT burden would be an additional EUR 122.3 billion until 2030.

Generous FIT schemes made Germany, a country with one-third the average solar irradiation of India, a renewable energy powerhouse in terms of installed capacity. It added 7.4 GW of solar capacity in 2010 alone. In addition to solar heat and power, Germany is also among the top five in the world for installed wind and biomass power capacity.[86] The manufacturing industry, however, faces competition from cheaper Chinese supplies, although the German Renewable Energy Federation (BEE) maintains that its competitive advantage lies in the quality of its products.

In addition to the EEG, Germany introduced a Renewable Energies Heat Act (EEWärmeG) in 2009. The Act aims to increase the share of RE for heat production to 14 per cent by 2020 and makes the use of RE for space and water heating mandatory for new buildings. The share of RE depends on the technologies deployed (solar thermal installations, 15 per cent; gaseous biomass, 30 percent; other biomass installations, liquid biomass and geothermal heat, 50 per cent). Funding to support the transition in existing buildings increased from EUR 130 million in 2005 to EUR 350 million in 2008, further rising to EUR 500 million from 2009.[87]

*India*

India has a long history of promoting RE. It was one of the first countries to establish a dedicated Ministry of Non-Conventional Energy Sources, now renamed the Ministry of New and Renewable Energy (MNRE). The Indian Renewable Energy Development Agency (IREDA) was incorporated in 1987. However, the trajectory shifted from 2003, when the Electricity Act liberalised the power market and called on each state's Electricity Regulatory Commission (SERC) to institutionalise the minimum purchase of RE electricity or RPOs; non-compliance would invite penalties. Furthermore, 'open access' provisions allowed RE power generators to access distribution systems and transmission lines for a nominal fee.[88]

---

[84] Vaughan and Harvey, 'Solar subsidies to be cut by more than half'.  [85] *Ibid.*
[86] See n. 61.    [87] International Energy Agency (2011d).
[88] Meisen, Quéneudec, Avinash and Timbadiya (2010): 22.

The objectives of the Electricity Act were further strengthened in 2006 with a national tariff policy, under which SERCs had to specify RPOs with distribution companies and set time limits for implementation. The policy also introduced preferential tariffs and competitive bidding to select projects. By April 2010, eighteen states had established or drafted RPOs for 1–15 per cent of total electricity generation.[89]

In December 2009, the government introduced generation-based incentives (GBIs) for wind power projects with a minimum capacity of 5 MW. The GBI (1 US cent/kWh against the average wind power price of 6 US cents/kWh) was an additional incentive for the SERCs' preferential tariffs. The overall programme budget was INR 3.8 billion (USD 81.6 million).[90]

FITs have also been used for small hydro and biomass projects. A small hydro scheme for 2009–2010 offered capital subsidies for new plants with capacity ranging from 100 kW to 25 MW. For 100–1000 kW projects, the subsidy was USD 500/kW, but it was double that amount for 'special category' and northeast states. For 1–25 MW plants, capital subsidies amount to USD 1 million for the first MW and an additional USD 100,000 for each extra MW for the targeted states.[91]

Perhaps the most exciting developments have been in the solar sector. Since it was launched in 2010, the JNNSM has been a key part of the National Action Plan on Climate Change. The JNNSM offers attractive FITs after projects have been selected by a reverse auctioning process. The overall objective is to install 22 GW of solar power (grid and off-grid) using both PV and concentrated solar power (CSP) technologies by 2022. In the first phase (2010–2012), targeting 1000 MW, the Central Electricity Regulatory Commission (CERC) announced twenty-five year-long PPAs with FITs of 36 US cents/kWh for PV projects and 31 US cents/kWh for CSP ones. The rates are expected to be revised downward for each batch of auctioning (the lowest bids had already dropped to about 15 US cents/kWh by end-2011).[92] For the mission's first phase, LCR rules are applied to solar PV modules and cells, although solar thermal projects are spared (these provisions are not present in state-level solar missions in India). Annual module production is expected to exceed 2500 MW by 2015 while JNNSM targets 4000–5000 MW of capacity by 2022 (or one-fourth of the expected installed capacity by that date).[93]

---

[89] Arora et al. (2010).
[90] Sunil Raghu, 'India announces incentives for wind power generation', *Wall Street Journal*, December 17, 2009.
[91] See n. 89: 61.   [92] See n. 55.   [93] Sutrave (2010). See also Arora et al. (2010): 43.

Notwithstanding the partial requirements for local manufacturing, India's policies are primarily aimed at increasing generation capacity. India's target is that, by 2012, 10 per cent of all new capacity in the power sector will be in renewables. With energy access for millions of potential consumers currently not connected to any modern source of electricity a key political imperative, it is unlikely that LCR requirements will be a reason to hold back on rapid deployment of RE capacity.

*Spain*

Spain has been promoting RE in several sectors, with a succession of laws passed since the mid 1990s. More recently, the Sustainable Economy Act set the goal of meeting the European target of a 20 per cent share for RE in total energy consumption. The Act aims to stimulate R&D and innovative projects and offers tax deductions as incentives.[94]

Another ACT (RD 2818/1998), passed in December 1998, was significant because it allowed RE producers to feed all power generated into the grid and receive a premium over the wholesale price. The premiums were updated annually. In March 2004, a new Act (RD 436/2004) made the system more predictable by publishing prices and premiums in advance as a fixed percentage of the average electricity price (the premium for RE over the average electricity tariff was 40 per cent, but for solar it was 250 per cent).[95]

Regional governments have also had a role. The government of Navarre introduced an Energy Plan in 1995 to support energy efficiency and investments in RE capacity. From 2002, these measures were supported by a dedicated training centre and funding support lines for biomass, solar PV and thermal and wind.[96] In addition to promoting RE and diversifying sources of energy, job creation was a key objective of the programme.[97] The Andalusian government's Programme for the Promotion of Solar Thermal Energy Installations (PROSOL) offered direct grants and subsidised loans to install projects. In 2007, the grant component was 40 per cent of the cost of solar panels. This was reduced to 30 per cent in 2009, but wind energy was included among the beneficiary sectors.[98]

Numerous laws have also helped increase the use of RE in buildings. At the Federal level, these have included upgrading building codes and

---

[94] International Energy Agency (2010d).  [95] Ragwitz and Huber (2005): 15.
[96] Renewable Energy for Europe (2003).
[97] European Renewable Energy Council (2004),
[98] Edmund Sykes, 'Photovoltaic (PV) solar & wind electricity generation in Spain – a personal experience', *Ezine Articles*, 2010.

Governing Clean Energy Subsidies 173

certifications for the energy performance of buildings. The government also provided EUR 1 billion (USD 1.28 billion) for refurbishing buildings during 2008–2012. Regional governments in Catalonia, Extremadura, Madrid, Murcia, Navarre and Valencia have promoted the use of solar PV, solar thermal and biomass in buildings.[99]

The biofuels sector has received support in the form of regulations for the mandatory commercialisation of biofuels. The National Energy Commission (*Comisión Nacional de la Energía*) was tasked with issuing biofuel certificates, managing the certification process and monitoring compliance (ITC/2877/2008). Minimum targets for biofuel use were set at 2.5 per cent for biodiesel and bioethanol by 2009 and 3.9 per cent by 2010. By 2009, production capacity of biofuels had reached 4 million tonnes of oil equivalent (mtoe).[100] The government has also supported the sector by providing subsidies to biodiesel and bioethanol R&D projects, often involving several Spanish companies. For biomass, the government recognised that imported products were sharply driving down prices and adversely impacting domestic industry. As a consequence, it launched a strategy to develop forest residue biomass and linked it to the development of rural areas.[101]

For energy efficiency, Spain developed an action plan for 2008–2012 involving EUR 2.4 billion of public investment.[102] The expectation was to stimulate private investments worth EUR 22.2 billion, with a focus on efficiencies in the transport, buildings and industrial sectors.

Finally, the Renewable Energy Plan (REP), which ran from August 2005 to December 2010 (building on an earlier programme in place since 2000), set the target of RE at 30 per cent of electricity consumption and 12 per cent of primary energy consumption by 2010. The rationale for the policy was to reduce dependence on oil imports, meet international commitments under the Kyoto Protocol and to phase out nuclear power. Thus, wind power targets were raised from 13 GW to 20.15 GW by 2010. For solar, the target was 1200 MW by 2010 (up from 400 MW by 2007). In order to meet these goals, EUR 23.6 billion was budgeted (mostly from private sources while EUR 4.956 billion was set aside for FITs.[103] Solar thermal projects were awarded investment subsidies to cover 37 per cent (in some cases 50 per cent) of total project cost.[104]

A special regime for FITs was introduced in 2007 (Royal Decree 667/2007) to cover RE facilities up to 100 MW in capacity. Up to 50 MW, project operators could either choose a FIT or a feed-in premium

---

[99] Gobierno de España (2010): 30–31.   [100] *Op. cit.*: 18.   [101] See n. 99.
[102] Instituto para la Diversificación y Ahorro de la Energía (2011).
[103] International Energy Agency and International Renewable Energy Agency (2010).
[104] European Renewable Energy Council (2009b): 4.

over the market price. FITs were guaranteed for twenty-five years for PV, tidal and small-hydro projects, for twenty years for wind and geothermal and for fifteen years for biomass projects. For projects between 50 MW and 100 MW, a bonus was promised for the electricity generated. The exception was solar PV, which was not subject to the cap on project size to qualify for the FIT benefits. In 2009 the cap was raised to 500 MW for PV projects, although the FIT was reduced to 32 cents/kWh for ground-mounted projects and 34 cents/kWh for rooftop systems (earlier, the tariff level could go up to EUR 0.44/kWh).[105]

Unlike Germany, however, Spain's policies to support RE projects are more threatened in the long term. When it launched its FIT scheme, the aim was to install 400 MW of RE capacity by 2010. By September 2008, Spain had installed 344 MW with an expected FIT cost burden of EUR 53 billion over 25 years (75 per cent more than the cost of conventional power).[106] The economic crisis, rising public debt and competition from cheaper imports have made Spain's support policies unviable. Thus, in December 2010, FITs for solar PV were cut retroactively, partly to discourage speculative investors who had little experience in RE (this resulted in a 'crisis of confidence' for existing industry players).[107] FITs for wind projects were competitive with fossil fuels by 2010. The unsustainability of generous support measures over the long term constitutes a policy risk, as described above. As RE projects need longer recovery periods, owing to the high upfront cost of capital, the policies and cost burdens have to be calibrated from the early stages as well.

Moreover, the purpose of the subsidy also matters. In 2010 the government estimated that with all the efficiency measures and policies to promote RE capacity at home, Spain would have surplus energy to sell into the wider European market. To that extent, Spain's policies were also geared toward promoting RE exports. However, with the persisting economic crisis in Europe, there is no guarantee that domestic RE capacity will find the markets to justify the subsidies received by local producers. In fact, even for 2020, with a planned 40 per cent RE share in electricity generation, Spain's policies are contingent on grid connections to Central European markets.[108]

---

[105] International Energy Agency (2011b).
[106] Jeffrey Ball, 'Clean energy sources: sun, wind and subsidies', *Wall Street Journal*, January 15, 2010.
[107] Peter Wise, 'Renewable energy: subsidy cuts cause crisis of confidence', *Financial Times*, 5 October 2011: SR1-4.
[108] Gobierno de España (2010).

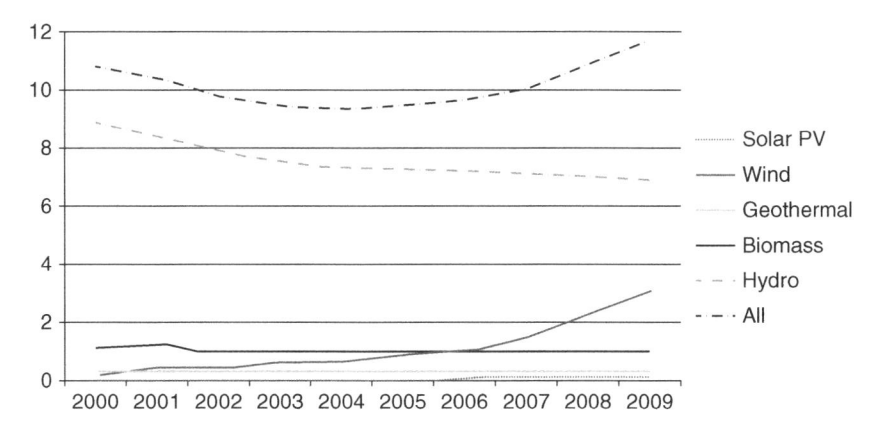

Figure 4.4: RE's Share in Total Electricity Capacity in the United States (per cent)
Source: Gelman (2010): 24.

*United States*

The United States has been promoting RE for more than two decades. If hydropower is included, the Department of Energy (DOE) estimates that RE's share in total electricity capacity is about 12 per cent, although the share of RE in total energy production is lower (8.7 per cent in 2000; 10.6 per cent in 2010) (see Figure 4.4).

The Federal Energy Policy Act 1992 established the Renewable Energy Production Incentive (REPI) to offer support for RE projects using solar, wind, geothermal, biomass, landfill gas, livestock methane and ocean resources. A FIT rate of 2 US cents/KWh applied for the first ten years of a project's operation. REPI was reauthorised in 2006, subject to the availability of annual appropriations.[109]

More recently, the DOE's SunShot Initiative has focused on leveraging the potential of solar power in the United States (the highest in the industrialised world) to reduce the costs of utility scale solar installations by 75 per cent (approximately, 6 US cents/kWh) to make them cost competitive with other sources of energy. In this way, solar power could account for 15–18 per cent of US electricity generation by 2030. The objective is also to 're-establish American technological leadership, strengthen US economic competitiveness in the global clean energy race, and lead to America's secure energy future'.[110] Thus, the initiative has a strong manufacturing component, with emphasis on R&D to reduce PV

[109] Campbell (2010): 20.   [110] US Department of Energy (2011h).

module costs and increasing PV manufacturing at home. The DOE has invested USD 60 million since 2007 for a PV incubator programme to attract over USD 1.3 billion of private investment. Smaller grants support non-hardware concepts, while larger grants (up to USD 5 million) – with 50 per cent cost shares over eighteen months – are used to demonstrate and deploy technologies at scale.[111]

Specific policies for large-scale wind projects are also available. A Production Tax Credit (PTC) was introduced in 1992. Currently, a tax credit of 21 US cents/kWh is available for electricity generated from utility scale wind turbines under the PTC. In order to stimulate investment during the recession, the American Recovery and Reinvestment Act (ARRA) of 2009 extended the credit, giving projects coming online during 2009–2012 the choice between an investment tax credit of 30 per cent or a 30 per cent grant.[112]

Some programmes have had an explicit jobs creation focus, even if the results are mostly in the form of political pay-outs. The Volumetric Ethanol Excise Tax Credit (VEETC) was set up by the American Jobs Creation Act, 2004; running until December 2011, the policy gave eligible blenders or retailers a 45 US cent/gallon tax credit on pure ethanol blended with gasoline.[113] In 2006, that policy alone accounted for 54.6 per cent (USD 2.6 billion) of the Federal ethanol subsidy and 41.6 per cent of total RE subsidies.[114] Ethanol production has also received large subsidies from the Department of Agriculture, accounting for up to 20 per cent of the corn harvest in 2006.[115] In August 2011, a new USD 510 million initiative was launched to promote next-generation biofuels.[116] In addition, a rule proposed by the US Department of Agriculture (but not yet implemented) suggests that production subsidies might be available only to bio-refineries with at least 51 per cent US ownership. If implemented, this would be a discriminatory measure and potentially disputed by trade partners like Brazil, Canada, China, Germany and India.[117]

State-specific programmes have been equally important in the United States. Connecticut combined a zero emissions REC scheme with a reverse auction to select companies that needed the least subsidy. This was combined with fifteen-year-long contracts to guarantee policy certainty.[118]

---

[111] US Department of Energy, 'Energy Department announces $7 million to reduce non-hardware costs of solar energy systems', Energy.Gov, November 15, 2011.
[112] See Campbell (2010): 15–16. [113] US Department of Energy (2011j).
[114] Texas Comptroller of Public Accounts (2011). [115] *Ibid.*
[116] Tildy Bayar, 'US biofuel industry prepares for life without subsidies', Renewable Energy World.com, September 9, 2011.
[117] Global Trade Alert (2010).
[118] Murray, 'Carrot and stick approach supports but can distort'.

The California Solar Initiative (CSI), launched in 2009, provides financial incentives with a total budget of USD 3.2 billion over ten years, of which USD 216 million is devoted to increasing solar PV access for low-income households.[119] Another CSI programme for solar water heating has a budget of USD 350 million to offer rebates to replace natural gas and electric heaters.[120] The California Energy Commission's New Solar Homes Partnership works with builders to encourage uptake of solar installations in new construction. With a budget of USD 400 million over ten years, it hopes to install 400 MW of solar technologies. Subsidies for standard housing (USD 2.25/W) are raised in the case of affordable housing projects (USD 3.15/W).[121] Californian utilities also use FITs over multi-year contracts (ten, fifteen or twenty years). Until 2016, property taxes are also exempted for up to 75–100 per cent of the value of solar energy systems.[122]

Colorado has used both concessional loans and tax exemptions to promote RE projects. In 2006 Boulder offered a tax rebate of about 15 per cent on the sale and use of solar installations (amounting to 1 per cent of the average system cost or USD 50,000 for 500 kW PV installations).[123] The state also used USD13 million of ARRA funds in 2009 to offer loans for RE and energy efficiency projects.[124]

In Texas, the LoanSTAR revolving loan programme was initiated by the state's Energy Office in 1988. Approved by the DOE, by 2007 USD 240 million had been given as low-interest loans. Much of the focus has been on energy saving, emissions reduction and building retrofits.[125] Loan maturity periods were also increased from four to eight years in the mid 1990s and then again to ten years in 2002.

The development of wind energy in Texas showcases how a combination of support measures for more than a decade have made the state host the largest wind power capacity in the United States. Texas has a potential wind power potential of 524,800 MW (or 493 per cent of current electricity consumption).[126] By end 2010, 10085 MW of wind capacity had been already installed. The key trigger for this development was the 1999 Texas Renewable Portfolio Standard (RPS). The mandated use of RE was raised periodically, resulting in USD 1 billion of investments in wind power; the ten-year goal for the policy was met within six years.[127] The RPS was complemented by a federally supported production tax credit, offering 1.5 cents/kWh of electricity generated, adjusted for

[119] US Department of Energy (2011a).   [120] US Department of Energy (2011b).
[121] US Department of Energy (2011b).   [122] US Department of Energy (2011g).
[123] US Department of Energy (2011d)   [124] US Department of Energy (2011d).
[125] Texas State Energy Conservation Office (2011a).   [126] Infinite Power (2011).
[127] Texas State Energy Conservation Office (2011b).

inflation.[128] The Texas Tax Code also offers 100 per cent property tax exemptions on the value of on-site solar, wind or biomass power-generating devices.[129]

Nevertheless, subsidy programmes have also resulted in perverse outcomes thanks to over-capacity and the viability of grid connections. In Texas, an investment of USD 3 billion in grid capacity to connect wind projects in dispersed locations to urban consumers was found to be viable only if coal power projects were also included in some of the transmission line corridors.[130]

## 4  What Role for Trade Rules in Governing Clean Energy Subsidies?

In light of the country examples discussed above, this section will look at how the WTO's rules on subsidies and exemptions apply to disputes on clean energy subsidies. It will examine how vulnerable some of the policies and measures might be to legal disputes at the WTO. It will, then, revisit the typology of subsidies set out earlier to present an alternative framework that could reconcile the tensions between ensuring that domestic policies are not discriminatory against trading partners and the different motives for supporting clean energy.

### Subsidies at the WTO

According to the SCM Agreement, a 'financial contribution' from the government is considered a subsidy if it confers a 'benefit' on the recipient (Article 1). Financial contributions include: direct transfer of funds; when the government forgoes revenue; when it provides goods or services; or if the government delegates a programme to a private body that it would normally have followed. A benefit is conferred only if the government's contribution is more favourable than what would be available to the RE project developer or manufacturer in the open market.

This broad treatment of subsidies would cover many of the measures noted in the typology developed in this chapter, namely direct financial transfers; preferential tax treatment; regulations (in the form of government procurement policies or the establishment of special economic zone, SEZs); infrastructure support (such as subsidising access to the grid or offering land at below-market prices) and some forms of trade

---

[128] Texas State Energy Conservation Office (2011c).
[129] Texas State Energy Conservation Office (2012).
[130] Jeffrey Leonard, 'Get the energy sector off the dole', *Washington Monthly*, January–February 2011.

restrictive policies, such as LCRs (since these are prohibited; see below). But for the SCM Agreement to apply, the measure must also confer a benefit. So, if the subsidies simply cover some of the costs of acquiring RE systems or setting up plants in remote locations, they might not necessarily be treated as a benefit.[131] In these situations, it could be argued that government support is only a compensation to encourage actions, either through climate mitigation or energy access measures that may not have occurred otherwise. Further, it might be preferable to measure net subsidy (the difference between the gross cost to the government funds allocated for the subsidy and the revenue resulting from the measure) to get a more accurate assessment of the conferment of benefit. But the SCM Agreement does not specify this.

The SCM Agreement further classifies subsidies as either prohibited (Article 3) or actionable (Article 5).

*Prohibited Subsidies*

The SCM Agreement, Article 3, prohibits export subsidies and measures favouring domestic over imported goods, i.e. local content requirements. These subsidies are assumed to be damaging to other countries and, therefore, must be 'withdrawn without delay' (Article 4.8).

One dispute that has emerged is over the LCR provisions in Ontario's Green Energy and Green Economy Act 2009. The Act requires that 60 per cent of materials in RE projects be locally sourced. Japan initiated consultations at the WTO's Dispute Settlement Mechanism (DSM) in 2010, and the EU followed in August 2011.[132] The EU has claimed that, as a 'significant' exporter of wind power and solar PV equipment (EUR 300-600 million during 2007–2009), it was harmed by Ontario's LCRs. In response, the Ontario Energy Minister defended the provisions on the grounds that they were necessary to create jobs:[133] '[W]e will [stand up] against anybody outside of Ontario that wants to threaten our efforts to create jobs.' In other words, LCR provisions remain contentious as they have as much to do with generating employment as with RE capacity installation.

---

[131] Howse and Eliason (2009): 265.
[132] See www.wto.org/english/tratop_e/dispu_e/cases_e/ds412_e.htm for details about the dispute. Only Germany's FIT programme had been disputed before this case although the European Court of Justice (ECJ) had ruled it legal in 2001.
[133] International Centre for Trade and Sustainable Development, 'EU joins Japan in contesting Ontario renewable energy plan', *Bridges Trade BioRes*, 11(15), 5 September 2011.

### Actionable Subsidies

Actionable subsidies, although not prohibited, may be challenged either through the DSM or through countervailing action, if they cause 'adverse effects to the interests of another Member'.[134] Actionable subsidies may be contested if they are 'specific', that is directed toward a particular industry (Article 2). If a specific subsidy causes adverse effects for foreign firms, it is actionable under the WTO (Article 5). Even if the subsidy is actionable, the complaining party has to prove the harm caused (the subsidy is not automatically considered illegal). Unlike prohibited subsidies, therefore, the violating country has to only remove the adverse effects of the measure rather than the measure itself (Article 7.8).[135]

### Adverse Effects

Assessing adverse effects means, first, determining that the financial contribution confers a benefit and, second, that there is a connection between a particular policy and subsequent commercial losses. As noted earlier, if the subsidy recipient enjoys contributions more favourable than were otherwise available, then the measure is considered a benefit.[136] Establishing whether harm was caused will, of course, vary case by case.

The US solar manufacturing industry was jolted in August 2011 when three firms – Solyndra, Evergreen Solar and Spectra Watt – filed for bankruptcy. The industry, having faced stiff competition from Chinese suppliers, claims that subsidies unfairly promote Chinese firms while eroding market share for other firms. (China produces 60 per cent of all solar panels but exports 95 per cent of its production. The US solar market alone is worth USD 6 billion annually, and China had exported USD 1.6 billion of panels to the country during January–August 2011.) Rising competition, on the one hand, and lack of clarity on subsidy impacts, on the other, can result in distorted arguments. US Senator Wyden from Oregon argued that if the demand for solar products was rising but American production was continuing to fall, China must be violating trade rules, a claim that does not necessarily follow but has political traction.

Consequently, in October 2011, seven US-based solar panel manufacturers (led by German subsidiary, SolarWorld) filed a case with the US Commerce Department to protest against Chinese solar subsidies

---

[134] 'Adverse effects' include: injury to the domestic industry of another Member; nullification or impairment of benefits accruing directly or indirectly to other Members under the GATT; and/or serious prejudice to the interests of another Member.
[135] Wilke (2011): 7.   [136] *Ibid.*; 9.

and demand countervailing tariffs of more than 100 per cent of the price of Chinese panels.[137] By December, the US International Trade Commission (USITC) had unanimously decided that there was evidence that US firms had indeed been injured by subsidised Chinese imports, raising the trade dispute to a higher level that could lead to countervailing and anti-dumping duties.[138]

But in March 2012, the US Department of Commerce's preliminary findings suggested the imposition of low tariffs as countervailing duties (CVD): 4.73 per cent on imports from Trina Solar, 2.9 per cent from Suntech and 3.59 per cent from all other remaining Chinese manufacturers. On May 17, the Commerce Department also imposed higher anti-dumping (AD) duties: 31.14 per cent on panels from Suntech, 31.22 on panels from Trina Solar, 31.18 per cent on all other companies that requested individual duty determinations and nearly 250 per cent to all other Chinese manufacturers, including state-controlled companies. These duties were applied ninety days' retroactively.

In retaliation, China announced its own investigation into US subsidies for solar, wind and hydroelectric industries.[139] The China Chamber of Commerce for Import and Export of Machinery and Electronic Products, the Chinese Renewable Energy Industries Association and its National Development and Reform Commission have contested that Chinese firms benefit from economies of scale while US subsidies to its own firms are much larger, often in the hundreds of millions of dollars.[140]

Also in October 2011, the US notified nearly 200 Chinese subsidy programmes to the WTO SCM Committee, alleging that they violated trade rules. At a minimum, the Americans claimed that no updates of Chinese subsidies had been provided since 2001 while notifications were due every two years.[141]

The mutual dependence of the two countries in solar industry trade could either escalate or mitigate an outright trade war. The US exports more than USD 800 million of polysilicon, a key ingredient in solar panels, to China annually. While some Chinese firms have already demanded anti-dumping measures against these imports from the United States, firms dependent on Chinese components within the latter

---

[137] Keith Bradsher, '7 US solar panel makers file case accusing China of violating trade rules', *New York Times*, October 20, 2011: B1.
[138] Keith Bradsher, 'Trade war in solar takes shape': B1.
[139] Hook 'China to probe US clean energy subsidies'.
[140] *Xinhua*, 'China concerned over US ruling on solar panel probe', *China Daily*, 4 December 2011.
[141] USTR (2011). The US Trade Representative also notified 50 Indian subsidy programmes.

have formed a Coalition for Affordable Solar Energy (CASE) to oppose trade disputes with China.[142]

It should also be noted that subsidies for production are not prohibited outright, unless they have adverse effects on foreign firms in the domestic or other markets. US federal solar subsidies amounted to USD 1,134 billion in 2010, when it was only USD 179 million in 2009. But, the United States claims that, since much of this production is for domestic use rather than exports (unlike China), the production subsidies should not violate WTO rules.[143] In the case of bioethanol exports, however, in October 2011 the European Commission initiated anti-dumping and countervailing duty investigations against US exports. This was due to a complaint from the European Producers Union of Renewable Ethanol Association (ePURE) that federal production subsidies to US firms had allowed them to export to the EU market, which had had an adverse impact on the European industry.[144]

However, not all subsidy-related concerns have resulted in legal disputes.[145] In October 2009, at the 20th meeting of the United States–China Joint Commission on Commerce and Trade, China agreed to remove a provision for 70 per cent LCR for wind projects (Brazil, India and Germany had also been affected by this measure). In June 2011, China terminated its policy of giving USD 6.7–22.5 million to wind turbine makers who substituted Chinese components for foreign ones. Some measures, of course, take a long time to be corrected. China had imposed anti-dumping duties on Korean polyester films (the main component in solar panels) in August 2000, a policy that was stopped only in December 2010.

*Seeking Exceptions*

Although the SCM Agreement restricts certain subsidies, the GATT 1994 allows for certain exceptions to the rules under Article XX, which is particularly important in the case of measures related to the environment. It allows restrictions on trade to 'protect human, animal or plant life or health' (Article XX(b)) and for the 'conservation of exhaustible natural resources' (Article XX(g)), as long as the measures are applied in a non-discriminatory manner. It must be noted, however, that Article XX

---

[142] Hook and Crooks, 'China's rush into renewables': 10.
[143] Bradsher, '7 US solar panel makers file case': B1.
[144] Global Trade Alert (2011). The EC had also applied anti-dumping duties on US exports of biodiesel in 2009, which were extended in May 2011.
[145] For more information on shifting trade policies and their implications, see www.globaltradealert.org/.

exceptions apply only to the GATT 1994 and not to other WTO Agreements, such as on SCM, intellectual property (TRIPS) or investment measures (TRIMS). There is no clear ruling on whether Article XX may be invoked to justify FITs and other forms of government support directed at clean energy.[146] With a growing number of disputes surrounding subsidies as well as disputes related to process and production methods, some have called for widely applying the Article XX exceptions, although there is no certainty how these justifications would be treated by the WTO dispute settlement panels.[147]

But, even if the Article XX exceptions were applied to clean energy access and generation, it is unlikely to justify exceptions for the primary purpose of equipment manufacture. Some might argue for supporting domestic equipment manufacturing on the grounds that in the long run it could result in lower costs and thus help the environment. But this depends on whether the subsidy is successful, for which there is no guarantee. It is difficult to envisage that a WTO panel would admit such an argument for manufacturing subsidies (as they will need to be linked to the objectives of the GATT, Article XX), and any country could claim such an exception to afford incentives to its domestic manufacturing sector.

In the case of subsidies aimed at increasing clean energy access, these measures could cause adverse effects for goods and services based on their *design* (and also perhaps for cross-border electricity exports) but, even so, it may be easier to provide an environmental justification here rather than for manufacturing subsidies. Once again, the motivation of the subsidy programme matters.

The SCM Agreement, Article 8, included a specific list of non-actionable subsidies, including those for R&D and environmental protection. However, this provision lapsed in 2008, and a new list has not been agreed to date. Nevertheless, the precedent suggests that it is indeed possible to add more specificity within the WTO SCM Agreement to give more clarity on the use of subsidies to promote clean energy.[148]

Existing energy markets have distribution and retail networks that favour fossil fuels.[149] Thus, as argued previously, clean energy subsidies for addressing climate change are interventions in markets that are otherwise distorted by subsidies to fossil fuels. If developing countries use clean energy support measures to bolster efforts at climate mitigation or increase energy access, the threat of trade sanctions for the use of such subsidies is inimical to climate goals.[150] This is one reason that developing countries have opposed linking trade and climate issues.

---

[146] Wilke (2011): 20.   [147] Condon (2009): 903–906; Howse (2010): 17–18.
[148] See n. 145: 21.   [149] Howse (2010): 6.   [150] Jha (2011): 15.

Another critical issue is the scale of the subsidy, and whether it is used as a transitional measure. In India, low-carbon technology is targeted primarily for the purposes of addressing energy scarcity and providing access to unconnected households. In fact, rather than emphasise local manufacturing, India has focused on building solar capacity, which is why, in January 2011, it reduced import tariffs for components used to install solar projects to 5 per cent. By contrast, China's low-carbon technology is driven by 'a desire to become world leader in clean energy'.[151] Thus, if subsidies are used to boost domestic manufacturing capacity, it is important to assess whether it is primarily for creating jobs and promoting exports, or whether it is to boost energy access and a transition to a low-carbon pathway. But, as argued above, existing provisions in the SCM Agreement do not explicitly allow for exceptions on environmental grounds. If such exceptions were to be introduced in a new Agreement (see the case for a SETA below), it is unlikely that a job-creation imperative will suffice to permit subsidy measures, even if the jobs were in clean energy sectors. Such clarifications on the justification and end-use of subsidy measures are currently missing in existing legal provisions.

While debates continue at the multilateral level, regional agreements have also established some links between energy, trade and the environment. Ontario's FIT programme is already under threat of challenge under the North American Free Trade Agreement (NAFTA). The Texas-based Mesa Power Group claimed that the policy violated NAFTA's government procurement provisions, which are stiffer than those of the WTO.[152] But NAFTA also allows states to take measures to ensure that investments are 'sensitive to environmental concerns' (Article 1114(1)). Both the Energy Charter Treaty and the Economic Community of West African States (ECOWAS) Energy Protocol, while recognising state sovereignty over energy resources, expect parties to minimise any harmful environmental impacts of energy-related activities (Articles 18(1) and 19(1)). In addition, in 1994 the Energy Charter Treaty (ECT) introduced an associated Protocol on Energy Efficiency and Related Environmental Aspects (PEEREA), which lets a country hosting foreign investors and traders apply environmental conditions on grounds of 'necessity'. It can do so to protect its environment even if there is uncertainty about the future (say, with regard to the impacts of climate change)[153] (see Ghosh 2011a: 114). These provisions could serve as

[151] Levi et al. (2010): 9.
[152] International Centre for Trade and Sustainable Development, 'EU joins Japan in contesting Ontario renewable energy plan', *Bridges Trade BioRes*, 2 September 2011.
[153] Ghosh (2011a): 114.

Governing Clean Energy Subsidies 185

a basis to subsidise domestic firms under regional agreements (but not under the WTO) in order to promote clean energy (and protect the environment from foreign entities with lower environmental standards). Eventually, disputes will have to be adjudicated both on the basis of adverse trade impacts and the purpose of the measure. One could even argue that promoting costlier local manufacture could slow down the process of meeting environmental goals at a lower cost and sooner. But, there could also be a case that a low-cost and reliable domestic manufacturing base might be needed if a large clean energy infrastructure base is to be sustainable and well maintained over the long term.[154]

### *A New Framework for Trade Rules on Clean Energy Subsidies*

The country cases, emerging disputes and lack of clarity with respect to exceptions to WTO rules underscore the tension between maintaining non-discriminatory trade practices (a primary objective of the trade regime) while also promoting greater and faster adoption of clean energy (a key strategy in response to climate change as well as deficits in energy access) (see Figure 4.5).

Figure 4.5 captures the tension between maintaining non-discriminatory trade practices, on the one hand, while promoting greater and faster adoption of clean energy, on the other. Along the horizontal axis is essentially the concern of the trade regime (in particular the WTO SCM Agreement, but the GATT 1994 and other free-trade agreements (FTAs) would also be applicable). Firms in the clean energy sector are worried about the subsidies offered in other countries that are expected or not expected to have an adverse impact on their commercial interests. The vertical axis depicts the spectrum of rationales driving government support to clean energy. As discussed before, these include providing energy access, building clean energy generation capacity, creating a domestic manufacturing base, or promoting exports of clean energy products and services.

Subsidies marked in black boxes are clearly prohibited under the WTO rules. Measures with the explicit purpose of promoting exports or sourcing local content are not permitted. They are automatically assumed to have adverse impacts on foreign trade interests. However, other subsidies – in areas such as the provision of a stimulus during a recession, creating jobs or addressing market failures (light grey boxes) – may be actionable only if they are proved to have an adverse impact. Note that the subsidies listed in

---

[154] For instance, there have been concerns about the quality of solar panels imported into India, i.e. whether they would meet efficiency standards over a long period given different weather conditions.

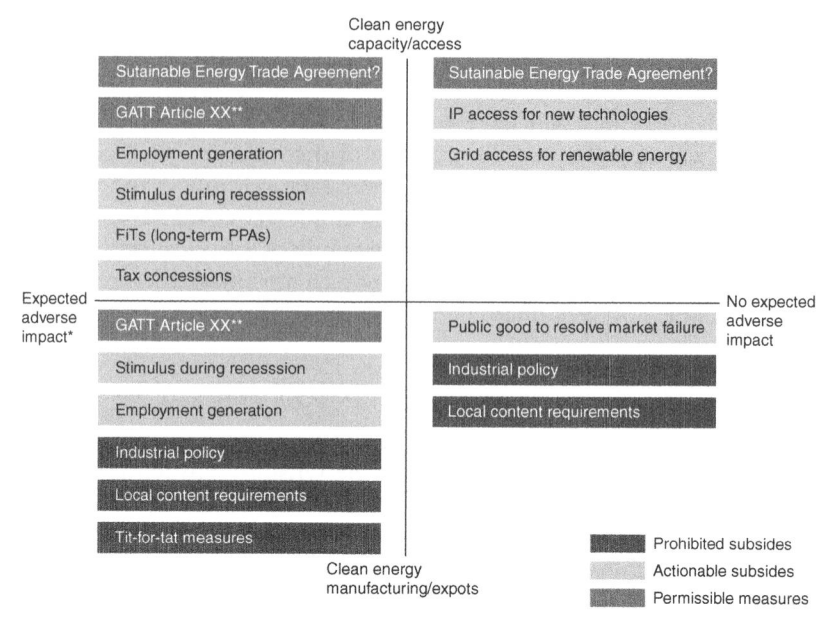

Figure 4.5: Clean Energy Subsidies Have Multiple Rationales, Impacts and Counteractive Reactions
Source: Author.

the four quadrants are neither an exhaustive list nor exclusive to specific quadrants in every case. Figure 4.5 is a schematic representation of how different types of subsidies for clean energy could have different rationales, impacts and counteractive reactions from other countries.

In the bottom left quadrant, most of the measures that have clear mercantilist purposes would be prohibited. In addition, measures found to have adverse impacts could invite challenges. In these cases, exceptions under the GATT, Article XX, may be also invoked, although it is debatable whether its provisions can be applied for clean energy subsidies. Thus, in the top left quadrant, although energy access or capacity installation might be the motive (thereby invoking possible exceptions), the expectation of adverse impact would be problematic. This could happen if FITs and tax breaks were found to be discriminatory or if stimulus during a recession (say in the form of producer subsidies) resulted in greater exports at the cost of other firms.

The bottom right quadrant is particularly interesting because, while promoting manufacturing or exports might seem mercantilist, not all measures need have an adverse impact. In fact, if the financial contribution is not greater than market value, or if it only is a means to correct market failure, the measure is not automatically prohibited. Some might even argue that a domestic manufacturing base is a precursor to establishing robust capacity for generating clean energy. Of course, if industrial policy relies only on LCRs and export subsidies, it will be prohibited irrespective of the motive.

Finally, the top right quadrant suggests that motives for clean energy subsidies matter. If subsidies were used, for instance, for extending grid connections to RE sources (whether project developers were domestic or foreign), ideally they should not be challenged. Again, if subsidies were offered to acquire intellectual property (IP) for emerging clean energy technologies, no adverse impact is caused if a country is able to expand its clean energy generation capacity. Currently, however, such exceptions are not explicitly permitted under the WTO rules and until these issues are resolved such policies might continue to attract trade disputes. Similarly, as discussed above, certain clean energy support measures might have adverse impacts even if their motive is to expand energy access (top left quadrant). With the current uncertainty over whether such policies could be contested, the incentives to invest in these sectors could be reduced.

Thus, there might be a case for clarifying rules under future trade-related initiatives for sustainable energy, including possibly a separate SETA, which could set out the key principles for what would be permissible subsidies. If the measures are for non-mercantilist purposes, such as increasing clean energy generation capacity or offering energy access, then a SETA could potentially carve out the policy space for countries to pursue such goals.

## 5  Conclusions and Policy Recommendations

Clean energy subsidies are complex instruments that reflect the multiple motivations of governments, industry and non-governmental actors. By emphasising the importance of motives, this chapter has extended the debate over subsidies beyond the realm of a legal interpretation of WTO law. Instead, it has provided a framework to understand who the supporters and opponents would be, depending on which rationales are used to offer the subsidies. This report has also developed a typology of subsidies, covering financial transfers, taxes, regulations, infrastructure and trade rules, that is broader than the definitions usually applied. This typology is further refined by categorising subsidies by their end-use, whether to extend energy access, install capacity, develop manufacturing, or promote exports. This chapter also describes the major policies in some of the leading RE powers in the world. Since some of these policies have been in place for more than two decades, this approach improves our understanding of clean energy subsidies despite the difficulties in cross-country comparisons. Finally, the chapter identifies gaps in existing trade law to argue that merely legal analyses would not be sufficient to deliver policy clarity and attract greater investments in this crucial sector. In short, if the two imperatives of expanding clean energy access and responding to the threat of climate change are to be taken seriously, trade law would have to respond to the challenges outlined in this chapter. If clean energy investments are encumbered by the WTO law, due to restrictions on subsidy measures, other legal provisions would have to be developed.

## *What Can Be Done? Five Proposals for Policy and Legal Clarity*

At least five aspects of the governance of clean energy subsidies need attention at the national and international levels. First, international institutions with rules governing trade, energy flows and climate change need greater coordination.[155] For the trade regime, subsidies are subsidies, whether for fossil fuels or for clean energy. From the climate perspective, there is a clear case for investing in clean energy sources, especially since fossil fuels developed on the back of government support as well. From the perspective of energy access and energy security, once again there is a need to diversify sources while delivering energy to hundreds of millions of poor people who are not being served by the market. Correcting for these market failures does not mean that new subsidy programmes should be developed without considering trade discrimination, distortion and rent-seeking. This is why coordination across regimes is warranted.

One way to accomplish such coordination is through various Sustainable Energy Trade Initiatives that can be pursued both multi-laterally as well as in regional or bilateral settings, including through a new international Agreement – a SETA – which would draw on rules from multilateral and regional trade, energy, environmental and climate-specific institutions. Proposals for a SETA have already emerged, with pros and cons for different architectures (plurilateral, within or outside the WTO, open membership and accession rules, etc.).[156] Whatever the

---

[155] Ghosh (2011a): 106–119.
[156] International Centre for Trade and Sustainable Development (2011).

pathway for a SETA, the first step in the process must be to undertake legal and economic analyses to clarify existing rules on how clean energy subsidies might be treated. The interesting aspect of subsidies is that any reform a country might undertake autonomously or as part of any trade initiatives or agreement (even a restricted one) would usually affect all its trade and investment partners, whether or not they are party to the agreement or initiative. This is because tariffs could be tailored according to trading partners as part of a bilateral or regional agreement while domestic subsidy-related policies would have the same impact on all trading partners.

Second, common metrics to count subsidies can help increase transparency. With trillions of dollars of projected investments in the clean energy sector, the time is ripe to develop a common country reporting format, establish the frequency of such reporting, and designate institutions or agencies that can collate and distribute the information (for example, through an online portal). In the absence of reporting at the WTO, non-governmental sources (such as the Global Trade Alert, which serves as an independent information source for trade measures and their impact on countries) could fill the information gaps.[157] Unless clean energy subsidies are measured in a transparent manner, there will be greater danger of misinterpretation and potentially more trade disputes arising. Such a situation cannot bode well for a relatively fledgling industry.

The United Nations Sustainable Energy for All (SE4All) initiative is one possible route to bring together the relevant international and national agencies to develop the common definitions and metrics for clean energy subsidies. Among the international agencies involved would be the World Bank, since it manages many of the clean energy-related climate funds; the International Monetary Fund (IMF), since it reviews its member countries in terms of their subsidy policies and their fiscal impact; the Organisation for Economic Cooperation and Development (OECD), since it has developed the Rio Markers methodology for assessing development finance, which could have some relevance for measuring clean energy subsidies as well; the IEA, a part of the OECD that can integrate assessments of all sorts of energy support measures; regional development banks; the WTO; and other regional or plurilateral trade agreements (NAFTA, the ECT, etc.).

Third, the relationship between rationalising fossil-fuel subsidy programmes and the use of subsidies to promote clean energy sources should be further investigated. Without the former, promoting clean energy

---

[157] Ghosh (2011a): 411. See also Ghosh (2010b): 451.

subsidy programmes would have little impact on the energy mix, especially where fossil-fuel electricity generation or consumer tariffs are subsidised. Instead, the programmes would take on a more mercantilist flavour, whereby subsidies would be used to promote manufacturing and exports rather than increase domestic uptake of renewable energy.

The G20, which has already served as a forum for reviewing and suggesting reforms for fossil-fuel subsidies, would be an ideal location to discuss how subsidies for clean energy compare and how they could benefit from subsidy rationalisation.

Fourth, there is a need to establish the purpose of government support. While retaining policy flexibility is important, subsidies to increase energy access or energy generation capacity would have different impacts from those geared primarily toward promoting manufacturing and exports. The pursuit of policy clarity would allow countries to review their policies and justify those that have limited mercantilist impacts.

Currently, no forum exists where governments could discuss their reasoning for clean energy support programmes. As a result, there is the risk that more trade and investment disputes might arise at the multilateral level or through bilateral arbitration channels. So long as subsidies for clean energy are viewed only as mercantilist instruments, rather than measures to promote energy access, countries will tend to dispute each other's policies. A forum for discussion could ensure that a host of countries are able to explain the purpose of their programmes. Combined with a SETA, these review sessions could potentially ease the pressure on countries that are seeking to increase energy access rather than manufacturing or exports. The Rio + 20 Sustainable Development Summit could have served as a platform for this conversation to begin, allowing governments to both describe and clarify their subsidy programmes, while enabling others to learn policy lessons (successful and unsuccessful) from each other's experiences. However, the limited outcomes of that meeting mean that the quest for an appropriate forum will continue.

Therefore, fifth, independent assessments of the alleged adverse impacts of subsidy policies could reduce the threat of unilateral trade sanctions or other penalties. Establishing the adverse impact of subsidies cannot be left to national agencies alone. Even without raising disputes at the WTO, it is conceivable that relevant WTO committees could debate the nature, purpose, scale and impact of different types of clean energy subsidies. Such a debate could help clarify individual country measures – say, at the WTO Trade Policy Review (TPR) meetings – but they could also offer greater policy clarity for clean energy subsidies as a whole. This would reduce the chances of disputes and give investors more legal clarity than currently exists.

In addition to reviews at the WTO's TPR Body, the Committee on Regional Trade Agreements could include discussions on the impacts of clean energy subsidies in terms of regional trade flows and even the integration of regional electricity networks. The United Nations Industrial Development Organization (UNIDO) should also undertake an economic assessment of the scale and impact of subsidies in promoting clean energy research, development, deployment and commercialisation. Independent country assessments could be undertaken by a host of trade and sustainable development-focused research institutions and think tanks. The greater the research into clean energy subsidies, the deeper would be the understanding of their purpose and impacts and, therefore, the lesser would be the potential for dispute. If disputes are eventually linked to the purpose of the subsidy programmes, then the confidence of investors seeking to focus on energy access or generation capacity could also increase.

Investors, energy consumers and government policy makers would all benefit from a more certain trade and investment environment for clean energy. The UN Year of Sustainable Energy for All (2012) was an ideal opportunity to focus international attention on the issue. If the transition to a low-carbon green economy is going to be a long-haul objective, the aim must be to offer policy and legal clarity regarding supporting measures over the long term.

### References

Arora, D. S. et al. (2010) *Indian Renewable Energy Status Report: Background Report for DIREC 2010*. NREL, GTZ, REN21 and IRADe, NREL/TP-6A20-48948

Bridle, Richard and Christopher Beaton (2011) 'The cost-effectiveness of solar PV deployment subsidies', *NCCR Trade Regulation Working Paper*, No. 2011/31

Campbell, Richard (2010) *China and the United States: A Comparison of Green Energy Programs and Policies*. Congressional Research Service R41287

China Direct (2011) *China's Twelfth Five Year Plan (2011–2015) – the Full English Version*

Condon, Bradly J. (2009) 'Climate change and unresolved issues in WTO law', *Journal of International Economic Law*, 12 (4): 895–926

Council on Energy, Environment and Water (CEEW) and Natural Resources Defense Council (NRDC) (2012) *Laying the Foundation for a Bright Future: Assessing Progress under Phase 1 of India's National Solar Mission*. Interim Report, New Delhi

Dooley, J. J. (2008) *US Federal Investments in Energy R&D: 1961–2008*. Paper prepared for the US Department of Energy, PNNL-17952, Washington, DC

European Renewable Energy Council (2004) *Renewable Energy Policy Review: Spain*. EREC, Brussels

(2009a) *Renewable Energy Policy Review: Germany*. EREC, Brussels

(2009b) *Renewable Policy Review: Spain*. EREC, Brussels

Gelman, Rachel (2010) *2010 Renewable Energy Data Book*. United States Department of Energy, Washington, DC

Ghosh, Arunabha (2010a) *Climate, Trade and Global Governance in the Midst of an Economic Crisis*. Briefing at a Public Hearing on Global Governance, Special Committee on the Financial, Economic and Social Crisis, European Parliament, Brussels, 25 March

(2010b) 'Developing countries in the WTO Trade Policy Review Mechanism', *World Trade Review*, 9 (3): 419–455

(2011a) 'Seeking coherence in complexity? The governance of energy by trade and investment institutions', *Global Policy*, 2 (3) (Special Issue)

(2011b) 'Strengthening WTO surveillance: making transparency work for developing countries', in Carolyn Deere-Birkbeck (ed.), *Making Global Trade Governance Work for Development*. Cambridge University Press

Global Trade Alert (2010) *United States of America: US Citizenship Requirements for Biofuel Subsidies*, St Gallen

(2011) *EC: Initiation of CVD Investigation Concerning Imports of Bioethanol Originating in the US*, St Gallen

Gobierno de España (2010) *Spain's National Renewable Energy Action Plan 2011–2020*. Madrid

Howse, Robert (2010) *Climate Mitigation Subsidies and the WTO Legal Framework: A Policy Analysis*. IISD, Geneva

Howse, Robert and Antonia Eliason (2009) 'Countervailing duties and subsidies for climate mitigation: what is, and what is not, WTO-compatible?', in Richard B. Stewart et al. (eds.), *Climate Finance: Regulatory and Funding Strategies for Climate Change and Global Development*. New York University Press, New York and London

Infinite Power (2011) *Texas' Renewable Energy Resources*. Austin, TX

Instituto para la Diversificación y Ahorro de la Energía (IDAE). *Energy Saving and Efficiency Strategy Action Plan 2008–2012 in Spain*. IDAE, Madrid

International Centre for Trade and Sustainable Development (ICTSD) (2011) *Fostering Low Carbon Growth: The Case for a Sustainable Energy Trade Agreement*. ICTSD, Geneva

International Energy Agency (1999) *Looking at Energy Subsidies: Getting the Price Right*. IEA, Paris

(2010a) *Luz Para Todos (Light For All) Electrification Programme*. IEA, Paris

(2010b) *Programme for Incentives for Alternative Electricity Sources: Programa de Incentivo a Fontes Alternativas de Energia Elétrica – PROINFA*. IEA, Paris

(2010c) *PROINFA: Program to Foster Alternative Sources of Electric Power*. IEA, Paris

(2010d) *Sustainable Economy Law*. IEA, Paris

(2011a) *Clean Energy Progress Report: IEA Input to the Clean Energy Ministerial*. IEA, Paris

(2011b) *Feed-in Tariffs for Electricity from Renewable Energy Sources (Special Regime)*. IEA, Paris

(2011c) *Golden Sun Programme*. IEA, Paris
(2011d) *Renewable Energies Heat Act*. IEA, Paris
(2011e) *World Energy Outlook 2011*. IEA, Paris
(2012) *2012 Amendment of the Renewable Energy Sources Act – EEG*. IEA, Paris
International Energy Agency and International Renewable Energy Agency (2010) *Renewable Energy Plan 2005–2010*. IEA, Paris
International Institute for Sustainable Development (2012) *Fossil Fuel Subsidy Reform: Building Momentum at Rio and Beyond*. Side Event Summary, 26 March
International Renewable Energy Agency (IRENA) (2011) *Renewable Energy Jobs: Status, Prospects & Policies*. IREN
Jha, Vyoma (2011) *Cutting Both Ways? Climate, Trade and the Consistency of India's Domestic Policies*. Council on Energy, Environment and Water (CEEW), New Delhi
Levi, Michael et al. (2010) *Energy Innovation Driving Technology Competition and Cooperation Among the United States, China, India, and Brazil*. Council on Foreign Relations Press, New York
McCrone, Angus et al. (2011) *Global Trends in Renewable Energy Investment: Analysis of Trends and Issues in the Financing of Renewable Energy*. UNEP and Bloomberg New Energy Finance, New York
Meisen, Peter, Eléonore Quéneudec, H. N. Avinash and Palak Timbadiya (2010) *Overview of Sustainable Renewable Energy Potential of India*. Global Energy Network Institute, San Diego
Mez, Lutz (2005) *Renewable Energy Policy in Germany: Institutions and Measures Promoting a Sustainable Energy System*. Environmental Policy Research Centre, Berlin
Moerenhout, Tom, Tilman Liebert and Christopher Beaton (2012) *Assessing the Cost-Effectiveness of Renewable Energy Deployment Subsidies: Wind Power in Germany and China*. NCCR Working Paper, No. 2012/03. Swiss National Centre of Competence in Research, Berne
Pokharel, Govind Raj (2002) *Subsidy for Renewable Energy Technologies in Developing Countries: Some Useful Discussions*. Appropriate Technology Forum
Ragwitz, Mario, and Claus Huber (2005) *Feed-In Systems in Germany and Spain and a Comparison*. World Future Council, Fraunhofer Institute Systems and Innovation Research, and Energy Economics Group, Hamburg
REEEP (2009–2010) *Corporate Clean Energy Investment Trends in Brazil, China, India and South Africa*. Renewable Energy Efficiency Partnership and the Carbon Disclosure Project, Kenilworth
REN21 (2009) *Renewables Global Status Report: Update*. REN21 Secretariat, Paris
(2011) *Renewables 2011 Global Status Report*. REN21 Secretariat, Paris
Renewable Energy for Europe (2003) *Energy Planning in Navarre*. <publication details at proof>
Robins, Nick, Robert Clover and D. Saravanan (2010) *Delivering the Green Stimulus*. HSBC Global Research, New York

Seligsohn, Deborah and Angel Hsu (2011) 'How Does China's 12th Five-Year Plan address energy and the environment?', WRI, Washington, DC

Spain Ministry of Industry, Tourism & Trade (2010) *Spain's National Renewable Energy Action Plan 2011–2020*, Ministry of Industry, Tourism & Trade and Instituto para la Diversificacion y Ahorro de la Energía (IDAE), Madrid

Sutrave, Lakshman Rao (2010) *Future of Indian Solar PV Industry*. Frost & Sullivan, New Delhi

Texas Comptroller of Public Accounts (2011) *Government Financial Subsidies*. Window on State Government. Austin, TX

Texas State Energy Conservation Office (2011a) *LoanSTAR Revolving Loan Program* Austin, TX

(2011b) *Texas Renewable Portfolio Standard*. Austin, TX

(2011c) *Wind Energy Incentives*. Austin, TX

(2012) *Texas Tax Code Incentives for Renewable Energy*. Austin, TX

United Nations Environment Programme (UNEP)(2009) *Global Green New Deal: Policy Brief*. UNEP, Nairobi

US Department of Energy (2011a) *California Solar Initiative – PV Incentive*. Database of State Incentives for Renewables & Efficiency, Washington, DC

(2011b) *California Solar Initiative – Solar Water Heating Rebate Program*. Database of State Incentives for Renewables & Efficiency, Washington, DC

(2011c) *CEC – New Solar Homes Partnership*. Database of State Incentives for Renewables & Efficiency, Washington, DC

(2011d) City of Boulder – Solar Sales and Use Tax Rebate. Database of State Incentives for Renewables & Efficiency, Washington, DC

(2011e) *Direct Lending Revolving Loan Program*. Database of US Department of Energy State Incentives for Renewables & Efficiency, Washington, DC

(2011f) 'Energy Department announces $7 million to reduce non-hardware costs of solar energy systems', *Energy.Gov*

(2011g) *Property Tax Exclusion for Solar Energy Systems*. Database of State Incentives for Renewables & Efficiency, Washington, DC

(2011h) *SunShot Initiative*. Energy Efficiency & Renewable Energy, Washington, DC

(2011i) *Volumetric Ethanol Excise Tax Credit*. Alternative Fuels & Advanced Vehicles Data Center, Washington, DC

US Energy Information Administration (2011) *Direct Federal Financial Interventions and Subsidies in Energy in Fiscal Year 2010*. Washington, DC

USTR (2011) *United States Details China and India Subsidy Programs in Submission to WTO*. Washington, DC

Wilke, Marie (2011) *Feed-in Tariffs for Renewable Energy and WTO Subsidy Rules*. ICTSD, Geneva

World Bank (2006) *Summary Note on Literature Review on Energy Subsidies*. Energy and Water Department, Washington, DC

(2008a) *Designing Sustainable Off-Grid Rural Electrification Projects: Principles and Practices*. The Energy and Mining Sector Board, Washington, DC

(2008b) *Operational Guidance for World Bank Group Staff: Designing Sustainable Off- Grid Rural Electrification Projects: Principles and Practices*. Washington, DC

(2008c) *REToolkit: A Resource for Renewable Energy Development*. Washington, DC

(2010) *Background Information Note on the Role of Subsidies in Renewable Energy Promotion*. Climate Investment Funds Meeting of the SREP Sub-Committee, SREP/SC.4/Inf.2. Washington, DC

World Economic Forum (2011) *Green Investing 2011: Reducing the Cost of Financing*, Geneva, April

World Resources Institute (2011) *Programme of Incentives for Alternative Electricity Sources (PROINFA)*. WRI, Washington DC

Ying, Liu (2009) *Renewable Energy Status and Policies in China*. Chinese Renewable Energy Industries Association, Beijing

## Further Reading

UN (2012) *The Future We Want*. Final Outcome of the Conference. Rio + 20 United Nations Conference on Sustainable Development, Rio de Janeiro

UN General Assembly (2010a) *Objective and Themes of the United Nations Conference on Sustainable Development*. Preparatory Committee for the United Nations Conference on Sustainable Development Second Session, A/CONF.216/PC/7. New York

(2010b) *Resolution Adopted by the General Assembly: Implementation of Agenda 21, the Programme for the Further Implementation of Agenda 21 and the Outcomes of the World Summit on Sustainable Development*. A/RES/64/236. New York

Van der Graaf, Thije, Sovacool, Benjamin K., Ghosh, Arunabha, Kern, Florian and Klare, Michael T. (eds.) (2016) *The Palgrave Handbook of the International Political Economy of Energy*. Palgrave Macmillan, Basingstoke

# 5 Trade Law Implications of Procurement Practices in Sustainable Energy Goods and Services: September 2012

*Alan Hervé and David Luff*

Competition is growing among countries hoping to capture important new markets for clean energy technologies and products. Meanwhile, domestic policy measures in this area often attempt to address multiple policy goals, including job creation.

Countries are now often turning to government procurement as a means of creating demand for both clean energy and goods and services related to clean energy and energy efficiency. Governments, as large consumers of goods and services, can leverage their purchasing power to create or further expand existing markets for goods and services. At the same time, procurement policies can discriminate against foreign suppliers by favouring domestic suppliers in either a *de jure or a de facto* manner. Many governments use procurement policies as a tool for promoting domestic sustainable energy capacities and industries; while this aids domestic industry, it also means that countries might not be choosing among the most competitively priced equipment and services available globally.

Existing international instruments address the tension between the promotion of Sustainable Energy Goods and Services (SEGS) in public procurement and its discriminatory effects. One of the most important of these instruments is the WTO Government Procurement Agreement (GPA), which contains rules that provide a useful framework for openness, non-discrimination and transparency. Given its fragmented nature, however, the international framework fails to address the existing issues raised by national policies promoting SEGS.

This raises the need for effective incentive schemes that encourage clean energy solutions. This chapter suggests that a Sustainable Energy Trade Agreement (SETA) would provide an opportunity to actively promote SEGS in public procurement while ensuring that they will not be used simply to give preference to domestic suppliers.

The first section of this chapter discusses the existing international regulations on public procurement. The second section presents the policy landscape surrounding the promotion of SEGS in public procurement, with an overview of the field's existing policies and instruments. The third section assesses the compatibility of those policies and practices with existing international instruments, with an emphasis on the GPA. Finally, the fourth section examines how a SETA could improve procurement practices.

## 1  Government Procurement: Regulatory Frameworks, Issues and WTO Disputes

### *What Constitutes Government Procurement?*

The term 'government procurement' generally refers to government purchases of goods and services for the government's own use. Such goods and services range from office equipment, transport vehicles, and cleaning and transport services to advanced technology goods such as weapons systems. The terms 'public procurement' and 'government contract or public contract',[1] used by the United States, describe the same activity. The United Nations Commission on International Trade Law (UNCITRAL) defines government procurement as 'the acquisition of goods, construction and services by a procuring entity'.[2] The GATT, Article III:8, defines it as 'procurement by governmental agencies of products purchased for governmental purposes and not with a view to commercial resale or with a view to use in the production of goods for commercial sale'.

In this chapter, the terms 'public procurement' and 'government procurement' will apply to all acquisitions by any means of goods, works, or services by public procuring entities, such as central government ministries, municipalities, public schools, hospitals, or even state enterprises. The suppliers are generally from the private sector, although in some cases a procuring entity may purchase goods and services from another public body related to the state (for example, a state-owned enterprise or SOE).

Public authorities are major consumers of goods and services in both developed and developing countries. According to the European Commission, European public authorities spend approximately 2 trillion Euros each year (equivalent to 19 per cent of the entire EU Gross Domestic Product or GDP).[3] In most countries, public procurement

---

[1] Arrowsmith (2010).
[2] UNCITRAL Model Law on Public Procurement, Official Records of the General Assembly, 66th Session, Supplement No. 17 (A/66/17), Annex I, 07.2011, Article 2 j.
[3] The European Commission also emphasises that for most public authorities, construction and renovation works and the running cost of buildings represent a major share of annual

accounts for a significant proportion of GDP: around 10 to 15 per cent in the OECD countries, up to 25 per cent in developing countries, and even more in countries in transition.[4]

Traditionally, governments have used public procurement as a policy tool – mainly to favour domestic industry, foster development of certain regions, or create jobs. Discriminatory practices take a wide variety of forms, from explicit requirements that domestic suppliers be preferred over foreign suppliers to procurement procedures that are *de facto* discriminatory.[5] This favouritism towards internal suppliers constitutes an obvious barrier to trade, one that has been partially addressed by international regulation, especially within the WTO framework.

## Existing Government Procurement Regulatory Frameworks at the Regional and International Levels

### Non-Binding International and Regional Instruments Regulating Public Procurement

The Model Law on Procurement of Goods, Construction and Services, negotiated within the UNCITRAL, provides a template for the design and development of public procurement regulations based upon best practices. Initially designed to provide guidelines to developing countries, the Model Law has inspired the procurement legislation of various Central and Eastern European (CEE) countries, and has lately showed an increasing influence in Asian and African states as well.

The UNCITRAL Model Law was revised by the UN General Assembly in July 2011.[6] This revision was prompted mainly by new technological developments – most notably, the use of electronic communication in public procurement. Despite these developments, the basic features of the Model Law have not changed.

Its main objectives are to create standardised approaches to public procurement and help states to achieve domestic procurement objectives, including value for money, efficiency and probity, among others. Unlike the GPA, the UNCITRAL Model Law is not a legally binding instrument. It is simply intended to provide a regulatory blueprint for public procurement, for adoption by individual countries as they choose.

The Model Law contains procedures addressing 'standard procurement, urgent or emergency procurement, simple and low-value procurement, and

---

expenditure, in some case over 50 per cent. EU Commission, *Buying Green! A Handbook on Green Public Procurement*, 2nd edn., 2011, http://ec.europa.eu/environment/gpp/pdf/handbook.pdf.
[4] McCrudden (2004): 257.  [5] Dischendorfer (2000).
[6] UNCITRAL Model Law on Public Procurement, Annex I, 07.2011.

large and complex projects', stating that all procedures should be subject to rigorous transparency mechanisms and should promote competition and objectivity. It also provides that potential suppliers should be able to challenge 'all decisions and actions taken in the procurement process'. As noted by UNCITRAL, 'while government purchasers should have discretion to decide what to purchase and how to conduct the procurement, that discretion is subject to safeguards that are consistent with other international standards – notably, those imposed by the United Nations Convention Against Corruption'.[7] The Model Law allows the enacting State to pursue its domestic policy objectives, such as promoting economic development through supporting small and medium-sized enterprises (SMEs), as well as environmental goals, as discussed below.

In addition to the Model Law, other regional non-binding instruments have been developed in recent years. In 1995, the Government Procurement Expert Group (GPEG) of the Asia-Pacific Economic Cooperation Forum (APEC) was created to encourage the voluntary liberalisation of procurement markets in the Asia-Pacific region. The GPEG developed a set of non-binding procurement principles, which included transparency, 'value for money', fair dealing, accountability and due process.[8] These non-binding principles do not contain specific rules concerning types of contracts or entities. The APEC countries decide for themselves how to implement the principles in their own domestic systems.[9]

### The WTO Government Procurement Agreement

*Presentation of the Revised GPA*  The WTO GPA came into effect on January 1, 1996, one year after the completion of the Uruguay Round. It was intended initially to apply to all WTO Members but this proved impossible; thus, the GPA constitutes one of the few plurilateral agreements within the WTO legal framework. It is not included in the Uruguay Round's 'Single Undertaking', under which the signatories are required to assume the rights and obligations arising from all the Agreements contained in Annexes 1 to 3 of the WTO Agreement. The GPA creates obligations and rights only for those WTO Members that have signed it.[10]

---

[7] www.uncitral.org/uncitral/en/uncitral_texts/procurement_infrastructure/2011Model.html.
[8] APEC Government Procurement Experts Group on Non-Binding Principles on Government Procurement.
[9] *Ibid.*   [10] See n. 5.

Only forty-two of the WTO's 157 members have signed the GPA, with developed countries constituting the majority of the GPA parties; ten countries are currently in the process of acceding to the agreement. The accession process requires negotiations on coverage issues (in particular, regarding the entities to be covered and other aspects of coverage described below) and on verification that the acceding party's national legislation is consistent with the Agreement.

In accordance with the GPA, Article XXIV (7b), the parties negotiated a revision in consideration of special and differential treatment for developing countries 'with a view to improving this Agreement and achieving the greatest possible extension of its coverage among all parties on the basis of mutual reciprocity'.[11] These negotiations 'were not part of the Doha Round negotiations in the WTO, which are multilateral rather than plurilateral and which are related to a range of different topics (agriculture, non-agricultural market access, services, intellectual property ...)'.[12]

In December 2006, the parties reached a provisional agreement on the text of a revised agreement to replace the existing one.[13] The revised GPA text, however, could not enter into force until the parties had reached a final agreement on the issue of coverage. In December 2011, this revised version was formally approved by the parties meeting at ministerial level in Geneva, and it has now been submitted to the parties to the GPA for internal ratification. The revised text 'entails a complete revision of the Agreement to simplify its structure, modernize the text and make it easier to understand and more user-friendly'.[14] According to the WTO, it constitutes 'a historic opportunity to improve the disciplines for this key sector of the economy and expand market access coverage, valued at between 80 to 100 billion dollars a year'.[15] Moreover, 'the negotiations have resulted in a significant extension of the coverage of the Agreement (which will be effective after the entry into force of the revised Agreement). These gains in market access result from lower thresholds and the addition of new entities and sectors to the existing Parties' current commitments.'[16] The scope of the commitments from the parties is thus significantly extended to new sectors, including local government and sub-central entities, services and other areas of public procurement

---

[11] See Arrowsmith (2002); Anderson (2007).   [12] Anderson and Arrowsmith (2011).
[13] Report of the WTO Committee on Government Procurement to the General Council, GPA/89, 11 December 2006.
[14] Anderson (2007): 13. The revised text entered into force on 6 April 2014 was was ratified by forty-three WTO Members.
[15] WTO Brief on Government Procurement (last updated in November 2012)   [16] *Ibid.*

activity in the current agreement. The discussion in this chapter reflects the revised version.[17]

The GPA provides the principal contractual obligations determining how governments frame and implement procurement legislation and regulations. The stated aims of the Agreement are 'the establishment of a multilateral framework for government procurement, with a view of achieving greater liberalisation and expansion of, and improving the framework for, international trade', in order to eliminate discriminatory treatment favouring domestic suppliers, goods and services.

Whereas the former GPA tended to focus merely on eliminating discrimination in public procurement, the revised GPA included in its Preamble new provisions referring to other goals.

In this respect, the third recital now states:

Recognizing that the integrity and predictability of government procurement systems are integral to the efficient and effective management of public resources, the performance of the Parties' economies and the functioning of the multilateral trading system.

Moreover, the Preamble now refers to a set of other horizontal policy objectives, including the need to account for the development, financial and trading needs of developing countries, especially the least-developed countries (LDCs). It emphasises the importance of transparency and the fight against corruption in public procurement. As noted by Anderson and Arrowsmith, 'the GPA now pursues not only the objective of non-discrimination but also best value for money (the "efficient and effective management of public resources") and the avoidance of corruption and conflict of interest – and moreover, these objectives are pursued in their own right and not merely as ancillary to trade objectives'.[18] However, the recitals of the revised GPA remain silent with respect to the environmental or social objectives that could be pursued through public procurement.

The two main legal measures aimed at abolishing discriminatory trade practices are the Most- Favoured Nation (MFN) and National Treatment (NT) obligations. Those principles, also embodied in the other main WTO Agreements, are adapted to the proper rationale of government procurement. The signatories are authorised to make explicit derogations from the non-discrimination principles vis-à-vis other parties if the latter do not grant similar access to their own

---

[17] Ministerial Level Meeting of the Committee on Government Procurement (15 December 2011), Decision on the Outcome of the Negotiations Under Article XXI:7 of the Agreement on Government Procurement, GPA/112.
[18] See n. 13.

markets. While ensuring strict reciprocity of rights and obligations, this provision constitutes a significant derogation from the traditional MFN obligation, as it gives to the GPA 'an appearance of a series of bilateral arrangements under a common umbrella rather than a genuine bilateral agreement'.[19] Therefore, this reciprocity limits the benefits of concessions only to the parties that are able to make offers of interest to others.

According to certain authors, reciprocity-based obligations contribute to the extension of the coverage of the GPA, since the classical application of MFN treatment would have limited the commitment of the parties to the lowest common denominator.[20] Conversely, other authors consider that the inability of a party to grant reciprocity to another party might exclude it in practice from the benefits of the agreement in a given economic sector. As such, developing countries, especially those with a small procurement market that have often been unable to formulate offers of interest to developed countries, have *de facto* been barred from (or have chosen not to participate in) the negotiations. This could be different if the existing GPA parties reduced the thresholds or coverage to smaller-sized contracts, so that developing countries would have the capacity to make offers and benefit from reciprocal concessions.[21]

The real scope and coverage of the GPA depend largely on the entities covered in the parties' Annexes to Appendix I of the Agreement. Each party must specify which central and sub-central government entities (and other entities) will be covered by the obligations imposed by the Agreement. Three other Annexes refer to the coverage of goods, services and construction services. To this point, a 'negative list' approach has been followed concerning the coverage of goods, meaning that all goods are covered unless listed in the Annexes. The approach with respect to services is positive, meaning that only services scheduled in the Annexes are covered. The scope of coverage of the GPA is therefore narrower for services than for goods (Box 5.1).

---

[19] See n. 5: 23.  [20] De Graaf and King (1995): 446.
[21] Guimarães De Lima e Silva (2008). See also Anderson and Osei-Lah (2011): 13. The authors also point out two structural problems for developing countries aiming to join the GPA. These are the sensitive issue of SOEs (indeed, the question of whether or not the purchasing of SOEs should be seen as 'government procurement' is currently one of the most important issues in the GPA accession of China) and, secondly, the belief that GPA accession could potentially conflict with existing social and policy programmes (such as Black economic empowerment in South Africa or the preferences accorded to indigenous minorities in Malaysia). The latter issue is discussed below.

> **Box 5.1. Sectors for Sustainable Energy Goods and Services Procurements Potentially Covered by the Revised GPA[22]**
>
> | Sectors | Coverage under the revised GPA[23] | Examples of SEGSs |
> |---|---|---|
> | Goods | Above 130,000 SDRs: all goods are covered except those expressly mentioned by the Parties | All kinds of RE products such as alternative fuel vehicles |
> | Services | Above 130,000 SDRs: services are listed positively and negatively by the parties. | Pre-construction power plant services, design and engineering services, energy performance contracting services.[24] |
> | Construction | Above 5 million SDRs | Smart buildings using RE or renewable materials |

Some parties to the GPA expressly maintained non-application provisions with regard to domestic policy considerations in their Annexes and general notes. These include set-asides for small and minority business, single tendering procurement and set-asides for SMEs, and contracts to be awarded to cooperatives or associations.[25] No party seems to have included in its schedule explicit non-application provisions for the promotion of SEGS. In this respect, reference must be made to the sectors, sub-central authorities and other agencies that are still excluded from the parties' commitments and to the general exception provisions of the GPA pertaining to horizontal policy objectives.

---

[22] Table 5.1 is based on information collected in the lists of commitments of the parties to the GPA. Special drawing rights (SDRs) are the supplementary foreign exchange reserve assets defined and maintained by the International Monetary Fund (IMF).

[23] Coverage also depends on the entities (central and sub-central) mentioned in the parties' commitments. When concluding the revised GPA, the parties generally accepted to extend the scope of the GPA to sub-central entities.

[24] See also Monkelbaan (2013).

[25] See Anderson and Osei-Lah (2011): 13. See also the US General Notes to Appendix 1 of the revised GPA which states that '[n]otwithstanding the above, this Agreement will not apply to set asides on behalf of small and minority business', WT/Let/672, 19 March 2010. The Canadian General Notes contain a similar exclusion.

For instance, Annex 3 of the US commitments mentions 'the waiver of Buy American restrictions on financing for all power generation projects'.[26] Canada excludes from its commitments 'urban rail and urban transportation equipment, systems, components and materials incorporated therein' whereas this potentially constitutes one important sector for SEGS promotion.[27] The EU's schedule is extremely diversified, as each EU Member State has filed different commitments and exclusions. The EU, however, expressly covers in its schedule the 'making available or exploitation of fixed networks destined to supply a service to the public in the field of production, transportation or distribution of electricity or the supply of electricity to these networks'.[28] There are as many commitments, entities covered, and goods and services subject to the GPA as there are parties to it. It is recommended, therefore, to check each schedule individually, as a general synthesis is almost impossible.

The revised GPA, Article III, allows parties to derogate from their commitments based on national security considerations or on a number of general exceptions. Coupled with Article II.3, this new language incorporates exclusions to coverage, which until now had been contained in the Annexes of the individual parties, sometimes in different ways. The ability to derogate from the general provisions of the GPA is particularly relevant with respect to the promotion of SEGS, and will be discussed in detail below.

The GPA also outlines a set of procedural disciplines aimed at implementing the principle of transparency. To this end, the Agreement provides a large number of detailed rules on the conduct of award procedures. These procedural requirements cover a number of matters, including the publication of information on the procurement system and detailed notices of intended procurements, information on the conditions for participation in a procurement, qualifications of suppliers, technical specifications and tender documentation, time limits for tender and delivery, treatment of tenders and awarding contracts, transparency for procurement information, disclosure of information and domestic review procedures.

Finally it is worth mentioning that the final provision of the revised GPA refers to the commitment of the Members of the Committee on Government Procurement to pursue negotiations concerning the future treatment of 'sustainable procurement' following the entry into force of the new legal framework.[29]

---

[26] See WT/Let/672, 19 March 2010.   [27] See WT/Let/454, 9 December 2003.
[28] See Annex 3 of EU's Schedule, doc. WT/Let/330, 1 March 2000.
[29] This clause appears in revised GPA., Article XXII:6

*GATT/WTO Disputes Involving the GPA* Given its effectiveness and quasi-judicial nature, the WTO Dispute Settlement system could have played an important role in the interpretation of the GPA. However, only three out of the more than 400 complaints filed with the WTO have involved the GPA, and only one of these led to the adoption of a panel report.[30] Moreover, no complaint involving the GPA has been filed with the WTO since 1999.

The small number of cases surely results in part from the limited number of parties to the GPA. Additionally, the economic importance of government procurement notwithstanding, the parties to the agreement have until now made few commitments, thus limiting the possibility of challenging discriminatory measures in this field. Finally, given its delays and the fact that compensation for past harm is unavailable, dispute settlement may be of little value in the context of government procurement.

The first complaint involving the GPA 1994 was brought by the European Communities (EC), triggered by a procurement tender published by the Ministry of Transport of Japan for the purchase of a multi-functional satellite for air traffic management. The EC contended that the specifications in the tender were not neutral, as they referred explicitly to US specifications. This meant, according to the EC, that European bidders were effectively barred from participating in the tender. The EC alleged that the tender was inconsistent with Annex I of Appendix I of Japan's GPA commitments and violated of the GPA, Articles VI:3 and XII:2.[31] As mentioned previously, a panel finding of a GPA violation would not have provided compensation to the potential European suppliers. Therefore, the 'mutually satisfactory arrangement' eventually signed by the parties to this dispute was unquestionably the best option for both sides.[32]

---

[30] See *Japan – Procurement of a Navigation Satellite*, Request for Consultations by the European Communities, DS73/1, 26 March 1997; *United States – Measures Affecting Government Procurement (Massachusetts State Law Prohibiting Contracts with Firms doing business with or in Myanmar)*, Request for Consultations by the European Communities, DS88/1, 20 June 1997, and Japan, DS95, 18 July 1997; *Korea – Measures Affecting Government Procurement (Procurement Practices of the Korean Airport Construction Authority)*, Panel Report, WT/DS163/R, 1 May 2000.

[31] *Japan – Procurement of a Navigation Satellite*, DS73/1.

[32] See the Notification of Mutually-Agreed Solution, WT/DS73/5, 3 March 1998. '[T]he European Commission and the Ministry of Transport of Japan have reached a settlement through the establishment of cooperation between the European Tripartite Group (consisting of the European Commission, the European Space Agency and Euro control) on the one hand and the MOT on the other in the field of interoperability between the MTSAT Satellite-Based Augmentation System (MSAS) and the European Geostationary Navigation Overlay Service (EGNOS). This cooperation is aimed at jointly contributing to the implementation of a global seamless navigation service for

The second complaint involved the famous *Myanmar Case*. In 1997, both Japan and the EC requested consultations with the United States, following the adoption of a local law by the State of Massachusetts that essentially prohibited the public authorities of Massachusetts from procuring goods or services from anyone doing business with Burma.[33] This was the first and, so far, only WTO case involving secondary policy objectives pursued in connection with the award of public contracts. The EC and Japan, concerned that other US states might pass similar legislation, argued that the Massachusetts law limited market access for European and Japanese companies and was inconsistent with several provisions of the GPA. Interestingly, the main arguments raised in the request for consultations were based on the possibility of introducing policy objectives as part of the conditions imposed on tendering companies. The WTO panel did not have an opportunity to rule on the validity of secondary policy objectives in public procurement, as its work was suspended following a ruling by the US Supreme Court that the Massachusetts law was incompatible with the Commerce Clause of the US Constitution.[34]

The third case – the only one that led to the adoption of a panel report – involved a US complaint against the procurement practices of the Korean Airport Construction Authority (KACA), relating to qualifications for bidding as a prime contractor, domestic partnering, and the absence of access to challenge procedures.

The main issues before the panel were (i) whether the procuring entity for the project at issue was covered by Korea's list of commitments, (ii) whether the procurement practices were compatible with the GPA and, finally, (iii) whether the benefits reasonably expected to accrue under the agreement, or in the negotiations resulting in Korea's accession to the GPA, were nullified or impaired by measures taken by Korea (whether or not in conflict with the provisions of the GPA) within the meaning of the GPA, Article XXII:2.

The panel emphasised that each country's schedule of commitments formed an integral part of the GPA, and was therefore to be interpreted in the same way as the latter,[35] in accordance with the Vienna Convention

---

aeronautical end-users through the interoperability among MSAS, EGNOS and other equivalent systems. It has also been agreed that the requirements for interoperability will be mentioned in MSAS and EGNOS documentation for all future procurement in and after 1998, on condition that both sides reach the conclusion that the interoperability is feasible.'

[33] *United States – Measure Affecting Government Procurement*, Request for Consultations by the European Communities, WT/DS88/1, 26 June 1997.
[34] See *Crosby v. National Foreign Trade Council*, 530 US 363, 2000.
[35] See *Korea – Measures affecting Government Procurement*, WT/DS163/R, 1 May 2000, § 7.9.

on the Law of Treaties. The panel concluded that the procuring entity was not expressly included in Korea's schedule of GPA commitments. The issue, then, was whether an entity not expressly mentioned in the list of commitments of a party to the GPA could be subject to the agreement because it was controlled by other entities that were included in the schedule. After a thorough analysis of the history of the Korean accession to the GPA, the panel concluded that the KACA was not sufficiently related to a covered entity to be subject to the rules of the GPA.[36]

In the fourth section, we will address the *Feed-in Tariff [FIT] Case*.[37] This case is now settled. While it predominantly involves other WTO Agreements than the GPA, it could have a systemic relevance concerning the WTO-compatibility of measures aimed at promoting SEGS.

## Free-Trade Agreements and Government Procurement

Almost all Members of the WTO – as well as many non-Members – are parties to various bilateral and plurilateral free-trade agreements (FTAs). More than 300 FTAs are currently in effect, and, due in part to the apparent failure of the WTO's Doha Round, many more are in the discussion or negotiation process. These FTAs contain 'WTO plus' obligations – rules and disciplines on which the full membership of the WTO cannot agree upon under the single undertaking principle. Public procurement is one of the items on which the major players, such as the United States and the EU, seek additional liberalisation and disciplines.

The North American Free Trade Agreement (NAFTA) – concluded in 1992 between the United States, Canada and Mexico – contains a chapter dealing with public procurement. This chapter, whose provisions are practically identical to those of the 1996 GPA, follows a 'negative list' approach, meaning that all goods and services are covered except for those specifically exempted by the parties. The coverage is also based on the principle of reciprocity.[38]

Besides the exceptions allowed in the field of national security procurements, NAFTA authorises the parties to adopt or maintain measures 'necessary to protect human, animal or plant life or health', provided that 'such measures are not applied in a manner that would constitute a means of arbitrary or unjustifiable discrimination between Parties where the same conditions prevail or a disguised restriction on trade between the

---

[36] *Ibid.*, § 7.89. The United States also failed to demonstrate that benefits reasonably expected to accrue under the GPA, or in the negotiations resulting in Korea's accession to the GPA, had been nullified or impaired by measures taken by Korea.

[37] DS412 (Japanese complaint) and DS426 (EU complaint).

[38] See www.nafta-sec-alena.org/en/view.aspx?conID=590&mtpiID=140#An1001.1a-3.

Parties'.[39] In relation to the GPA, therefore, NAFTA allows through its exceptions the possibility to promote secondary policy objectives, though it remains difficult to include SEGS-related procurement within these exceptions.

Central entities are covered widely by NAFTA; sub-central entities, however, usually remain outside the scope of application of the agreement.[40] Goods and services are all covered by NAFTA with some limited exceptions. Exclusions in the services sector are more important than in goods one. There are exclusions in some sectors that are particularly relevant for the development of sustainable procurements such as 'services with reference to transportation equipment'.[41]

More recent US FTAs contain rules that are closer to the revised GPA. They also tend to encourage secondary policy objectives, including environmental initiatives. The FTA signed with Morocco in 2004, for instance, takes the 'negative list' approach and indicates that the article on technical specification 'is not intended to preclude a procuring entity from preparing, or applying technical specifications to promote the conservation of natural resources or to protect the environment'.

FTAs completed by the EU also contain provisions on public procurement. The EU–Korea FTA, for instance, stresses the commitment of the parties to liberalise public procurement and promote the application of the revised GPA, and it creates a joint committee aimed at fostering cooperation in the field of government procurement. Because both countries are also party to the GPA, however, the FTA states that:

Nothing in this Chapter shall be construed to derogate from either Party's rights or obligations under the GPA 1994, or from an agreement which replaces it.

The FTAs concluded by the EU with States not party to the GPA contain much more detailed provisions. For instance, Title VI of the FTA between the EU and its Member States and the members of the Andean Community (Colombia and Peru) provides a detailed framework under which each signatory, including its procuring entities, shall accord – immediately and unconditionally – treatment on goods, services and suppliers of other signatories no less favourable than treatment accorded to domestic goods, services and suppliers. As in the GPA, this liberalisation of government procurement is based upon both a 'positive' and a 'negative list' approach. Moreover, like the provision of the GPA whose discussion ensues, exceptions to this part of the agreement can be justified on the grounds of the necessity 'to protect human, animal or plant health'.

---

[39] See NAFTA, Article 1018.  [40] See NAFTA, Chapter 10, Annex 1001.1a–3.
[41] See NAFTA, Chapter 10, Annex 1001.1b–2.

Procurement Practices in SEGS 209

The EU–Colombia and Peru FTAs explicitly added to this sentence the words 'including the respective environmental measures'.

Finally, a limited number of the South–South FTAs also contain provisions on government procurement. Those provisions are usually limited to promoting the liberalisation of public procurement as an objective, but without coverage commitments.[42]

These FTAs, particularly those signed with non-parties to the GPA, pave the way for future accession to the GPA. They also allow for the introduction of 'GPA plus' provisions, which raises the possibility of promoting SEGS in government procurement. The provisions of the SETA could indeed be integrated within future FTAs that would then allow for a progressive 'multilateralisation' of this agreement.

## 2   Policy Context and Landscape

### *Motivation for Procurement Practices for SEGS*

Governments cannot be considered simply as market participants. Purchasing entities play an active part in the markets as important consumers, significantly influencing markets through both their actions and inactions. That is why 'sustainable procurements' (also called 'green procurements') and, specifically, the acquisition of SEGS in public procurements, can fulfil a set of secondary environmental and sustainable policy objectives.

Sustainable or green procurements provide a tool to limit the impact of procurement on human health and the environment, providing an opportunity to mitigate over-exploitation of scarce resources. The promotion of SEGS within public procurements also provides a means of complying with the international obligations imposed by the UN Framework Convention on Climate Change (FCCC) and the Kyoto Protocol.

As major consumers, governments can influence the development of private markets. Public sector demand can be used strategically to influence the behaviour of private actors in the production of SEGS. Procurement promoting SEGS can be 'a major driver for innovation, providing industry with real incentives for developing green products and services – particularly in sectors where public purchasers represent a large share of the markets',[43] such as public transport or construction. Governments have an important role to play by offering green innovators

---

[42] See Anderson and Arrowsmith (2011).
[43] See Commission, *Buying Green! A Handbook on Green Public Procurement*, 2nd edn., 2011.

a guaranteed market for their products, thereby generating economies of scale and lower costs.

Sustainable procurement offers a more comprehensive approach to the costs and outcomes associated with procurement decisions. A narrow approach based solely on price fails to take account of the full life cycle cost of a contract. For instance, low-energy products will, in the long run, allow significant reductions of utility bills. Use of the lowest price as the sole criterion is to be replaced by the concept of 'best value for money'.[44]

Social benefits can also be expected from sustainable procurement. Even if the phenomenon remains difficult to assess quantitatively, health benefits can be expected from procurements promoting SEGS.[45] Moreover, in many cases sustainable procurements provide a good stimulus for local business, which will often have greater capacity to fulfil energy-saving criteria and benefit from the promotion of domestic products and services.

## Overview of Procurement Practices and Instruments for SEGS

The following discussion focuses on the European and Chinese promotion of SEGS. Among the developed countries, the EU and its Member States have, for more than a decade, developed a set of political instruments and incentives aimed at promoting sustainable procurements (usually called 'Green Procurements' within the EU). China developed a sustainable procurement policy more recently.

### SEGS Promotion In Developed Countries: The EU and its Member States

*The EU Legal Framework*  Public procurements at the EU level are regulated by three main Directives: the Public Sector Directive[46]

---

[44] The concept of 'best value for money' was first developed by the UK Treasury, which defined it as 'the optimum combination of whole life cost and quality (or fitness for purpose) to meet the customer's requirement' (www.hm-treasury.gov.uk/d/government_procurement_pu147.pdf). As noted by Brammer and Walker (2007), 'through the focus on whole life cost, the definition of the best value of money gives scope to public bodies to take social and environmental policy objectives into account in their procurement activities'.

[45] Long-term health benefits can be expected, for instance, from reduced carbon emissions. From the social point of view, the promotion of sustainable procurement is a way to stimulate the private sector and make an economy more innovative and competitive.

[46] The Treaty on the Functioning of the European Union (TFEU) defines Directives as instruments having a binding legal nature with respect to the Member States (TFEU, Article 288).

Procurement Practices in SEGS 211

(Directive 2014/24/EU[47]), which defines the procedure for awarding most major contracts by public bodies (national governments, regional and other public entities) and the Utilities Directive (Directive 2014/25/EU[48]), which regulates the procedures for awarding major contracts by bodies engaged in certain activities in the sectors of water, transport, energy and postal services, and the Concession Contracts Directive (2014/23/EU[49]), which regulates the award of concession contracts. The principle of transparency and equal treatment of bidders, best value for money, and free movement of goods and services form the basis of the three Directives. These instruments allow the application of sustainable procurement principles but do not force the European procuring entities to do so.

The provisions concerning sustainable procurements contained in both instruments are quite similar. Both Directives, for example, require major contracts to be advertised through the EU's official journal, to publicise the contracts to all interested parties and to regulate the criteria that can be used to tender and award contracts. Several recitals of Directive 2004/18 illustrate the EU's approach towards sustainable procurements and the way in which environmental considerations can be taken into account by European purchasers.[50]

Since the adoption of the earlier version of these Directives in 2004, the European Commission adopted a proactive approach toward sustainable procurement through regular communications and staff working documents. Its updated handbook on green procurements, *Buying Green!*, provides useful guidelines for public purchasers who want to introduce sustainable considerations into their tendering procedures.[51] SEGS can be promoted at different stages of the European procurement procedures:

---

[47] Directive 2014/24/EU of the European Parliament and of the Council of 26 February 2014 on Public Procurement and Repealing Directive 2004/18/EC (OJ L 94, 28 March 2014:65).
[48] Directive 2014/25/EU of the European Parliament and of the Council of 31 March 2004 Coordinating the Procurement Procedures of Entities Operating in the Water, Energy, Transport and Postal Services Sectors and Repealing Directive 2004/17/EC (OL J 94, 23 March 2014: 243).
[49] Directive 2014/23/EU of the European Parliament and of the Council of 26 February 2014 on the Award of Concession Contracts (OJ L 94. 28 March 2014: 1).
[50] See in particular the 1st, 5th, 29th, 44th and 46th recitals of the EU Directive 2004/18. The same can be observed concerning the so-called Utility Directive 2004/117. For a deeper analysis, see Arrowsmith (2010). Moreover, several rulings from the European Court of Justice (ECJ) deal with the issue of sustainable procurement. See for instance Case C-448/01, *EVN AG and Wienstrom GmbH* v. *Austria* ('EVN') [2003] ECR I-14527.
[51] See n. 44.

Deciding which Procedure is Applicable
The preparatory stage of a public procurement is crucial, especially when it comes to choosing the procedures that could facilitate the introduction of sustainable development. For instance, an *open procedure*, in which any operator may submit a tender, allows access to the maximum choice of environment-friendly solutions but does not require tenders to be selected solely on the basis of environmental considerations. In a *restricted procedure* (with a limited number of operators invited to tender) or through a *negotiated and competitive dialogue* (used in particularly complex procurements), the environmental technical capacity of the tenders may be assessed at an early stage.

Defining Contract Requirements
Once the subject of a contract is defined (with a possible reference to the use of sustainable energy), technical specifications, which are included in the contract notice or tender documents, are crucial when it comes to introducing sustainability considerations. Within the EU, technical specifications may be formulated by reference to European, international, or national standards, as well as in terms of performance or functionality (Directive, Article 23) or in terms of environmental performance levels of a material, product, supply or service.[52]

## Technical Specifications by Reference to Standards

Environmental standards containing characteristics such as energy use may be included in the specifications. The procurement Directives refer to European or national standards as a means by which specifications can be defined. Indeed, 'standards are useful in public procurements as they are clear, neutral, and usually developed using a process which includes a wide range of stakeholders, including national authorities, environmental organizations, consumer associations and industry'.[53]

## Technical Specification by Reference to Performance or Functional Requirements

A performance specification describes the desired result in terms of the outputs that are expected – for example, with respect to quality, quantity and reliability. The bidding documents ask the tenderers to achieve certain results, but do not specifically address how they should be achieved, thus allowing more scope for market creativity. For instance, in the construction sector, a purchasing authority may indicate that the heating system should guarantee a constant temperature of 20°. In that

---

[52] The European Commission gives the examples of a computer that should not consume more than a certain amount of energy per hour, and a vehicle that does not emit more than a certain quantity of pollutant.

[53] See n. 44: 26. See also *Integration of Environmental Aspects of European Standardization*, COM(2004) 130, 30 January 2004.

case, 'suppliers may opt for innovative heating and ventilation systems which reduce dependence on fossil fuels'.[54]

Under the procurement Directives, technical specifications can include references to sustainability-related materials and production.[55] All technical specifications, however, should be related directly to the subject matter of the contract, including only requirements pertaining to the production of the goods or services purchased. This also holds when specific production and process methods are required; a tender, for instance, may indicate that electricity should be produced from renewable resources. In line with the GPA, EU law aims at avoiding discrimination by prohibiting purchasing authorities from insisting upon a production method that is proprietary or available only to one supplier, unless such a requirement appears to be justified by exceptional circumstances and is accompanied by the words 'or equivalent'.

Finally, Directive 2004/18, Article 23.6, encourages the use of European ecolabels in the tender documents.[56] An ecolabel can be used to facilitate the assessment of compliance with technical specifications, but given the voluntary nature of such labels, tenderers cannot be required to register under a certain ecolabel scheme. Equivalent means must always be accepted by the purchasing entities when it comes to assessing compliance with the requirements relating to the products or services being acquired.

### Selecting the Suppliers and Awarding the Contract

Selection criteria focus on a tenderer's capacity to carry out the contract. During the selection of suppliers, European purchasers are allowed to take into account their experience and competence related to environmental matters.[57]

Award criteria indicate the characteristics that will enable the purchaser to make a choice. They must be distinguished from the selection criteria

---

[54] Commission, *Buying Green!*: 27.
[55] See Annex VII of the Directive 2014/24/EU and Annex VIII of the Directive 2004/17/EC.
[56] The European Ecolabel is a voluntary scheme, established in 1992 to encourage businesses to market products and services that are environmentally friendly. Products and services awarded the Ecolabel 'carry the flower logo, allowing consumers – including public and private purchasers – to identify them easily. Today, the EU Ecolabel covers a wide range of products and services, with further categories being continuously added. Categories covered include cleaning products, appliances, paper products, textile and home and garden products, lubricants and services such as tourist accommodation. Unlike standards, Ecolabels are voluntary.' See the Communication on *Sustainable Consumption and Production and Sustainable Industrial Policy Action Plan*, COM(2008) 397 final, 16 July 2008: 4.
[57] Operators who have violated environmental laws can be excluded.

and also from the technical specifications mentioned above, which aim to indicate a set of minimum requirements. The EU legal framework allows considerable scope for the use of environmental criteria, as the award of the contract can be chosen not only on the basis of the lowest price but also on the 'most economically advantageous tender'.[58] This last notion allows the introduction of secondary policy objective criteria, including sustainability.

*Implementation Policies* With its legislation and policy aimed at encouraging green procurement, the EU provides an interesting study in the promotion of SEGS in public procurement. The European Commission strongly encourages the creation of networks and the exchange of good practices at the national and the local levels, such as the Local Governments for Sustainability (ICLEI), which is dedicated to introducing new instruments, mechanisms and tools for municipal management in order to ensure the implementation, effective monitoring and continual improvement of sustainable development policies.[59] Other initiatives include the Local Authority Environmental Management Systems and Procurement (LEAP), which established a toolkit and developed an array of tools and guidance for public authorities to deal with green procurement as part of an environmental management system.[60]

Although a complete analysis of the policies of various EU Member States is beyond the scope of the present chapter, it may be noted that several governments have set ambitious targets with a strong emphasis on SEGS. In 2008, the EU Commission noted that 'the Dutch Government has set a 100% Sustainable Procurement target to be reached by 2010; the Austrian Government has identified different targets to be met by 2010 for five product groups: IT: 95%; electricity: 80%; paper: 30%; cleaning products: 95%; and vehicles: 20%. In France, 20% of the vehicles purchased by the central government should consist of "clean" vehicles, 20% of new construction should be compliant with HQE16 standards or equivalent, and 50% of all wood products should be sustainable by 2010. In the UK, the Sustainable Procurement Action Plan is closely linked to a series of sustainable operations targets for the Government office estate, including a pledge to become carbon neutral by 2012 and to reduce carbon emissions by 30% by 2020.'[61]

---

[58] This concept applies the 'best value for money' approach at the award stage.
[59] For an introduction to ICLEI, see www.iclei-europe.org/index.php?id=procurement.
[60] For an introduction to LEAP, see www.laep-gpp-toolkit.org/.
[61] See the Commission's Communication, *Public Procurement for a Better Environment*, COM/2008/0400 final, 2 July 2008.

Some studies, however, are less optimistic about sustainable procurement on the part of European institutions and member states. The *Green Public Procurement in Europe 2006 Report*, produced by four NGOs with support from the European Commission, assessed the state of green public procurement in the then-twenty-five EU Member States.[62] The report identifies among the main barriers to the development of green procurement the high cost of green products, the lack of environmental knowledge on the part of purchasers, the absence of managerial and political support, and the lack of information and appropriate training of the bidders. Other studies conducted by EU Member States, such as Sweden, have confirmed this conclusion regarding the factors that limit tenderers from promoting SEGS.[63]

### SEGS in China

In the past decade China, in parallel with its economic development, has officially introduced a set of concrete initiatives designed to achieve energy conservation and emissions reduction. In 2006, the 11th Five-Year Plan for National and Economic Development set binding targets for the period from 2006 to 2010, directing the GDP unit target consumption to be reduced by 20 per cent and the total sum of the main pollutant emissions to be cut down by 10 per cent. The Chinese State Council published a *Decision for Strengthening the Work on Energy Conservation* with a series of measures and policies to promote energy conservation and reduce emissions. A November 2009 decision includes concrete targets for reducing greenhouse gas (GHG) emissions.[64]

However, despite the many references to sustainable procurement in Chinese regulations, the country still faces important problems that limit

---

[62] The data for this study came from two sources. One was from 865 responses to 8787 questionnaires, and the other was from a survey of 1000 tender documents. See http://ec.europa.eu/environment/gpp/pdf/take_5.pdf.

[63] Lina Carlsson and Fredrik Waara conducted a survey in Sweden before the implementation of Directive 2004/18 through interviews with twenty-nine procurement officers in eight Swedish municipalities, one county and one region in Sweden. Three types of limitations to the integration of environmental concerns were identified: *the lack of administrative resources* (including environmental know-how), *legal concerns* (suppliers refrained from using environment-related award criteria because it could result in bid protests from unsuccessful bidders) and *lean budgets* (some purchasers considered that environmentally friendly goods and services were too expensive). See Carlsson and Waara (2006).

[64] Among the many implementation activities involving SEGS in China were the Green Olympics. Green procurement was used in acquiring construction materials, designing the facilities, and providing services.

the implementation of this policy and the promotion of SEGS in procurement.

*The Chinese Legal Framework* The liberalisation of Chinese public procurement started in the early 1980s, after which the country passed two primary laws on public procurement – the Bidding Law in 1999[65] and the Government Procurement Law in 2002.[66] These laws were supplemented by a series of implementing measures taken by different government agencies.[67]

Sustainable procurement is not a legal concept as such within Chinese law. Still, in recent years, a great number of laws have been enacted to implement sustainable policies in public procurement.[68] The Chinese Government Procurement Law, Article 9, states that 'government procurement shall be conducted in such a manner as to facilitate achievement of the economic and social development policy goal of the state, *including but not limited to environmental protection'*. And the Bidding Law, despite its lack of secondary policy objectives, implicitly allows the consideration of sustainable development policies at different stages of the procurement (such as the qualification process, technical specifications and the award stage).[69]

The Clean Production Promotion Law of the People's Republic of China of 2002 indicates that 'governments at all levels, in their procurement, *should give priority to the products that are environment friendly and resource-conserving'*. This law also states that *'all levels of government should use advocacy and education to encourage the public to purchase and to use environment friendly and resource-conserving products'*.[70]

A third important legal provision is the Circular Economy Law of the People's Republic of China, Article 47 of which provides that entities and individuals purchasing goods with public funds should give preference to energy-saving, water-saving, material-saving and environmentally friendly and recycled products.[71]

This set of legal instruments refers to two lists that have a critical impact on the use of sustainable products and of labels that specify exactly which products should be preferred for environmental reasons:

---

[65] Bidding Law of China, www.lawinfochina.com/display.aspx?lib=law&id=1014
[66] Government Procurement Law of China, www.gov.cn/english/laws/2005-10/08/content_75023.htm
[67] See Fuguo (2008). [68] See Fuguo et al. (2011) See also Ju et al. (2009).
[69] See n. 67: 329.
[70] Clean Production Promotion Law of the People's Republic of China, http://english.mep.gov.cn/Policies_Regulations/laws/envir_elatedlaws/200710/t20071009_109966.htm.
[71] Standing Committee of the 11th National People's Congress of the People's Republic of China, Circular Economy Law of the People's Republic of China, 2008.

- The *Labelling List* (established in 2006 by the Ministry of Environmental Protection) lists products quality-verified by third-party verification agencies, who attest *inter alia* that they are energy-efficient or contain recyclable material. This system of green labelling is voluntary, and suppliers are allowed to give other kinds of evidence aside from the label to attest that their products comply with the green technical requirements and specifications of a public contract. The list contains twenty-one categories of products, such as light vehicles, photocopiers, computers, water-based paints and furniture.
- The *Energy-Saving List* covers both energy-saving and water-saving products. It is promulgated by the Government Procurement Supervision and Administration Department under the Central Government, or at the provincial level jointly with the provincial department.[72] It contains more than twenty-five categories of energy-saving products, such as air conditioners, refrigerators, water heaters, computers, and seven categories of water-saving products, such as toilets, showers and faucets.

Chinese regulations also require that the State Council and provincial governments give priority to products and equipment that have an attestation certificate for energy conservation in preparing the lists.[73] The list and accreditation system are crucial for suppliers, as the eligibility for a green procurement contract is entirely dependent on them. In this respect, the 2008 Chinese Law on Energy Conservation requires all public entities to procure products and equipment that are on the energy-saving list.[74] Moreover, procurement of energy-consuming products and equipment that have been explicitly eliminated by the state is prohibited. A public institution that considers procuring energy-saving products and equipment but fails to procure a product or equipment included in the lists is subject to sanctions.[75]

*Difficulties in Implementing Sustainable Procurement Policies* China's promotion of sustainable procurement is still affected by a set of legal, institutional and socio-cultural constraints.

---

[72] See Chinese Law on Energy Conservation, Article 51.   [73] *Ibid.*, Article 64.
[74] *Ibid.*, Article 51: 'When a public institution purchases energy-consuming products and equipment, it shall purchase those products and equipment that have been incorporated into the government procurement inventory of energy-saving products and equipment.' For an English translation, see http://faolex.fao.org/docs/texts/chn76322E.doc.
[75] *Ibid.*, Article 81.

**The country's legal environment could be considerably improved:** Too often, Chinese laws consist only of general statements, lacking precise rules and regulations concerning the implementation of the sustainable procurement policy. For example, Qiao and Wang observe that 'the ninth provision of [the] Government Procurement Method states that [the] government should give priority to high-tech products and eco-friendly products, but it does not define eco-friendly products and does not specify the importance of green products'.[76] Another example is the absence of provisions that give the purchaser the opportunity to provide performance-based specifications, which can play a crucial role in green energy market innovation.[77] Finally, the provisions of existing Chinese regulations on public procurement do not explicitly cover construction and services, despite these sectors' potentially crucial importance to sustainable procurement efforts. This said, the 2008 Chinese Law on Energy Conservation requires all public entities to procure products and equipment that are on the energy-saving list.

**The institutional perspective:** China's sustainable procurement policy suffers from a lack of unity. China 'does not have a single designated agency that is charged with managing green procurements. Several agencies and ministries are involved in green procurement management, including the Environment Protection Ministry, the Finance Ministry, the National Development and the Reform Committee, as well as the various procurement centres at the provincial and local levels.'[78] Different, often rival, actors pass regulations 'either jointly, or on their own, causing policy overlaps, management duplication, and even conflicts amongst agencies'.[79]

**China still lacks a real market for sustainable procurement:** Technological investments in this field are low and many barriers to trade still impede access to the Chinese market. The priority given in Chinese law to national products and suppliers certainly limits the purchasing entities' choice of SEGS.[80] The compulsory and exclusive nature of the energy-saving list is in part responsible for this phenomenon.

**The development of sustainable procurement in China is affected by the weak human and financial resources of the public**

---

[76] See Fuguo et al. (2011).   [77] *Ibid*.   [78] *Ibid*.
[79] Some authors therefore propose the creation of a unified 'green procurement agency'. See Qiao and Wang (2011): 1040.
[80] The Government Procurement Law implements a buy-national policy in Article 10, which provides that 'the *government shall procure domestic goods*, works, and services except where: (1) Goods, works, or services to be procured are not available within the territory of People's Republic of China or though available, cannot be acquired on reasonable commercial terms and conditions. (2) Items to be procured are for use abroad. (3) Otherwise provided by laws and administrative regulations.'

**authorities:** As noted by Qiao and Wang, 'the government does not have trained green procurement professionals. Those involved in green procurement are from the finance department or are management personnel. Many of them do not have procurement experiences and know very little about market analysis, procurement cost control, supplier assessment and management, procurement contract management, negotiation, or communication. They have even less understanding and knowledge about green procurement ... Therefore they tend to use their subjective judgment in deciding the bid.'[81] Coupled with insufficient funds appropriated by public procurement, Chinese purchasers logically tend to award procurement based solely on the lowest price.

### *Concrete Illustrations of SEGS Procurements*

SEGS-related procurements are increasing in many developed and developing countries. While an exhaustive overview of such practices falls beyond the scope of the present chapter, some concrete cases can be addressed.

#### *US Strategies Developed at the Federal and State levels*

The United States has recently developed several initiatives to promote SEGS-related procurement, following the historical willingness of the American authorities to use public procurement as a tool for developing strategic policies.[82]

At the Federal level, President Obama signed an Executive Order in 2009 aimed at establishing 'an integrated strategy towards sustainability in the Federal Government and to make reduction of greenhouse gas emission a priority for federal agencies'.[83] The Executive Order states that Federal agencies must immediately increase energy efficiency, reduce their GHG emissions from direct and indirect activities, conserve and protect water, eliminate waste, recycle and prevent pollution, among other initiatives. To that end, very detailed and precise objectives are

---

[81] See Qiao and Wang (2011): 1041.
[82] Besides the promotion of SMEs already mentioned, one can observe that the US Department of Defense played a critical role in promoting R&D and innovation through public procurements. For instance, 'in the 60s, when the technological options were far ahead of civilian applications in the semi-conductor business, the US defense sector represented the only customer for the American sector industry. With its high-level technological requirements, the public sector created a strong demand for innovation in order to satisfy the specifications imposed by military applications. The Department's willingness to pay almost any price for compact, lightweight electronics for its missile programmes stimulated the infant semi-conductor industry.' See Nyiri et al. (2007).
[83] The Executive Order is available at www1.eere.energy.gov/femp/pdfs/eo13514.pdf.

given to the agencies. For example, Section 2h of the Executive Order directs heads of agencies to advance sustainable acquisition by ensuring that 95 per cent of new contract actions are purchased through green-certified and labelled programmes.

Moreover, several initiatives have been taken at the sub-Federal level. In the field of electricity supply, for instance, twenty-nine states and the District of Columbia have implemented the so-called mandatory renewable portfolio standards (RPS), i.e. standards that encourage production of energy from renewable energy (RE) sources, including wind, solar, biomass and geothermal.[84] Although their designs differ considerably from one state to another, RPS policies usually imply an obligation for utilities or load-serving entities to procure a certain proportion of RE by a specific date. Most RPS requirements carry through to 2020 or even longer.

### Other Examples of SEGS Procurements in the United Kingdom and Morocco

There are myriad other national and local examples of public procurement involving SEGS. The UK government, which aims to be a leader in the EU strategy for sustainable procurement, has formally recognised that sustainability should be a core component of public procurement.[85]

Several on-going projects promote both SEGS and innovation.[86] The UK government has used public procurement for the development of a carbon capture and storage (CCS) pilot power plant.[87] The aim was to help private developers overcome the technical and commercial risks and uncertainties in the development and deployment of CCS technologies. The issued tender contains funding for research on CCS technology and the arrangement of pilot CCS sites. Clearly defined criteria include a provision that the pilot plants should use post-combustion capture technology and store the sequestered $CO_2$ in offshore geological sites. This technology should be able to sequester 90 per cent of $CO_2$ and to cover the whole project cycle (capture, transport, and storage), while reaching an electrical output of at least 300 MW. Finally, the project should be built in the United Kingdom.

---

[84] See Kreycik et al. (2011), www.nrel.gov/docs/fy12osti/52983.pdf.
[85] See also Kenrick (2011).
[86] For a deeper presentation of these projects see Brenton et al. (2011), http://graduateinstitute.ch/ctei/.
[87] UK Department for Business Enterprise and Regulatory Reform, 'Competition for a Carbon Dioxide Capture and Storage Demonstration', Project Information Memorandum, 2007: 8.

In the area of stimulating green transport innovations, it was estimated that in 2007 the Department for Transport spent £5 million per year on grants designed to support UK-based low-carbon road vehicle technologies at the research and pre-competitive development stages.[88] The Department for Transport also provides grants for the testing and demonstration of infrastructure for alternative fuels and vehicles, including infrastructure for biofuels, electric vehicles and hydrogen. In 2007, grant funding for infrastructure projects was estimated at around £0.5 million per annum.[89]

The United Kingdom also launched a 'hydrogen fuel-cell and carbon abatement technology fund' in 2006.[90] Technology demonstration of fuel-cell and hydrogen technologies received £15 million of this funding, and part is allocated to transport-related applications. The UK government also provides 'funding of an initial £20 [million] to support a new programme aimed at accelerating the market penetration of lower carbon vehicles and reducing the barriers faced by companies in moving from prototype demonstrations of lower carbon technologies to full commercialisation. This programme provides financial support for the public procurement of fleet demonstrations of lower carbon vehicles [and, where appropriate, supporting infrastructure].'[91] An additional £10 million research and development (R&D) fund, designed to accelerate growth in low-carbon transport technologies and support the emergence of green auto manufacturers in the United Kingdom, was launched in March 2011. The new fund was part of a package of government measures intended to encourage domestic entrants into the low-carbon vehicle sector and overcome the financial difficulties experienced by a number of green car start-ups.

Recently, the United Kingdom decided to extend this strategy to the energy procurement of schools and higher education establishments, 'until now a sector which had not moved from a fixed price, fixed term contract, to a flexible, risk-managed contract as recommended by the British government'.[92] It is noted that, '15% of public sector carbon emissions arise from activities in the English schools system and about a third of this is directly from energy usage in school buildings. In light of this, the Government aims to make all schools "sustainable schools" by 2020, by not just promoting sustainability through teaching methods but also by encouraging schools to participate in local authority carbon reduction commitment opportunities and other initiatives in order to reduce their energy consumption.'[93] Schools face a problem of expertise in this field and are thus encouraged to collaborate with experts in energy procurement, often from the private sector.

---

[88] www.energysavingtrust.org.uk/Transport.   [89] *Ibid.*   [90] *Ibid.*   [91] *Ibid.*
[92] Kenrick (2011).   [93] *Ibid.*

This example illustrates the importance of public–private partnerships (PPPs) in the development of sustainable procurements and, particularly, SEGS procurement. A report by Colverson and Pereira addressed this issue.[94] One of the case studies in their report concerned a global rural electrification programme developed in Morocco.[95] With the aim of 'improving the living conditions of its rural population, the Moroccan Government set itself the target of improving access to electricity from its level of 12 per cent in 1994 to 97 per cent by 2007/2008'.[96] The National Electricity Office (ONE) made the choice to utilise photovoltaic (PV) solar power, 'for the more remote households physically beyond connection to the grid',[97] about 10 per cent of the rural population. TEMASOL,[98] successful in its application after a call for tenders, signed a service contract with ONE in 2002 'to supply solar power to 16,000 homes across four provinces'.[99] In 2004, this joint venture obtained a new procurement for 42,500 homes across twenty-five provinces. Moreover, 'the contract included not only the supply and installation of the PV kits but also their operation and maintenance over the 10-year life of the customer contracts'.[100] The whole project was financed mainly through public and donor funds and a reduced contribution from user fees. In the beginning, the French furnisher 'experienced some problems with late payment of fees',[101] caused by the low income of the consumers. TEMASOL considered that it did not receive financial returns proportional to its investments, particularly in some difficult geographic areas. Finally, the consortium had to deal with the twin challenges presented by a lack of knowledge regarding solar energy in Morocco and the difficulty of finding expertise in this field at the local level at all stages of the operation (sales, installation and service).[102] These difficulties were eventually overcome thanks to a strategy based on the involvement of local employees, regional branches and local authorities – not only in collecting fees, but also in dealing with the capacity and knowledge gap.[103] The project

---

[94] See Colverson and Pereira (2012). The report includes a set of instructive case studies related to the PPP dimension of sustainable procurements, particularly in the field of SEGS. It draws several lessons from these practical experiences on the opportunity of involving private actors in the elaboration and implementation of public sustainable procurements projects.
[95] Data and information utilised by Colverson and Pereira (2012) were retrieved from the UNDP Growing inclusive database and can be found in the report *TEMASOL: Providing Energy Access to Remote Rural Households in Morocco*, UNDP, 2011.
[96] Ibid.   [97] Ibid.
[98] TEMASOL is a joint venture between the French oil and electricity companies TOTAL and Électricité de France (EDF). It was created especially for this procurement.
[99] See n. 95.   [100] Ibid.   [101] Ibid.   [102] Ibid.
[103] According to Colverson and Pereira (2012), 'Local staff were specifically chosen and trained to bridge the trust gap that inevitably exists between international private sector service providers and rural populations, and introduction of new technologies'.

eventually became profitable in 2008, and the outcome seems to be positive overall. A developing country like Morocco draws non-quantifiable but obvious social benefits from increased access to RE, including benefits in public education.

These cases show that the development of SEGS necessitates not only a strong political drive at the highest level but also large cooperation networks established at national and local levels. Additionally, if an efficient regulatory framework establishing clear guidelines and securing the use of SEGS in public procurement is essential, the proactive cooperation of the private sector, experts and the public also conditions the success of public procurement policies in SEGS.

We now turn to the question of potential trade distortions caused by SEGS and related procurement policies. Equilibrium must be found between the promotion of SEGS and the requirement of non-discrimination in international trade.

## 3 Trade Implications of Procurement Policies and Instruments for SEGS

### *Trade Distortions Caused by Green Procurement Practices*

The promotion of SEGS with respect to public procurement and policies intrinsically bears the risk of discriminating amongst potential suppliers. Non-sustainable products or services will be excluded from the different stages of the procurement process (through specifications, selection criteria, the timing of awarding the contract, and contract performance clauses that integrate environmental concerns). *De jure* and *de facto* discriminations can be identified in the European and the Chinese legal instruments and policies with respect to public procurement.

The compulsory nature of the energy-saving list imposed on the tendering entities by the 2008 Chinese Law on Energy Conservation, for example, clearly excludes suppliers that cannot be registered under this list. Moreover, the compulsory list excludes other certification agencies (among them international organisations). Thus, non-Chinese as well as domestic suppliers may be excluded from tenders without being able to demonstrate the sustainable nature of their products.

In practice, measures promoting sustainable procurement have the potential to be *de facto* discriminatory, as countries do not expressly state the required country of origin for green products and services or the nationality of the qualified service supplier. The major issue of contention concerning the practices and policies promoting the use of SEGS is whether the latter constitute 'non-tariff barriers' (NTBs, i.e. non-tariff measures that have a protectionist intent). Such situations may arise when authorities use standards and ecolabelling to define the characteristics of goods and services to be procured, and such standards and ecolabels are already met by a clearly identifiable category of operators, to the exclusion of others.

Furthermore, standards and ecolabelling are potentially trade-restrictive when they are based on process and production methods (PPMs) that are not apparent from the product itself. Although crucial when it comes to assessing the likeness of SEGS with other products or services, the discriminatory nature of PPMs is still a sensitive and unresolved issue.[104]

### *Compatibility with Multilateral Rules*

The existing non-binding international and regional instruments on government procurement are sufficiently flexible to allow SEGS promotion, but the compatibility of the previously described practices with the WTO GPA is more uncertain.

#### *Compatibility of Practices and Measures Promoting SEGS-Related Procurement with Multilateral and Regional Guidelines*

The UNCITRAL Model Law on Public Procurement has incorporated several provisions that can be interpreted as encouraging states to favour the use of SEGS in public procurement.

While the preamble of the Model Law does not refer to green procurement or to the protection of resources, Article 2 refers to 'social-economic policies' that cover, *inter alia*, environmental policies.

Concerning SEGS-related procurement more specifically the Model Law, Article 11, indicates that evaluation criteria related to procurement may include 'the characteristics of the subject matter, and the terms of payment and of guarantees in respect to the subject matter of the procurement, such as the functional characteristics of the subject matter'. By the same token, Article 11 allows a margin of preference for the benefit of domestic suppliers of domestically produced goods. Local producers or

---

[104] A requirement concerning the way the product should perform or how the service should be delivered in terms of energy efficiency must be distinguished from a situation where a standard or an ecolabel indicates how much energy was used and/or saved in producing the product or the service, or takes into consideration the type of energy used within the process (for example, if it was produced using a renewable source or a conventional fossil-based source). In the latter case, the standard or eco-label directly imposes a PPM. See Malumfashi (2010): 163–164.

service suppliers, who may be more efficient energy consumers than their foreign counterparts, may thus benefit from positive discrimination at the award stage of the procurement procedures.[105]

Another SEGS-friendly provision is found in the Model Law, Article 43.3, which introduces the concept of the 'most advantageous tender' (as opposed to the 'lowest tender price') in the examination and evaluation of tenders. The notion of 'best value for money' – which allows the introduction of sustainable selection criteria and development-related specifications in tenders – is also encouraged by regional organisations, including the APEC 1999 Non-Binding Principles on Government Procurement discussed earlier.

*WTO Compatibility of Practices and Measures Promoting SEGS-Related Procurement*

Whether procurement practices and policies promoted by countries are WTO-compatible raises several questions.

*Is the Country a Party to the GPA?* If a country is not a party to the GPA, the compatibility of the measure with WTO rules must be assessed with respect to the GATT and the GATS. Public procurement appears to be excluded from the scope of application of these agreements.

First, the GATT, Article III:8(a) (National Treatment) and Article XVII(2) (State Trading Enterprises), include an explicit exception with respect to public procurement. The GATS, Article XIII(1) also exempts 'law, regulations and requirements governing the procurement by governmental agencies or services' from the GATS disciplines on MFN treatment, market access and NT. These exemptions in the GATT and the GATS Agreement apply only to purchases for governmental purposes and not to purchases of goods or services for resale, as is done by state trading enterprises who sell products or services on a commercial basis, or for the production of goods for resale.

In short, for non-parties to the WTO GPA, discrimination favouring SEGS in public procurement cannot, in principle, be successfully challenged before the WTO.

In 2010/2011, Japan and the EU initiated WTO disputes – DS412 and DS426, respectively – against Canada in relation to the province of Ontario's local content requirements in a FIT procurement scheme.

---

[105] See UNCITRAL Secretariat, *Revised Guide to Enactment to Accompany to UNICITRAL Model Law on Public Procurement*, A/CN.9/WG.I/WP.79 cong. (W. G. I),13 June 2012, § 23–43.

The outcome of these challenges is likely to have systemic consequences regarding SEGS procurement policies and measures not covered by the GPA.

The case concerned the Ontario Green Energy and Economy Act, which empowers the Ontario Power Authority (OPA) to develop programmes to encourage the use of RE. Under this regime, the OPA developed a FIT scheme that allowed the buying of RE (solar and wind electricity) at an above-market price in order to compensate for the higher production costs. In order to benefit from this incentive programme, the OPA set domestic content requirements (for solar, initially 40–50 per cent of the costs to develop a project, rising to 60 per cent for projects after 2011; and for wind, initially 25 per cent, rising to 50 per cent after 2012).[106]

Japan and the EU maintained that the procurement scheme unfairly discriminated against foreign RE products through its 'domestic content' clause.[107] They argued that the FIT scheme was a subsidy under the WTO Agreement on Subsidies and Countervailing Measures (SCM Agreement) and that domestic content requirements were prohibited under Article 2. The complainants also claimed that the measures violated the NT clauses of Article III of the GATT and the Agreement on Trade-Related Investment Measures (TRIMs). For its part, Canada maintained that the FIT programme was a government procurement scheme intended only to promote use of clean energy in Ontario. It also contended the only relevant WTO agreement in the dispute was the GPA, not the GATT national treatment provision, nor the TRIMS or SCM Agreements. Canada held that since the OPA was not inscribed in Canada's schedule, it was not subject to GPA provisions.[108]

*If the Country is Party to the GPA, is the Procurement Covered by its Schedule of Commitments?* Whether the procurement is covered can be analysed based on the methodology laid down by the WTO panel that decided the previously mentioned *Korea Case*:[109]

[106] This information was partly extracted from the EU Commission website, http://trade.ec .europa.eu/doclib/docs/2007/may/tradoc_134652.pdf?&lang=en_us&output=json.
[107] See a first analysis of the case, http://ictsd.org/i/news/bridgesweekly/129972.
[108] In September 2011 (Japan) and January 2012 (the EU) requested the establishment of a WTO panel in order to determine whether Ontario's measures were consistent with WTO rules. Panel rulings on the twin disputes were issued in December 2012. Canada appealed the verdicts, but in reports released on 6 May 2013, the Appellate Body ruled that the local content requirements in Ontario's FIT scheme did indeed violate the NT principle of the TRIMS and the GATT Agreements. Canada has agreed to bring its measures into compliance by 24 March 2014.
[109] See *Korea – Measures Affecting Government Procurement (Procurement Practices of the Korean Airport Construction Authority)*, § 7.9. For a description of this case, see pp. 206–207.

Table 5.1 *Main Measures Promoting SEGS in EU Procurement Policies and Relevant WTO Provisions*

| Stages of the Procurement Procedure Promoting SEGS | Means of Promoting SEGS | Relevant Provisions of the GPA WTO Agreement (GPA 1994 and Revised GPA) |
|---|---|---|
| **Choice of the applicable procedure** | Restricted procedure or negotiated and competitive dialogue selecting suppliers able to provide SEGS | GPA 1994, Articles VII, X, XIV and XV<br>Revised GPA, Articles VII, XII and XIII |
| **Definition of the requirements of the contract** | *Technical specifications:*<br>– Reference to standards related to SEGS<br>– Reference to performance or functional requirements related to SEGS<br>– Through Ecolabels | GPA 1994, Article VI<br>Revised GPA, Article X |
| **Selection of the suppliers and service providers** | Selection criteria mentioning SEGS<br>Award criteria mentioning SEGS | GPA 1994, Article XIII<br>Revised GPA, Article XV |
| **General exceptions under the GPA** | | GPA 1994, Article XXIII<br>Revised GPA, Article III |

(a) Procurement is not covered by the GPA List of Commitments

In such cases, procurement is exempt from the GPA obligations, and the general provisions of the GATT 1994 and the GATS exempting public procurement will be applicable. The Canada *FIT Case*, however, shows that the application of other WTO Agreements, such as the SCM and TRIMs Agreements, may have to be considered.

(b) Procurement is covered by the GPA List of Commitments

In order to properly assess the compatibility of measures taken by a party to the GPA, reference is made to the way that SEGS can be promoted by purchasers. For ease of reference, the European legal framework and practices will serve as the basis for the assessment. This section also addresses the non-discrimination requirement as laid down in the revised GPA and its possible conflict with SEGS policies (Table 5.1).

## The Existence of a Set of SEGS Procurement-Friendly Provisions

Choice of Procedure (Especially Restricted Procedure and Negotiated Dialogue)   The provisions of the GPA – both the 1994 and the revised GPA – leave the parties room to manoeuvre concerning the choice of the applicable procedure, provided that the tendering procedures are applied in a non-discriminatory manner or, following the formulation of the revised GPA, in a way that 'protects domestic suppliers'. The discrimination may occur when a restricted procedure or a competitive dialogue is applied by a European purchaser who wants to select only those European suppliers that appear to have the technical capacity and experience to provide SEGS. In such a case, the measures may fall within one of the General Exceptions under the GPA, especially in its revised version (see below).

Use of Technical Specifications   Difficulties may arise from the contractual requirements and the possible use of technical specifications related to SEGS. The technical specification stage constitutes a key feature of procurement, providing an opportunity for public purchasers to include eco-friendly requirements for the goods and services demanded through reference to environmental standards or ecolabels. 'Technical specifications' can refer to the product or service itself at the consumption level, or to the process and methods of production.

The revised GPA, Article X (the GPA 1994, Article VI) regulates the use of technical specifications in relations to goods, services and their processes. It requires procuring entities to respect the principles of non-discrimination and transparency, stating that:

A procuring entity shall not prepare, adopt or apply any technical specification or prescribe any conformity assessment procedure with the purpose or the effect of creating unnecessary obstacles to international trade. [110]

The GPA encourages the use of 'standards' in the technical specifications related to goods and services.[111] More precisely, parties are encouraged to refer to international standards (where they exist) or to 'technical regulations' or 'regional standards'. Technical specifications may also include labels or other non-mandatory instruments. The revised GPA, Article X:2(a), specifies that:

---

[110] See GPA 1994, Article VI and revised GPA, Article X.
[111] See GPA 1994, Article VI:2 and revised GPA, Article X:2.

Where design or descriptive characteristics are used in the technical specifications, a procuring entity should indicate, where appropriate, that it will consider tenders of equivalent goods or services that demonstrably fulfil the requirements of the procurement by including words such as 'or equivalent' in the tender documentation.

Moreover, the revised version of the GPA contains two new provisions facilitating the inclusion of specifications related to SEGS:

First, Article I of the revised GPA stipulates that a standard means a document approved by a recognized body that provides for common and repeated use, rules, guidelines or characteristics for goods or services, or related processes and production methods, with which compliance is not mandatory. It may also include or deal exclusively with terminology, symbols, packaging, marking or labelling requirements as they apply to a good, service, process or production method.

Therefore, a process and production methods requirement can be included in standards or labels. Second, the revised GPA, Article X:6, indicates that:

For greater certainty, a Party, including its procuring entities, may, in accordance with this Article, prepare, adopt or apply technical specifications to promote the conservation of natural resources or protect the environment.

Both provisions help facilitate the use of specifications related to the sustainability of a product or service, especially when it comes to PPMs. This would be particularly useful, for instance, when a standard or a label specifies that a good or a service must be produced through energy-saving methods.

The Use of Sustainable and Award Criteria  Through the concept of the 'most economically advantageous' tender instead of the single 'lowest price' tender, procuring entities may take into account secondary policy objectives when awarding a contract. This practice seems compatible with the GPA, confirming the possibility of selecting not only the lowest price tender but also the one considered as most advantageous depending on the specific evaluation criteria set forth in the tender notice.[112]

**SEGS-Related Procurement versus Non-Discrimination**
Non-discrimination is one of the major principles laid down in the revised GPA. Article IV states the following:

With respect to any measure regarding covered procurement, each Party, including its procuring entities, shall accord immediately and unconditionally to the goods and services of any other Party and to the suppliers of any other Party

---

[112] See GPA 1994, Article XII.4(b) and revised GPA, Article XV.5.

offering the goods or services of any Party, treatment no less favourable than the treatment the Party, including its procuring entities, accords to:

(a) domestic goods, services and suppliers; and
(b) goods, services and suppliers of any other Party.
(c) With respect to any measure regarding covered procurement, a Party, including its procuring entities, shall not:
(d) (i) treat a locally established supplier less favourably than another locally established supplier on the basis of the degree of foreign affiliation or ownership; or
(e) (ii) discriminate against a locally established supplier on the basis that the goods or services offered by that supplier for a particular procurement are goods or services of any other Party.
(f) The main question concerning SEGS-related procurement is whether the regulatory provisions allowing such procurement could be considered to be introducing *de facto* discrimination between local and foreign suppliers of goods and services. A SEGS-related provision, for instance, could possibly indirectly favour regional suppliers of renewable-energy and related goods and services.[113]

A parallel exists, in this regard, between the revised GPA, Article IV, and the NT rules described in the GATT, Article III:4[114] and the Agreement on Technical Barriers to Trade (TBT), Article 2.1.[115] The main issue under these provisions, however, relates to the ordinary meaning of the term 'like product' and, unlike these provisions, the revised GPA, Article IV, does not contain any reference to likeness. This concept does not fit well with the rationale of public procurement provisions, which are mostly addressed to suppliers and procuring entities of countries. This does not necessarily mean that no difference should be made between products and services. An argument could be made, for instance, that energy provided by a supplier through solar PV is very different from that provided through a traditional thermal power station. In this case, treating them differently would not amount to discrimination. A dispute involving the GPA, Article IV, therefore, would necessarily have to deal

---

[113] See Davies (2011): 431–433.
[114] GATT, Article III:4 states that: 'The products of the territory of any contracting party imported into the territory of any other contracting party shall be accorded treatment no less favourable than that accorded to like products of national origin in respect of all laws, regulations and requirements affecting their internal sale, offering for sale, purchase, transportation, distribution or use.'
[115] TBT, Article 2.1 indicates that: 'Members shall ensure that in respect of technical regulations, products imported from the territory of any Member shall be accorded treatment no less favourable than that accorded to like products of national origin and to like products originating in any other country.'

with a likeness criterion. Some lessons about this can be drawn from case law related to the TBT Agreement.

In a report dealing with the TBT Agreement, Article 2.1, the Appellate Body used a methodology based predominantly on the competitive relationship between the two products in order to establish 'likeness'. In the *Clove Cigarettes Case*, Indonesia was the complainant against a US provision of the Family Smoking Prevention Tobacco Control Act of 2009 that bans clove cigarettes. Indonesia alleged that Section 907, signed into law on 22 June 2009, prohibits the US production or sale of cigarettes containing certain additives, including clove, but does not ban the production and sale of cigarettes with other additives, such as menthol. The Appellate Body disagreed with the panel that 'like products' in the TBT Agreement, Article 2.1, should be interpreted based on the regulatory purpose of the technical regulation at issue. It ruled that the determination of whether products are 'like' within the meaning of the TBT Agreement, Article 2.1, is a determination about the competitive relationship between the products, based on an analysis of the traditional 'likeness' criteria – namely, physical characteristics, end-uses, consumer tastes and habits, and tariff classification. Further, according to the Appellate Body, the regulatory concerns underlying a measure – such as the health risks associated with a product – may be relevant to the determination of 'likeness' to the extent that they have an impact on the competitive relationship between the products. Based on this interpretation of the concept of 'like products', the Appellate Body agreed with the panel that clove cigarettes and menthol cigarettes were to be considered as 'like products' within the meaning of the TBT Agreement, Article 2.1.[116] Hence, in a case involving public procurement, the main criterion to be used to establish discrimination in a given sector is the one based on the competitive relationship between the products and the services at issue.

Should the promotion of SEGS in procurement procedures be considered to be discriminatory, justification can be sought under the 'general exceptions' of the GPA agreement:

Subject to the requirement that such measures are not applied in a manner which would constitute a means of arbitrary or unjustifiable discrimination between countries where the same conditions prevail or a disguised restriction on international trade, nothing in this Agreement shall be construed to prevent any Party from imposing or enforcing measures: necessary to protect ... safety, human, animal or plant life or plant health.

---

[116] The competitive relationship between two products criterion has also been used as a tool to establish likeness by a panel in *US – Tuna*, (WT/DS381/R), September 15, 2011, and confirmed by the Appellate Body (WT/DS381/AB/R), May 16, 2012.

This provision mirrors the general exception provision of the GATT, Article XX. A WTO panel would, therefore, undoubtedly refer to the case law developed on the basis of that article (in particular, the 'chapeau' conditions and the necessity test). In this context, a dynamic interpretation of the expression 'necessary to protect safety, human, animal or plant health' can be proposed to justify the promotion of 'sustainable energy goods and services'. Examples of similar dynamic interpretations, which also referred to sustainable development, are found in the *US – Gasoline*[117] and *US – Shrimp*[118] cases, and despite the tenuous link between SEGS and the protection of human, animal or plant safety, a possible justification under this exception could exist. A difficulty, however, would arise due to the requirement that SEGS prescriptions based on PPMs *are not applied in a manner which would constitute a means of arbitrary or unjustifiable discrimination between countries where the same conditions prevail or a disguised restriction on international trade*. PPMs do require a party to the GPA to adopt the manufacturing processes of the procuring party in order to benefit from its GPA-outlined rights. Case law has repeatedly asserted the principle that unilateral PPMs do not meet the conditions related to the lack of *arbitrary or unjustifiable discrimination* unless the Member adopting them has proactively engaged in prior consultations with the other Members whose trade interests are affected. These consultations must be conducted with a view to accommodate the trade and environmental interests of all parties involved.[119]

This stated, given the lack of clear WTO case law addressing sustainable procurement policies, it is worth examining how they have been handled by the Court of Justice of the EU.

*Justification on the Basis of Developmental/Environmental Grounds in National Policies: The Example of the Case Law of the Court of Justice of the EU*

Since the end of the 1990s, several cases have been decided by the European Court of Justice (ECJ) concerning secondary policy objectives in public procurement. While stressing the necessity to respect the principles of EU law – in particular, the principles of fair treatment and non-discrimination flowing from the right of establishment and the freedom to

---

[117] *United States – Import Prohibition of Certain Shrimp and Shrimp Products*, WT/DS58/AB/R, 10 August 1998, § 144.

[118] *United States – Standards for Reformulated and Conventional Gasoline*, WT/DS2/AB/R: 20–21, April 29,1996. In these cases, the WTO Appellate Body found that clean air and turtles were an 'exhaustible natural resource' under the GATT, Article XX(g).

[119] See n. 73.

provide services – the Court held that social and environmental objectives in procurement decisions were not, as such, contrary to EU Law.[120]

The *Concordia Buses* judgment[121] clearly demonstrated the possibility of promoting SEGS within the EU legal system. The Helsinki City Council had specified in a tender for buses for the Helsinki urban bus network that the selection would be based on the 'most economically advantageous tender', reflecting several criteria. One of these was the quality of the vehicle fleet, and points were awarded for the use of buses with low nitrogen oxide emissions and low noise levels. The ECJ faced the decision of whether this provision was compatible with EU legislation relating to procedures applying to the award of public service contracts.[122]

In response, the ECJ considered that the EU Directive 'was to be interpreted as meaning that where the contracting authority decides to award a contract to the tenderer who submits the economically most advantageous tender, it may take into consideration ecological criteria such as the level of nitrogen oxide emissions or the noise level of the buses', provided that:

– they are linked to the subject matter of the contract;
– they do not confer an unrestricted freedom of choice on the authority;
– they are expressly mentioned in the contract documents or the tender notice;
– they comply with all the fundamental principles of Community law, in particular the principle of non-discrimination.[123]

In that case, the criteria related to the nitrogen oxide emissions were considered to be effectively linked to the subject of the contract.[124] Furthermore, the points system used to measure the extent to which environmental criteria had been applied did not confer unrestricted freedom of choice on the contracting authority, since it had required tenderers to meet specific and objectively quantifiable environmental requirements.[125]

---

[120] See especially, Case C-225/98, *Commission* v. *France ('Nord-pas-de-Calais')* [2000] ECR I-7445.

[121] Case C-513/99, *Concordia Bus Finland Oy Ab* v. *Helsingin Kaupunki and HKL-Bussiliikenne ('Concordia Buses')* [2002] ECR I-7213.

[122] Council Directive 92/50/EEC of 18 June 1992 relating to the coordination of procedures for the award of public service contracts, OJ 1992 L 209/1. Article 36(1)(a) of this Directive dealing with 'Criteria for the Award of Contracts' was indeed very similar to the now applicable Article 53 of Directive 2004/18.

[123] See ECJ ruling, para 64.   [124] See ECJ ruling, para 66.

[125] See also C-448/01, *EVN AG and Wienstrom GmbH* v. *Austria* [2003] ECR I-14527. In that case, the Austrian authorities required electricity suppliers to supply the Federal offices with electricity generated from RE sources, subject to any technical specifications, and in any case, not knowingly supplying those offices with electricity generated by

The case law of the ECJ is, as in this instance, often quite straightforward.

That is not the case for the WTO. Considering the general reluctance of the WTO system to accept unilateral measures based on prescriptions that are deemed discriminatory, such as PPMs, and also because WTO rules do not sufficiently differentiate the products and services subject to government procurement based on environmental criteria, recourse to an international agreement addressing the issue could prove useful. This issue could be addressed by a SETA.

## 4 Addressing Trade-Restrictive Practices within a SETA Mindful of Sustainable Development Objectives

The objectives listed in the Preamble to the WTO Agreement, as well as the substantive provisions of other WTO Agreements, especially those of the GPA, do not explicitly authorise the promotion of SEGS by public purchasers. WTO panels and the Appellate Body might consider such promotion to constitute indirect discrimination that limits trade in an unjustifiable and arbitrary manner. However, if a party could rely on the existence of a proper international definition of SEGS and the appropriate standards to be applied in tenders, potentially discriminatory measures could perhaps be justified under the existing exception to the rules on public procurement. A SETA could constitute an appropriate framework to define SEGS in an objective and neutral way.

Moreover, the negative effects of the multiplicity of FTAs and the dispersion of rules governing public procurement could be contained by such an Agreement. States wishing to pass laws allowing SEGS promotion could use the SETA framework as a relevant international standard when negotiating the provision of FTAs.

A SETA could also provide an opportunity to change the current approach to SEGS in public procurement. At the international level, and especially within the WTO, this is still a controversial issue, as SEGS and sustainable procurement can most often be justified only as exceptions to the multilateral trade rules. A response to this issue should consist of a positive and proactive approach that would encourage and facilitate tender requirements based on SEGS. The revised GPA specifies

---

nuclear fission. The ECJ considered that these criteria were in conformity with the EU legislation on public procurement, in the context of the assessment of the most economically advantageous tender.

Table 5.2 *SEGS Prescriptions Based on PPMs and Gaps in Current Multilateral Trade Law*

| Gaps in the Current Multilateral Trade Law | Could a SETA Fill the Gap? |
|---|---|
| Legitimate SEGS are not defined in any WTO Agreement. | A SETA should provide a definition of SEGS |
| SEGS prescriptions based on PPMs can be considered as discriminatory if they do not sufficiently relate to the physical characteristics of the products concerned | A SETA could contain an acknowledgement by its parties that products and services complying with SEGS requirements that are consistent with the SETA are different from products and services not complying with these prescriptions |
| If considered discriminatory, SEGS prescriptions based on PPMs cannot be 'applied in a manner which would constitute a means of arbitrary or unjustifiable discrimination between countries where the same conditions prevail or a disguised restriction on international trade' | A SETA could specify that SEGS prescriptions that are consistent with the SETA are assumed not to be 'applied in a manner which would constitute a means of arbitrary or unjustifiable discrimination between countries where the same conditions prevail or a disguised restriction on international trade' |
| Traditional case law is not sympathetic to unilateral PPMs that have not been negotiated beforehand with the affected trade partners | |

that sustainable procurement should be one of the subjects for future GPA negotiations.[126] These negotiations could result in provisions linked to a future SETA Agreement, and conversely – if a SETA is negotiated first – future GPA provisions could also refer to those SETA provisions.[127]

Furthermore, the relationship between government procurement and SEGS – not only from an 'enabling environment' perspective but also from a 'promoting' perspective[128] – may merit examination. A SETA could constitute the legal basis to both allow and promote SEGS-related procurement. An advance group of 'like-minded' parties to the SETA could, for instance, agree to liberalise sustainable

---

[126] See revised GPA, Article XX:8(a)(i).
[127] Concerning the general structure of a SETA, see *Fostering Low Carbon Growth: The Case for a Sustainable Energy Trade Agreement*, ITCSD, Geneva, November 2011.
[128] These expressions are those of one of the reviewers of this article, Marie Wilke.

procurement on a reciprocal basis to develop an international market in this field. Such a proposal, of course, is potentially highly controversial and against the current tendency of promoting non-discriminatory free trade. As observed above, however, liberalisation in the field of public procurement has so far been conditioned by reciprocity. To this end, a SETA could outline the principle of reciprocity for the liberalisation of SEGS, while the parties to the GPA would translate this principle into their respective schedules. Similarly, SEGS-related procurement could then be on a much stronger legal ground than is currently the case.[129]

Finally, the contents of the SETA as they relate to public procurement and the interface with WTO rules, both in terms of content and dispute settlement procedures, should be clearly specified ().

One recommendation would be to try to avoid 'forum shopping' among the various agreements. The inclusion of the SETA within the WTO may be the best option in this regard.[130]

The enforceability of the procurement-related provisions of a SETA could vary. Soft-law provisions, for instance, could adequately address SEGS-related requirements and the exchange of best practices between the parties to the SETA. Existing standards and labels related to SEGS in the technical specifications and in the awards of the parties to the agreement could be developed and promoted similarly, as this requires flexibility and should not limit private sector innovation. In other areas, quantitative objectives could be imposed on the parties to the agreement. For instance, a SETA could require certain proportions of SEGS-certified products in some key sectors (building, construction, transport, etc.), and it could assess such objectives through a peer review mechanism. The requirements could vary based on the level of development of the contracting parties, encouraging broad participation in a SETA.

Table 5.3 is designed to summarise different options with respect to provisions on procurement that could be included in a SETA.[131]

---

[129] Non-parties to the GPA but parties to the SETA could also participate in this reciprocal commitment for SEGS.
[130] See *EC – Approval and Marketing of Biotech Products*, WT/DS291/R, 29 September, 2006, § 7.76–7.89. See also Matthew Kennedy, *Legal Options for a Sustainable Energy Trade Agreement*, ICTSD, Geneva, July 2012: 34–37.
[131] See n. 127.

Table 5.3 *Possible Options for Procurement-Related Provisions of a SETA*

| Characteristics of SETA Provisions on Government Procurement | Pros | Cons |
|---|---|---|
| **Type of Agreement** | | |
| Within the scope of the WTO | Greater legal certainty<br>Non-discriminatory nature of sustainable procurement could be promoted with respect to all WTO Members<br>Avoids 'forum shopping'<br>Could facilitate coordination with the Committee on Public Procurement Activities and the negotiations on Public Procurement<br>More efficient when it comes to justification of SEGS procurement before WTO adjudicatory bodies | Excludes non-WTO Members<br>Difficulties deriving from the limited membership to the GPA will not necessarily be solved |
| Outside the scope of the WTO | Can include non-WTO Members<br>Negotiations of the provisions will not be suspended until resolution of other WTO issues<br>May provide useful lessons that could be replicated within the WTO | Risk of 'forum shopping'<br>Possible conflicts between GPA and SETA provisions (especially before WTO adjudicatory bodies) |
| **Membership** | | |
| Universal | Could allow universal promotion of SEGS in public procurement | Long negotiations<br>Limited results |
| Limited | Could allow group of like-minded countries to develop tools aiming at promoting public procurement | The obligations contained in the agreement could preclude other parties from accepting them |
| **Scope and Content** | | |
| Soft promotion of SEGS in public procurement (with exchange of good practices) | Allow proactive approach regarding SEGS in procurement instead of the current defensive approach | Weak added value considering that non-binding instruments and recommendations already exist at the international level |
| Quantitative objectives imposed | Real incentive to develop SEGS in public procurement that is | Resistance from some countries could limit membership |

Table 5.3 (cont.)

| Characteristics of SETA Provisions on Government Procurement | Pros | Cons |
|---|---|---|
| on the parties (varying with development levels), e.g. 50 per cent of developed-country procurement should use SEGS by 2020 | still lacking at the international level | Difficulty of establishing subjective criteria on which to determine quantitative objectives Requires a proper legal definition of SEGS |
| **Link with Other SETA Provisions** | | |
| Provision on public procurement binding on all SETA members | Greater coherence of entire agreement | Future SETA members could advocate limited provisions concerning public procurement, stressing their discriminatory nature |
| Provision on public procurement binding for some SETA members (SETA *à la carte* approach) | Could allow group of like-minded countries to develop an efficient and detailed legal framework | Difficulties inherent in limited membership |

# References

## *ICTSD Publications*

ICTSD (2011) *Fostering Low Carbon Growth: The Case for a Sustainable Energy Trade Agreement.* Geneva

Kennedy, Matthew (2012) *Legal Options for a Sustainable Energy Trade Agreement.* Geneva

Monkelbaan, Joachim (2013) *Sustainable Energy Services in a SETA.* Geneva

## *Books and Reports*

Anderson, Robert and Sue Arrowsmith (eds.) (2011) *The WTO Regime on Government Procurement – Challenge and Reform.* Cambridge University Press

Arrowsmith, Sue (ed.) (2010) *Public Procurement Regulation: An Introduction.* EU Asia Inter University Network for Teaching and Research in Public Procurement Regulation, Nottingham

Malumfashi, Garba Ibrahim (2010) *'Green' Public Procurement Policies, Climate Change Mitigation and International Trade Regulation: An Assessment of the WTO Agreement on Government Procurement.* University of Dundee

## *Articles, Discussion Papers and Book Chapters*

Anderson, Robert (2007) 'Renewing the WTO Agreement on Government Procurement: progress to date and ongoing negotiations', *Public Procurement Law*, 16 (4): 255–273

Anderson, Robert and Kodjo Osei-Lah (2011) 'Forging a global procurement market: issues concerning accessions to the agreement on government procurement', in Robert Anderson and Sue Arrowsmith (eds.), *The WTO Regime on Government Procurement – Challenge and Reform.* Cambridge University Press

Arrowsmith, Sue (2002) 'Reviewing the GPA: the role and development of the plurilateral agreement after Doha', *Journal of International Economic Law*, 5 (4): 761–790

Brammer, Stephen and Helen Walker (2007) *Sustainable Procurement Practice in the Public Sector: An International Comparative Study.* University of Bath

Brenton, James, Kamala Dawar and Jan-Cristoph Kuntze (2011) *Issues and Tensions in Public Procurement of 'Green Innovation': A Cross-Country Study.* Working Paper of the Centre for Trade and Economic Integration, Geneva, June

Carlsson, Lina and Fredrik Waara (2006) 'Environmental concerns in Swedish local government procurement', in Khi Thai and Gustavo Piga (eds.), *Advancing Public Procurement.* PrAcademics Press, Boca Raton, FL

Colverson, Samuel and Oshani Pereira (2012) *Harnessing the Power of Public–Private Partnerships: The Role of Hybrid Financing Strategies in Sustainable Development.* International Institute for Sustainable Development (IISD), Geneva

Davies, Arwell (2011) 'The national treatment and exceptions provisions of the Agreement on Government Procurement and the Pursuit of Horizontal Policies', in Robert Anderson and Sue Arrowsmith (eds.), *The WTO Regime on Government Procurement – Challenge and Reform.* Cambridge University Press

De Graaf, Gerard and Matthew King (1995) 'Towards a more global procurement market: the expansion of the GATT Government Agreement in the context of the Uruguay Round', *International Lawyer*, 29 (2): 435–452

Dischendorfer, Martin (2000) 'The existence and development of multilateral rules on government procurement under the framework of the WTO', *Public Procurement Law Review*, 9: 1–38

Fuguo, Cao (2008) 'China's government procurement policy and framework: history, structure and operation', in Khi V. Thai (ed.), *International Handbook of Public procurement.* CRC Press, Boca Raton, FL

Guimarães De Lima e Silva, Valéria (2008) 'The revision of the WTO Agreement on Government Procurement: to what extent might it contribute to the expansion of current membership?', *Public Procurement Law Review*, 17 (2): 61–98

Ju, Meiting et al. (2009) *Public Green Procurement in China:* Development Course, Program Management and Technical Methods, *Sustainable Public Procurement in Urban Administration in China: An Action under Europe Aid's SWITCH-Asian Program Paper*, No. 02-EN/CH

Kenrick, Victoria (2011) 'Sustainable energy procurement recommendations for schools', 2 February 2011, www.environmentalleader.com/2011/02/02/sustainable-energy-procurement-recommendations for schools/

Kreycik, Claire et al. (2011) *Procurement Options for New Renewable Electricity Supply*. NREL, Washington, DC

McCrudden, Christopher (2004) 'Using public procurement to achieve social outcomes', *Natural Resources Forum*, 28: 257–267

Nyiri, L. et al. (2007) *Public Procurement for the Promotion of R&D and Innovation in ICT*. European Commission, Brussels

Qiao, Yuaha and Wang, Conghu (2011) *China Green Public Procurement Program: Issues and Challenges in Its Implementation*. China Agricultural University

## Further Reading

UK Department for Business Enterprise & Regulatory Reform (2007) *Competition for a Carbon Dioxide Capture and Storage Demonstration, Project Information Memorandum*, www.ucl.ac.uk/cclp/ccsdedlegnat-UK.php

## GATT and WTO Documents

*Japan – Procurement of a Navigation Satellite*, Request for Consultations by the European Communities, DS73/1, 26 March 1997

*Korea – Measures Affecting Government Procurement (Procurement Practices of the Korean Airport Construction Authority)*, Panel report, WT/DS163/R, 1 May 2000

Report of the WTO Committee on Government Procurement to the General Council, GPA/89, 11 December 2006

*United States – Import Prohibition of Certain Shrimp and Shrimp Products*, WT/DS58/AB/R, Appellate Body Report, 10 August 1998

*United States – Measures Affecting Government Procurement (Massachusetts State Law Prohibiting Contracts with firms doing Business with or in Myanmar)*, Request for Consultations by the European Communities, DS88/1, 20 June 1997, and Japan (DS95), 18 July 1997

## Other Documents

### European Union

Council Directive 92/50/EEC of 18 June 1992 relating to the coordination of procedures for the award of public service contracts, OJ 1992 L 209/1

Directive 2004/17/EC of the European Parliament and of the Council of 31 March 2004 coordinating the procurement procedures of entities operating in the water, energy, transport and postal services sectors, OJ L 134, 30 April 2004: 1

Directive 2004/18/EC of the European Parliament and of the Council on the coordination of procedures for the award of public works contracts, public supply contracts and public service contracts, OJ L 134, 30 April 2004: 114

### EU Commission

*Buying Green! A Handbook on Green Public Procurement*, 2nd edn., 2011, http://ec.europa.eu/environment/gpp/pdf/handbook.pdf

(2004) *Integration of Environmental Aspects of European Standardization*, COM (2004) 130, 30 January 2004

*Public Procurement for a Better Environment*, COM/2008/0400 final, 2 July 2008

*Sustainable Consumption and Production and Sustainable Industrial Policy Action Plan*, COM(2008) 397 final, 16 July 2008

### Case Law

Case C-225/98, *Commission* v. *France ('Nord-pas-de-Calais')* [2000] ECR I-7445

Case C-448/01, *EVN AG and Wienstrom GmbH* v. *Austria* ('EVN') [2003] ECR I-14527

Case C-513/99, *Concordia Bus Finland Oy Ab* v. *Helsingin Kaupunki and HKL-Bussiliikenne ('Concordia Buses')* [2002] ECR I-7213

EC – Approval and Marketing of Biotech Products, WT/DS291/R, 29 September 2006

Standing Committee of the 11th National People's Congress (NPC) of the People's Republic of China, *Circular Economy Law of the People's Republic of China*, 29 August 2008

UNCITRAL Model Law on Public Procurement, Official Records of the General Assembly, 66th Session, Supplement No. 17 (A/66/17)

UNCITRAL Secretariat, *Revised Guide to Enactment to Accompany to UNICITRAL Model Law on Public Procurement*, A/CN.9/WG.I/WP.79 cong. (W. G. I), 13 June 2012

US Supreme Court, *Crosby* v. *National Foreign Trade Council*, 530 US 363, 2000

## 6 Selling the Sun Safely and Effectively: Solar Photovoltaic Standards, Certification Testing and Implications for Trade Policy: December 2013

*Sunny Rai and Tetyana Payosova*

This chapter presents an overview of the mandatory technical regulations and voluntary standards for solar photovoltaic (PV) equipment in the main world markets. Specific examples are provided to illustrate how standards can affect international trade in solar PV equipment and consequently the development of the industry. It also seeks to explore which World Trade Organization (WTO) disciplines are applicable to technical regulations and standards, and whether they can accommodate industry concerns related to harmonisation. Finally, the chapter offers recommendations from the industry perspective, that are reflected in the suggested legal solutions within the framework of a possible future Sustainable Energy Trade Agreement (SETA).

This chapter is one of eight studies commissioned by the International Centre for Trade and Sustainable Development (ICTSD), which launched idea of a SETA with a scoping paper entitled *Fostering Low Carbon Growth: The Case for Sustainable Energy Trade Agreement* in 2011 (ICTSD 2011), a report identifying technical regulations and standards as one of the key trade barriers to be dealt with in the SETA. Gary Hufbauer and Jisun Kim in Chapter 2 in this book have also underlined the importance of international harmonisation of technical requirements (both regulations and standards) for sustainable energy technologies.

This chapter deals only with one specific type of sustainable energy technology – solar PV – and, more particularly, the red tape involved in getting products to market. While the safety and protection of human life and health is the over-riding goal of standard-setters and manufacturers alike, one of the great concerns of the solar industry is the potential of standards and regulations to obstruct trade. Those concerns are addressed in the final section.

# 1 Rationale for Standards in the Solar PV Equipment Market

Safety and performance standards exist for virtually all commercially available electrical products, from air conditioners, toasters[1] and wind turbines and, of course, for solar PV technologies and the balance of system components used to operate a PV system connected to the utility grid. Standards for electrical products have developed for a combination of social and economic reasons voiced by a variety of stakeholders, from advocacy organisations to insurance companies and investors attempting to evaluate risk, to government regulators ensuring public safety and fair-trade practices and, finally, to the actual manufacturers who have a vested interest in selling their products and protecting their industries.

This chapter provides information on the current state of standards applied in the PV industry and explores how these may impede the trading of PV products. Examples are provided to help illustrate the effect of standards on the PV industry and suggestions are provided on how to address these issues. It should be noted that in industry terminology, the word 'standards' refers to both mandatory and voluntary product requirements, while in trade law only voluntary requirements are called 'standards' whereas mandatory requirements are referred to as 'technical regulations'. The meaning of this differentiation will be further explored in the final section.

## *Primary Reasons That Stakeholders Want Standards*

Each key stakeholder in the solar PV market has a primary reason for imposing standards requirements on PV equipment, and many have secondary reasons. Figure 6.1 represents the main reasons. Governments' primary concern is public safety; however, they are also interested in protecting the market, making life better and easier for their producers, creating competitiveness, taking advantage of economies of scale and reducing costly litigation. Consumers and investors are most concerned that the product will be reliable and perform as expected, but they also want to make sure that they have a safe installation – and that it has been certified as such – to protect them from legal problems. Manufacturers want to cover all these bases, but the primary driver in most instances is that they are required to conduct the testing, and would open their companies to legal liability if they tried to work around the system. Harmonisation of standards across the globe helps manufacturers

---

[1] American Society for Testing and Materials (ASTM), *Standard Test Method for Performance of Conveyor Toasters*. West Conshohocken, PA, 2010.

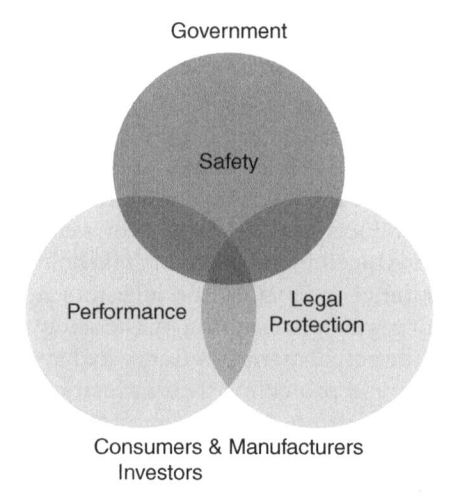

Figure 6.1: Stakeholders and Rationale for Standards

create a competitive environment and take advantage of economies of scale. It also helps keep costs and prices down.

### *Benefits of Employing Standards and Other Quality Assurance Measures*

Quality assurance methods, such as certification that products meet industry standards, are an integral part of all established industries. As described above, industry stakeholders must be able to verify critical aspects of the equipment, the overall system, the installation methods and personnel, operation, performance, safety and many other elements. Without this ability, investment in the industry would be difficult to obtain; prices for critical services, such as financing and insurance, would be prohibitively high; and regulators would be hesitant to approve equipment, installations and work practices. These and other factors would produce a drag on the industry, limiting its growth.

This is the situation faced by the renewable energy (RE) industry for much of its existence. Eliminating barriers like these has led to the development of quality assurance methods, such as product standards that specify safety and performance requirements; inspection regimens to verify the quality of production and installation; and standards for the production and testing of equipment to ensure that measurements are accurate and representative of real world conditions. While many of these efforts started out in the PV industry as loosely organised initiatives with little oversight, most have now developed into structured programmes, employing consensus standards and accredited testing bodies, inspection agencies and certifiers. This provides a level of independence in the testing, inspection, evaluation and certification services that enhance the value of quality assurance methods to industry stakeholders.

### *Rationale for Safety Standards*

The importance of standards with respect to safety of equipment that conducts electrical current goes back to Thomas Edison's first flip of the switch in 1882 at Pearl Street Station in New York. Shortly thereafter came the first victim of electric shock that led to concerns about the safety of this new technology.[2] Given the potential for electric shock and the arc-flash hazards that could lead to fires, it was not long before various attempts to develop safety procedures and standards for electrical devices were developed, resulting in the first National Electric Code (NEC) in the United States in 1897. The advancement of this technology globally gave rise to similar codes and requirements in various regions of the world to promote public safety and guidelines for installers and inspectors of electrical systems.

Solar PV technology produces direct current (DC), and typical installations involve PV modules in a string that operate at values of up to 600–1,000 volts or higher. These installations are routinely exposed to the outside environment and attached to structures where people live or work. These installations therefore pose a significant safety concern both during installation and over the lifetime of their operation.

### *Electrical Safety Hazards and Possible Result of Failures*

The two main electrical-related dangers with respect to solar PV are electric shock and fire.[3] Electric shock can occur during installation, maintenance, or as a result of improper installation, improper grounding, or access to the system by unqualified individuals. Fires generally occur as a result of loose, worn or faulty connectors that create a high-power plasma discharge using partially ionised air as a conductor.[4] Research is underway on the incidence of arc-fault/flash in solar PV[5] systems and,

---

[2] http://brainfiller.com/attachments/ec-history-of-electrical-safety-may2009-pdf.21.
[3] Oregon Solar Energy Industries Association (OSEIA), *Solar Construction Safety*, 2006, Portland, OR: 91–99.
[4] CivicSolar, *Arc Fault Detection*. Boston, MA, 2003.
[5] Jason Strauch, *PV Module Arc Fault Modeling and Analysis*, Slides 1–16, Sandia National Laboratories, Livermore, CA, 2011.

based on that research, changes are being proposed to the NEC to mitigate these dangers.[6] Few products designed to address arc-fault/flash concerns are currently available on the market. This is a good example of how very rational safety standards for a certain market can impact product design for manufacturers with an existing product design based on markets without such requirements.

Given the potential seriousness of harm to installation personnel, the public and property that can result from the improper design, installation, or maintenance of solar PV systems, there is a strong rationale for requiring adherence to installation codes and certifying products to standards in order to ensure their safe installation, operation and protection from possible unintended consequences. However, given the differences in electrical grid characteristics and electrical code requirements across continents, nations, regions and even sometimes localities, these standards can create difficulties for manufacturers of solar PV products.

### *Rationale for Performance Standards*

Beyond the safety concerns that lead to a very understandable rationale for standards for solar PV, the performance of these systems is also an important factor in protecting consumers and investors. It ensures that faulty products and/or misleading claims of product performance do not affect the industry.

Even the design qualification and type approval test standards, like IEC 61646 for thin-film and IEC 61215 for crystalline silicon modules, are primarily designed to predict 'infant mortality', i.e. the likelihood of an early failure in the field. There are several on-going efforts to develop other testing protocols that will predict performance over time. Attempts to create accelerated lifetime tests are difficult to validate to real world conditions owing to the time it takes to perform the testing, the concurrent advancement of materials used in solar PV products and the many different combinations of failures that can occur given the variety of environmental conditions where solar PV systems are installed.

### *Protect Investments with Reliable Information*

There is a need on the part of investors or those offering investments to developers in the solar PV market for more accurate power production, reliability and lifetime data based on environmental integrity and long-term performance modelling. This growing trend in the solar PV investment community is often referred to as 'bankability', and used as a

[6] *Ibid.*

description assigned to a solar PV product or project and the ability of a developer to convince an investor to fund a project.

## *Protect Markets against Faulty Products and Unsubstantiated Performance Claims*

Another important reason for performance standards is to ensure that inferior products and claims do not negatively affect those in the industry who offer quality products, marketed in an honest manner. The solar thermal industry (mainly solar water heaters) in the late 1970s and early 1980s in the United States is an example of the potential for a market to get out of hand when products and claims are poorly regulated. During this period, there was a severe energy crisis in the United States, and solar thermal was less expensive than solar PV. This made it a much better choice (some would argue that it is better value even today). The result was that many 'fly-by-night' companies took advantage of customers by installing inferior systems and making false claims. Much of the damage from this exaggerated marketing and poor products took place before the Solar Rating & Certification Corporation (SRCC) was established in 1980.[7] However, by then it was too late, and significant damage had been done. When the next US Administration took office in 1980, RE incentives were no longer a focus of policy.

Other industries have several examples of organisations that advocate for industry or consumer interests on product claims, performance and safety. A good example is the Air-Conditioning, Heating, and Refrigeration Institute (AHRI).[8] The AHRI is a US industry organisation that has developed performance standards to set the bar for entry into the market and to protect the industry from inferior products and false claims.

## *Rationale for Installation Standards*

Although not directly related to the product standards that affect manufacturers, in order for a PV system to go live there are various requirements their installation, and implementation will vary depending on nation, state (or province), even down to the county and town level in some cases. Solar PV installation training is not a requirement in every location, and this leaves open a high potential for faulty installations and possible harm to installers or system owners. Indirectly, this has

---

[7] Solar Rating and Certification Corporation, www.solar-rating.org/about/general.html.
[8] Air-Conditioning, Heating, and Refrigeration Institute, *Certification Programs*, AHRI, Arlington, VA, 2013.

consequences for manufacturers of solar PV products and components, because it affects the still nascent market in which they are selling. Therefore, manufacturers often find themselves in the position of assisting the growth of the market, particularly in the early stages of development.

### Installation by 'Qualified Professionals'

In markets without specific solar PV training requirements, where few installations have occurred, there is a concern that solar equipment will be installed incorrectly by inexperienced but otherwise qualified electricians. This can lead to installations that do not appropriately to take into account the design concerns specific to solar PV and can cause the system to under-perform, damage the structure where it is placed, or result in potential health hazards.

### Familiarity with DC Voltage and Solar PV Technology

Fortunately, many states in the United States and locations globally do have requirements to ensure that installers have sufficient training. They often have incentive programmes, requiring that the installation be conducted by a recognised or certified installer. For example, the North American Board of Certified Energy Practitioners (NABCEP) provides testing to verify solar PV installer competence. The Interstate Renewable Energy Council (IREC) is the body that certifies training institutions to the Institute for Sustainable Power Quality (ISPQ) international standards for solar training. Those wishing to take the NABCEP examination are required to train at an ISPQ-accredited institution and have a certain amount of installation experience prior to sitting the NABCEP installer exam, which is widely recognised in the United States and in some cases is a requirement for incentives (e.g. New York).

Even with training, there is still room for error by the installer, and given the relative newness of the technology, it may not be picked up by code inspection officials as they, too, have been exposed to few of these systems and have little training in solar PV installations.

Many manufacturers have developed programmes to provide training; however, this is usually limited and specific to their products. Therefore, such programmes may not be viewed by the IREC as sufficiently comprehensive for certification. While there is not a lot that a manufacturer can do to influence the factors of training and inspection directly, this variability certainly does impact a manufacturer's ability to enter markets where there is still a high learning curve for the installation industry. For example, certain states within the United States and certain countries have a relatively small installation base in comparison with a state such as California, or a country like Germany, where the installation base is large, and the learning curve is much further along. The larger installed base means that there is more general consumer knowledge of who is qualified and a higher likelihood of local, regional, or national qualification standards for installers. This is an area that is likely to improve over time, but one way to accelerate this improvement is for manufacturers to work with national and international organisations focused on enabling the solar PV market and supporting manufacturers' efforts to communicate and educate installers and inspectors.

## 2  Overview of Global PV Product Standards

Market commercialisation of a product is often an unexpected difficulty for manufacturers that have never before dealt with the process. Many solar PV products are developed within the structure of new companies that do not have in-house experience with the standards process related to other products. This relative inexperience has created somewhat more difficulty than would generally be expected with the arrival of new or innovative products from manufacturers with existing products and experience in how the general landscape of standards, testing and certification works.

For those not initiated into the world of global standards, these issues can be extremely confusing. This is especially true for those who understand their home market and, quite rationally, believe that the processes related to standards, testing and certification will be similar when expanding to the global market. These individuals will find that the process in other locations can be completely different from their home market. One effort to address this problem is an industry–academic partnership called the Center for Evaluation of Clean Energy Technology (CECET). A subsidiary of the global testing and certification organisation, Intertek, CECET includes several universities and private companies with a specific focus on clean technology that requires significant assistance in preparation for commercialisation. This assistance includes guidance to companies whose technologies do not fit the existing standards. CECET works with them to develop appropriate testing protocols and collecting data. This information is shared with standard-writing bodies with a view to accelerating the adoption of new standards or revisions to existing ones, thus helping speed the entry of innovative new technology to the market. A whole book could – and should – be written on the historical development and current state of testing standards across the globe, but this is beyond the scope of the present chapter. This chapter focuses on the main markets, primarily in Europe and North America, with some discussion of other major solar PV markets. It is impossible within the framework of this discussion to do more than just to scratch the surface of the way standards are developed and recognised by multiple nations and how organisations are accredited to test and certify that products do indeed meet these standards.

### *Standard-Writing Bodies*

Several organisations have been involved in writing PV-related regulatory compliance and performance-based standards. Some of the original standards published in the 1970s have been revised to include certain newer technologies. A number of new standards – mostly related to components and support systems – have been developed in the last decade. For details and the specific history for each standard and associated writing body, please reference the websites of the individual organisations discussed below.

#### *International Electrotechnical Commission (IEC)*

The IEC is a standard-writing body and publisher of international standards for electrical, electronic and related technologies. IEC International Standards form the basis of many national standards that are adopted either wholly or with national deviations, including EN standards for the EU, as well as some standards for Canada, the United States and many other markets around the world. The deviations are added by the adopting country to address any concerns and differences in local power conditions, grid configurations, or other issues related to the use of the equipment.

#### *Institute of Electrical and Electronics Engineers (IEEE)*

An international professional society, the IEEE Standards Association (IEEE-SA) oversees the development of standards via an accredited consensus-based standard development process. IEEE standards 'address a range of issues, including but not limited to various protocols to help maximise product functionality and compatibility, facilitate interoperability and support consumer safety and public health'.

#### *American Society for Testing and Materials (ASTM) International*

The ASTM develops international voluntary consensus standards with over 12,000 current standards published worldwide.

*Underwriters Laboratory (UL)*

The UL is an American National Standards Institute (ANSI) accredited standards-writing body for US electrical safety standards. It has published over 1,000 standards developed through an accredited consensus-based standards development process.

*Canadian Standards Association (CSA)*

The CSA is a not-for-profit membership-based association that develops safety and health standards through a consensus-based standards development process. CSA is one of four standards development organisations accredited by the Standard Council of Canada to develop standards for the Canadian market.

### Major Standards Testing Related to Solar PV

Table 6.1 lists the standards most often used by the industry for solar PV modules, components and major balance of system (BOS) products. The requirements for testing to these standards vary depending on countries and sometimes within countries.

The IEC International Standards are by far the most recognised globally. They allow for differences based on variations in electrical grids and other factors on a country-by-country basis. These are mostly listed as country deviations towards the end of the standards.[9] The country variations are usually fairly minor (examples of some of these variations are given in the specific country descriptions below).

The IEC Technical Committee related to Solar PV Energy Systems is TC 82. The working groups of TC 82 are constantly updating existing standards, as well as meeting to discuss the adoption of new standards. The working groups are focused on building consensus, and contributions from members are entirely voluntary. Therefore, it is not uncommon for a change to take three or more years to be adopted and included in a revised standard. The timeframes are even longer for the adoption of new standards. This presents a huge difficulty for manufacturers of truly unique products in categories that may not fit the testing standards.

The IEC International Standards with local variations make up the majority of the global market for recognised and therefore effectively 'required' standards. The only market-significant global region that does not follow a variation of the IEC International Standards is North America, where UL standards are currently those recognised by

---

[9] IEC International Electrotechnical Commission, www.iec.ch/standardsdev/how/.

Table 6.1 *PV Testing Standards*

| Standard | Testing Overview |
| --- | --- |
| IEC 62109 | Safety standard for inverter technology |
| UL 1741 | Safety standard for inverter technology |
| UL 1703 | Safety standard for solar PV modules |
| IEC 61730 | Safety standard for solar PV modules |
| IEC 61215 | Performance standard for Crystalline Silicon Technology |
| IEC 61646 | Performance standard for Thin-Film PV Technology |
| UL 8703 | Performance standard for Concentrated PV |
| UL 2703 | Safety standard for Rack Mounting Systems and Clamping devices |
| UL 3703 | Safety standard for Solar Tracking Systems |
| UL 1699B | Arc-Fault standard for Solar PV systems |
| IEC 62093 | Performance standard for PV systems |
| UL 61701 | Salt Mist Corrosion Testing |
| UL 790 | Fire Test |

*Source:* Intertek.

government agencies and, more importantly, the code officials who conduct on-site inspections. The code officials or authorities having jurisdiction (AHJs) are more important because, in almost all cases, they have the full authority to accept or deny the installation of equipment within their jurisdictions. In practice, they generally follow closely the NEC in the United States or the Canadian Electrical Code (CEC) in Canada and any regional, state, or local variations in the respective codes. For example, in Canada, the province of Ontario follows the Ontario Electrical Safety Code, which takes precedence over, but is strongly aligned with, the CEC.[10] California also has its own Electrical Code, not to be confused with the Canadian CEC, called the California Electrical Code, which is part of the Title 24 Building Code Requirements[11] and is very closely aligned with the NEC, just as the code in Ontario is closely aligned with the CEC.

### *Certifying the Certifiers: The Process of Approving Testing and Certification Organisations*

To verify that the organisations that are doing testing and certification based on standards are doing so in a consistent and accurate manner, the

---

[10] Electrical Safety Authority, Government and Regulations, *The Ontario Electrical Safety Code*.
[11] State of California Building Standards Commission, *Codes*, 2013, www.bsc.ca.gov/codes.aspx.

majority of markets use a national or international governmental or government-recognised organisation to provide certifier accreditations. They usually require the companies to get accreditation against the ISO 17025 standard to serve as a testing laboratory and/or to provide certifications based on test results. There are three main systems and many local variations, largely due to the historical context of markets that were not as global as they are today. The three main organisations are the Worldwide System for Conformity Testing and Certification of Electrotechnical Equipment and Components (IECEE); [12] the Occupational Safety and Health Administration (OSHA) in the United States;[13] and the Standards Council of Canada (SCC). In the majority of Europe and most other markets outside of North America that use the IEC International Standards, the testing and certification organisations are members of the IECEE.

In the United States, OSHA, which is a Federal governmental agency, designates which organisations are recognised to conduct testing to recognised standards. These testing organisations are called Nationally Recognised Testing Laboratories (NRTLs).[14] The OSHA model requires that the testing organisation provide both testing and certification, and that the product standards be developed by an 'appropriate' US standards organisation. Although global companies are now included, there is still a historical dominance of the UL standards in the United States.

Canada follows a model similar in some ways to OSHA; however, instead of requiring organisations to provide both testing and certification, the SCC allows for, but does not require, separation. Another difference is that while OSHA just specifies that the standards be developed by a US standards organisation, the SCC recognises only four major standards organisations.[15] Of those, only the UL and the CSA are used for solar PV.

The IECEE requires the separation of testing from certification and designates the certification bodies as National Certification Bodies (NCBs). Member countries then designate the Certification Body (CB) Testing Laboratories (CBTLs) that, in association with the NCBs, will be responsible for recognising and issuing the CB Test Certificates. The result

---

[12] IEC/IECEE. List of Photovoltaics Members, www.iecee.org/pv/html/pvcntris.htm.
[13] United States Department of Labor, Occupational Safety & Health Administration, *Typical Registered Certification Marks*, Washington. DC.
[14] United States Department of Labor, Occupational Safety & Health Administration. *Nationally Recognised Testing Laboratories (NRTLs)*, Washington, DC.
[15] Standards Council of Canada. *Directory of Accredited Standards Development Organizations*, Ottawa, www.scc.ca/en/accreditation/standards.directory-of-accredited-standards-development-organizations.

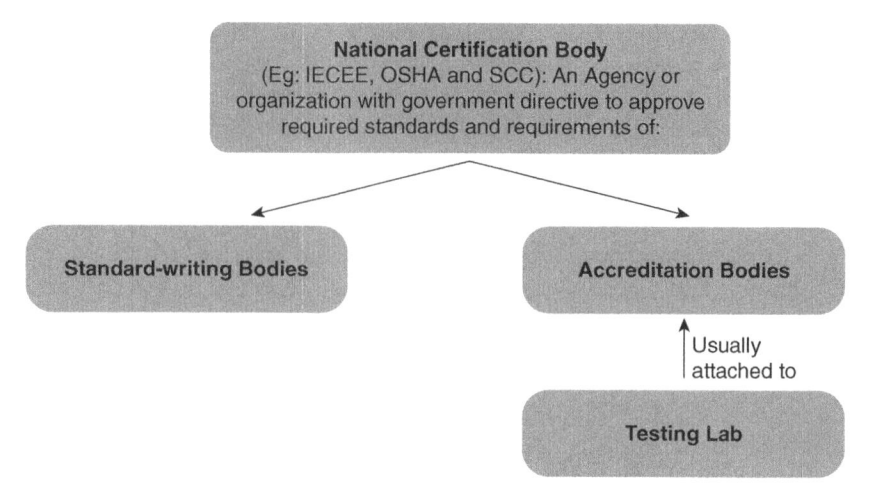

```
                    National Certification Body
              (Eg: IECEE, OSHA and SCC): An Agency or
              organization with government directive to approve
                   required standards and requirements of:

        ↙                                              ↘
 Standard-writing Bodies                        Accreditation Bodies
                                                        ↑ Usually
                                                        | attached to
                                                    Testing Lab
```

Figure 6.2: The Role of National Certifying Bodies
Source: Intertek.

is much the same as in the model used by the United States, except that there are multiple countries involved and therefore multiple NCBs. Unlike the United States under OSHA, the testing and certification is done by separate entities. It should, however, be noted that even though the entities are separate, often the mother organisation for a NCB and a CBTL is the same. The IECEE approach also entails an effort toward harmonisation of standards, making it an approach well suited for companies working in a global market. Unlike OSHA, IECEE results are recognised in countries other than that in which the testing was done, so that only one test is needed for many markets (see Figure 6.2).

### *Hypothetical Example of Testing Differences in Germany and the United States*

There are several different standards related to solar PV, and multiple standards for the same general purpose, across countries and regions. The following example illustrates the case of a hypothetical German PV module manufacturer who wants to have an existing PV module already certified to sell in the German market to be certified for sale in the North American market as well.

German manufacturer JaVolt has been selling a 72 cell monocrystalline 230 w solar PV module for feed-in-tariff (FIT) projects throughout Germany for several years. Its initial product was tested by VUT

Nordest (a hypothetical certification organisation recognised by the German accreditation body DAkkS to certify test results from recognised testing providers). For the actual testing, JaVolt worked with Herrnhopfner (a hypothetical non-profit testing organisation recognised by DAkkS for providing testing services). Herrnhopfner conducted the IEC 61730 and IEC 61215 International Standard tests for safety and performance. Although JaVolt went through a few re-tests, it was able to pass and sell its product in Germany and, from its understanding, across the European and other markets that had adopted IEC International Standards for solar PV.

A few years ago, JaVolt expanded into the French and Italian markets. It realised that in these markets, even though they are in Europe and the company had already performed the IEC testing, JaVolt would need to undertake further verification and testing order to qualify for FIT incentives. In France, the incentive had significant focus on the building-integrated photovoltaics (BIPV) market, and the demand was for a product that was approved for BIPV installations. This required JaVolt to change the design of the frame, and the changes required that JaVolt re-test a significant portion of the product under IEC International Standards for the design revision. For Italy, JaVolt found out that, in order to qualify, its modules needed to be recognised by the International Association of Plumbing and Mechanical Officials (IAMPO).

After another year of success in these new markets, JaVolt Business Development Director, Friedrich Sonnenbaum, realised that, to maintain market growth, the company would need to enter the North American market. Friedrich hired a friend from college, Richard Schwarzenegger (no relation to Arnold), who spoke excellent English to go to the United States to prepare for this market expansion. While in the United States, Richard began speaking with wholesalers and distributors of solar PV and learned that in order to sell JaVolt PV modules, he would need to get UL 1703 certification. Upon further research, Richard found that IEC testing was in many ways similar and, if anything, more extensive than UL testing, but in discussions with representatives of US testing laboratories, he learned that regardless of the results of IEC testing, the company would need to undergo re-testing to get UL 1703 certification. As VUT Nordest was not able to conduct this testing, Richard contacted a NRTL based in the United States. Working with this group, he learned that he could receive certification only by conducting additional tests that were required for the UL 1703 standard and not already covered in IEC testing.

## 3 Country Experiences

### *Global Solar PV Production: Historical Overview*

Solar PV was largely invented and first commercialised in the United States, and up until the twenty-first century the United States both manufactured and installed more solar PV systems than any other country. Manufacturing dominance moved to Japan in the mid 1990s and remained there until the European market began to increase its installation base, owing to the introduction of FITs, which resulted in manufacturing ramping up in Europe from 2000 to 2006. During this period, China also increased its PV manufacturing, and eventually took the lead position in 2007. It has since significantly dominated the market (Figure 6.3).[16]

### *North America*

In the beginning of solar PV, North America had the largest number of installations, although most were off-grid. Following the growth of installations in Japan and the introduction of grid-connected FIT incentives in Germany in 2000, North America quickly lost this position.

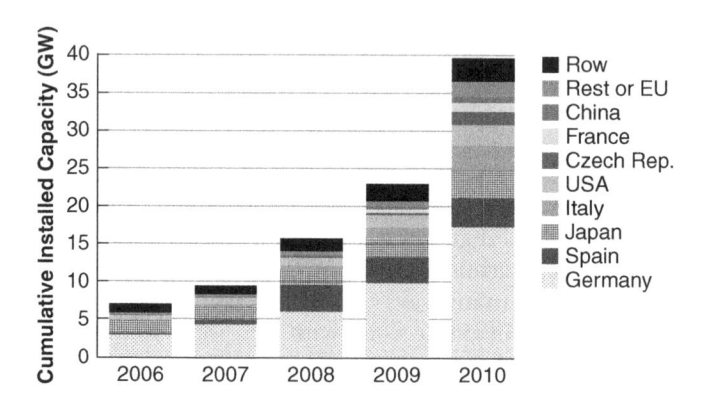

Figure 6.3: Cumulative Installed PV Capacity in the Top Eight Countries
*Source:* US Department of Energy, *Solar Technologies Market Report*, 2011.

---

[16] Earth Policy Institute, Climate, Energy, and Transportation, Data Center, www.earth-policy.org/data_center/C23.

*United States*

In the United States, certification to standards adopted by the ANSI is required to demonstrate compliance for most elements of a solar PV installation. The United States continues to use the UL standards for solar PV: UL 1703 and UL 1741 for interconnection equipment, such as inverters, combiner boxes and other BOS components. Efforts are underway to replace or harmonise current solar PV and inverter standards with US versions of the IEC International Standards. Although the standards will not be identical, having a common framework and many common requirements will provide a clearer path to compliance across markets.

Standards are also being developed related to rack-mounting systems and clamping devices (UL 2703) and tracking systems (UL 3703). Certification to these standards must be conducted by a nationally recognised testing laboratory accredited by OSHA. As of the time of writing, only three NRTL organisations provided PV module testing in the United States: Intertek/ETL, UL and the CSA.

*Canada*

Canada uses standards very similar to UL 1703, but there is an industry expectation that the country will move toward adoption of the IEC International Standards in the future. In Canada, the province of Ontario has experienced the most activity with regard to solar PV, due to its FIT scheme. While there were many difficulties in getting the programme off the ground, the scheme has attracted several manufacturers to Ontario, where they must comply with the domestic content rules of the FIT programme. The rules are too complex to cover for the purposes of this chapter; however, the intent is to ensure that the funds spent on the FIT programme also bring jobs and manufacturing to the province. The domestic content rules effectively require that the majority of assembly for all components of a PV system (inverter, racking and PV modules) be conducted in the province of Ontario. Due to this requirement, many companies have moved manufacturing to Ontario, or developed contract manufacturing agreements so that they can have a presence in this market.

*Mexico*

Growth in PV installations in the Mexican market has been very modest in comparison with the rest of the world. However, several PV manufacturers have assembly lines in Mexico; among the more well known are

SunPower, Kyocera and Sanyo. Manufacturers undertake operations in Mexico to take advantage of lower labour costs and the close proximity to the rest of the North American market.[17] Mexico has some incentives, including tax breaks for renewable energy investments and funds available via the country's participation in the Kyoto Protocol. Also, Mexico plans to have 8 per cent of power produced by renewables. The majority of installations appear to be wind power or other technologies, but there is certainly a good solar potential in Mexico, although growth may be seen more in off-grid or small to medium installations that use the majority of the power produced.[18]

## South and Central America and the Caribbean

The electrical systems in many of the countries in Central and South America, and the Caribbean island nations are diverse. This is likely due to the historical development of electricity and the governments in power at the time electric technology became prevalent in these locals.

The mixture of distribution networks is based both on the 50 Hz and 60 Hz frequencies at voltages similar in range to the US 110 v or the predominant European 230 v. Certain countries, like Brazil, do not have a common voltage across the country. Instead, there are '27 Brazilian federative units and their respective voltages'.[19]

### US Virgin Islands

Many of the island territories and nations are highly motivated to adopt solar PV, because they currently depend on fuel oil (diesel) for power generation.[20] The costs make solar PV an attractive option even without incentives. Given the environment and the weather that often occurs in the Caribbean, PV installations require more planning and design for high wind loads, power fluctuations and capacity concerns.

### Brazil

Despite the twenty-seven different federative units in the country, the Brazilian market offers a great opportunity and interest. The units are

---

[17] Michaela Platzer, *US Solar Photovoltaic Manufacturing: Industry Trends, Global Competition, Federal Support*, Congressional Research Service, 2012: 1–32.
[18] Allan Marks, 'Mexico offers diverse opportunities for investment in renewable energy', *World Energy Magazine*, 10 (4), 2008: 104–108.
[19] 19. World Standards. *Electricity around the World*, www.worldstandards.eu/electricity.htm.
[20] Virgin Islands Water and Power Authority (WAPA).

based on differences in the electric grids of various localities. While this is not a problem for solar PV *per se*, it does present challenges for the inverter technology used to convert the DC power created by PV modules into AC power for the grid. Given the variations in grid structure, there is a potential need for twenty-seven different configurations for inverters. Although a very modest amount of PV has been installed in Brazil (around 30 MW) so far, discussions are underway on policies that could lead to a solar boom in Brazil.[21] Given the large size of the country and the fact that the majority of the population is concentrated in only a few of the twenty-seven units, installations may pose only a modest challenge to manufacturers of inverters and, even if they need to provide multiple types of units in the market or conduct additional testing, they are likely to make the effort to do so. This is an example of how the actual interconnection and safety standards, even when complicated, do not keep companies from entering a market.

### Central America

While there is great potential for solar PV in Central America given its reliance on high-cost fossil fuels[22] and the lack of utility grid infrastructure, progress in the region has not matched that of the rest of the world.[23] Economics is of course a major factor in this slow growth. A very large portion of the population of Central America does not have access to electricity. This makes for a great opportunity for distributed generation and mini-grid sources based on renewable technologies, including solar PV.[24]

### Europe

At the turn of the millennium, Germany and other European nations adopted energy policies aligned with meeting their commitments under the Kyoto Protocol. Germany was the first to offer a FIT, and Spain, France, Italy and others followed. The installations also attracted a manufacturing ramp-up, and Germany gained a fairly dominant position in global solar PV with many manufacturers of solar PV and related BOS.

---

[21] 'Brazil to issue regulations supporting solar energy, Aneel says', *Bloomberg*, March 14, 2012.
[22] Martin and Carlos (2012): 1–4.  [23] Worldwatch Institute (2013).
[24] Elzinga et al. (2011): 1–64.

### Germany

Germany is often cited as a model for implementing solar PV, and its production in this field now supplies about 3 per cent of total power output.[25] In 2012, Germany reached 25 per cent of its power production from RE sources, including wind, biomass, hydro, PV and other technologies.[26] With the announcement that Germany will be phasing out nuclear power,[27] which accounts for about 20 per cent of power production, there is still a promising future for more RE and, in particular, solar PV production in Germany owing to the successful FIT programme. However, as is the case for any incentive scheme, some still oppose the transition to renewable sources and, as in other traditional centralised power organisations, are fighting to retain the status quo. Although the high tariffs of the programme, combined with significant reductions in the cost of solar PV equipment, have resulted in a very attractive investment for some, others blame it for rising energy costs for many while filling the pockets of only a few. However, for the most part, the German incentive programme has continued to reduce its tariffs and kept on-going market changes reflected in updates to the FIT policy.

Germany, like most of Europe, has adopted the IEC International Standards as a requirement for products intended for installation of solar PV. Many of the companies providing testing services, including TUV, VDE and Fraunhofer, are based in Germany. Even though competition from Chinese manufacturers has had an impact on the PV manufacturing sector in Germany, it is still clearly the global leader in installations and experience in implementing a successful solar PV energy policy.

### Spain

Spain, like Germany and the majority of Europe, uses the IEC International Standards. The rush to install solar PV in Spain, leading up to the peak in 2008, shows how many barriers such as standards, codes and other issues do not halt progress when the investment is attractive enough. The boom-and-bust experience in Spain is a case worthy of further evaluation to understand how standards and certifications did or did not impact installations under the initial FIT policy.

---

[25] 'German solar power output up 60 per cent in 2011', *Reuters*, December 19, 2011.
[26] Bdew, *Energie.Wasser.Leben*, Bundesverband der Energie-und Wasserwirtschaft, Berlin, 2012.
[27] 'A revolution for renewables: Germany approves end to the nuclear era', *Spiegel Online International*, 30 June 2011: 1–2.

## Italy

Italy has experienced excellent growth in solar PV since 2008. The Italian FIT is doubly attractive given the prevailing high cost of electricity in the country. As in other countries, there are complaints about subsidies for RE production. In light of waning public support and a real reduction in the cost of solar PV, tariffs have been reduced to adjust to these market realities. However, the growth curve for Italy is strong, and if nothing happens to cause a boom-and-bust scenario like that in Spain – always of concern with the constant changes in Italian politics and economy – an optimist would say that there is great hope that PV will continue to grow strongly.

Italy also uses the IEC International Standards; however, it has additional requirements for 'made in EU' content to qualify for the higher percentage of government rebate, including specific variations for connecting to the grid and additional requirements based on the climate, type of technology and location of use. The laws are quite complicated and shrouded in technical terminology that makes the programme difficult to understand even for a native Italian. Major changes to Italy's FIT programme are set to take effect in 2013.[28]

## Middle/Near East and Africa

The Middle East and Africa obviously have great solar potential. However, the combination of political instability, varying conditions of electrical transmission grids and investor hesitancy risk creating barriers above and beyond standards and trade issues. Nevertheless, the region is hosting many events for solar PV in Egypt and Abu Dhabi,[29] where the very impressive plans for Masdar City have been progressing despite several hiccups.[30] Even the oil-producing nation of Saudi Arabia is investing in solar.[31]

In some cases, the instability of surrounding nations is causing countries to look into solar. For example, Jordan relies heavily on natural gas from the Arab Gas Pipeline, which fuels 88 per cent of its electricity. Instability has interrupted this supply and caused the country to use costly diesel and oil generators. Jordan, unlike its neighbours, does not have significant oil in the ground, so this is not a sustainable solution, and the

---

[28] *Gazzetta Ufficiale della Repubblica Italiana*, 23 February 2007.
[29] World Future Energy Summit, Abu Dhabi, 20–24 January 2014.
[30] 'Abu Dhabi's $600m solar power plant to be completed this year', *Gulf News*, 29 March 2012: 1–2.
[31] 'Why Saudi Arabia is taking a shine to solar', *Christian Science Monitor*, October 28, 2012: 1–2.

government has recognised this. Fortunately, Jordan is in a great location for solar PV and has ambitious goals to reach renewable targets. It is unclear what the specific requirements are for solar installations in the country, but an initiative by a New Jersey-based company with ties to Jordan is looking to assist the government with developing a solar PV strategy, and it is likely that Jordan will adopt international standards similar to the IEC for any proposed programmes.[32] Turkey has similar concerns to those of Jordan and is starting to develop incentive programmes and host events for solar PV as well.[33]

South Africa has become a hot topic, as the government has begun some pilot programmes to develop more energy independence. Given South Africa's excellent solar potential, solar PV is a good candidate. It is unclear at this time what specific standards the government will require for solar products, but it is likely, as for the Middle Eastern countries, that some variation of international standards will be the choice.[34]

The rest of the African continent holds lots of potential for micro-grids and smaller local off-grid applications of solar; however, the lack of transmission will make the approach to supplying the majority of African nations quite different from that of most other markets. The interesting aspect of this vast continent is that given the lack of existing transmission infrastructure, there is huge potential for distributed and micro-grid generation with storage technology. This kind of development could bring to many in Africa a dependable source of electricity that they have never had access to before.

This kind of development could allow the African continent to leap ahead of the rest of the globe that, for many reasons, will be reliant on a grid delivery structure for the foreseeable future. Cell phone use offers a parallel: in Africa, growth in the use of mobile phones makes perfect sense as it costs much less than installing infrastructure and traditional land-line telephony and, as the technology advances, it is not as dependent on a physical infrastructure. We may see a similar progression with solar power in Africa, where economic growth is spurred by access to electricity that was previously infeasible. One day, we may look at Africa as the model to follow, instead of continuing to invest in a large and expensive physical electrical grid structure. For much of Africa, there is a complete lack of standards or, if standards do exist, there is minimal enforcement. In

---

[32] 'Jordan turns to solar for energy security', *PV Insider*, June 28, 2011: 1–2.
[33] Eyigün Sevnur and Önder Guler, *Turkey Solar Potential and Viability of Solar Photovoltaic Power Plant in Central Anatolia*: 94–99, Paper written for the International Renewable Energy Congress, Tunisia, November 2010.
[34] Department of Energy of South Africa, *Renewable Energy Independent Power Producer Procurement Programme*, Pretoria.

addition, if you are installing an off-grid or micro-grid application, you are not interacting with a utility grid and can therefore install technology with minimal concern for regulatory standards. Off-grid installations may have some limitations, but there are significantly fewer barriers than for grid-tied installations. While some may see regulatory standards as market barriers, these standards help ensure safety and reliability, in particular for products that are expected to last more than twenty years in the field and where their use often powers critical loads.

### *Asia and Oceania*

Mainland Asian countries, Australia, and large island nations are all excellent candidates for a growing solar PV market. However, even more than elsewhere in the world there are many different electrical utility requirements if we look at the region as a single entity.

#### *China*

China has without question had a significant impact on the solar industry. The country has become the undisputed leader in the manufacture of solar PV in a short time, surpassing Japan around 2007 and then skyrocketing 500 per cent in total production over three years. This rapid growth has led to claims of unfair competitive practices on the part of Chinese companies and accusations that the Chinese government is subsidising the industry to be able to sell at less than cost. Whether or not these claims are valid, the PV industry outside of China has been severely impacted by the rapid emergence of Chinese dominance (see Figure 6.3).

Beyond the subsidies provided by the government to manufacturers that sell in an export market, China has a huge potential market of its own.[35] With a growing middle class and large rural electrification projects under way, solar PV is an important focus of the country's energy policy. China has its own certification process for products; the [China Compulsory Certificate] CCC mark requires testing to standards that 'correspond largely to [Economic Commission for European] ECE and other international standards. Nevertheless, the Chinese authorities do not recognise any tests except the tests conducted in Chinese test labs during the certification.'

'[36] Unlike many other countries, which recognise testing to international or national standards conducted in other countries as long as it is

---

[35] 'China strives to develop new, renewable energy: white paper', *Xinhua*, 24 October 2012.
[36] China Certification, *CCC Made Easy*, 2013: 6.

done by an approved internationally recognised and accredited testing organisation, China requires domestic testing labs and auditors to conduct all testing in the country, effectively providing no other market entry option than physically testing products in China. The testing process can take from four to eighteen months, and often requires the additional cost of a consultant.

However, at least for solar PV, it appears that this requirement is not stringently enforced. If it were systematically enforced, it would create an additional barrier to entry in an already difficult market to penetrate for solar PV manufacturers. Some of the other challenges that solar PV manufacturers from outside China may experience are intellectual property (IP) protection, different tax treatment for foreign companies conducting business in China, restrictions on foreign direct investment (FDI), and general issues related to the relationships between government officials and foreign companies.[37] In spite of the challenges, solar PV companies with headquarters and the majority of their manufacturing based outside of China (i.e. SunPower and First Solar) have signed major deals in anticipation of China's becoming the next big market for solar PV installations.[38] However, most manufacturers will find the combination of additional requirements and affordable alternatives produced in the home market to be a significant barrier to entering the Chinese market.

### India

India is similar to China in that it also has a rapidly growing middle class and broad needs for electrification updates throughout the country. The government has begun to initiate programmes to attract both foreign investment and support for local manufacturing of PV. According to the Government of India Ministry of New and Renewable Energy:

The Jawaharlal Nehru National Solar Mission [JNNSM] was launched on the 11th January 2010 by the Prime Minister. The Mission has set the ambitious target of deploying 20,000 MW of grid connected solar power by 2022 is aimed at reducing the cost of solar power generation in the country through (i) long term policy; (ii) large scale deployment goals; (iii) aggressive R&D; and (iv) domestic production of critical raw materials, components and products, ... to achieve grid tariff parity by 2022. [The m]ission will create an enabling policy framework to achieve this objective and make India a global leader in solar energy.[39]

---

[37] US Department of Commerce, *An Exporter's Guide to China*, 2008: 1–100.
[38] 'First Solar, SunPower expanding in booming China market', *Bloomberg*, December 2, 2012.
[39] Government of India. www.mnre.gov.in/solar-mission/jnnsm/introduction-2/.

The basic requirements for solar PV installation in India are conformity with the 'latest edition' of the IEC safety and performance standards. There are additional requirements to conduct a salt/mist corrosion test for coastal environments, and a unique obligation for radio-frequency identification (RFID), which may be the most significant barriers to trade, given the logistical hurdles they present for manufacturers of solar PV based outside of India.

The requirement to include RFID traceability in each module is a very useful method for administering a programme, ensuring against counterfeit products and reducing the likelihood of theft. The regulation requires, *inter alia*, that each PV module must use an radio frequency [RF] identification tag (RFID) with the following information: name of the manufacturer of PV module or solar cell; month and year of manufacture; country of origin; I-V curve for the module; peak wattage, Im, Vm and FF; unique serial and model numbers; date and year of obtaining IEC PV module qualification certificate; name of the test lab issuing the certificate; and other relevant information on the traceability of solar cells and modules as per ISO 9000 series.[40]

Until March 2013, the RFID could be placed inside or outside the module laminate, but since April 2013 the identification tag must be placed inside.

Innovative as this idea is, it is not a normal process for most manufacturers of solar PV. It is an example of policies that may have unintended consequences that may impact trade. As stated above, until April 2013, RFIDs could be attached outside the module, which meant increased costs and a logistical challenge for manufacturers wishing to enter the Indian market. The fact the identification tags must now be placed inside the laminate is a direct change to the manufacturing process and could have impacts on the industry. There may be benefits if all manufacturers adopt this process, and if it does not increase costs significantly; however, if only some companies adopt the measure – or if the cost of doing so is prohibitive – it can create indirect trade barriers.

### Japan

Solar PV has had more impact and market visibility in Japan than perhaps any other country except Germany. Prior to Germany's FIT and rise as a global leader of installed PV, Japan held the lead in annual installed capacity from 1998 to 2004 (see Appendix I). Japan also held a significant percentage of the markets for manufacturing and installations of solar PV. Until 2004, when installation prominence was taken over by Germany,

---

[40] Government of India, *Minimal Technical Requirements/Standards for SPV Systems/Plants to be Deployed During F.Y. 2012–2013*: 1–5.

Japan was the undisputed leader in PV manufacturing, with 50 per cent of the market (see Figure 6.3).

Japan also follows the IEC standards, with specific variations adapted for the market and electrical grid. A Japanese Industrial Standardization (JIS) certification is required for safety, as well as a Japan Electrical Safety and Environment Technology Laboratories (JET) PVm certification, which includes factory inspection, and is a widely recognised voluntary certification often used in the Japanese market for installations.[41]

### Republic of Korea

The rapid growth of Korean companies in the electronics industry makes the country a very logical place for adoption of PV manufacturing and installations given the established and growing middle class. Samsung, LG and Hanwha are some examples of large Korean manufacturers in the solar PV market. These manufacturers have a significant amount of cell manufacturing related to the semiconductor industry on the peninsula.

The market has been slow to adopt and enforce standards; therefore, there are a lot of uncertified products in the market. The PV industry and government have realised the need for adherence to international standards and are implementing requirements in order to receive attractive incentives for installations.[42]

The Korean solar PV requirement for safety is testing to the KSCIEC 61730-1 and KSCIEC 61730-2 standards. For performance testing, the crystalline silicon standard is KSCIEC61215, and the thin-film standard is KSCIEC 61646. You may notice that the numbers follow the IEC International Standards and are a clear indication of the related similarity. This is a common practice when the International Organization for Standardization (ISO) or IEC International Standards are adopted across the world. As many other countries, Korea has adopted the IEC International Standards with specific variations for the local market. There are emerging efforts in research and development (R&D) aimed at creating a market with recognised standards in the country. With a global presence, strong financial backing and a significant domestic market, Korea could be a future leader in solar PV from both a manufacturing and installation perspective.[43]

---

[41] Japan Electrical Safety & Environment Technology Laboratories. *Certification of Photovoltaic Modules*: 1–2, http://www.jet.or.ip/en/products/solar/index.html.
[42] 'Korean solar industry speaks', *InterPV*, 2010: 1–3.
[43] Korean Institute of Energy Research, *New and Renewable Energy Technology*, Global Kier, 2011: 1–2.

*Australia*

Australia has also recognised the potential that it has in adopting solar technologies. Major investments by the government into R&D in the sector and collaboration with the United States are indications of the intention to promote solar PV in the region.[44]

PV modules installed in Australia must be certified and approved to AS/NZS 5033 PV installations regardless of whether a rebate is sought. This standard is referred to in the Australian/New Zealand Standard for Wiring Rules (AS/NZS 3000:2007), which is legislated in each state. Section 4 states that modules shall be compliant with IEC/EN 61730 and either IEC/EN 61215 or IEC/EN 61646. This was enforced effective from 1 June 2009. PV systems above 50 V (open circuit) or 240 W must meet Application Class A of IEC/EN 61730.

The Clean Energy Council (CEC) maintains a database and website listing of AS 5033-compliant PV modules. The modules are always listed under the certificate holder's name with the model numbers shown on the certificate. Importers or distributors branding original equipment manufacturer (OEM) products must obtain a certificate in their own name, showing their product numbers as offered for sale in Australia.

## 4   Diversity of Regional and Local Requirements and Challenges to a Harmonised Global System

In addressing the challenges related to adapting products for global markets, it is important to understand the underlying needs for these variations and to keep in mind that the intent is not to impede trade, but to ensure safe and effective performance of solar PV system installations. In order to gain a deeper understanding of these needs, this section discusses some of the major differences between regions or specific locations that lead to the diverse requirements. By understanding this diversity better, solar PV technology stakeholders can learn ways to develop products and strategies to more easily bridge the differences with minimum additional design complications and costs.

### *Global Electrical Grid Characteristics*

While there is some commonality in voltage and frequency use among different regions, there is a very evident difference between the Americas and Europe and Asia. Even though this could lead one to believe that

[44] 'Australia takes a shine to solar energy research', *Sunday Morning Herald*, 13 December 2012: 1–2.

designing for these three major markets is sufficient, there are several other local factors beyond the grid interface that require product variations to install in these markets. We will cover a few examples in the following sections.

### Voltage Limits

Until recently, the US NEC limited the maximum PV system voltage to 600 VDC, while in Europe 1,000 VDC has been the standard voltage. Significant advantages exist in conversion efficiency, system cost and design related to higher DC voltage; these advantages have prompted the adoption of 1000 V systems in the United States and a move to 1500 V in the European market.

The varying voltage requirements affect modules, inverters, combiner boxes, DC disconnects and many other BOS components making it difficult to produce a single model for use in both the North American and European markets.

### Ride Through

An important safety feature of grid-connected solar PV inverters is that they will shut themselves down upon sensing a loss of the grid (anti-islanding). In some installations, particularly larger 'utility scale' solar installations, there is an advantage to keeping the system online to 'ride through' momentary abnormal grid conditions, such as voltage sags. The desire for this feature and the level and type of grid abnormality that the system will be required to ride through varies between utilities, making a standard set of requirements difficult to institute and leaving manufacturers of inverter equipment at a loss for how to design their products to fit all needs while still providing the required safety features. It is important to note here that only a few of these differences are justified by any objective criteria, and in large part are the result of history or decisions taken when each utility was a world unto itself. Consequently, it would make enormous sense for international standards to be written, and then adhered to, for connecting PV installations to the grid (including, but not limited to, ride-through rules), just as it makes sense to have international standards for the PV modules themselves.

### To Ground or not to Ground

Grounding is another example of a major difference between the North American and European markets that causes confusion and complication for manufacturers looking to manufacture products for a global market. Up until recently system grounding (grounding either the positive or negative output of the PV module) was mandatory by the NE; however, in the European system of grounding (or earthing, as they call it), it is not generally used.

One example of a complication regarding grounding is the general requirement for a transformer in power conversion technology in the US market that is not needed in the European market. This adds weight and cost, and reduces the efficiency of power conversion. Changes are currently underway and 'transformerless' inverters will become more common on the market, including in the United States, due to changes in the electrical code initiated by the industry. Bonding requirements and safety grounding for racking and PV modules will still be required. System grounding of the PV string through the inverter will not be required. These changes allow for the use of transformerless or 'non-isolated' inverters. These inverters save both the consumer and manufacturer significantly with regard to weight and cost of product, and as long as the equipment is properly installed in accordance with the code, it will work within a grounded utility grid connection.[45]

Ungrounded and transformerless systems require different ground fault detection mechanisms compared with isolated and grounded systems. In the United States, the current inverter standard, UL 1741, does not cover the requirements for ungrounded and non-isolated systems, except through an unofficial UL Certification Requirement Decision (CRD), which has not been adopted as part of the ANSI standard. Slightly different requirements for detecting and responding to ground faults are contained in the IEC International Standards, which the United States is moving toward adopting. In the interim, manufacturers of inverters, developers and system installers are left with uncertain requirements and little guidance.

### *Building Codes Characteristics*

Building codes have developed in the same manner as electrical codes, in response to national and local concerns in the form of requirements. In the United States, the AHJ has the final word for code enforcement, and although the AHJs use certification marks as a method of recognising and verifying that the product is properly tested, in the end an AHJ has the right to allow or not allow a project. This authority structure appears to be

---

[45] Wiles (2010): 1–6.

similar globally,[46] and the levels at which this authority is appropriate or over-enforced is likely to create variations all over the world. The enforcement officers are therefore highly dependent on access to information, and new technologies are always going to be a little more complicated for them to accept due to lack of experience. It is likely that only time and exposure to many solar PV systems will reduce these barriers, but efforts made by interested parties to educate and provide the necessary information can speed the process along.

### Fire Testing

PV modules in North America are required under the UL 1703 standard to pass the spread-of-flame test and the burning-brand test under the Standard Test Methods for Fire Tests of Roof Coverings, UL 790. UL 790 has additional tests that are used for BIPV modules. Some US states require changes in the fire-testing methodology and want the PV modules to be tested together with the roof racking system. The reason for this is that certain combinations of PV modules with racking will cause more fire hazards depending on the module racking combination. Some combinations can trap significantly more heat and cause a greater fire danger, but it is difficult to plan for every combination. Standard UL 1703 and UL Subject 2703 (rack mounting systems) are currently being modified to include a system-testing approach, which may lead to safer installations, but also more difficulty in testing for racking manufacturers and consumers with regard to choosing module/racking combinations.

Germany is also developing new standards in its building codes that will require solar PV to be tested for fire. The DIN EN 13501–1 will mandate testing for solar PV installed in the German market. The requirements for this testing, while similar in intent to UL 790, have variations that are likely to impose testing for each standard required for a manufacturer wishing to sell the same product in both markets. This trend toward more extensive fire testing is likely to continue making it challenging for PV module and racking system manufacturers to test their products for all markets and harder still as the requirements often call for all PV module and racking combinations. While the rationale for more stringent testing requirements is valid, the result can lead to unintended consequences on the market and create obstacles to manufacturers preparing their products for a global market.

---

[46] PV Legal, *Reduction of Bureaucratic Barriers for Successful PV Deployment in Europe Final Report*, 2012, www.pvlegal.eu.

## Climate and Location of Use-Based Variations and Requirements

It may not seem instantly obvious, but there are many climate conditions that will put additional strain on a PV system, and in these environments additional tests are often required. Established manufacturers are making attempts to prepare their products to pass as many climate- and location-based tests as possible, but this leads to additional costs in materials that may never really be necessary for the environment where the module is intended to be used. Also, changes intended for adherence to one location may cause it to be less effective in another, especially when considering differences in hot, humid cold, or dry climates. It can be a significant burden for a manufacturer to cover all the bases with one product, and yet it may be too difficult or costly to justify creating multiple versions of a PV module. As the market and volumes grow, this may be more justified, much as it has for building materials and other products that are subjected to different climate environments. Nevertheless, we are still in a relatively early stage in the development of solar PV technology, and this makes the 'one product for all situations' approach rather onerous.

Appropriateness for the installation environment is an important issue for consumers who want some indication of the ability of the solar PV products they select for their installation to be able to withstand the natural environment where it will be placed. The recognition of this has led to national and international efforts to define logical categories to assist both consumers and manufacturers in selecting the right product for the intended place of use.

### Approach from the IEC

Below is an excerpt from an IEC working draft to create location-based standards and a proposed indication of where additional testing would be required. Although these additional tests are not yet provided, research into appropriate additional testing is currently being evaluated:

The purpose of this international Standard is to define a rating system and associated tests to give a comparative indication of the long-term reliability of PV modules in multiple climates and use conditions. The commercial success of PV technologies is based on long-term reliability. As the industry grows into markets around the world, there will be an increasing need to understand how climates and specific use conditions around the world affect long-term reliability. In some cases, a single design may prove to be robust in all locations of the world. In other cases, designing for a limited use environment may enable lower product cost. As the community explores these questions, it will be useful to have a test standard that can compare and differentiate PV modules according to their expected reliability in specific use environments.

*Summary of the Rating System* Table 6.2 shows the various tests used to rate PV system performance. The rating stress label is given in the first column. The corresponding tests for each rating stress label are listed in the chart in the second through the fifth column. All four tests must be successfully completed to receive a 'pass' for each rating stress label. The first row is given for completeness to describe the present IEC 61215 Design Qualification Test.

In addition to passing the tests indicated in Table 6.2, we anticipate that modules will be required to pass a test for potential-induced degradation. Also, when diode failure is better understood, an additional test may be required to test PV modules for expected reliability when exposed to prolonged shading.

The rating system will give a comparative index that can be used to differentiate PV module products that have passed the tests. The comparative index will include consideration of:

- Electrical performance (module efficiency) retained after successful completion of each test leg.
- Low leakage current in both the dry and wet conditions to indicate better electrical isolation than the minimum required for successful completion of the test.
- Absence of evidence of change to the module that could lead to failure such as development of hot spots or cracked cells observed by infrared or electroluminescent imaging.

The comparative index will be defined in each of the IEC 62XXX–Y standards and will be communicated through a more detailed report. All the tests and indices will use test experience to give confidence that the

Table 6.2 *PV Performance Rating Tests*

| Label/Stress | Humidity | High Temperature | Thermal Cycling | UV |
| --- | --- | --- | --- | --- |
| IEC 61215 | IEC 61215 | IEC 61215 | IEC 61215 | IEC 61215 |
| Temperate | IEC 61215 | IEC 61215 | IEC 62XXX – 2 | IEC 62XXX – 3 |
| Hot and dry | IEC 61215 | IEC 62XXX – 4 | IEC 62XXX – 2 | IEC 62XXX – 3 |
| Hot and humid | IEC 62XXX – 5 | IEC 62XXX – 4 | IEC 62XXX – 2 | IEC 62XXX – 3 |

*Source:* Intertek.

accelerated test results will have predictive value for long-term performance in real-world conditions.

Source: International PV Module Quality Assurance Task Group, formed July 2011.

### NREL-led Approach

The Evaluation of Value and Structure of Rating System for PV Quality Assurance Comparative Testing Task Group 6, is working on a similar approach to the one indicated above. This group puts an emphasis on adding extended UV exposure, which leads to discoloration and delamination in PV laminates, and thermal, shading and mechanical stresses, which lead to accelerated 'wear-out' of interconnections, solder bonds and diodes. While other environmental or location-use conditions such as wind, marine and snowy environments are also considered, standards for these already exist, and the focus is on heat and humidity requirements beyond the existing IEC testing for solar PV. There is planned communication of these efforts to the IEC Technical Committee 82 Working Group 2, which is intended to harmonise the development of the approaches.

### Ammonia Test

The PV modules are electrical devices intended for continuous outdoor exposure during their lifetime. Highly corrosive wet atmospheres, such as in the environment of stables of agricultural companies, could eventually degrade some of the PV module components (corrosion of metallic parts, deterioration of the properties of some non-metallic materials – such as protective coatings and plastics – by assimilation of ammonia), causing permanent damage that could impair their functioning and safe operation. This standard describes test sequences useful to determine the resistance of PV modules to ammonia ($NH_3$).

### Salt Spray Test

Highly corrosive wet atmospheres, such as marine environments, could eventually degrade some PV module components (corrosion of metallic parts, deterioration of the properties of some non-metallic materials – such as protective coatings and plastics – by assimilation of salts, etc.), causing permanent damage that could impair their functioning and safe operation. Temporary corrosive atmospheres are also present in places where salt is used in winter periods to melt ice formation on streets and roads. This standard describes the test sequences useful to determine the resistance of different PV modules to corrosion from salt mist containing sodium chloride (NaCl, MgCl2, etc.).

### Snow Load and Wind Load Testing

While the existing IEC and UL standards both have requirements for mechanical load testing that are intended to indicate operational and safety continuance after being subjected to specified mechanical stresses, many in the industry commonly believe that these requirements are not sufficient for PV module installation in many high-wind, high-snow environments. Efforts are underway to develop new standards, and possibly testing protocols, to verify preparedness for extreme wind and snow load installations.

Evaluating only the current requirements created for mechanical load, PV modules that are already IEC certified may also be certified with regard to the mechanical loading test under Standard UL1703, since the IEC testing subjects the PV module to 2400 Pa for 1 hour and the UL 1703 standard requires a minimum of 30lbs/ft$^2$ for 30 minutes.

It is clear that the IEC is more stringent in this respect, and therefore the specific test of the IEC International Standards should also be valid for UL 1703 certification, but not the other way around. However, most manufacturers with a UL certification seeking entry into an IEC market will have to conduct a significant amount of testing anyway to sell into that market.

The following is part of the IEC 61215 standard for the mechanical load test, which indicates that for heavy snow and ice conditions, during the last cycle of this test the load is increased from 2400 Pa to 5400 Pa or approximately 113 lb/ft$^2$:

2400 Pa corresponds to a wind pressure of 130 km2 h–1 (approximately ±800 Pa) with a safety factor of 3 for gusty winds. If the module is to be qualified to withstand heavy accumulations of snow and ice, the load applied to the front of the module during the last cycle of this test is increased from 2400 Pa to 5400 Pa.

However, even this additional pressure is questioned, because wind gusts and uneven snow distribution can cause more severe mechanical stresses. There are several taskforces and researchers working on this. For example, the International PV Module Quality Assurance Task Force QA Forum, which was initially held in July 2011 in San Francisco, encourages international participation in the development of a rating system that meets the needs of all countries and customers so that PV manufacturers will need to complete only a single test. A Task Group was formed in April 2012, to investigate developing standards or test methods for severe weather. According to the group:

The proposed scope for Wind involves developing test methods for modules installed in hurricane and cyclone prone areas require additional testing for high wind loadings. This includes both static uplift forces, and dynamic forces from buffeting. Installations in areas subject to Category 5 winds should be able to withstand gusts of 317kph (200mph). This group will focus on testing required to demonstrate whether PV modules are suitable for high wind regions. For Snow they will concentrate on studying non-uniform snow loads. For Hail they will work on studying a more severe hail test which will involve larger hail balls of 30 mm or greater where 25 mm is the common used currently for testing.

### *Incentive-Based Variations*

Many of the variations in requirements in North America take into account different conditionalities to receive incentives. Nationally, the United States has offered a Federal tax credit; however, most states with active solar installations also have state or utility-based programmes, which have varying requirements for equipment, installers and manufacturing content. Below are a few examples of how some of these regional or state incentive programmes impact trade for manufacturers.

#### *California SB1 Guidelines-Compliant PV Modules*

This list is compiled by the California Solar Initiative and requires California Energy Commission (CEC) testing that goes beyond the standard UL 1703 testing. The list is also used for other state programmes (e.g. Texas), and the required additional testing is based primarily on performance testing (based on the IEC International Standards for PV module design qualification and type approval) of PV modules under more 'realistic' environmental conditions than just the standard testing condition (STC) used in the UL 1703 standard. Many manufacturers outside the United States are surprised that they need to conduct this additional testing, and it can slow their entry into the US market. But, generally, it is a matter of inconvenience and time, rather than a financial barrier, as the cost is not that great, especially if the testing is conducted in conjunction with UL 1703 testing. Given California's prominence in solar energy, most manufacturers quickly learn what is necessary and get on the CEC list.

#### *Florida Solar Energy Center (FSEC)*

Florida has its own set of requirements called the FSEC Standard 202–10.[47] While Florida will accept results from outside testing organisations, it still requires some testing to be done at the FSEC facility.

---

[47] Florida Solar Energy Center (2010): 1–5.

### Harmonisation of Standards

There are multiple ongoing efforts to harmonise standards for the solar PV industry, and the general trend is toward using the IEC International Standards as the common denominator.

#### Solar ABCs

In the United States, the Solar America Board for Codes and Standards (ABCs) is a major collaborative effort funded by the US Department of Energy (DOE). The Solar ABCs says it was 'formed to identify current issues, establish a dialogue among key stakeholders, and catalyze appropriate activities to support the centralized development of codes and standards that facilitate and accelerate the installation of high quality, safe photovoltaic (PV) systems'.[48] To date, this group has published several papers and conducted many meetings in an effort to harmonise solar PV standards and discover the gaps in standards with regard to their application for installations in the United States.[49]

Solar ABCs have specific on-going efforts with several standards organisations to improve standards from both the standpoint of better addressing safety and quality concerns and harmonising standards to create reasonable measures that do not impede the market.

The recommendation of Solar ABCs is to adopt IEC performance standards to ensure that product claims and quality are ensured in the market. This would in a sense also partially harmonise standards for many module manufacturers that are already using the IEC as their standard.[50]

#### Natural Resources Canada

The adoption of IEC International Standards in Canada came about as part of the research and efforts of the Canmet ENERGY group under the National Research Council (NRC). While driven more directly by a government entity and less involved in the direction of actual improvement of standards than Solar ABCs, it serves essentially the same purpose.

---

[48] Solar America Board for Codes and Standards (ABC), *About Solar ABCs*, University of Central Florida, 2011.
[49] Solar America Board for Codes and Standards (ABC), *Codes and Standards*, University of Central Florida, 2011.
[50] Solar America Board for Codes and Standards (ABC), *Policy Recommendation/Recommended Standards for PV Modules and System*, University of Central Florida, 2011.

*European Harmonisation Efforts*

Several efforts are underway in Europe and by European-based organisations to harmonise standards. Given that the IEC is the recognised standard for the European market, most of these efforts are geared toward implementing this standard on a global scale. Research is being conducted to validate existing standards and propose new testing protocols by institutes such as Fraunhofer[51] in Germany, SUPSI[52] in Switzerland and other centres throughout Europe.

*Asia, Oceania and Other Regions*

Similar efforts to those taking place in the US and European markets have been undertaken to harmonise standards and facilitate the trade process for solar PV across markets. For instance, the members of Asia-Pacific Economic Cooperation (APEC) have established a Solar Technology and Conformance Initiative to discuss the solar market's future development. APEC has also set up a standards working group for solar PV.

## 5 Private Sector Impact on Trade Caused by Standard Variability

### Reasons for Failure

Intertek published a presentation to inform solar PV customers of the most common reasons products fail to receive certification. Although there are certainly issues related to variations in markets, the number one reason in the majority of cases is inappropriate or incomplete documentation. This could be an area of focus for facilitating the process if programmes were promoted to assist manufacturers in the preparation of installation manuals and technical documentation materials for their products in the language of the country where certification testing will take place, or at least in a language that is well known. However, it is not only a problem of language *per se*; cultural understanding and technical jargon that varies widely between countries (even in the same language) are equally important. The United Kingdom and the United States are good examples of this.

### Inappropriate Installation Instructions

In over 85 per cent of the modules evaluated by Intertek, installation instructions were incomplete or contained errors. Lack of clear

---

[51] Fraunhofer ISE, www.ise.fraunhofer.de/en.
[52] SUPSI, *Studio e professine*, www.supsi.ch/isaac.html.

instructions hampers the ability of the evaluating engineer to determine how the module should to be installed. Installation documentation is particularly important as numerous sections in UL 1703 are related to assessing the product after it has been installed in the intended manner (as specified in the instructions).

Instructions should be written with an understanding of the technical skill level of the intended reader. Some of the most common issues are a lack of explanation of symbols used on the equipment or diagrams; instructions that are worded in a way that does not adequately address intended use. If the instructions are over-simplified or vague, they will not provide sufficient information for the end-user during the installation process. In addition, a common deficiency is instructions that do not adequately describe the grounding method intended to be employed. Another common occurrence is that the instructions are simply adapted from another piece of equipment and do not properly match the equipment being tested.

### *Samples Provided are not Representative of All Products Intended for Certification*

Another source of delay in testing is the failure to clearly define the proper model scheme during the initial process setup. In some cases, a manufacturer will add bigger and higher power modules than those available at the beginning of the testing process, mainly because the added modules were not available before the evaluation and testing began. This change in scope after the testing has begun may require additional evaluations and thus result in significant delay.

### *Testing Without Initial Construction Evaluation*

Temperature cycling testing takes several weeks to complete, and it is not uncommon for manufacturers to request starting this test as early as possible. However, without first performing the construction evaluation it may be necessary to repeat the test. As part of the test setup, assumptions must be made that cannot be confirmed without the construction evaluation. Key areas will be identified during the construction evaluation that will lead to proper sample preparation and ensure that the right samples are chosen with the right components.

### *Insufficient, Inaccurate or Incomplete Bill of Materials*

Not providing a complete bill of materials is another common issue that causes delays.

Missing component certification ratings leads to tests being performed without checking to make sure that the ratings of different materials are within those of the tests being performed. If any of the components or materials do not have the required ratings, it may not be possible to allow alternative components to be included in the product listing.

*Incorrect Component Ratings*

Many substrates for solar PV modules in the market today do not have the minimum relative thermal index (RTI) value required by UL 1703 clause 7.3, which states:

7.3 A polymeric substrate or superstrate shall have a thermal index, both electrical and mechanical, as determined in accordance with the Standard for Polymeric Materials – Long Term Property Evaluations, UL 746B, not less than 90°C (194° F). In addition, the thermal index shall not be less than 20°C (36°F) above the measured operating temperature of the material. All other polymeric materials shall have a thermal index (electrical and mechanical) 20°C above the measured operating temperature. The measured operating temperature is the temperature measured during the open-circuit mode for Temperature Test, Section 19, or the temperature during the short-circuit mode, whichever is greater.

If the outer layer is a spray-on thin coating it will not be considered for the RTI requirement. If the middle layer of the substrate has an RTI of less than 90°, it will not be acceptable. The inner layer mentioned is often the same as the encapsulate for the cells in the module. Since encapsulates do not need to meet the RTI requirement in clause 7.3, the engineering rationale has been that this layer of the laminate does not need to have an RTI of 90°. An inner layer of the PV substrate laminate or film that does have the required RTI as measured by an NRTL is considered acceptable.

### *Additional Costs and Design Challenges*

Manufacturers of solar PV and related components/equipment are impacted by the need to adapt their product to different market standards. While this is a common experience for nearly all manufacturers of electronic products, it can have a greater impact on solar PV companies that do not have a lot of other products. Larger, more established industries in consumer electrical products are able to absorb adaptation costs much more easily than most solar PV companies that only produce solar PV. Also, the fact that solar PV products are intended not to consume, but to supply power, creates a more challenging level of adaptation than is the case for products that consume electricity.

### Emerging Technologies Not Covered by Existing Standards

While there is a great push toward the development of new and innovative products in the solar PV market, little effort is being made to improve communication on these budding technologies to the standard-writing bodies. There is a need for concerted communications to ensure that when innovative technologies are ready, applicable standards are in place to test them. Sometimes even innovations to existing products can make them non-compliant with existing standards. Although the engineers that test a product have some ability to provide an engineering judgement, this is often limited and requires updates to the standards. Communication about 'up and coming' technologies may be conducted in the home country of the innovation and adopted in that market only to realise later that a hurdle still exists in every other market. Given the global nature of business today, in particular for solar PV-related products, this can be a major impediment to trade.

## 6 Review and Recommendations from the Industry Perspective

### Review

In some cases, at least with regard to safety and grid connectivity, there may be no easy way to harmonise standards beyond a certain point. Hence, one strong point of the structure of the IEC is that each country acts as a member and is required to have its own reviewing body to make sure that the standards adopted are appropriate for its systems.

### Recommendations from the Industry Perspective

Policies promoting training and informational support for inspectors and installers
- Support for organisations such as Solar ABCs and others working to harmonise standards while recognising the need to allow variations for the purposes of grid interactivity, safety and specific policy compliance. At the same time, for ease of facilitating trade, international standards should be written, and enforced, for connecting PV installations to the grid (including, but not limited to, ride-through rules). It thus makes just as much sense as having international standards for the PV modules.

- Policies promoting assistance in the preparation of product documentation for countries that face barriers of language and cultural communications.
- Greater harmonisation and mutual recognition of testing procedures so that solar PV products do not need to go through unnecessary re-testing to enter multiple markets except for justifiable reasons, such as differences in climate, etc.
- To encourage continuous innovation in solar PV products and a faster introduction to the market, there should be speedier facilitation of acceptability with regard to 'up and coming' technologies and products involving regulatory authorities, standard-setting, testing and certification bodies and organisations, such as the CECET.

A number of other useful recommendations related to renewable energy standards in general have been put forward in a report by the International Renewable Energy Agency (IRENA) on international standardisation in the field of renewable energy.[53] Many of these recommendations would be useful to the solar PV industry as well. For instance, one recommendation is to investigate ways to facilitate the search for, and access to, required standards. This could include working with the standard-setting bodies to explore possibilities for a collaborative, interactive database that provides a searchable standards portal that includes a hierarchical tree of normative standards for the technologies being researched. Another is the use of the latest communication technologies for engagement in standardisation.

One of the conclusions of the report is the importance of understanding which standards are the same, or just slightly modified, and those that interconnect with other standards through normative references. The IRENA report proposes the creation of a web-based information platform in consultation with relevant stakeholders to bridge this information gap. This would provide easy access to information on international standards for renewable energy, including existing standards and those under development, as well as providing information on the benefits of standardisation and the development process and the use and application of standards.[54] This will certainly be useful for the industry as well. The concordance table in Appendix II provides one such illustrative example of the inter-relationship between international solar PV standards and additional requirements at the national level.

---

[53] International Renewable Energy (2013). [54] *Ibid.*

## 7 Does the WTO Help the Sun Shine Brighter? WTO Disciplines and Relevant Case Law Applicable to PV Standards

### Introduction

Standardisation is an important part of the national policies aimed at achieving certain legitimate public goals, including the protection of life and health of humans, animals and plants, as well as the environment or national security.[55] By setting certain product standards, including size, form and performance, countries can control which products of what quality enter their markets. At the same time, however, standards can pose a serious threat to free-trade flows, especially if they are applied in an unjustifiably restrictive or discriminatory manner. Therefore, technical regulations and standards must not be more trade restrictive than necessary for fulfilling certain legitimate policy objectives, and they should not serve protectionist intents through less favourable treatment of foreign products compared with like domestic products. Compliance of a product with certain safety/quality standards serves as a signal of its quality, thereby increasing its competitiveness and reducing transaction costs.[56]

In recent years, the growing negative impact of non-tariff barriers (NTBs), including technical regulations, has been widely recognised.[57] Unlike trade restrictions caused by tariffs, the effects of NTBs, including subsidies, government procurement and certain quality or performance requirements, are less transparent and more difficult to curb.

While climate change-related standards, particularly those concerning carbon emissions and energy efficiency, have already become a prominent topic for discussions within the WTO framework[58] PV standards, which apply to one of the most promising RE technologies, have received only limited consideration. A general mandate within the WTO to deal with standards for environmental goods and services (EGS) is indicated in the Doha Declaration, para 31. It provides for the reduction or elimination of NTBs to EGS, which clearly encompasses technical regulations and standards for solar equipment such as solar panels.[59] Within the Technical Barriers to Trade (TBT) Committee, concerns were raised about the standards for thin-film solar panels in Korea in effect since 2007. Moreover, since 2008, Korea has required solar panels to be sold in

---

[55] International Centre for Trade and Sustainable Development (2001).
[56] WTO (2012): 143.   [57] WTO (2005) and (2012).
[58] See WTO, *Concerns Raised about Tobacco and Environmental Measures*, Technical Barriers to Trade: Formal Meeting, June 15 and 16, 2011.
[59] Hufbauer and Kim (2012): 21–22.

the country to be certified by the Korean Management Energy Cooperation. The United States raised this issue for the first time at the June 2010 meeting of TBT Committee,[60] and in six subsequent meetings.[61] Although the Korean representatives maintain that the standard and the related certification system are not mandatory, the concerns remain unaddressed, as confirmed by reports from TBT Committee meetings.[62] The possible solution offered by the United States was for Korea to apply a relevant international IEC standard.

Unlike energy efficiency and carbon standards, which mostly aim at controlling and reducing greenhouse gas (GHG) emissions, the main objective of the current PV standards is to ensure the safety and reliability of this RE technology. And, although solar PV standardisation aims at positive societal effects, industry's evaluation of its trade effects on the PV market shows that there are many drawbacks to the existing system.

This chapter focuses on the existing WTO law disciplines applicable to PV standardisation. It evaluates their suitability to address solar PV industry needs with respect to both solar PV exporting and importing countries. Based on this analysis, possible solutions within a SETA that could be negotiated by WTO Members (and possibly outside the WTO) will be suggested. The SETA will address key issues and challenges at the interface of trade policy and sustainable energy, including providing predictability for traders of sustainable energy goods and services, as well as a trade-supportive enabling environment for sustainable energy scale-up.

### *PV Standards versus PV Technical Regulations*

Most of the PV standards described so far are related to goods, and thus would be covered by the TBT Agreement. The TBT Agreement is one of the WTO Multilateral Agreements on Trade in Goods[63] and is applicable to (mandatory) technical regulations and (voluntary) standards, as well as to conformity assessment procedures for goods.[64] The Agreement aims to

---

[60] Committee on Technical Barriers to Trade, *Minutes of the Meeting of 23–24 June 2010*, G/TBT/M/51, paras. 33–35.
[61] Committee on Technical Barriers to Trade, *Eighteenth Annual Review of the Implementation and Operation of the TBT Agreement*, Note by the Secretariat, 27 February 2013, G/TBT/33: 22.
[62] Committee on Technical Barriers to Trade, *Minutes of the Meeting of 10–11 November 2011*, Note by the Secretariat, G/TBT/M/55, 9 February 2012, paras. 108–110; Committee on Technical Barriers to Trade, *Minutes of the Meeting of 20–21 March 2012*, Note by the Secretariat, G/TBT/M/56, 16 May 2012, paras. 93–95.
[63] Annex 1A to the Final Act Marrakesh Agreement Establishing the World Trade Organization, 1994.
[64] TBT Agreement, Annex 1.

ensure that the introduction and implementation of standards does not create unnecessary obstacles to international trade, striking a balance between the various legitimate objectives of WTO Member States and the pursuit of further trade liberalisation.[65] The harmonisation of standards at the international level, which WTO Member States are encouraged to use whenever they are appropriate and efficient to achieve the legitimate objective pursued, provides a coherent framework and equal opportunities for stakeholders to participate in the standard-setting process.[66]

While the TBT Agreement is a specific agreement within the WTO framework to deal with standardisation issues, the application of the General Agreement on Tariffs and Trade (GATT) – the core building block of WTO disciplines on trade in goods – to standardisation is not excluded *per se*. The relationship between the two agreements is determined by the *lex specialis derogat legi generali* rule in the General Interpretative Note to Annex 1 of the WTO Agreement. According to this rule, in case of conflict between a GATT provision and a provision of another, more specific covered Agreement (e.g. the TBT Agreement, Agreement on Agriculture, etc.) applicable to trade in goods, the latter should prevail. However, as determined by the panel in *EC – Asbestos*, and confirmed in *US – COOL* and *US – Tuna II (Mexico)*,[67] this rule does not lead to the mutual exclusiveness of the GATT and the TBT Agreement. In a dispute settlement procedure, a panel would first have to scrutinise a measure under the TBT Agreement. If the measure is found to be consistent with the *lex specialis*, the panel will proceed with the analysis under the relevant GATT provisions according to the terms of reference.[68]

It is important to emphasise again that, unlike in the industry terminology, the notion of 'standards' has a very specific meaning in WTO law. The TBT Agreement differentiates between mandatory product-related requirements, which are referred to as 'technical regulations', and voluntary product-related requirements, which are known as 'standards'.[69] A technical regulation is defined as a 'document that lays down product characteristics or their related processes and production methods, including the applicable administrative provisions, with

---

[65] TBT Agreement, Preamble. [66] TBT Agreement, Articles 2.4 and 2.5.
[67] Panel Report (PR), *EC – Asbestos*, paras 8.16–8.17; PR, *US – COOL*, para 7.73; PR, *US – Tuna II (Mexico)*, paras 7.39–46.
[68] PR, *EC – Asbestos*, para 8.16; PR, *EC – Sardines*, paras 7.14–7-19; Appellate Body Report (ABR), *US – Tuna II (Mexico)*, paras 7.39–46.
[69] This is also important in the light of the definition in the ISO/IEC Guide 2, according to which standards can be both voluntary and mandatory.

which compliance is mandatory'. Such product characteristics may include packaging and labelling requirements, among others. On the other hand, a standard is a 'document approved by a recognised body that provides for common and repeated use, rules, guidelines, or characteristics for products or related processes and production methods with which compliance is not mandatory'.[70] The TBT Agreement generally gives preference to technical regulations and standards that are based on performance requirements, rather than on design or descriptive characteristics, as the latter have a higher probability of constraining trade. In addition to these two categories, the TBT Agreement deals with conformity assessment procedures, which cover a broad range of procedures to determine whether the requirements of technical regulations and standards have been fulfilled.

Based on this differentiation, various provisions and obligations of the TBT Agreement apply to PV standardisation. These are addressed below.

### *Technical Regulations for PV*

Most of the 'standards' described above, namely the safety and performance requirements, would likely fall within the category of technical regulations under the TBT Agreement since they are of a mandatory character and constitute a condition for market access. It should be noted that the PV standards subject to analysis in this chapter can be characterised as product-related; thus, the much discussed non-product-related production and processing methods (NPR–PPM) issue does not arise here.[71] Since mandatory technical requirements impose stronger effects on trade in the goods concerned, the TBT Agreement provides for a set of strict rules on the preparation, adoption and application of such provisions.[72] The most important of these include the obligations on non-discrimination and the avoidance of unnecessary barriers to trade, the use of international standards, transparency, and mutual recognition of technical regulations They also allow for emergency exceptions. It should be noted that these rules apply not only to technical regulations issued by central government bodies (e.g. national laws providing for certain fire safety standards that also apply to PV equipment), but also to regulations

---

[70] TBT Agreement, Annex 1.
[71] In the case of PV, such production and processing methods (PPMs) could include energy-efficient technologies, or the use of 'green electricity' in the process of PV equipment production. These would not be reflected in product characteristics and would not affect performance or safety, and thus are not directly product-related. For further information on this issue, see Conrad (2014).
[72] TBT Agreement, Articles 3 and 4.

adopted by local government bodies and even non-governmental entities. While compliance with the TBT Agreement, Article 2 for the former is unconditional, for the latter technical regulations, according to the TBT Agreement, Article 3.1, Member States are just required to take reasonable measures to ensure their compliance.

Until recently, the definition of technical regulation was considered to be fairly straightforward. In practice, however, this question has been raised several times in WTO dispute settlement proceedings. A three-tier test has been established by the Appellate Body in *EC – Asbestos*,[73] *EC – Sardines*,[74] and confirmed in *US – Clove Cigarettes*[75] and *US – Tuna II (Mexico)*.[76] According to this test, a measure is a technical regulation if:
- it applies to an identifiable product or group of products;
- it lays down one or more characteristics of the product (intrinsic or product related); and
- compliance with these product characteristics is mandatory.

Moreover, the Appellate Body in *US – Tuna II (Mexico)*[77] confirmed that the determination of whether a certain measure constitutes a technical regulation has to be made 'in the light of the characteristics of the measure at issue and the circumstances of the case'.

Furthermore, technical regulations may be expressed in positive or negative wording, providing either that certain characteristics should not be present in a product (the PV cell should not contain a certain type of silicon), or that a product should possess certain characteristics (e.g. the PV module must be tested for snow and wind load).

Most of the technical requirements described so far would easily qualify under the first two criteria. First, they apply to PV equipment, which is an identifiable group of products; and second, they set certain safety or performance requirements, which in turn are reflected in technical characteristics of the equipment.

Finally, the question of the mandatory character of the measure would depend on the specific case. In *US – Tuna II (Mexico)*, the Appellate Body had to weigh the question of whether the US 'dolphin-safe' labelling scheme for tuna was mandatory. Unlike previous cases, including *EC – Sardines*, where only the one species of sardine would qualify as 'preserved sardines' and only it would be admitted for

---

[73] PR, *EC – Asbestos*, paras 66–70.   [74] ABR, *EC – Sardines*, para. 176.
[75] ABR, *US – Clove Cigarettes*, para 87.
[76] PR, *US – Tuna II (Mexico)*, paras 7.53–7.54. The three-tier test has been reiterated in ABR, *US – Tuna II (Mexico)*, paras 178–199. The only issue appealed by the United States under the 'technical regulation' criterion was the determination of a labelling scheme as mandatory by the panel.
[77] ABR, *US – Tuna II (Mexico)*, para 199.

importation and sale as a 'sardine',[78] or *EC – Asbestos*, where the ban on products containing asbestos or asbestos fibres prevented them from entering the EC market,[79] the labelling scheme in *US-Tuna II (Mexico)* was not a mandatory pre-condition for market access. Confirming the panel's approach, the Appellate Body in the latter dispute adopted quite a low threshold for 'mandatory'.[80] Because 'any producer, importer, exporter, distributor or seller of tuna products must comply with the measure at issue in order to make any "dolphin-safe" claim' and this measure 'sets out a single and legally mandated definition of a "dolphin-safe" tuna product', the labelling scheme was determined to be mandatory and thus to constitute a technical regulation under the meaning given in the TBT Agreement, Annex 1, para 1.[81]

Applying this approach to the PV sector in practice would lead to the conclusion that the labelling, certification and testing schemes, etc. for PV equipment – confirming certain performance characteristics desirable for electricity generators, but not mandatory for market access – could, in certain cases, qualify as technical regulations and thus be subject to the stringent rules of the TBT Agreement, Article 2.

### *When PV Technical Regulations are Discriminatory*

The TBT Agreement does not prohibit the application of technical regulations as such, but it aims at ensuring that these regulations are not enacted and applied in a discriminatory way.

The core principle of non-discriminate entails two main elements enshrined in the TBT Agreement, Article 21. On the one hand, a technical regulation shall not discriminate between like domestic and imported products (the national treatment or NT principle); on the other hand, it shall also treat the like imported products from one country no less favourably than from any other country (the most favoured nation or MFN principle). Importantly, as noted above, the non-discrimination obligation applies as an absolute obligation only to technical regulations adopted by central governmental bodies,[82] while for the same technical regulations adopted by local governments or non-governmental bodies, WTO Member States are just required to take 'reasonable measures as may be available'.[83]

---

[78] PR, *EC – Sardines*, paras 7.3–7.4, 7.20–7.35. This finding of the panel has been also confirmed by the Appellate Body, see the respective ABR, para 194.
[79] ABR, *EC – Asbestos*, paras 75–78.
[80] See Trujillo (2012): 25. The author suggests that through this determination of 'mandatory', the distinction between a 'technical regulation' and a 'standard' is blurred.
[81] ABR, *US – Tuna II (Mexico)*, paras 190–199. [82] TBT Agreement, Article 2.1.
[83] TBT Agreement, Article 3.1.

Thus, in a hypothetical example, if a US regulation required only imported PV equipment to comply with an 'infant mortality' performance standard, it would violate the NT obligation. On the other hand, if PV modules coming from China were required to comply with the above-mentioned regulation, and solar panels from Korea were not be subject to the same requirement, this would constitute a violation of the MFN principle.

Until recently, there has been no case law that would elucidate the due analysis of a measure under the TBT Agreement, Article 21.[84] However, the panel in *US – Clove Cigarettes*[85] suggested that the determination of inconsistency was based on three core elements:

- the measure at issue is a 'technical regulation' within the meaning of Annex 1;
- the imported and domestic products under consideration are 'like';
- and the measure results in a less favourable treatment of imported products compared with like domestic products.

The crucial point of the analysis under Article 2.1 lies in the notion of 'like products'. While the panel in *US – Clove Cigarettes* used the traditional four-elements likeness test,[86] well-established in WTO/GATT jurisprudence, it noted that 'it is far from clear that it is always appropriate to transpose automatically the competition-oriented approach to likeness under Article III:4 of GATT 1994 to Article 2.1 of the TBT Agreement', especially in the absence of a general principle as expressed in the GATT 1994, Article III:1.[87] The Appellate Body did not support the general approach of the panel as to the role of competitive relationship, and noted that '[i]n the light of this context and of the object and purpose of the TBT Agreement, as expressed in its preamble, we consider that the determination of likeness under Article 2.1 of the TBT Agreement, as well as under Article III:4 of GATT 1994, is a determination about the nature and extent of a competitive relationship between and among the products at issue'.[88] And, although the regulatory concerns cannot serve as the basis for a 'likeness' determination, they may play a role for the analysis, especially if they are reflected in the competitive relationship of the products.[89] Thus, competitive relationship would play a crucial role for establishing the 'likeness' of PV equipment.

---

[84] Van den Bossche (2008).    [85] ABR, *US – Clove Cigarettes*, paras 104–120.
[86] Namely, the physical characteristics, end-uses, consumer tastes and habits and tariff classifications of the products concerned.
[87] *Ibid.*, para 104.
[88] *Ibid.*, para 120. See also an overview of the panel's decision in Voon (2012).
[89] ABR, *US – Clove Cigarettes*, para 119.

As for the less favourable treatment element, the Appellate Body in *US – Clove Cigarettes* also clarified that any distinction, even if based on particular product characteristics or their PPMs, should not directly lead to a conclusion that imported products are accorded less favourable treatment.[90] The Appellate Body came to the conclusion that 'the context and object and purpose of the TBT Agreement weigh in favour of reading the "treatment no less favourable" requirement of Article 2.1 as prohibiting both *de jure* and *de facto* discrimination against imported products, while at the same time permitting detrimental impact on competitive opportunities for imports that stems exclusively from legitimate regulatory distinctions'.[91] While *de jure* discrimination cases are more obvious and easy both to reveal and to scrutinise under Article 2.1, in cases of *de facto* discrimination, the panel would have to 'carefully scrutinise the particular circumstances of the case, that is, the design, architecture, revealing structure, operation, and application of the technical regulation at issue, and, in particular, whether that technical regulation is even-handed' to weigh a legitimate regulatory distinction against the discrimination.[92] More simply put, the TBT Agreement allows members to enact trade restrictive technical regulations, but these should not lead to arbitrary or unjustifiable discrimination or a disguised restriction of international trade.[93]

Every specific case for PV technical regulation would have to be subject to the tests outlined above to establish whether the regulation leads to discrimination among types of PV equipment.

Finally, the TBT Agreement, unlike the GATT, does not have a general exceptions clause.[94] Although the TBT Agreement, Article 2.2, has very similar wording to the GATT, Article XX, it is not an exception, but rather an obligation with which every technical regulation must comply. This 'positive approach' also constitutes the fundamental difference between the GATT and the TBT Agreement. Thus, even if there is no violation of the non-discrimination obligations under Article 2.1, a measure can still violate Article 2.2 if it does not comply 'with the substantive obligations set out in the non-economic justification provisions'[95] and cannot be justified otherwise. Notably, through the introduction of the notion of 'even-handedness' into the 'less-favourable treatment' analysis under the TBT Agreement, Article 21 – which, in fact, corresponds to the chapeau of the GATT, Article XX – the Appellate Body, in the three cases

---

[90] *Ibid.*, para 169.   [91] *Ibid.*, para 175.   [92] *Ibid.*, para 175.
[93] Zhou (2012): 1100 *et seq.* at 1075.   [94] Vranes (2009): 302–303.
[95] Zleptnig (2010): 118–119.

(*US – Tuna*, *US – COOL* and *US – Clove Cigarettes*), tried to cure the lack of general exceptions in the TBT Agreement.[96]

### *The Necessity Test for PV Technical Regulations*

It should be noted that countries have the right to set product requirements for fulfilling their legitimate domestic policy objectives (regulatory autonomy), but they must do this in a manner that is not more trade restrictive than necessary.

This key obligation is contained in the TBT Agreement, Article 2.2: 'technical regulations shall not create unnecessary obstacles to trade.' The second sentence further explains that the technical regulation 'shall not be more trade-restrictive than necessary to fulfil a legitimate objective, taking account of the risks non-fulfilment would create'.

It had already been established in *EC – Sardines* that the panels need to decide whether the objective pursued by a Member State is legitimate, as well as whether the technical regulation is suitable and necessary to achieve this objective.[97]

The TBT Agreement, Article 2.2, provides an illustrative list of legitimate objectives, including 'national security, protection of human life and safety, animal or plant life and health, or the environment'. The *US – COOL* and *US – Tuna II (Mexico)* disputes extensively addressed the notion of 'legitimate'. First, the open character of the list in the TBT Agreement, Article 2.2, was confirmed. In practice, it means that Member States are not limited to the objectives enumerated in this provision. Second, the panel in *US – COOL* rejected a suggestion that an objective pursued not mentioned in the list must be 'linked in nature' to those explicitly mentioned in the TBT Agreement, Article 2.2.[98] Finally, the Appellate Body recalled that the Preamble of the TBT Agreement and other covered Agreements could also be used as 'guidance in determining what is "legitimate"'.[99]

Most PV standards are related to the protection of health and life of people, while some others may be enacted to ensure the security and stability of the electricity grid, or the protection of the environment (e.g. from fire). These would fall within the category of legitimate objectives.

In a second step, it shall be established whether a specific technical regulation is more trade restrictive than necessary. Similar to the provisions of Article 2.1, the 'necessity' element was interpreted by WTO adjudicating bodies only recently. While the wording of Article 2.2

---

[96] Marceau (2013) and Arcuri (2012). [97] ABR, *EC – Sardines*, para 286
[98] ABR, *US – COOL*, para 453. [99] *Ibid.*, para 445.

suggested that legal analysis of 'necessity' would follow the established practice under the GATT, Article XX, a question about the specific context of the TBT Agreement remained open.[100] Finally, the panel in *US – Clove Cigarettes* clarified this issue and held that GATT jurisprudence was relevant for the 'necessity' analysis, but it did not suggest that it should be applied in its entirety to interpret the TBT Agreement, Article 2.2.[101] The decisions in *US – Tuna II (Mexico)* and *US – COOL* disputes confirmed this approach.[102]

The most recent case law on the TBT Agreement, Article 2.2, suggests that the necessity analysis is a 'relational analysis of the trade restrictiveness', based on examination of several factors.[103] First, the panels need to consider:

(i) the trade restrictiveness of the technical regulation;
(ii) fulfilment of the objective pursued at the level chosen by a Member, which does not entail any minimum threshold and simply refers to a 'degree of contribution to the achievement of objective'[104]; and
(iii) the risks of non-fulfilment, with due regard to the importance of the interests and values at stake. Notably, the TBT Agreement, Article 2.2, allows for some trade restrictiveness of the technical regulation. However, it shall not be excessively trade restrictive and shall not constitute an unnecessary obstacle to trade.[105]

Second, in order to ensure proportionality between the trade restrictiveness and the need to employ the challenged measure, the latter may be compared with a possible alternative measure. Here, the panel shall answer the following four questions:

(i) Is an alternative measure less trade restrictive?
(ii) Would it make an equivalent contribution to the legitimate objective?
(iii) Is it reasonably available?
(iv) What are the risks of non-fulfilment of the legitimate objective?[106]

All these factors shall be examined and constitute a single 'necessity' analysis under the TBT Agreement, Article 2.2.[107]

In practice, applying this analysis to a specific PV standardisation scheme could mean, for instance, that for a very narrowly construed performance requirement, a less trade-restrictive measure can be found

---

[100] Mitchell, Andrew (2008): 201. [101] PR, *US – Clove Cigarettes*, paras 7.353–7.369.
[102] ABR, *US – COOL*, para 374; ABR, *US – Tuna*, paras 7.457–7.458, 7.471, with references to *Brazil – Measures Affecting Imports of Retreaded Tyres* and *China – Measures Affecting Trading Rights and Distribution Services for Certain Publications and Audiovisual Entertainment Products*, as well as the SPS Agreement, Article 5.6.
[103] PR, *US – Clove Cigarettes*, paras 7.356–7.418. [104] ABR, *US – COOL*, para 468.
[105] *Ibid.*, para 338. [106] *Ibid.*, para. 461; ABR, *US –Tuna II (Mexico)*, paras 312–322.
[107] ABR, *US – Tuna II (Mexico)*, para 322; ABR, *US – COOL*, para 455.

in a more flexible performance requirement or even in equivalence recognition of performance requirements in (some) other WTO Member States. The condition here would be that an alternative measure should contribute in an equivalent manner to the achievement of the legitimate objective. So, if a PV module safety standard (which constitutes a technical regulation under the TBT Agreement) is too low to protect human health and life from PV installation and exploitation-related threats, it will not be recognised by a panel as a viable substitute for another technical regulation that is more stringent but achieves the pursued objective considerably better.

### International Standards and Their Role in PV Standardisation

The PV industry supports the harmonisation of technical regulations and standards at the international level since it can decrease the variety of requirements in export markets and thus to make it easier for producers to sell their products not only domestically, but also abroad. Another reason for industry support is the positive effect that harmonisation will have on trade through the reduction of trade restrictions.[108] Due consideration of the importance of international standards is also given in the WTO.

International standards are adopted by international standardisation bodies. The TBT Agreement lacks definition as to what qualifies as such a body. However, it provides a definition of an international body or system. According to the WTO, this is understood as 'a body or system whose membership is open to relevant bodies of at least all Members'.[109] The Appellate Body in *EC – Sardines* held that international standards, in order to qualify under the TBT Agreement, do not necessarily have to be adopted by consensus.[110] However, the TBT Committee noted that: 'In order for international standards to make a maximum contribution to the achievement of the trade facilitating objectives of the TBT Agreement, it was important that all Members *had an opportunity to participate* in the elaboration and adoption of international standards' (emphasis added). This is aimed at ensuring the transparency of international standards and the participation of both developed and developing countries in the process. It is also intended to provide a framework for legitimacy.

Recent jurisprudence on the TBT Agreement, namely the Appellate Body ruling on *US – Tuna II (Mexico)*, provides an extensive interpretation of the term 'international standard'. The Appellate Body first

---

[108] WTO (2005): 35 et seq. See also Zleptnig (2010): 379.
[109] TBT Agreement, Annex 1.4.
[110] ABR, *EC – Sardines*, para 227; see also the Explanatory Note to the TBT Agreement, Annex 1.2.

addressed the definition of 'standard' in the TBT Agreement itself. An explanatory note to Annex 1.2 on the term also refers to the ISO/IEC Guide 2, which includes an additional prerequisite for standards, i.e. that they are adopted by consensus. It is, however, noted that the TBT Agreement also covers documents that are not based on consensus. The Appellate Body thoroughly analysed the relationship between the definition of 'standard' in the TBT Agreement, Annex 1.2, and the ISO/IEC Guide 2. It came to the conclusion that 'international standards' are adopted by international standardising bodies with recognised activities in standardisation and whose membership is open to the relevant bodies of at least all WTO Members.[111] However, standardisation does not have to be the principal function of such bodies, although WTO Members should be 'aware or have reason to expect, that the international body in question is engaged in standardisation activities'.[112] The Appellate Body further found interpretative guidance in the TBT Committee Decision on Principles for the Development of International Standards, Guides and Recommendations with Relation to Articles 2 and 5, as well as Annex 3 to the Agreement, which stipulates the principles and procedures for the development of international standards.[113] One of the key principles mentioned in the decision is 'openness', meaning that the relevant bodies of the WTO should be able to participate, without any discrimination (*de jure or de facto*), at the policy level and at every stage of standards development. The Appellate Body found that a standardising body can be considered international only where such openness is granted 'at every stage of standards development'.[114]

Both the International Organization for Standardisation (ISO) and the IEC would qualify as international standardising bodies for the purposes of the TBT Agreement.[115] Most of the standardisation organisations listed earlier in this chapter would need to be scrutinised according to

---

[111] ABR, *US – Tuna II (Mexico)*, para 359.   [112] *Ibid.*, para 362.

[113] *Ibid.*, para 366. Decision of the Committee on Principles for the Development of International Standards, Guides and Recommendations with relation to Articles 2, 5 and Annex 3 of the Agreement, WTO document G/TBT/1/Rev.10, Decisions and Recommendations adopted by the WTO Committee on Technical Barriers to Trade since 1 January 1995, 9 June 2011: 46–48. The Appellate Body in *US – Tuna II (Mexico)* found that the Decision by the TBT Committee had a status of subsequent agreement between the parties regarding the interpretation of the treaty or the application of its provisions within the meaning of the Vienna Convention on the Law of Treaties, Article 31(3): see para 371 of the Appellate Body report.

[114] ABR, *US – Tuna II, (Mexico)*, para 373.

[115] The International Electrotechnical Commission (IEC) is the world's leading organisation that prepares and publishes International Standards for all electrical, electronic and related technologies. The IEC is one of three global sister organisations (IEC, ISO, ITU) that develop International Standards.

the set of prerequisites set by the Appellate Body in *US – Tuna II (Mexico)*. Many of these bodies would fail to meet the 'openness' requirement mentioned above, given that they are national standard-setting bodies.

Further, the TBT Agreement, Article 2.4, requires the technical regulation to be based on relevant international standards. According to TBT case law, there should be 'a very close and very tight' relationship between the technical regulation and the relevant international standard.[116] Since this requirement is construed very narrowly,[117] the TBT Agreement, Article 2.4, allows for policy space through an 'ineffectiveness and inappropriateness' clause. The use of international standards is not required, specifically where these would be ineffective or inappropriate to achieve the legitimate objective pursued by the Member, for instance, due to climatic and geographic factors, or fundamental technological problems. The Appellate Body in *EC – Sardines*[118] confirmed the findings of the panel that the 'effectiveness' of a measure relates to its capacity to accomplish the stated legitimate objectives (as a result of the application of an international standard), while 'appropriateness' means the measure's suitability to achieve legitimate goals (the nature of an international standard). Ineffectiveness and inappropriateness are not mutually exclusive, and the determination of one does not automatically lead to the determination of the other. Special consideration in PV standardisation cases will have to be given to the 'relevance' of an international standard due to differing climatic and geographical conditions, as well as rapidly developing technology.

Importantly, according to the TBT Agreement, Article 2.5, the compliance of a technical regulation with an international standard creates a (rebuttable) presumption that the measure does not create unnecessary obstacles to trade and consequently is necessary in light of the legitimate objective pursued (TBT Agreement, Article 2.2).[119] Surely, this makes the vulnerability of a technical regulation in disputes much lower.

Finally, the role of international standards for standards and conformity assessment procedures is slightly different. Paragraph F of the Code of Good Practice (Annex 3) provides for the use of an international standard whenever it is existent or its adoption is imminent, unless it appears to be ineffective or inappropriate. Although the TBT Agreement, Article 5.4, contains similar provisions for the use of guides and recommendations of international standardisation bodies, it includes only the

[116] ABR, *EC – Sardines*, paras 166, 245.
[117] See Zleptnig (2010): 383, n. 95, suggesting that such a narrow wording almost in every case leads to the obligation under the TBT Agreement, Article 2.4.
[118] ABR, *EC – Sardines*, para 289.   [119] See Vranes (2009): 308, n. 94.

inappropriateness condition, but not the ineffectiveness one. We will address this issue in more detail below.

*Equivalence of Technical Regulations*

Given that many sectors still lack international standards, the TBT Agreement encourages states to recognise foreign technical regulations. Article 2.7 provides that:

Members shall give positive consideration to accepting as equivalent technical regulations of other Members, even if these regulations differ from their own, provided they are satisfied that these regulations adequately fulfil the objectives of their own regulations.

Thus, the sovereignty of states to adopt their own regulation is preserved. However, even where regulations are not the same, there is a possibility to recognise them as equivalent in their effectiveness to fulfil the objectives pursued. Notably, mutual recognition is not a strict obligation, but rather a recommendation that leaves a lot of policy space, including for abuse. Taking into consideration the lack of international standards for some solar PV equipment and the speed with which the technology develops, mutual recognition of the equivalence of technical regulations would be a viable option to enhance trade in solar PV equipment.

*Transparency*

In order to level the playing field and ensure that there are no disguised restrictions on trade, the TBT Agreement imposes a transparency obligation (Article 2.9), specifically in cases where (i) there is no relevant international standard, or (ii) the technical regulation did not follow the international standard, and (iii) it may have a significant effect on trade.

Since the 'significant effect on trade' is a vague concept, the TBT Committee clarified that, in order to assess such an effect, 'the Member concerned should take into consideration such elements as the value or other importance of imports in respect of the importing and/or exporting Members concerned ..., the potential growth of such imports, and difficulties for producers in other Members to comply with the proposed technical regulation'.[120]

The transparency obligation includes a duty to (i) publish a notice on a new technical regulation; (ii) notify WTO Members about its coverage;

---

[120] Committee on Technical Barriers to Trade, Decisions and Recommendations adopted by the Committee since January 1995, Note by the Secretariat, Revision, G/TBT/1/Rev. 8, dated 23 May 2002.

and (iii) conduct consultations on request.[121] Derogation from this duty is allowed in cases of emergency (urgent problems of safety, heath, environment, etc.). However, even then, the state has a duty to notify the technical regulation and ensure that it is published without delay once it has been adopted.[122]

In general, the TBT Agreement requires that there should be a reasonable interval between the publication of a technical regulation and its entry into force, so that WTO Members can adapt to the new requirements. The TBT Committee issued an interpretation of the 'reasonable period', which 'mean[s] normally a period of no less than six months, except when this would be ineffective in fulfilling the legitimate objectives pursued'.[123] The six-month interval has been confirmed by the Appellate Body. However, WTO Member States are exempted from this obligation in cases of emergency.

There is a great variety of technical regulations for PV equipment, and compliance is often very difficult and costly for foreign PV producers. Thus, reinforcing the transparency requirement would play a crucial role in the preparation, adoption and implementation of PV technical regulations. It would also be beneficial to industry while helping to meet the public policy objectives pursued.

### *PV Standards and the TBT Agreement*

Standards differ from technical regulations in their non-binding, voluntary nature.[124] All kinds of requirements for PV that have a non-mandatory character, such as those dealing with labelling and certification, would fall within the category of standards under the TBT Agreement.

The TBT Agreement also provides for a separate set of rules for standards. Unlike technical regulations and conformity assessment procedures, the substantive obligations for standards are contained in Annex 3 to the Agreement, which is called the Code of Good Practice for the Preparation, Adoption, and Application of Standards. Paragraph D of the Code of Good Practice provides for a non-discrimination obligation (both national treatment and MFN) similar to the TBT Agreement, Article 2.1. Although the Code of Good Practice does not fully incorporate wording analogous to the TBT Agreement, Article 2.2, concerning legitimate objectives, para E requires that standards are not prepared,

---

[121] TBT Agreement, Article 2.9.   [122] TBT Agreement, Article 2.10.
[123] See n. 120 and Van den Bossche (2008).
[124] Notably, if a technical requirement is by its legal nature non-binding, but it is applied as if it were a compulsory requirement, it would qualify as a *de facto* technical regulation under the TBT Agreement.

adopted, or applied in a way that would create unnecessary obstacles to trade. WTO Member States must ensure, according the TBT Agreement, Article 4, that their central government standardisation bodies accept and comply with the Code of Good Practice, and that their local governmental and non-governmental standardisation bodies take reasonable measures to comply.

Importantly, the TBT Agreement recognises the possible negative effects of parallel and overlapping standardisation efforts within the same country, or between national and international standardisation bodies, and calls for coordination among them.[125] Finally, paras J–P of the Code of Good Practice provide for comparable transparency obligations.

Following the description of standard-setting organisations, such as ANSI in the United States and CSA in Canada, it is clear that the stringent obligations imposed by the TBT Agreement, Article 4, would not apply, since the standard-setting organisations in question are private. However, the United States and Canada should take reasonable measures to ensure that these organisations comply with the Code of Good Practice and do not adopt standards that are discriminatory (in nature and/or through application). They should also notify the WTO about the adoption of new standards. This obligation is not a strong one, but rather a secondary obligation, and it is thus unlikely that it would be invoked in dispute settlement.

So far, no disputes have arisen on the issue of non-compliance of certain standards with the TBT Agreement. However, Article 14.4 gives any Member a right, in cases where its interests have been significantly affected due to the non-compliance of another Member with Article 4 (and by reference also with the Code of Good Practice), to resort to dispute settlement.

### *Conformity Assessment Procedures*

If a dispute arises over technical requirements for PV, it would make sense for a country whose exporters are affected by them to claim that the measures are technical regulations and thus fall under the more stringent obligation contained in the TBT Agreement, Article 2.1. It should be noted, however, that the disciplines on conformity assessment procedures also offer certain protection for PV exporters. These disciplines can be seen as a safety net in cases where technical regulations and standards have been adopted in a due manner and comply with the relevant TBT

---

[125] Code of Good Practices, para H; TBT Agreement, Annex 3.

disciplines, since they prevent the remaining negative trade effects and discrimination.

A 'conformity assessment procedure' is defined as 'any procedure used, directly or indirectly, to determine that relevant requirements in technical regulations or standards are fulfilled'.[126] In practice, it would cover such procedures as testing, verification, certification and inspection. While technical regulations and standards are mentioned separately in the TBT Agreement, Annex 1, they are also included in the definition of conformity assessment procedures. The panel in *EC – Trademarks* held that this suggests that technical regulations, standards and conformity assessment procedures are distinct and mutually exclusive.[127] Conformity assessment procedures allow an exporter to prove compliance with the local technical regulations and standards of the importing country. However, they usually impose certain costs and time constraints and, if administered in a discriminatory and non-transparent manner, may be used for protectionist purposes.[128] Given that most countries (except the United States) base their requirements on the IEC International Standards with minor national variations, it is evident that conformity assessment disciplines would play a key role in ensuring market access to foreign-produced, tested and certified PV equipment.

The TBT Agreement contains a separate group of rules for the conformity assessment procedure (Articles 5–9). The MFN and NT obligations apply to access to an assessment of conformity, its duration, strictness and applicable fees. A distinct obligation relates to the confidentiality of information provided for the procedure. Similar disciplines apply to harmonisation efforts and the use of relevant international guides and standards, as well as transparency. Transparency could be ensured by creating a database containing information on variations of national PV standards based on the IEC International Standards or divergent standards, including data on the international or national conformity assessment procedures available for foreign-certified PV equipment.

Whenever possible, WTO Members must ensure that the results of conformity assessment procedures in other Members are accepted (mutually recognised). Even if procedures in other countries differ from those used in the domestic market, but they still provide an equivalent assurance of conformity, these shall be accepted. However, countries may have concerns about technical competence in the country where the initial conformity assessment was carried out. Verified compliance

---

[126] TBT Agreement, Annex 1.
[127] PR, *EC – Trademarks and Geographical Indications for Agricultural Products and Foodstuffs (Australia)*, paras 7.512–7.513.
[128] Clarke (1996): 31–34.

(such as accreditation) with international guides and recommendations may serve as an indication that the level of technical competence is adequate and thus reliable.[129] Here, the use and recognition of international systems for conformity assessment procedures (for instance, the IECEE system of conformity assessment schemes for PV equipment) would be highly recommended.[130] This is particularly important, since the TBT Agreement, Article 5.4, requires WTO Member States to use the international guides or recommendations already issued by international standardising bodies, or the completion of which is imminent. A Member State may decide not to use these if it considers them inappropriate, for instance due to national security concerns or fundamental climatic or geographical factors.

There may be instances where no guides or recommendations are available, or where the technical content of the proposed conformity assessment procedure differs markedly from that of available guides. In such circumstances, the procedure may have a significant effect on the trade of other WTO Members (this would apply to most PV conformity assessment procedures). In those cases, the country concerned must publish the conformity assessment procedure it wishes to apply, notify the WTO about its coverage, and allow for comments and discussions from other WTO Members. In any case, the procedure should be published promptly, and enter into force within six months of publication.[131]

Again, similar to technical regulations and standards, less stringent compliance obligations apply to TBT provisions for local governments and non-governmental bodies. Articles 7 and 8 encourage WTO Members to take reasonable measures upon their availability to ensure compliance with the TBT Agreement. In many countries, conformity assessment procedures are carried out by non-governmental bodies accredited by governmental authorities, but the (rather lenient) TBT provisions may not be strong enough to ensure compliance by all such organisations.

## *PV Installation Standards and Domestic Regulation under the GATS*

Most measures related to the safety and performance of PV equipment fall under the categories of technical regulations, standards, or conformity

---

[129] TBT Agreement, Article 6.
[130] IEC/*IECEE PV Certification: The Sure Way to Safety, Quality and Performance, System of Conformity Assessment Schemes for Electrotechnical Equipment and Components*, IECEE, 2010
[131] TBT Agreement, Articles 5.6 – 5.9.

assessment procedures for goods. However, some of the standards deal with technical and qualification requirements for services – for instance, PV installation standards – and thus fall outside of the scope of the GATT and the TBT Agreement. One of the key questions in this regard is the training of installers, as a pre-condition for access to installation services. In many cases, this appears to be one of the key obstacles to dissemination of solar PV technology.

The General Agreement on Trade in Services (GATS) does not contain comparable detailed rules for technical requirements. These are usually classified under the category of domestic regulation, addressed in the GATS, Article VI. In relation to qualification requirements and procedures, technical standards and licensing requirements, it calls for development of additional disciplines. For this purpose, a Working Party on Domestic Regulation has been established to develop coherent horizontal disciplines.[132] So far, it has developed special rules for the accountancy services sector. This sectoral approach could be followed further, and specific disciplines for technical requirements for renewable energy services, including PV services, could be developed within the framework of the SETA.

The qualification requirements and procedures, technical standards and licensing requirements under Article VI aim to ensure the desired quality of services, and thus are of a qualitative, rather than a quantitative, nature.[133] The requirements for PV installers would fall mainly within the category of qualification and licensing requirements, which contain a number of requirements (experience, education, etc.), upon completion of which a service provider can be granted official permission to perform PV installation services. It is important that disciplines under the GATS, Article VI, be further advanced in a horizontal manner to all services sectors where commitments have been made.

### *Possible Solutions Provided by the SETA*

Special rules on technical requirements for goods and services would constitute an important part of the SETA. The TBT Agreement, Article 10.7, recognises the right of Members to conclude bilateral or plurilateral agreements on issues such as technical regulations, standards and conformity assessment procedures. These must be notified to the WTO, with a description of the main disciplines and coverage. Several issues related to technical requirements could be taken up in the SETA, specifically for solar PV, or in general for other RE equipment and services that face similar problems.

---

[132] Delimatsis (2010): 647–648.   [133] Delimatsis (2008): 365–408.

Suggested new disciplines of a SETA within the WTO framework can include, but are not limited to:

- Disciplines on the preparation, adoption and application of technical regulations by local governments and non-governmental bodies could be strengthened through an unconditional obligation to comply with the TBT Agreement, Articles 2.1 and 2.2, since the effect of technical regulations that are not adopted by central governments still may have a crucial negative impact on trade in PV products.
- Ensuring a high level of security during the installation and use of PV (as well as other RE) equipment. When this cannot secured through cooperation efforts at the international level, a new Article XX-like general exceptions clause, with a focus on requirements necessary to protect human and animal life and health, could potentially be introduced into the TBT Agreement. The need for such a clause should be clarified in light of the recent and up-coming TBT case law, and could be further discussed with industry representatives.
- Enhanced mutual recognition and transparency obligations would form an important part of the SETA provision on standardisation. Transparency could be enhanced by the creation of a database on national PV standards based on the IEC International Standards or divergent standards, including data on international or national conformity assessment procedures available for foreign-certified PV equipment. This would ensure that countries and companies are aware of the standards adopted abroad, which in turn can positively influence decisions in mutual recognition procedures. In addition, international systems for conformity assessment procedures, such as the IECEE system of conformity assessment schemes for PV equipment, would be highly recommended. Better accountability should also be sought in the standardisation activities of local governments and non-governmental bodies, as this issue is not sufficiently addressed in current TBT disciplines. Furthermore, as noted by Hufbauer and Kim, instead of trying to negotiate a single international standard for PV equipment, a more realistic solution would be to use mutual recognition as a first step.[134] Oversight of the functioning of bilateral mutual recognition mechanisms is usually carried out by a joint committee, which also approves national conformity assessment bodies.[135] A specialised sub-committee could carry out similar coordination functions within the SETA framework.
- New disciplines for PV-related services, where the commitments will be made through a SETA, are needed.[136] Domestic regulation

---

[134] Hufbauer and Kim (2012): 9.   [135] 135. See, for instance, Beynon (2003): 239.
[136] *Ibid.*

requirements following the mandate in the GATS, Article VI, could be elaborated on a sector-specific basis (PV or sustainable energy in general) to address the qualification, licensing and technical requirements of specific importance for PV-related services. This could follow the model adopted for the accountancy services sector, or a broader approach that goes beyond the disciplines in domestic regulation.

- It would also be advisable to create a special sub-committee for sustainable energy technical requirements. The sub-committee could coordinate notifications and communication between WTO Members in relation to standards for PV and other RE equipment, as well as qualification, licensing and technical requirements for related services. The sub-committee could also ensure coordination and cooperation between the WTO and key international standardising organisations. The procedural issues of such cooperation could be addressed in a separate agreement between the organisations (similar to the 1996 IMF–WTO Agreement), but would find a general reference in the SETA.
- Moreover, a special information system for sustainable energy can be created based on the proposed WTO/ISO Standards Information System and the old ISO/IEC Information Centre.[137]

Finally, there is a possibility to include technical regulation provisions in a SETA concluded outside the WTO framework. In this case, the disciplines on technical regulations, standards and conformity assessment procedures, as well as domestic regulation provisions for PV-related service providers would be binding only for the SETA parties.[138] As required by the TBT Agreement, Article 10.7, the countries joining the SETA would have to notify it to the WTO Secretariat. It would also be necessary to make a reference to the selected existing disciplines of the TBT Agreement, thus incorporating them into the SETA, and enhancing them through additional provisions that go beyond the obligations provided for in the TBT Agreement.

Thus, there are several options to enhance free-market access for, and non-discriminatory treatment of, PV equipment in the SETA through more coherent and transparent standards requirements for this equipment – both within and outside the WTO framework. The analysis above shows that standardisation activities by non-governmental bodies is insufficiently addressed through existing disciplines, and there is a need to enhance overall transparency and cooperation at the international level, as well as mutual recognition of the international, regional and national standards applied by various stakeholders in the PV sector.

---

[137] WTO/ISO Standards Information System, and ISO/IEC Information Centre.
[138] Kennedy (2012): 31–33.

# Appendix I Solar PV Capacity

Table 6A.1.1 *Annual Installed Solar PV Capacity in Selected Countries and the World, 1998–2010 (megawatts, MW)*

| Year | GER | ITA | Czech Rep | Japan | US | FRA | China | Spain | Others | World |
|---|---|---|---|---|---|---|---|---|---|---|
| 1998 | 10 | n.a. | n.a. | 69 | n.a. | n.a. | n.a. | 0 | 76 | 155 |
| 1999 | 12 | n.a. | n.a. | 72 | 17 | n.a. | n.a. | 1 | 112 | 197 |
| 2000 | 40 | n.a. | n.a. | 112 | 22 | n.a. | 0 | n.a. | 106 | 280 |
| 2001 | 78 | n.a. | n.a. | 135 | 29 | n.a. | 11 | 2 | 76 | 331 |
| 2002 | 80 | n.a. | n.a. | 185 | 44 | n.a. | 15 | 9 | 138 | 471 |
| 2003 | 150 | n.a. | n.a. | 223 | 63 | n.a. | 10 | 10 | 125 | 581 |
| 2004 | 600 | n.a. | n.a. | 272 | 90 | n.a. | 9 | 6 | 142 | 1,119 |
| 2005 | 850 | n.a. | n.a. | 290 | 114 | n.a. | 4 | 26 | 155 | 1,439 |
| 2006 | 843 | n.a. | n.a. | 287 | 145 | 8 | 12 | 102 | 174 | 1,581 |
| 2007 | 1,271 | 70 | 3 | 210 | 207 | 11 | 20 | 542 | 179 | 2,513 |
| 2008 | 1,809 | 338 | 61 | 230 | 342 | 46 | 45 | 2,708 | 589 | 6,168 |
| 2009 | 3,806 | 717 | 398 | 483 | 477 | 219 | 228 | 17 | 912 | 7,257 |
| 2010 | 7,408 | 2,321 | 1,490 | 990 | 878 | 719 | 520 | 369 | 1,934 | 16,629 |

*Notes:* n.a. = Data not available. Values include both grid-connected and off-grid PV systems.

*Source:* Compiled by Earth Policy Institute with 1998–1999 data for the world, Japan, and China, and with 1998–2005 data for all other countries, from European Photovoltaic Industry Association (EPIA), *Global Market Outlook for Photovoltaics until 2013* (Brussels, April 2009): 4; 2000–2010 data for the world, Japan, and China; 2006–2010 data for all other countries, from EPIA, *Global Market Outlook for Photovoltaics until 2015* (Brussels, May 2011): 9, 14–29.

## Appendix II Concordance of Various Standards for PV Modules (Tables 6A.2.1–Tables 6A.2.7)

Table 6A.2.1

| | | US Market | | |
|---|---|---|---|---|
| | | PV Modules | | |
| Key International Standards | Description | US Standards | Differences | General |
| IEC61215 ed2.0 (2005–04) | Crystalline silicon terrestrial PV modules Design qualification and type approval | UL61215 | Identical to IEC61215 Identical to IEC61646 | Key components like Jbox, wire, connectors need to be UL recognised components. UL requires 3 factory inspections per year UL allows voltages down to 600 V where IEC always requires 1,000 V Key additional test required is temperature co-efficient |
| IEC61646 ed2.0 (2008–05) | Thin-film terrestrial PV modules Design qualification and type approval | UL61646 | | |
| IEC 61730–1 ed1.2 Consol. with am1&2 (2013–03) and IEC 61730–2 ed1.1 Consol. with am1 (2012–11) | PV module safety qualification Part 1: Requirements for construction and Part 2: Requirements for testing | UL1703 | Different construction review requirements which require a full review for UL if IEC was done previously Different Safety Test which require additional testing if IEC was done previously | |
| | | CEC300 (California Energy Commission) needed for Renewable Energy Credits | Requires Listing to UL1703 plus additional tests from IEC61215 or IEC61646 | |

Table 6A.2.2

| Key International Standards | Description | US Standards | Differences | General |
|---|---|---|---|---|
| | | US Market | | |
| | | PV Modules | | |
| IEC 62109-1 ed1.0 (2010-04) and IEC 62109-2 ed1.0 (2011-06); IEC 62477-1 ed1.0 (2012-07) | IEC62109 Safety of Power Converters for use in Photovoltaic Power Systems Part 1: General Requirements and Part 2: Particular Requirements for Inverters and IEC62477-1 Safety Requirements for Power Electronic Converter Systems and Equipment Part 1: General | UL 1741 | 1. IEC62109 (-1,and-2) cover both transformer and transformerless inverter for both grounded and ungrounded PV application; UL1741 just cover inverter for grounded PV application. But it has several unofficial CRD to cover the above application.<br>2. IEC62109 and IEC62477 cover covers inverter connected to systems not exceeding maximum PV source circuit voltage of 1500DC. The equipment may also be connected to systems not exceeding 1000VAC; UL1741 does not specify the maximum working voltage. 600 V should be considered due to the rating limits of components and spacing table. Additional evaluation shall be applied for more than 600 V application.<br>3. IEC62109 and IEC62477 state the requirements on protection against environmental stresses such as the service conditions for operation, storage and transportation; UL1741 does not consider humidity; vibration; UV resistance; altitude.<br>4. IEC62109 and IEC62477 state the requirements on protection against the effect of sonic pressure; UL1741 does not have this kind of requirement.<br>5. IEC62109 and IEC62477 state the requirements on protection against the liquid and chemical hazards; UL1741 does not have this kind of requirement.<br>6. UL1741 refers to IEEE1547 and 1547.1 as a necessary grid connection standard for a utility-interactive inverter. IEC62109 does not include this kind of grid connection requirement, which is covered optionally by the other IEC International Standards. | |

Table 6A.2.3

## US Market

### PV Modules

| Key International Standards | Description | US Standards | Differences | General |
|---|---|---|---|---|
| EN50438 (VDE4105, G59, G83, CEI021, etc.) | Grid connection requirements for EU countries including Germany, UK, Italy | IEEE1547 and IEEE1547.1 | Significantly different testing requirements | |

## India Market

### PV Modules

| Key International Standards | Description | India Standards | Differences | Comments |
|---|---|---|---|---|
| IEC61215 | Crystalline silicon terrestrial PV modules Design qualification and type approval | BISIS14286 | Identical to IEC61215 | PV modules to be used in a highly corrosive atmosphere (coastal areas, etc.) must qualify Salt Mist Corrosion Testing as per IEC61701 |

Table 6A.2.4

| Key International Standards | India Market |||
| | PV Module |||
| | Description | India Standards | Differences | Comments |
|---|---|---|---|---|
| IEC61646 | Thin-film terrestrial PV modules Design qualification and type approval | IEC61646 | Identical to IEC61646 | PV modules must conform to the latest edition of any of the following IEC/equivalent BIS Standards for PV module design qualification and type approval. From April 1, 2013 RFID shall be mandatorily placed inside the module laminate. |
| IEC617730-1&-2 | PV module safety qualification Part 1: Requirements for construction and Part 2: Requirements for testing | BISIS/IEC 61730–1&-2 | Identical to IEC61730-1&-2 | |
| **PV Inverters** IEC62109-1&-2 | Safety of Power Converters for use in PV Power Systems – Part 1: General Requirements and Part 2: Particular Requirements for Inverters | BIS IEC61683 for Efficiency and IEC60068 for environmental requirements | BIS IEC616883 is Identical to IEC61683 | |

Table 6A.2.5

China Market

| Key International Standards | Description | China Standards | Differences | Comments |
|---|---|---|---|---|
| **PV Module** | | | | |
| IEC61215 | Crystalline silicon terrestrial PV modules Design qualification and type approval | SACGB/T9535 | The specifications are not identical: there are difference and SACGB/T 9535 is in Chinese only | A CNCA-designated test laboratory in China is needed. Initial Factory Inspection needed and follow-up certifications need to be conducted every 12 months. |
| IEC61646 | Thin-film terrestrial PV modules Design qualification and type approval | SACGB/T18911 | Identical to IEC61646 | |
| IEC617730-1&-2 | PV module safety qualification Part 1: Requirements for construction and Part 2: Requirements for testing | SACGB/T20047.1 for Part 1 | Part 1 is covered for construction requirements The specifications are not identical: there are difference and SACGB/T20047-1 is in Chinese only | |
| **PV Inverters** | | | | |
| IEC62109-1&-2 | Safety of Power Converters for use in PV Power Systems – Part 1: General Requirements and Part 2: Particular Requirements for Inverters | Inverter of wind and solar energy supply power system for off-grid – Part 2: Testing method (SACGB/T20321-2) | The Part 2 specifications are not identical: there are difference and SACGB/T20321-2 is in Chinese only | |

Table 6A.2.6

European Market

| Key International Standards | Description | EU Standards | Differences | CE Directive |
|---|---|---|---|---|
| **PV Module** | | | | |
| IEC61215 | Crystalline silicon terrestrial PV modules Design qualification and type approval | EN61215 | Identical to IEC61215 | LVD:2006/95/EC |
| IEC61646 | Thin-film terrestrial PV modules Design qualification and type approval | EN61646 | Identical to IEC61646 | LVD:2006/95/EC |
| IEC617730-1&-2 | PV module safety qualification Part 1: Requirements for construction and Part 2: Requirements for testing | EN61730-1 and -2 not adopted yet | 61730-1&-2 each country has its own version | LVD:2006/95/EC |
| **PV Inverters** | | | | |
| IEC62109-1&-2; IEC62477-1 | IEC62109 Safety of Power Converters for use in PV Power Systems – Part 1; General Requirements and Part 2: Particular Requirements for Inverters and IEC62477-1 Safety Requirements for Power Electronic Converter Systems and Equipment – Part 1: General | EN62109-1&-2 can not be purchased but it is identical to ISEN62109-1 and -2 | ISEN62109-1 is identical to IEC62109-1&-2 | 2006/95EC |
| EN50438 (VDE4105, G59, G83, CEI 021, etc.) | Grid connection requirements for EU countries including Germany, United Kingdom, Italy | EN50438 (VDE4105, G59, G83, CEI 021, etc.) | Grid connection requirements vary by country | |

Table 6A.2.7

Japan Market

| Key International Standards | Description | Japanese Standards | Differences | Comments |
|---|---|---|---|---|
| IEC61215 | Crystalline silicon terrestrial PV modules Design qualification and type approval | JISC8990 | Identical to IEC61215 | ET PVm certification service is a voluntary scheme operated by JET. Certificates are granted to each model of products after the successful completion of applicable tests based on the IEC/IEC harmonised JIS standards and the factory inspection of the quality management system at the manufacturing location. |
| IEC61646 | Thin-film terrestrial PV modules Design qualification and type approval | JISC8991 | Identical to IEC61646 | |
| IEC617730-1&-2 | PV module safety qualification Part 1: Requirements for construction and Part 2: Requirements for testing | JISC8992-1&-2 | Identical to IEC61730-1&-2 | Factory inspections consist of the initial factory inspection (carried out simultaneously with the conformity tests) and the regular factory inspections. Initial factory inspection is carried out to ensure the manufacturing system is suitable to keep manufacturing the products identical to as certified. Regular factory inspection is performed once a year for the purpose of ensuring that the products shipped with certification label are identical to as the tested/certified produc. |

## Appendix III

### *Glossary of Terms*

**Arc-fault/Flash** – A rapid release of energy due to an arcing fault between a phase bus bar and another phase bus bar, neutral, or a ground.

**Balance of Systems (BOS)** – Balance of systems, usually all remaining components of a PV system exclusive of the PV module.

**Bankability** – The ability to achieve a return on investment with minimal financial risk.

**Building-Integrated PV (BIPV)** – A term for the design and integration of PV technology into the building envelope, typically replacing conventional building materials. This integration may be in vertical facades, replacing view glass, spandrel glass, or other facade material into semitransparent skylight systems; into roofing systems, replacing traditional roofing materials; into shading 'eyebrows' over windows; or other building envelope systems.

**Concentrated PV** – A PV module or array that uses optical elements to increase the amount of sunlight incident on a PV cell. Concentrating arrays must track the sun and use only the direct sunlight, because the diffuse portion cannot be focused onto the PV cells.

**Crystalline Silicon** – Crystalline silicon cells are made of silicon atoms connected to one another to form a crystal lattice; the most prevalent bulk material for Solar PV cells (monocrystalline and polycrystalline)

**Feed-in-Tariff (FIT)** – The FIT system is a policy mechanism designed to accelerate investment in RE technologies. It achieves this by offering long-term contracts to RE producers, typically based on the cost of generation of each technology. Technologies such as wind power, for instance, are awarded a lower per-kWh price, while technologies such as PV are offered a higher price, reflecting higher costs. New programmes are being developed that allow payments to be based on the value provided by the use of the RE system instead of the cost.

**Inverter** – An inverter converts direct current (DC) power from the PV array/battery to alternating current (AC) power compatible with the utility and AC loads.

**Off-grid** – An autonomous or hybrid PV system not connected to a grid. May or may not have storage, but most systems require batteries or some other form of storage.

**PV** – Pertaining to the direct conversion of light into electricity.

**PV cell** – The smallest semiconductor element within a PV module to perform the immediate conversion of light into electrical energy (DC voltage and current).

**PV module** – A number of PV cells connected together, sealed with an encapsulant, and having a standard size and output power; the smallest building block of the power generating part of a PV array.

**Rack mounting** – A fixed PV array, using mounting hardware rather than roof membrane or other structures.

**Solar Tracking Systems** – A PV array that follows the path (track) of the sun to maximise the solar radiation incident on the PV array surface. The two most common orientations are (1) a single axis, where the array tracks the sun east to west and (2) dual-axis tracking, where the array points directly at the sun at all times. Tracking arrays use both the direct and diffuse sunlight.

**Thin-film** – A layer of semiconductor material, such as copper indium diselenide, cadmium telluride, gallium arsenide, or amorphous silicon, a few microns or less in thickness, used to make PV cells.

## References

### *Standards and Certification*

American National Standards Institute (ANSI), www.ansi.org

Civic Solar, *Arc Fault Detection*, Boston, MA, 2003

Earth Policy Institute, Climate, Energy and Transportation Data Center, www.earth-policy.org/data_center/C23

Electrical Safety Authority, *Government and Regulations*, The Ontario Electrical SafetyCode. Portland, OR

Florida Solar Energy Center (2010) *Testing Method of Photovoltaic Module Power Rating*. FSEC: 1–5

Go Solar California, www.gosolarcalifornia.org

IEC/IECEE, List of Photovoltaic Members, www.iecee.org/pv/html/pvcntris.htm

*PV Certification: The Sure Way to Safety, Quality and Performance, System of Conformity Assessment Schemes for Electrotechnical Equipment and Components*, 2010

International Centre for Sustainable Trade and Development (ICSTD) (2001) Submission of the UNFCCC, Information and News Relating to Modalities for the Operationalisation of a Work Programme and Possible Forum on Response Measures

International Electrotechnical Commission (IEC), www.iec.ch/about/ www.iec.ch/standardsdev/

International Laboratory Accreditation Cooperation (ILAC), www.ilac.org

International Renewable Energy Agency (2013) *International Standardisation in the Field of Renewable Energy*. IRENA, Masdar City

Japan Electrical Safety and Environment Technical Laboratories, Certification of Photovoltaic Modules, www.jet.or.jp/en/products/solar/index.html

Korean Institute of Energy Research, *New and Renewable Energy Technology*, Global Kier, 2011

SAI Global, www.saiglobal.com
Solar America Board for Codes and Standards, *About Solar ABCs*, 2011
*Codes and Standards*, 2011, www.solarabcs.org/about/index.htm
Solar Rating and Certification Corporation, www.solar-rating.org/about/general.html
Standards Council of Canada, *Directory of Accredited Standards Development Organizations*, www.scc.ca/en/accreditation/standards/directory-of-accredited-standards-development-organizations
State of California Building Standards Commission, Codes, 2013, www.bsc.ca.gov/codes.aspx
United States Department of Commerce (2008) *Clean Energy: An Exporter's Guide to China*, http://trade.gov/publications/pdfs/china-clean-energy2008.pdf
United States Department of Labor, Occupational Safety & Health Administration, *Recognised Testing Laboratories*. Washington, DC
*Typical Registered Certification Marks*, www.osha.gov/dts/otpca/nrtl/nrtlmrk.html
Wiles, John (2010). *Ungrounded Electrical Systems*, www.nmsu.edu/~tdi/pdf-resources/IAEI%20September-October%202010.pdf
World Standards, *Electricity Around the World*, www.worldstandards.eu/electricity.htm
Worldwatch Institute, The Way Forward for Renewable Energy in Central America. Washington, DC

### *Books and Articles*

Arcuri, Alessandra (2012) 'Back to the future: US – Tuna II and the New Environment–Trade Debate', *European Journal of Risk Regulation*, 3: 177
Beynon, Paul (2003) 'Community mutual recognition agreements, technical barriers to trade and the WTO's most favoured nation principle', *European Law Review*, 28 (2): 231–249
Clarke, John (1996) 'Mutual recognition agreements', *International Trade Law and Regulation*, 2 (2): 31–34
Conrad, Christiane (2014) *Processes and Production Methods (PPMs) in WTO Law: Interfacing Trade and Social Goals*. Cambridge University Press
Delimatsis, Panagiotis (2008) 'Determining the necessity of domestic regulations in services: the best is yet to come', *European Journal of International Law*, 19 (2): 365–408
(2010) 'Concluding the WTO services negotiations on domestic regulation: hopes and fears', *World Trade Review*, 9 (4): 643–673
Diller, Janelle (2012) 'Private standardisation in public international lawmaking', *Michigan Journal of International Law*, 33 (3): 481
Elzinga, David et al. (2011) *Advantage Energy*. IEA, Paris: 1–64
Fontanelli, Filippo (2011) 'ISO and Codex standards and international trade law: what gets said is now what's heard', *International and Comparative Law Quarterly*, 60 (4): 895–932
Hufbauer, Gary and Jisun Kim (2012) *Issues and Considerations for Negotiating a Sustainable Energy Trade Agreement*. ICTSD, Geneva

International Centre for Trade and Sustainable Development (2011) *Fostering Carbon Growth: The Case for a Sustainable Energy Trade Agreement*. ICTSD Platform on Climate Change. Geneva, November

Kennedy, Matthew (2012) *Legal Options for a Sustainable Energy Trade Agreement*. ICTSD, Geneva: 31–33

Marceau, Gabrielle (2013) 'The new TBT jurisprudence in US – Clove Cigarettes, WTO US – Tuna II, and US – COOL', *Asian Journal of WTO and International Health Law and Policy*, 8 (1): 1

Martin, Jeremy and Juan Carlos (2012) 'Central America's electric sector: the path to interconnection and a regional market', *Journal of Energy Security*, August 13: 1–4

Mitchell, Andrew (2009) *Legal Principles in WTO Disputes*. Cambridge University Press: 201

Trujillo, Elizabeth (2012) 'The WTO Appellate Body knocks down US "dolphin-safe" tuna labels but leaves a crack for PPMs', *ASIL Insights*, 16 (25), July

Van den Bossche, Peter (2008) *The Law and the Policy of the World Trade Organization: Text, Cases and Materials*, 2nd edn. Cambridge University Press

Voon, Tania (2012) 'Cigarettes and public health at the WTO: the appeals of the TBT labeling disputes begin', *ASIL Insights*, 16 (6), February

Vranes, Erich (2009) *Trade and the Environment: Fundamental Issues in International Law, WTO Law, and Legal Theory*. Oxford University Press

Zhou, Weihuan (2012) 'US-Clove Cigarettes and US – Tuna II (Mexico): implications for the role of regulatory purposes under Article III:4 of the GATT', *Journal of International Economic Law*, 15 (4): 1075

Zleptnig, Stefan (2010) *Non-Economic Objectives in WTO Law, Justification Provisions of GATT, GATS, SPS and TBT Agreements*. Martinus Nijhoff, Leiden/Boston, MA

## WTO Documents

Committee on Technical Barriers to Trade, Decision of the Committee on Principles for the Development of International Standards, Guides and Recommendations with Relation to Articles 2 and 5 and Annex 3 of the Agreement. Second Triennial Review of the Operation and Implementation of the Agreement on Technical Barriers to Trade, Annex 4, G/TBT/9, 3 November 2000. Geneva

Committee on Technical Barriers to Trade, Decisions and Recommendations adopted by the Committee since January 1995, Note by the Secretariat, Revision, G/TBT/1/Rev. 8, 23 May 2002. Geneva

Committee on Technical Barriers to Trade, Minutes of the Meeting of 23–24 June 2010, G/TBT/M/51. Geneva

Committee on Technical Barriers to Trade, Concerns Raised about Tobacco and Environmental Measures, Technical Barriers to Trade: Formal Meeting, 15 and 16 June 2011. Geneva

Committee on Technical Barriers to Trade, Minutes of the Meeting of 10–11 November 2011, Note by the Secretariat, G/TBT/M/55, 9 February 2012. Geneva

Committee on Technical Barriers to Trade, Minutes of the Meeting of 20–21 March 2012, Note by the Secretariat, G/TBT/M/56, 16 May 2012. Geneva

Committee on Technical Barriers to Trade, Eighteenth Annual Review of the Implementation and Operation of the TBT Agreement, Note by the Secretariat, 27 February 2013, G/TBT/33: 2. Geneva

WTO (2005) *World Trade Report 2005*. Geneva

WTO (2012) *World Trade Report 2012*. Geneva

## Case Law

Appellate Body Report. *European Communities – Measures Affecting Asbestos and Asbestos-Containing Products*, WT/DS135/AB/R, April 5, 2001, paras 66–70

Appellate Body Report. *European Communities – Trade Description of Sardines*, WT/DS231/AB/R, October 23, 2002

Appellate Body Report. *United States – Measures Affecting the Production and Sale of Clove Cigarettes*, WT/DS406/AB/R, April 4, 2012

Appellate Body Report. *United States – Measures Concerning the Importation, Marketing and Sale of Tuna and Tuna Products*, WT/DS381/AB/R, May 16, 2012

Appellate Body Report. *United States – Certain Country of Origin Labelling (COOL) Requirements*, WT/DS384/AB/R, WT/DS386/AB/R, June 29, 2012

*Brazil – Measures Affecting Imports of Retreaded Tyres*, WT/DS332

*China – Measures Affecting Trading Rights and Distribution Services for Certain Publications and Audiovisual Entertainment Products*, WT/DS363

Panel Report. *European Communities – Measures Affecting Asbestos and Asbestos-Containing Products*, WT/DS135/R, September 18, 2000

Panel Report. *EC – Trade Description of Sardines*, WT/DS231/R, May 29, 2002

Panel Report. *EU – Protection of Trademarks and Geographical Indications for Agricultural Products and Foodstuffs (Australia)*, WT/DS290/R, March 15, 2005

Panel Report. *United States – Measures Affecting the Production and Sale of Clove Cigarettes*, WT/DS406/R, September 3, 2011

Panel Report. *United States – Measures Concerning the Importation, Marketing and Sale of Tuna and Tuna Products*, WT/DS381/R, September 15, 2011

Panel Report. *United States – Certain Country of Origin Labelling (COOL) Requirements*, WT/DS384/R, WT/DS386/R, November 18, 2011

# 7 Addressing Local Content Requirements in a Sustainable Energy Trade Agreement: June 2013

*Sherry M. Stephenson*

## 1 Introduction

The financial crisis of 2008 heralded a more discrete, yet more pernicious form of protectionist trade policies: local content requirements (LCRs). At the same time, governments are placing greater importance on green policies to achieve sustainable economic growth. The global shift to green industrial growth was the overarching issue at the Rio + 20 summit in June 2012. However, stalled international climate negotiations and Doha trade talks within the World Trade Organization (WTO) are not fostering a transition to green growth. As noted by Joachim Monkelbaan, 'trade preferences for climate-related goods are supposed to increase exports of... products from developing countries and at the same time contribute to their dissemination in targeted developing countries due to building up or strengthening local environmental industries as well as cost and efficiency gains arising from economies of scale'.[1] Such anticipated effects both domestically and abroad are steps toward greener economies. However, the combination of the financial crisis and inadequate international policy momentum is contributing to the popularity of LCRs devised at the national level as green growth policy tools.

LCRs are policy measures that typically require a certain percentage of intermediate goods used in the production processes to be sourced from domestic manufacturers.[2] Such requirements in renewable energy (RE) policy tend to take one of two forms: a precondition to receive government support, such as tariff rebates; or an eligibility requirement for government procurement in RE projects. LCRs are usually coupled with other policy measures to encourage green growth.

Despite agreed WTO disciplines to promote freer trade, developed and developing countries increasingly use LCRs in their RE policies. Since the financial crisis, public financing for low-carbon energy policies has been

[1] Monkelbaan (2011): 25.   [2] Kuntze and Moerenhout (2013).

squeezed. At the same time, climate change and environmental degradation concerns are pressing. It is against this backdrop that the effectiveness of LCRs in achieving green industrial growth, as well as their legal aspects, must be addressed.

From the economic side, in the short term LCRs increase production costs, which then inflate retail energy prices. From the legal side, they are highly questionable under WTO law. Nevertheless, governments are continuing, and increasing, the use of these measures. This chapter outlines some basic conditions for LCRs' effectiveness in expanding local manufacturing, creating associated jobs and lowering retail energy prices, as suggested by Kuntze and Moerenhout (2013).[3] These conditions include: a stable and sizable market with potential for growth; a percentage of LCR that is not too restrictive; cooperation between governments and energy firms; and a baseline of current knowledge to facilitate further technology knowledge transfers.

Relatively little has been written on the topic of LCRs. In the existing literature, earlier work has tended to monitor the frequency of LCRs for renewable energy and their overall effectiveness. This chapter offers more pointed policy recommendations in the context of sustainable energy trade agreements. It draws and builds on the work by Kuntze and Moerenhout[4] as well as the work by Gary Hufbauer and other economists at the Peterson Institute for International Economics (PIIE).[5]

The first part of this chapter contains an overview of LCRs in RE and assesses the rationale and effectiveness of such measures. The second part presents two case studies of LCRs in RE, namely wind energy in two Canadian provinces (Ontario and Quebec) and solar energy in India. The third part analyses LCRs in the context of a proposed Sustainable Energy Trade Agreement (SETA), as well as alternatives to LCRs and lessons learned for future trade agreements.

## 2  The Breadth of LCRs

According to one set of estimates,[6] 107 new LCRs have been imposed since the onset of the financial crisis early in 2008. All together, they may have adversely impacted USD 2.7 trillion of world trade, or about 11 per cent of world commerce in goods and services. 'Impacted' does not mean reduced; however, the authors offer a speculative guess that the new LCRs may have reduced world commerce by USD 200-300 billion annually, about the same amount as the potential estimated gains from

---

[3] *Ibid.*: 5.  [4] See n. 2.  [5] See, in particular, Hufbauer et al. (2013).  [6] *Ibid.*

Doha Round trade liberalisation.[7] If this guess is near the mark, LCRs have significantly retarded world trade, and are partly responsible for the mediocre global trade performance in 2012 (under 4 per cent growth) and the sub-par outlook for years to come (3.3 per cent growth).[8]

Of course, the great majority of LCRs are aimed at sectors other than RE. Scanning the available data, it appears that perhaps twenty new LCRs affect the RE sector.[9] Most of these are contained in the broader mandates imposed on government procurement or government support of industry; a few of them single out RE. Collectively, LCRs in the RE space probably impact over USD 100 billion of trade annually, but the available data do not permit an estimate of trade stifled by LCRs.

Table 7A.1 in the Appendix, extracted from the database assembled by Hufbauer et al. (2013), identifies the LCRs that affect RE projects. Some are specifically targeted to wind turbines, solar panels or biomass. In other cases, RE projects are swept up in LCR measures that affect all government procurement or all government-financed projects. Very likely the original database missed many LCRs in the RE space, since it was designed to cover just new LCRs introduced since the recession of 2008–2009. However the examples cited in Table 7A.1 illustrate the flavour of LCRs that are applied in the RE area.

## 3  The Rationale for LCRS

### Arguments in Favour of LCRs

Moerenhout and Kuntze (2013) find that LCRs in green industrial policies are generally promulgated for four reasons. First, the political economy argument is made that LCRs augment public support for renewable energy projects. Second, proponents point to the classic case for protecting infant industries, especially in developing countries, until they can compete on the international market. Third and, quite importantly, the creation of 'green' jobs, especially in developed countries, is put forward as a justification for the use of LCRs. Fourth, proponents

---

[7] Hufbauer et al. (2010): 108. Note that the estimates in this volume are considered to be conservative and have been superseded by new research on the payoff from greater trade facilitation. A 2013 World Economic Forum–World Bank report entitled *Enabling Trade: Valuing Growth Opportunities* shows even larger estimates for payoff from a dramatic (but) feasible reduction in trade transactions costs, of a 5 per cent of global GDP and a 15 per cent gain in global trade, when these costs are brought down halfway to the level of the world's best practice.

[8] *Straits Times*, 'WTO cuts 2013 trade growth forecast to 3.3%', April 11, 2013: A24.

[9] This estimate is based on an examination of Appendix A in Hufbauer et al. (2013).

Local Content Requirements in a SETA 319

point to the potential environmental benefits of greater competition between renewable energy firms over the medium term.[10]

*Political Economy Argument* RE generally costs more, per kilowatt hour (KWh), than coal-fired power. One way to enlist public support for the extra cost is to tie RE projects to domestic innovation and job creation through LCRs. A worthwhile research project would investigate whether countries with 'strict' LCRs in fact use RE for a larger fraction of their power supply than countries with 'relaxed' LCRs (or no LCRs). The proposed investigation is well beyond the scope of this chapter, but the findings would make a valuable contribution to understanding the strength of the political economy nexus between LCRs and RE.

*Infant Industry Protection* Policy makers – usually in developing countries – contend that LCRs protect infant industries from foreign competition. LCRs present an attractive solution to allow infant industries to become internationally competitive in their renewable technology and manufacturing capability. Proponents argue that while the GATT 1947 reflected Keynesian precepts and was somewhat tolerant of government supervision of markets, the conclusion of the Uruguay Round and the creation of the WTO marked a shift towards neo-liberalism and a heavy preference for market outcomes. This framework, it is argued, is not amenable for developing countries to master advanced technologies and enjoy economies of scale.[11] By contrast, LCRs afford a certain 'policy space' to develop infant industries. In addition LCRs may counteract government subsidies in other countries. According to this line of argument, LCRs provide incentives for local firms to produce and eventually innovate in the most promising green energy sectors and to lower their production costs over time.[12]

*Green Job Creation* In industrialised economies, the same economic arguments for LCRs in terms of the infant industry rationale are not applicable. Instead, proponents of RE LCRs in developed countries point to the creation of green jobs. By requiring firms to use a certain percentage of local inputs, demand for domestic cleaner industries will increase, spurring green job creation in the short term.[13] In the long term,

[10] See n. 2.
[11] Gillian Moon, 'Capturing the benefits of trade? Local content requirements in WTO law and the human rights-based approach to development', University of South Wales, 2008.
[12] See n. 2.  [13] Farrell (2011): 28.

proponents argue that there are economic benefits to be gained from 'learning by doing' and from increasing the supply of RE.[14]

Policy makers find it politically compelling to push forward green industrial programmes with LCRs, given the anticipated economic gains in employment and growth of the green sector. Both developed and developing countries implement LCRs with the two-pronged goal of achieving a robust RE industry that will be competitive in international markets, and securing associated local job creation. In addition, it is sometimes argued that an expanded domestic manufacturing industry could entail a larger tax base for governments. An increased tax base allows governments to enjoy more revenues without raising tax rates.[15] However, it has not yet been demonstrated that the financial investment and incentives needed to expand the green manufacturing industry would be more than offset by the anticipated gains from a larger tax base.[16] Companies may also decide to settle in a location for reasons other than tax incentives, including the attractiveness of the overall business environment.

*Environmental Benefits* In the environmental arena, proponents of LCRs point to the positive spillover effects for the environment in the medium term. By increasing the number of players in the international market, proponents of LCRs contend that, in the medium term, greater competition will spur innovation in the RE sector and consequently lower green technology costs. Competition and innovation should reduce the time it takes for RE to compete with fossil fuels and nuclear energy.[17] Following this environmental line of argument, the medium-term benefits will compensate the short-term disadvantages in terms of greater production costs. In addition, proponents claim that, by promoting the transfer of technology, LCRs foster sustainable practices worldwide.

In theory, LCRs can also facilitate a transfer of technology from learning by doing and building local capacity. LCRs force firms to transfer technology so that the final quality of the product is maintained. As already mentioned, these positive spillover effects remain theoretical and have yet to be proven.[18] Thus, a LCR requirement may in fact make it even less interesting for companies to establish a productive base, thwarting the transfer of technology altogether.

*Arguments against LCRs*

Opponents of LCRs in RE policies point to the economic costs – inefficient allocation of resources, higher retail power prices, negligible employment

[14] See n. 2.   [15] Lewis and Wiser (2005).   [16] See n. 2.   [17] See n. 2.   [18] See n. 2.

gains and a negative impact on trade – and question the environmental gains in the medium term.[19]

*Inefficient Allocation of Resources* Opponents hold that LCRs lead to an inefficient allocation of resources by distorting the operation of comparative advantage: 'In practice, LCRs are discriminatory and can, like tariffs, constrain effective organization of sustainable energy supply chains.'[20] LCRs require or create incentives for enterprises to inefficiently invest their resources in local inputs to artificially improve the competitiveness of local products, making foreign products less attractive to potential buyers.[21] In the absence of LCRs, the same resources would be invested in other sectors more efficiently. The impact of LCRs is similar to that of subsidies – through an inefficient allocation of resources, local products become more competitive and foreign products less so.

Proponents of LCRs argue that they are a short-term policy, put in place to protect infant industries and businesses only for the amount of time needed to play 'catch up' with foreign economies by producing to scale the same products with greater efficiency, creating capacity to compete in international markets. In the long term, the need for LCRs will be obsolete once enterprises are able to compete with foreign firms. Opponents point out that, in reality, subsidies such as LCRs are politically sensitive. Once LCRs become a mainstay and expectation of businesses, withdrawal of government support will often be met with fierce resistance.[22] In addition, even if subsidies and LCRs are implemented with the intent of being temporary, it is possible that the relevant manufacturing sectors will never attain the level of efficiency necessary to operate without government support. Infant industries may never become competitive enough to export their RE products on the international market and instead require continuous government support. In such circumstances LCRs would become a policy of permanent protection.[23]

*Higher Power Prices* In the short term, LCRs inflate power costs. Since firms are required to purchase local inputs that are likely to be more costly than foreign ones, their manufacturing costs are increased. Producers eventually offset the higher manufacturing costs by passing

[19] See n. 2.
[20] International Centre for Trade and Sustainable Development (2011a): 41.
[21] Tomsik and Kubicek, (2006): 18.   [22] See n. 2 and n. 11: 13.
[23] It may also be true, however, that if no LCRs had been put in place, there may not have been any RE production at all, so the two objectives – economic efficiency and environmental sustainability – must once again be evaluated together.

these on in the form of increased power prices to domestic consumers.[24] LCR proponents contend that in the medium and long term, greater competition and innovation will eventually lower manufacturing costs, and hence consumer power prices, but this seems far from certain.[25]

*Green Job Creation Doubtful*   Although one of the intended benefits of LCRs is job creation in the green industrial sector, it is not certain that LCRs create additional jobs. Two opposing effects are at play. On the one hand is the output effect: LCRs increase the cost of RE production through higher input prices. As such, less RE is produced, resulting in zero job creation and possibly job losses in the green industrial sector. However, it is also possible that there is job creation but lower returns to other factors.[26] Since LCRs require firms to source components locally, employment will increase in the component industry. The net effect for job creation of higher input prices and hence less RE production combined with greater demand for component manufacturing is difficult to pinpoint.[27] The outcome depends on specific policies.

To make the story more complicated, countering the output effect is the substitution effect. The degree of local content required can affect whether jobs are lost or created. If the percentage of local content required is very high, then RE production will be reduced, accompanied with net job losses. However if the amount of local content required is not very high, then firms might increase their employment to offset higher prices for local material. This substitution effect assumes that labour can serve as a substitute for the local material.[28]

*Negative Impact on Trade*   LCRs negatively impact trade. They require firms to use material that is made locally and more expensive than foreign inputs. In the absence of LCRs, enterprises would opt for foreign-made material when that is the cheaper option. Like a subsidy, the effect of LCRs on trade is to discourage foreign imports and to stifle competition between domestic and foreign firms – by making locally produced material a requirement in the end product. The impact on trade of LCRs varies, depending on the percentage of local content required and the efficiency of existing firms. In an economy with inefficient firms, a high degree of required local content obviously thwarts competition. The LCR becomes a very high non-tariff barrier (NTB). However, the negative impact of high LCRs on competition is lessened in an economy with more efficient firms.

---

[24] Jie-A-Joen et al. (1998): 34.   [25] See n. 2.   [26] See n. 2.   [27] See n. 2.   [28] See n. 2.

In addition to the economic and environmental benefits and drawbacks set out above, LCRs may have a negative impact on the services portion of RE production, as well as innovation and quality.[29] In focusing on the manufacturing segment of the value chain, LCRs do not affect the services portion of the RE sector. In the short term, LCRs drive up production costs for RE, but might not encourage the services components of RE production. However, knowledge and technology transfers are essential for sustainable green growth. When they target manufacturing, LCRs do not take advantage of the employment gains to be made through investment in the services portion of the value chain, such as in engineering, installation and maintenance.[30] Given the critical nature of efficient services to the operation of value chains, any LCRs that target or impact services performance will have particularly detrimental cost and efficiency implications for the final products.

In addition, LCRs might hamper innovation and quality in the RE sector. With a restrictive LCR in place, investors might be deterred from investing in the sector owing to higher input prices. Meanwhile, the higher the LCR, the more the RE sector will be protected from foreign competition, resulting in lower quality and higher prices. Over time, this may impact the quality of foreign direct investment (FDI) attracted to the sector and encourage rent-seeking, less efficient FDI rather than cutting-edge, innovative FDI focused on both the domestic and world markets.

## 4  The Effectiveness of LCRs

Little has been written on the effectiveness of LCRs generally and empirical studies on their impact for RE are especially limited. Nevertheless, Kuntze and Moerenhout outline five agreed-upon preconditions for LCRs in RE production which have a beneficial impact for the domestic economy:
1. Stability and size of market;
2. Restrictiveness of LCRs;
3. Cooperation between government and firms;
4. Accompanying subsidies;
5. Technology and knowledge transfers.

First, LCRs in RE must be introduced in a stable and sizable market that has the potential for growth. In the absence of a stable market with growth potential, firms will be deterred from investing in RE manufacturing.[31] Ultimately, investors are concerned with whether the higher costs incurred to produce local material will be more than compensated for

[29] m Monkelbaan (2013).    [30] Brewer and Falke (2012).    [31] See n. 15.

through stable demand and industry growth. The larger the market, the more chance there is that welfare gains can be reached through LCRs. In addition, a large and stable market encourages transfers of knowledge and technology through learning by doing. If the existing industry and market potential are small, the LCR is unlikely to yield much in terms of welfare benefits to the host economy.[32]

Second, the impact of LCRs depends largely on the percentage of the local products required. To add value to the host economy, the LCR should be phased in gradually, and the percentage of local content required should not be too high. When LCR percentages are overly stringent, the LCR is more likely to damage the local economy. When the level is more appropriate, according to jurisdiction specifics, the increased cost of production may be offset by the gain in jobs or expansion of green manufacturing, for example. However, beyond a certain LCR percentage, the higher cost of production is greater than the value gained for the local economy. The appropriate LCR percentage depends on the size of the green industrial sector and the opportunity cost of capital.[33] Some economists have tried to model what this percentage would be, but there seems to be no definitive guide for this at present. However, it seems clear that an overly restrictive LCR is likely to be damaging.

Third, in setting the LCR rate, governments have much to gain from cooperating with local businesses.[34] Supply chains with numerous producers supplying components for intermediate goods are complicated. For example, solar photovoltaic (PV) cells consist of many components – cells, modules, etc. – that are potentially manufactured by different producers. Cooperation between governments and businesses increases information on both sides. This facilitates the determination of an appropriate LCR rate and the efficiently combination of various elements in the supply chain.

Fourth, Kuntze and Moerenhout find that a precondition for LCRs to be valuable to the host economy is ensuring that the subsidy to which the LCRs are to be coupled is sufficient to maintain market attractiveness.[35] There is still much research to be done on the appropriate type of subsidy for firms when introducing LCRs – for example, tax credits, soft loans, grants, capital subsidies – and which part of the value chain the subsidy should target.[36] This also depends on jurisdiction specifics, and the technology and value chain. Some form of financial support is intended to promote technology transfers. However, a large gap remains in the research concerning the best type of financial support.

Fifth, proponents of LCRs point to the positive spillover effects – in terms of greater efficiency – that come only with time and experience.

[32] See n. 20: 40.  [33] See n. 2: 8.  [34] Veloso (2001).  [35] See n. 20.  [36] See n. 2.

There is still much uncertainty as to whether the greater efficiency will be sufficient to offset the higher production costs in the long term.[37] Veloso shows that when there is already a certain level of local knowledge about the technology in question, the LCR will be more effective.[38] While they are less likely to bridge a wide knowledge gap between local and foreign businesses, LCRs may, however, bring added technology knowledge where a foundation has already been laid. An LCR will be more valuable if there is a high learning by doing potential. LCRs will yield more value to the host economy if they do not over-emphasise the manufacturing portions of the value chain, but also target training by doing to establish high-skilled workers.

Despite these five preconditions to gauge whether LCRs will have a positive welfare effect on the host economy, much remains uncertain. For instance, the appropriate LCR percentage will vary and depends on local market, technology, etc. There also remain questions in terms of the best subsidy – type, targeted value chain, duration and size. A misallocated subsidy could entail an additional cost and not an added benefit for the local economy. In addition, one of the chief and compelling arguments against LCRs is that they can easily become a permanent policy on account of the unwillingness of local firms to give up an obvious advantage. Thus an additional precondition is a clear timeframe for the term of the LCR, beyond which it would not be renewed.[39]

## 5  Two Illustrative Cases Involving LCR Use

The two illustrative cases discussed below provide a flavour of the type of products and services that governments are choosing to impose LCRs in the name of sustainable development and clean energy.

### *Wind Energy in Canada*

Wind energy capacity worldwide has grown at an incredible pace – doubling every three years according to the 2011 World Wind Energy Report. At the end of 2011, wind turbines accounted for roughly 3 per cent of the world's electricity consumption.[40]

Wind energy has grown especially in Canada, with the country becoming one of the top ten producers in 2010. Thanks to a wealth of natural resources, electricity prices in Canada are the fourth lowest among the Organisation for Economic Cooperation and Development (OECD) countries, following the United States, Mexico and Korea. After hydropower,

[37] See n. 34.   [38] See n. 2: 44.   [39] Hufbauer et al. (2013): 44.   [40] See n. 34.

which provides 60 per cent of Canada's electricity, wind energy is one of the major renewable energy sources in Canada. The Canadian Wind Energy Association (CanWEA) intends to provide 20 per cent of Canada's electricity with wind energy by 2025.[41]

Regulation over Canada's electricity market varies between its ten provinces and three territories, where different jurisdictions are in place. Wind energy projects almost all require municipal approval, while licensing and regulatory issues are handled at the provincial level. As a result of the federal system, the price of electricity in Canada varies considerably between regions. In Ontario, five separate legal bodies adjudicate matters concerning electricity generation, transmission and distribution. Nevertheless, the Province of Ontario indirectly provides over 70 per cent of Ontario's electricity through the Ontario Power Generation, Inc., which it owns.

*Ontario* In 2009, Ontario passed the Green Energy and Green Economy Act, aiming to expand the RE sector and create green jobs. Ontario's market for RE is sizable and has considerable growth potential. As already mentioned, one of the conditions for LCRs to be potentially beneficial to the local economy – in terms of job creation and for green industry expansion – is existing market size coupled with capacity for growth.

As part of the Green Energy Act, Ontario introduced a feed-in-tariff (FIT) programme to encourage investment in RE. To spur investment in local manufacturing, and hence create green jobs, the FIT programme is coupled with an LCR. Under the LCR, firms are required to use a certain percentage of locally manufactured material for wind and solar projects in order to receive government support. The level of the LCR varies depending on the type of renewable energy – wind, solar, etc. – as well as the size of the project. For wind and solar projects over 10 kilowatts (kW), the local requirement was 25 per cent and 50 per cent from 2009–11, respectively, and 50 per cent and 60 per cent from 2012 onward.[42] If the LCR is not met, firms are not eligible for FIT benefits.[43] In addition, the Act sets out specific percentage values for the activities and materials that can be used in fulfilling the LCR. For example, local steel used to produce turbine towers earns a value of 9 percentage points, and turbine towers that are made locally earn 4 percentage points.[44] The benefit of allocating

---

[41] Hufbauer et al. (2013).   [42] Wilke (2011): 5.
[43] Ontario Power Authority, 'Feed-In Tariff Program FIT Rules Version 2.0', 2010, http://fit.powerauthority.on.ca/august-10-2012-final-fit-20-program-documents-posted.
[44] See n. 2: 17.

different percentages to different activities and materials is that it gives policy makers flexibility in targeting green development or job creation.

This is the first time that an LCR has been introduced in Ontario, and it was not phased in gradually. As a result of Ontario's Act and the associated LCR, retail electricity prices increased by more than 17 per cent in 2010, with no expectation of a reduction in future years.[45] At the same time, however, Ontario's government said that the Green Energy Act had led to the creation of 20,000 jobs. Ontario's FIT scheme does not pay enough attention to investing in training to increase workers' skills or setting RE targets, which would incentivise investors based on perceived guaranteed demand.[46]

Japan, later joined by the EU, filed a WTO complaint against Canada's FIT scheme largely because of the LCR. The plaintiffs argued that the LCR connected to the FIT violated three different sets of WTO rules: the Agreement on Subsidies and Countervailing Measures (SCM), the national treatment requirement of the General Agreement on Tariffs and Trade (GATT), Article III, and the Agreement on Trade-Related Investment Measures (TRIMS).[47] Wilke writes: 'It is not the FIT programme as such, but a controversial "local content" provision of Ontario's FIT that landed Canada at the WTO. The "made-in-Ontario" requirement demands that up to sixty percent of all green energy project inputs (goods and services) be manufactured or provided for in the province.'[48] Canada defended Ontario's FIT programme, arguing that since the programme aims to expand renewable green energy, the GATT, Article XX (General Exceptions), comes into play and protects the programme from other GATT disciplines as well as the TRIMS. With respect to the SCM, Canada countered that the FIT programme involved government procurement, making it exempt from the agreement, as Canada had not included RE products among its list of procurement items.

In December 2012, the WTO panel concluded that Ontario's LCR, as a part of its FIT programme, was in violation of provisions in the GATT and the TRIMS. However, the WTO panel rejected the claim that the scheme constituted an 'actionable subsidy' under the SCM (the FIT programme being directed at domestic usage rather than exports). In early February 2013, the Canadian government appealed the decision, putting the case before the WTO Appellate Body. The Appellate Body in May 2013 agreed with the WTO panel and ruled that the LCR in Ontario's FIT programme for RE was inconsistent with WTO rules,

---

[45] Hao et al. (2010): 26.  [46] Fraser & Company (2009): 20.
[47] Hufbauer et al. (2013).  [48] Wilke (2011): vii, n. 42.

namely in violation of the national treatment (NT) obligation (GATT, Article III) and prohibited in the illustrative list of measures under the TRIMS, Article II (i.e. the policy requiring the purchase or use of products from domestic sources).[49]

*Quebec* While Ontario implemented its FIT programme, Quebec introduced a Request-for- Proposal (RFP) scheme. However, each scheme was coupled with an LCR.

In 2011, 97 per cent of Quebec's electricity was hydro-generated.[50] In its most recent energy strategy document, the government of Quebec made it a goal to build 4 gigawatts (GW) of wind power by 2015 as a complement to hydro energy. Although Quebec has a stable market, its wind energy market potential is considerably smaller than Ontario's – 4 GW compared with 24 GW.

In Quebec, new wind energy plants are built through RFPs. An LCR requirement for wind energy has been in place in the province since 2003. To date, Quebec has issued three wind energy RFPs. The first of these in 2003 required that the initial 200 megawatts (MW) of wind energy have 40 per cent local content, the next 500 MW have 50 per cent and the remaining 700 MW have 60 per cent. The second RFP in 2005 required that 60 per cent of the 2 GW of wind energy be sourced locally, with 10 per cent from the Gaspésie region. The third RFP in 2010 was almost identical in structure to the second. Despite these requirements, firms were not deterred from investing.[51]

In July 2012, the government in Quebec outlined the most recent RFP to increase wind energy capacity by 700 MW, coupled with an LCR. Under the LCR, 30 per cent of the turbine costs had to be spent in the Gaspésie and Matane municipalities, and 60 per cent of the overall costs had to be spent in Quebec. Both the 30 and 60 per cent LCRs had additional stipulations. For example, in the case of the 30 per cent requirement, the following costs were exempt: wind turbine warranties; transportation of wind turbines; building, testing and commissioning of

---

[49] The report of the Appellate Body supported the panel's conclusions that LCRs accord preferential treatment to products made in Ontario by requiring the purchase or use of products from domestic sources, which is prohibited in the illustrative list of the TRIMS Agreement, and therefore places Canada in breach of its NT obligation under the GATT, Article III, and the TRIMS Agreement, Article II. The WTO appellate judges also rejected Canada's rebuttal that the LCRs should be considered as 'government procurement which can be exempted from the national treatment obligation', the findings referring to one of the main arguments that Ottawa had made in its case. See *Bridges Weekly Trade News Digest*, 17 (16), 8 May 2013 and earlier (19 December 2012 and 20 February 2013).

[50] Hufbauer et al. (2013): 48.  [51] See Hao et al. (2010).

the turbines; and maintenance and operating costs.[52] Meanwhile, the 60 per cent LCR included the following costs: initial development costs; the cost of wind turbines; and construction and transportation costs. The LCR did not, however, include maintenance and operating costs, warranty coverage costs, or payments to landowners.[53] Reading beyond the fine print, Quebec's LCR programme points to the policy makers' goal of increasing capital investment in plant manufacturing and creating jobs.

Canada's approach to meeting growing electricity demand and expanding the RE industry has varied between provinces, with Ontario's FIT programme and Quebec's RFP approach. As with any trade barrier, LCRs in the cases of Ontario and Quebec increase the cost of producing RE. The higher cost of RE production from wind turbines will be passed on to consumers through higher electricity prices. In a study by economists at the PIIE, it is estimated that the LCRs in Canada resulted in an additional USD 386 per kW of installed capacity. At this rate, the additional cost incurred for the 800,000 kW of wind power installed in Ontario since 2009 amounts to over USD 300 million. For Quebec, where more than 500,000 kW have been installed since 2009, the additional costs amount to nearly USD 200 million.[54]

### Solar Energy in India

India's electrical infrastructure is outdated and unreliable: the existing energy resources have not kept pace with India's growing electricity demands. In July 2012, India experienced the largest power outage in its history: over 620 million people – roughly 9 per cent of the world population – were affected.

In 2010 India launched the Jawaharlal Nehru National Solar Mission (JNNSM).[55] The JNNSM, overseen by the Ministry of New and Renewable Energy, aims to increase solar power by installing 20GW of grid capacity by 2022 in three phases: Phase (1), 1,000 MW by the end of 2013; Phase (2), an additional 3,000 MW by the end of 2017; and Phase (3), an additional 16,000 MW by the end of 2022. India's solar subsidy programme was enacted to increase domestic manufacturing capacity and green jobs, promote sustainable growth, and reduce energy costs. Although

---

[52] 'Ressources naturelles', *Quebec Energy Strategy 2006 to 2015*.
[53] Kathryn Higgins, Matthew Sherrard and Thomas Timmins, 'Quebec announces procurement of 700 MW of wind energy', Gowlings Knowledge Centre, 2012.
[54] Hufbauer et al. (2013): 54.
[55] Ministry of Natural Resources and Environment, Government of India (MNRE), 'Jawaharlal Nehru Solar Mission: Towards Building Solar India', 2009.

the JNNSM scheme was devised in 2010, the 2012 power outage catalysed a momentum to establish India as a leader in solar energy manufacturing and deployment. As part of the Indian government's policy in the area of solar energy, an LCR was introduced in 2010.[56]

Like Canada's FIT and RFP programmes, subsidies are propelling the RE industry in India's JNNSM scheme. Solar energy is distinct from other types of RE industries worldwide in that it receives a disproportionate amount of subsidies. In 2011, subsidies towards solar energy constituted nearly 30 per cent of total global subsidies for RE.[57] Without subsidies, solar power is rarely viable and would not be the RE of first choice. As such, it is hard to estimate an international price for solar generated electricity, since each country has in place different subsidy schemes and incentives. Solar technologies, or PV systems, consist mostly of cells and modules.

An LCR is part of the JNNSM scheme, under which solar developers must purchase domestically manufactured crystalline silicon (CSi) modules. While the JNNSM mandates that solar producers purchase CSi modules that are manufactured domestically, solar developers using thin-film technology are exempt from the LCR. In response, the majority of solar developers in India have turned to imported thin-film technology. Worldwide, only 11 per cent of PV deployment uses thin-film, and the remaining 89 per cent is in CSi.[58] However, as a result of India's LCR, more than 70 per cent of solar developers have opted for cheaper imported thin-film technologies rather than local CSi modules.[59]

India is the only PV market in the world where thin-film is the dominant solar energy technology.[60] Solar developers tend to prefer CSi modules because of their efficiency – between 12 and 24 per cent of solar radiation is converted to electricity in comparison with between 4 and 12 per cent for thin-film.[61] However, in India, thin-film technology is preferred to CSi modules, because the LCR on domestically manufactured CSi modules makes the thin-film a much cheaper option.

An additional reason solar producers in India prefer thin-film technology is the better international financing options for solar energy projects not having an LCR, i.e. thin-film. For example, both the Export-Import Bank of the United States (EX-IM Bank) and the Overseas Private Investment Corporation (OPIC) have offered low-interest loans to solar energy firms provided they use thin-film produced in the United States. In 2010 and 2011, the EX-IM Bank lent USD 248 million to Indian firms

---

[56] The Government of India also introduced an LCR for thin film.
[57] Hufbauer et al. (2013): 71.   [58] Shiao (2012).
[59] Bridge to India, *The India Solar Handbook: November 2012 Edition*, 2012.
[60] Deign (2012).   [61] Dirjish (2012).

that bought thin-film modules.[62] Such a condition has increased Indian demand for US thin-film, while in India there has been an overproduction of silicone PV cells and modules. The EX-IM Bank financing distorts the impact of the LCR by lowering the cost of electricity through loans with low interest rates, and by shifting solar developers' module purchases away from domestically manufactured ones to imported ones.

The application of the JNNSM LCR exemption for thin-film has shifted the solar technology market in India from CSi to thin-film. As a result, domestic manufacturing has made negligible gains, and the LCR has slightly increased the cost of PV systems. Domestic manufacturers have scaled back the operations of their solar plants, operating below capacity or closing down altogether. In addition, the shift to thin-film deployment has undermined the anticipated economic and job growth from the JNNSM.

Without the LCR in place, solar developers in India would be able to import CSi modules and cells. Thin-film modules are slightly cheaper than CSi on a per-watt basis. However, thin-film modules have a lower efficiency, which translates to added costs for the developer. As such, even though thin-film modules might have a lower price per KWh than CSi modules, the overall cost might be higher once efficiency differences are taken into account. It has been suggested that an additional reason for the thin-film preference of Indian solar developers is that the hot climate provides ideal conditions to maximise thin-film efficiency. In the study by economists at the PIIE, it is estimated that India's LCR translates into a price increase of up to 12 per cent for PV modules and 3 per cent for PV systems for the solar developer. Meanwhile, the LCR has resulted in an estimated 3–7 per cent additional growth in domestic manufacturing of modules, compared with the market without the LCR.[63] It is expected that the price of CSi modules on the international market will decline due to technological advancements. As the price falls, the effect of the LCR will be greater.

Although global prices for CSi modules and cells continue to fall, owing to improved technology, Indian manufacturing competitiveness for CSi technology has not kept pace. The LCR is likely to discourage innovation in the solar energy industry and impede manufacturing competitiveness. The LCR might boomerang India's solar manufacturing and electricity goals.

Before the JNNSM programme was introduced, India's manufacturing sector for solar cells and modules was relatively small and relied on exports; between 70 and 80 per cent of locally manufactured solar

---

[62] Hufbauer et al. (2013): 79.   [63] Hufbauer et al. (2013): 69.

material was exported.[64] However, there are several obstacles hindering India from being a hub of renewable solar energy manufacturing and deployment and from being internationally competitive. First, there is an unfavourable business environment, in which Indian banks and international lenders are reluctant to finance solar energy projects that are perceived to be a high risk. Second, economies of scale are lacking. India's current infrastructure lacks the capacity to produce solar modules and cells to meet the JNNSM target goal. While foreign solar module manufacturing tends to produce 75 MW of capacity per line, India's infrastructure produces only 10–20 MW.[65] To increase solar energy production, India would first have to invest in fixed infrastructure.

Third, India's solar technology and knowledge is lacking in comparison with foreign competitors. Greater investment is needed in fixed costs – infrastructure – as well as the services portion of solar energy production, namely training of workers. India's LCR targets the manufacturing of solar modules rather than the services segment of the renewable solar energy sector. Manufacturing accounts for only 25 per cent of jobs on the solar electricity value chain. The majority of jobs require more training in installation and sales.[66] The narrow focus of the LCR limits its capacity to create jobs. Shifting the focus of the LCR more downstream to the services portion of the value chain would promote greater transfer of RE technology and knowledge, and increase the long-term capacity for green job growth.

The aim of the LCR is to help domestic solar producers overcome these domestic obstacles to develop solar energy infrastructure and technological capacity to eventually develop economies of scale, and become competitive enough to export solar technology on the international market. However, the LCR in India's solar technology area has resulted in higher costs for PV modules and cells, which have been passed on to the consumer.

In February 2013, the United States formally brought a complaint against India's subsidies before the WTO.[67] Indian policy makers contend that the JNNSM programme should not come under WTO scrutiny, because India is not a developed country like Canada and does not have a large market share in RE products. This is not unusual. Policy makers in developing countries have proposed special treatment with regard to regulation of NTBs for environmental goods and services, such as a greater time allowance for implementation and fewer reductions.[68] In addition, the Indian government defends the JNNSM – and its associated LCRs – on the

---

[64] Hufbauer et al. (2013): 72. [65] Jha (2011).
[66] Natural Resources Defense Council (2012).
[67] Inside US Trade, 'US Seeks WTO consultations with India over solar LCRs, subsidies', February 2013.
[68] Monkelbaan (2011): 4.

grounds that the programme consists of government procurement, since solar power is first purchased by the public National Thermal Power Corporation.[69]

## 6  Options and Alternatives for Dealing With LCRs

Although LCRs are prohibited under the WTO, both developed and developing countries use them in RE policies. Despite concerns about the consistency of LCRs, they need to be considered in the broader context, given the legitimate environmental concerns they are presumably set to address and the fact that RE is essential to mitigate climate change and environmental degradation. Monkelbaan writes: 'Environmental goods and services have become subject to special attention as sectors with potential win–win outcomes for both trade and the environment. Climate-friendly goods, technologies and related services can be a meaningful component of climate change mitigation strategies.'[70] However, further technological innovation for RE is costly, requiring considerable government support. To sustain a permanent shift towards green industry and RE, positive and well-directed incentives are needed.

In the interests of both the global economy and efficient RE production by developing as well as developed countries, less distorting options and alternatives for dealing with LCRs should be considered. These would put less stress on the multilateral trading system and would serve to address the legitimate concern that countries have when they try to stimulate employment while pursuing climate and energy policies. Such suggestions include, in particular, investment in infrastructure, promotion of government financing for infrastructure investment, creation of a better and more conducive business environment for firms in which to innovate toward more green technologies, as well as targeted and well adapted training programmes for workers to allow them to develop skills for the energy sector with environmentally friendly technologies. Addressing conditions that are hindering the development of competitiveness in RE manufacturing and services should be a high priority, together with providing a better enabling environment for firms to operate.

### Enhancing the Physical Infrastructure

An important factor in contributing to the progress of RE is the enhancement of the physical infrastructure. In this context, working toward the

---

[69] Hufbauer et al. (2013): 22.   [70] See Monkelbaan (2011): 2.

goal of achieving economies of scale, governments should prioritise infrastructure investment.

### Promoting Government-Sponsored Financing

In the context of the development of the physical infrastructure, financing is often a big constraint for developing countries. Regardless of the extent of renewable resources, or the potential for market growth, developing countries often lack the financial capacity to subsidise RE or the political capacity to impose carbon taxes – arguably the best policies to foster RE. Therefore, they resort to LCRs. To address this constraint, government-sponsored financing should be promoted, such as loan guarantees for developers of alternative, green energy.[71]

### Taking Better Advantage of Progress in Renewables

The advent of new technology and the rapid increase in production capacity in RE resources, such as solar and wind, have made them more competitive against conventional energy technology. Other sources of clean energy, including geothermal and biomass, are becoming more attractive and provide a huge potential for electricity production. Mandatory biofuels for transportation and less polluting energy use should be refined so as to be in balance with the need to preserve the biodiversity and lands needed to produce food resources. Policies such as FITs and other incentive mechanisms to stimulate investments in RE may be continued and enhanced as long as they are also required to ensure a healthy growth of renewable deployment that will further provide attractive returns to investors.

### Promoting Innovation and Training for Green Jobs

Focusing on innovation in green energy requires adapted training programmes for domestic workers. Such programmes should be designed with sustainable energy development and use in mind. To be most effective, these should be integrated with green industry needs, and periods of on-site training should be incorporated into the university curriculum or training programmes. Targeting all portions of the energy value chain rather than imposing an LCR aimed at domestic manufacturers should prove to be a better and less distorting way of expanding

---

[71] An Asian Development Bank (ADB) project is doing just this, offering to guarantee a maximum of USD 150 million in loans to Indian solar developers. See ADB, 'India Solar Generation Guarantee Facility', www.adb.org/site/private-sector-financing/india-solar-generation-guarantee-facility.

output in the green energy sector. It would have the added benefit of creating associated green jobs.[72]

*Focusing WTO Disputes on LCRs Outside Renewable Energy*

Since many LCRs have nothing to do with RE, countries that are rightly concerned with the use of this policy tool might focus their WTO disputes on LCRs outside the RE space. LCR disputes have squarely focused on wind and solar projects in Canada, India and the EU. Other LCR targets can certainly be found, which would lower trade tensions around 'green' or sustainable development issues and leave the question open as to the legitimacy and effectiveness of the application of LCRs in the area of RE.

## 7   Growing International Attention on LCRs

Governments have begun to examine more critically the merits of recourse to LCRs. A major step was taken in 2011 when leaders from the twenty-one members of the Asia-Pacific Economic Cooperation (APEC, which represents 54 per cent of world economic output, 40 per cent of world population and nearly half of world trade) highlighted the objective of advancing green growth in their Honolulu Declaration by 'speeding the transition toward a low-carbon economy in a way that enhances energy security and creates new sources of economic growth and employment'. In this context, they pledged that APEC economies will 'eliminate non-tariff barriers, including local content requirements that distort environmental goods and services trade'.[73] In an Annex of the same declaration, APEC Leaders also pledged to refrain from adopting new LCRs in the green energy area.

Subsequent work on LCRs within APEC featured them as the subject of a Trade Policy Dialogue conducted by the APEC Committee on

---

[72] It should be noted that if all countries adopt LCRs as part of their environmental policies, then these measures will simply negate each other, as to be effective LCRs rely on technology transfer from the more mature technology locations.

[73] More specifically, Annex C of the 2011 APEC Leaders' Honolulu Declaration was on 'Trade and Investment in Environmental Goods and Services'. In the paragraph devoted to LCRs, APEC economies commit to 'Eliminate, consistent with our WTO obligations, existing local content requirements that distort environmental goods and services trade in the region by the end of 2012, and refrain from adopting new ones, including as part of any future domestic clean energy policy'. See APEC Leaders' Honolulu Declaration of 13 November 2011, www.apec.org/meeting-papers/leaders-declarations/2011/2011_aelm .aspx.

Trade and Investment held in April 2013. The Dialogue provided an opportunity to better understand LCRs, the domestic policy objectives they try to address, their regional economic and commercial impacts, the impacts on economies using them, as well as the impacts on economies subjected to them and the ways that the APEC economies can seek to achieve domestic economic policy objectives through measures that achieve the same results but without distorting international trade and investment. Leaders of the APEC economies have agreed to further consider a list of alternative policies and measures to LCRs. APEC Members are also considering the possibility of having case studies carried out on how LCRs external to them are impacting their trade and investment interests.[74]

LCRs were also featured in the International Chamber of Commerce 8th World Trade Agenda Summit in Doha in April 2013, as part of the discussion on how to stimulate trade in environmental goods and services and what potential gains could be achieved from liberalisation in this area.[75]

## 8   Addressing LCRs within a SETA

A SETA presents an attractive solution to coordinating national policies with the aim of lowering the cost of RE policies.[76]

Negotiating a SETA would provide an excellent way to address RE concerns in a trade-friendly manner. To avoid the curse and cost of permanent protection, countries might agree within a SETA a non-renewable time limit, say of ten years, for their existing LCRs. To allow for an orderly transition and avoid litigation, signatories to a SETA could agree on a 'peace clause', so that they would not risk being taken to the WTO Dispute Settlement Body for existing LCRs during this agreed phase-out period.

---

[74] See the CTI Chair's Report, prepared for the Second Senior Officials' Meeting of APEC Economies, Surabaya, 18–19 April 2013, para 8, 2013/SOM2/016.

[75] The ICC World Trade Agenda Summit was held prior to the ICC 8th World Chambers Congress on 24 April 2013 attracting over 1,000 business leaders and representatives from 12,000 chambers of commerce worldwide. The participants approved a set of five recommendations for a meaningful interim Doha Round package, to be presented to the G20 Meeting in Saint Petersburg and the Bali 9th Ministerial Conference in December 2013. The ICC World Trade Summit discussed a Peterson Institute study showing that potential gains to be had from a WTO Agreement on liberalising trade in environmental goods and services could produce USD10.3 billion of additional exports.

[76] The idea of a SETA had already been proposed in the ICTSD study (2011a): xiii.

Governments might also consider agreement on a moratorium or standstill on the adoption of future LCRs within a SETA. This was done by the Group of Twenty (G20) with respect to trade protectionist measures during the economic recession of 2008–2010. To backstop such commitments, concerned countries might call upon the WTO Secretariat, through its Committee on Trade and Environment (CTE), to launch a surveillance programme of LCRs in the RE space. The programme would report on instances of adoption of LCRs and, where possible, assess their effectiveness. Reaching such an agreement in a SETA would reduce the risk for repeated trade disputes at the WTO and provide more clarity and certainty for business as well.

Under a SETA, countries might agree to include their partners in a 'regional content requirement' (RCR) rather than using LCRs, at least for scheduled projects during the agreed phase-out period in the renewable space. For example, the scheduled RCRs might refer to certain wind turbine components. This effective 'cumulation' of the LCR within the region constituted by the members of the SETA would effectively dilute the restrictive impact of the measure. Although an RCR might create some trade diversion, it would be less than that created by LCRs imposed purely at the national level. This trade diversion would have to be weighed against the environmental objectives.

Bearing in mind the scholarly literature on LCR effectiveness, countries might agree within a SETA to cap their LCR percentages at a moderate level, appropriate for the sector in question. Such a cap could be either maintained throughout the agreed phase-out period or could be agreed as a permanent deviation from existing rules. The former would likely be a more easily acceptable option. This limit might be best negotiated in the context of a SETA, against other trade-offs in the environmental area. Although such a cap could also be carried out unilaterally, a SETA would provide a vehicle to specifically address the cost-benefit analysis (CBA) of the recourse to LCRs in order to best evaluate how well such measures work in practice to meet the shared objective of moving toward a 'greener economy' in light of their trade distorting effects.

Moving forward with negotiating a SETA could serve to facilitate alternative or innovative approaches to liberalising sustainable energy goods and services (SEGS). It could provide a framework conducive to assessing the linkages between SEGS and serve as a useful 'laboratory', where rules and disciplines pertaining to sustainable energy could be clarified and take shape.

## Appendix

*LCRs and RE Projects*

Table 7A.1 *LCR Statistics*

| Country & Case # | Date Announced; Current Status | Affected Sectors | Size of Affected Domestic Sectors | Affected Trade | Description of LCR Measure |
|---|---|---|---|---|---|
| Aus-1 | June 2009; LCR rejected | Labour markets | | | The government of New South Wales included a 'Local Jobs First Plan' in its stimulus package, providing a price preference for Australian and New Zealand content |
| Aus-2 | July 2009; LCR remains in force | Government procurement | $A 32.6 billion in total federal government procurement in 2010 | Only $A 2.3 million in federal procurement contracts awarded to overseas vendors in 2010 | Australia provided $2.5 million over four years to apply the National Framework of the Australian Industry Program (AIP) to federal, state, and local governments The AIP program requires applicants for government tenders to give details on the participation of Australian companies in projects exceeding threshold levels |
| Bra-2 | 2010; LCR remains in force | Public procurement | | | Procurement Law (law no. 12.349/2010) establishes a 25 per cent margin of preference for manufactured goods and national services in compliance with Brazilian technical standards |
| Bra-4 | July 2010; LCR remains in force | Public procurement, ICT | | | The Buy Brazil Act (law no. 12.349/2010) establishes preferences for Brazilian goods and services in government contracts, to be determined by the president, though not in |

| ID | Date | Sector | Value | Description |
|---|---|---|---|---|
| | | | | excess of 25 per cent above the price of foreign goods and services |
| | | | | For strategic IT and communications technology contracts, tenders will be restricted to goods and services developed with national technology |
| | | | | Procurement rules were further tightened as part of the Brasil Maior plan (2011) |
| Can-2 | February 2009; LCR remains in force | All government procurement | $16 billion in federal procurement in 2010; $171.7 billion in products imported by Canadian government in 2010 | The Canadian Products Promotion Act (CPPA) alters public procurement decisions to favour goods that contain at least 50 per cent Canadian content (except for natural resources, for which the test is 75 per cent Canadian content) and limits the percentage of provincial purchases from abroad to no more than 50 per cent of the total amount spent on Canadian products in a given fiscal year |
| | | | | Products from North American Free Trade Agreement (NAFTA) countries are not counted as imports under these spending caps |
| Can-3 | October 2009; LCR remains in force | Electrical machinery | $1.0 billion in provincial imports in 2010 | Ontario's Feed-in-Tariff (FIT) programme requires developers to acquire a certain percentage of their project costs from Ontario goods and labour |
| | | | | The local content requirements differ by technology, project size and project timing |

Table 7A.1 (cont.)

| Country & Case # | Date Announced; Current Status | Affected Sectors | Size of Affected Domestic Sectors | Affected Trade | Description of LCR Measure |
|---|---|---|---|---|---|
| | | | | | For wind projects over 10 MW, the LCR is 25 per cent for a commercial operating date (COD) before January 2012, and 50 per cent with a COD after January 2012; for solar projects over 10 kW and less than 10 MW, the LCR is 50 per cent for a COD before January 2011, and 60 per cent with a COD after January 2011 |
| | | | | | The programme is now the subject of dispute-settlement in the WTO, after failed consultations with Japan |
| Chi-1 | October 2009; LCR remains in force | Wind turbines | $5.2 million in 2010 | $11.5 million in imports in 2010 | At the 20th US–China Joint Commission on Commerce and Trade meeting in October 2009, the Chinese government agreed to drop its local content requirement for wind turbines |
| | | | | | Previous to the agreement, the Chinese government demanded that local governments to source more than 70 per cent from domestic sources when planning wind power projects |
| | | | | | However, China requires wind turbine imports to meet local test certification by the National Energy Administration (NEA) |
| Chi-2 | November 2008; LCR remains in force | Energy | | | In November 2008, China implemented a $586 billion economic Stimulus Package, allocating a major portion of the government spending to RE projects |
| | | | | | A circular jointly released by nine government organisations requires that preference be given to domestic products |

| | | | | |
|---|---|---|---|---|
| Chi-3 | May 2009; LCR remains in force | Metal ores, textiles, basic chemicals, basic metals, fabricated metal products, machinery, office equipment, electrical machinery, communication equipment, precision instruments, transport equipment | $11 billion in imports in 2010 | This combination of measures virtually ensures a massive volume of sales of domestically manufactured RE equipment<br><br>The Ministry of Information Industry's planning Release entitled 'Restructuring and Revitalisation of Planning for the Equipment Manufacturing Industry' encourages state bodies to ensure that domestic industries meet the requirements of the national market, particularly with respect to power generation and capital equipment<br><br>The Ministry recommends measures that encourage the use of Chinese-made equipment, including insurance policies that favour local technologies and equipment<br><br>The Release also calls for an increase in the export tax rebates granted to producers of high-technology and high-value added equipment and the abolition of import tariffs on key components of these technologies and on related raw materials |
| Chi-4 | May 2009; LCR remains in force | Public procurement | Approximately $127.6 billion in central government procurement in 2010 | The National Development and Reform Commission (NDRC) implemented measures to ensure that local content would be prioritised in government contracts<br><br>The extent to which the Government Procurement Law governs procurement of RE services and equipment by state owned-enterprises (SOEs) is ambiguous<br><br>By its terms, the law applies to purchases of goods and services by numerous SOEs; the firms reportedly apply LCR principles when making procurement decisions |

Table 7A.1 (*cont.*)

| Country & Case # | Date Announced; Current Status | Affected Sectors | Size of Affected Domestic Sectors | Affected Trade | Description of LCR Measure |
|---|---|---|---|---|---|
| | | | | | The 'buy national' principles set out in the Government Procurement Law are most rigorously applied to procurement of equipment for projects that are funded by government investments |
| | | | | | Projects requiring imported products need prior approval from the relevant government authorities |
| Ind-1 | August 2009; LCR remains in force | Energy | | $295 million in wind turbines exported in 2010; $426.9 million imported; 20 per cent | To introduce newer wind turbine models (or to modify existing models), the new models have to be registered with the Centre for Wind Energy Technology (C-WET), which requires establishing an assembly facility in India |
| | | | | | Third-party certification is required in addition to the design assessment |
| | | | | | State agencies require C-WET certification for allowing connection to the grid |
| Ind-6 | | Energy | | $672.3 million in semiconductor devices used to generate solar power imported in 2010; 50 per cent | India's Ministry of New and Renewable Energy released guidance providing that project developers 'are expected to procure their project components from domestic manufacturers, as far as possible' as part of the country's Jawaharal Nehru National Solar Mission (JNNSM) |
| | | | | | For PV projects based on CSi technology, the guidelines require that all project developers uses modules manufactured in India; for such projects selected in FY 2011–2012, |

| | | | |
|---|---|---|---|
| | | | developers must use both modules and cells manufactured in India |
| | | | For projects based on solar thermal technology, the guidelines require 30 per cent local content in all plants and installations (under the JNNSM – Batch 1 and 2). |
| Indo-4 | December 2009; LCR remains in force | Energy | Indonesian regulation PTK No. 007 Revision-1/PTK/IX/2009 requires local and foreign bidders for energy service contracts to use a minimum of 35 per cent domestic content in their operations |
| Indo-8 | January 2011; Rule implementation | Government procurement | Presidential Decree 54/2010, Article 98, gives a public procurement preference to goods and services with a minimum of 25 per cent local content (even where the bid is 15 per cent higher in price) and applies to bids over $550,000 |
| | | | The Decree, Article 97, awards additional preference points to vendors with investments in Indonesia and partnerships with local small and medium-sized enterprises (SMEs) |
| Kaz-1 | May 2009; LCR remains in force | Public procurement construction | Kazakhstan adopted changes to the law on public procurement to include a 'local clause' in public procurement for goods (20 per cent) and services (15 per cent) |
| | | | Companies with more than 50 per cent foreign shareholding are considered foreign unless |

Table 7A.1 (*cont.*)

| Country & Case # | Date Announced; Current Status | Affected Sectors | Size of Affected Domestic Sectors | Affected Trade | Description of LCR Measure |
|---|---|---|---|---|---|
| Mex-1 | October 2010; LCR remains in force | Government procurement | | | they fulfill three criteria for qualifying as a 'national producer'. In October 2010, Mexico published new regulations to national content for government procurement. These regulations establish a minimum national content of 60 per cent for 2011 and 65 per cent for 2012 (but exceptions of 30–35 per cent for some light manufacturers and automobiles). The federal regulations only apply when federal funds are used, but the Mexican states develop their own rules |
| Par-1 | February 2009; LCR remains in force | Glass products, construction, public procurement | | | Paraguayan public bodies that spend national stimulus funds must give a minimum 70 per cent preference to national goods and services |
| Sau-1 | March 2010; LCR remains in force | Government procurement, transport, aviation | | | South Africa merged its National Industrial Participation Programme (NIPP) with its Competitive Supplier Development Programme (CSDP), which controls contracting by South Africa's nine SOEs. South African SOEs are now required to demand 30 per cent local purchases for any outlay of funds over US$10 million, |

| | | | | disproportion-ately affecting government contracts in the energy, rail, and aviation sectors |
|---|---|---|---|---|
| Tur-1 | December 2008; LCR remains in force | Government procurement | | Turkey's public procurement legislation allows for a 15 per cent price preference in favour of domestic suppliers when participating in tenders are set asides for Turkish goods and suppliers
A Prime Minister circular of December 2008 encouraged Turkish contracting authorities to apply those provisions more rigorously |
| Tur-2 | December 2010; LCR remains in force | Wind turbines | $1.3 billion in wind turbines imported in 2010; 25 per cent | Turkey implemented local content bonuses for different components of a wind turbine (tower, blade, mechanical, and electrical equipment)
The bonuses increase the wind FIT by up to 50 per cent |
| Ukr-1 | January 2012; LCR remains in force | Electricity derived from renewable sources | | Ukraine introduced LCRs for obtaining a specific FIT for electricity produced from renewables
The Law stipulates that government incentives for electricity production from alternative energy sources shall apply on condition that at least 15 per cent of the cost of the construction of the respective facility producing electricity must be comprised of materials, works, and services of Ukrainian origin |

Table 7A.1 (*cont.*)

| Country & Case # | Date Announced; Current Status | Affected Sectors | Size of Affected Domestic Sectors | Affected Trade | Description of LCR Measure |
|---|---|---|---|---|---|
| USA-11 | June 2011; LCR rejected | Government procurement | | | The US House of Representatives passed the 'Department of Homeland Security Appropriations Act, 2012' (H.R.2017) by a vote of 231–188 on June 2, 2011 The Act provides $40.6 billion for operations of the Department of Homeland Security (DHS) in Fiscal Year 2012. In the course of its debate, the House rejected two efforts to attach 'buy-American' provisions to the Act The Act was signed by the president on November 18, 2011 and became Public Law No.112–33 |

*Source:* Measures – Global Trade Alerts, www.globaltradealert.org/). Size of Domestic Market and Affected Trade – Authors' own calculations.

## References

Deign, Jason (2012) *What is behind India's Love Affair with Thin Film?* PV Inside, 63

Dirjish, Mat (2012) *What's the Difference between Thin Film and Crystalline Silicon Solar Panels?* Electronic Design, 16 May

Farrell, John (2011) *Maximising Jobs from Clean Energy: Ontario's 'Buy Local' Policy*. The New Rules Project, Washington, DC

Fraser & Company (2009) *Developing Ontario's Green Energy and Green Economy Act to its Full Potential*. Fraser & Company, Ontario

Hao, May et al. (2010) *Local Content Requirements in British Columbia's Wind Power Industry*. Faculty of Business – Pacific Institute for Climate Solutions, University of Victoria

Hufbauer, Gary et al. (2010) *Figuring out the Doha Round*. Peterson Institute for International Economics, Washington, DC

(2013) *Local Content Requirements: A Global Problem*. Peterson Institute for International Economics, Washington DC

International Centre for Trade and Sustainable Development (2011a) *Fostering Low Carbon Growth: The Case for a Sustainable Energy Trade Agreement*. ICTSD, Geneva

(2011b) 'Japan challenges Canadian sustainable energy incentives at WTO', *Bridges Trade BioRes*, 10 (17). ICTSD, Geneva

Jha, Vyomi (2011) *Cutting Both Ways: Climate, Trade and the Consistency of India's Domestic Policies*. CEEW Policy Brief, Council on Energy, Environment and Water, New Delhi

Jie-A-Joen, Clive, René Belderbos and Leo Sleuwaegen (1998) *Local Content Requirements, Vertical Cooperation, and Foreign Director Investment*. Faculty of Economics and Business Administration, Maastricht University

Kuntze, Jan-Christoph and Tom Moerenhout (2013) *Local Content Requirements and the Renewable Energy Industry – A Good Match?* ICTSD, Geneva

Lewis, Joanna and Ryan Wiser (2005) *Fostering a Renewable Energy Technology Industry*. Environmental Technologies Division, Ernesto Orlando Lawrence Berkeley National Laboratory, University of California

Monkelbaan, Joachim (2011) *Trade Preferences for Environmentally Friendly Goods and Services*. ICTSD, Geneva

(2013) *Services in a Sustainable Energy Trade Agreement*. ICTSD, Geneva

Natural Resources Defense Council (2012) *Laying the Foundation for a Bright Future: Assessing Progress under Phase 1 of India's National Solar Mission*. Council on Energy, Environment and Water (CEEW); Natural Resources Defense Council (NRDC), New Delhi

Rivers, Nic and Randy Wigle (2011) *Local Content Requirements in Renewable Energy Policies: Ontario's Feed-In Tariff*. Paper presented at International Association for Energy Economics, Stockholm

Shiao, M.J. (2012) *Thin Film Manufacturing Prospects in the Sub-Dollar-Per-Watt Market*. Greentech Solar, Cayman Islands

Tomsik, Vladimir and Jan Kubicek (2006) *Can Local Content Requirements in International Investment Agreements Be Justified?* NCCR, Berne

Veloso, Franc (2001) *Local Content Requirements and Industrial Development: Economic Analysis and Cost Modeling of the Automotive Supply Chain*. Massachusetts Institute of Technology, Engineering Systems Division, Boston, MA

Wilke, Marie (2011) *Feed-in Tariffs for Renewable Energy and WTO Subsidy Rules*. ICTSD, Geneva

## Further Reading

Brewer, Thomas and Andreas Falke (2012) 'International transfers of climate-friendly technologies: how the world trade system matters', in David Ockwell and Alexandra Mallet (eds.), *Low Carbon Technology Transfer: From Rhetoric to Reality*. Routledge, New York

Cosbey, Aaron (ed.) (2008) *Trade and Climate Change: Issues in Perspective*. International Institute for Sustainable Development, Winnipeg

Inside US Trade (2013) 'US seeks WTO consultations with India over solar LCRs, subsidies', February 7

International Centre for Trade and Sustainable Development et al. (2008) *Liberalisation of Trade in Environmental Goods for Climate Change Mitigation: The Sustainable Development Context*. ICTSD, the German Marshall Fund of the United States, and the IISD, Geneva

Janssen, Ron (2010) *Harmonising Energy Efficiency Requirements: Building Foundations for Co-Operative Action*. ICTSD, Geneva

Jha, Veena (2008) *Environmental Priorities and Trade Policy for Environmental Goods: A Reality Check*. ICTSD, Geneva

(2009) *Trade Flows, Barriers and Market Drivers in Sustainable Energy Supply Goods*. ICTSD, Geneva

Meléndez-Ortiz, Ricardo (ed.) (2013) *Global Challenges and the Future of the WTO*. ICTSD Programme on Global Economic Policy and Institutions, Geneva

Moon, Gillian (2009) *Capturing the Benefits of Trade? Local Content Requirements in WTO Law and the Human Rights-Based Approach to Development*. University of New South Wales Faculty of Law, Sydney

Selivanova, Julia (2007) *The WTO and Energy: WTO Rules and Agreements of Relevance to the Energy Sector*. ICTSD, Geneva

UNCTAD (2010) *World Investment Report 2010: Investing in a Low-Carbon Economy*. UNCTAD, Geneva

World Bank (2007) *International Trade and Climate Change: Economic, Legal, and Institutional Perspectives*. World Bank, Washington DC

(2010) *World Investment Report 2010: Investing in a Low-Carbon Economy*. UNCTAD, Geneva

# 8 International Technology Diffusion in a Sustainable Energy Trade Agreement: September 2012

*Thomas L. Brewer*

Policy makers and researchers widely acknowledge that international diffusion of sustainable energy technologies has the potential to make substantial contributions to climate change mitigation as well as sustainable development. Facilitating such diffusion is an objective of proposals to negotiate a Sustainable Energy Trade Agreement (SETA), which could be developed in multilateral, regional and/or other venues.

It is particularly important to address the issues associated with *all* the modes of technology transfer used by firms, namely international direct investments, licensing, and trade in services and goods. Because international direct investments and international services transactions are integral to technology diffusion processes, a SETA agenda should include *non-tariff barriers* (NTBs) to these modes of international technology diffusion, in addition to tariffs on goods and barriers to licensing.

From a macro as well as a micro perspective, international technology diffusion is inherently embedded in international trade, investment and licensing flows. Further, trade, investment and licensing together with technology diffusion are central to sustainable development processes. Together, they represent a tightly integrated economic package.

A SETA should also take into account the following key features of sustainable energy technologies:

– Energy efficiency technologies are among the most cost-effective ways to reduce greenhouse gas emissions (GHGs), as well as gain other benefits from reducing dependence on fossil fuels. A SETA agenda should therefore include the numerous, diverse and expanding lists of energy efficiency technologies that could make a significant contribution to sustainable development.

– Government procurement practices are important factors in the demand for sustainable energy technologies and their international diffusion. Although some sustainable energy technologies are covered for some countries in the existing World Trade Organization (WTO) Agreement on Government Procurement (GPA), there are significant gaps in its coverage in terms of both technologies and countries.

– Standards and testing, which are also inherently problematic in the context of trade policy issues because of concerns about disguised protectionism, are even more problematic for sustainable energy technology diffusion issues because the technologies themselves are rapidly evolving.

– Government subsidies of sustainable energy projects by technology exporting and importing countries can be justifiable on economic efficiency grounds because of market failures. A SETA agenda should therefore not only be about trade liberalisation; it should also be about finding a balance between the roles of governments and markets. Achieving such a balance is one of the most analytically and politically challenging topics for any SETA dialogue. A new paradigm about the role of government in economies, including international trade, is needed in order to adequately accommodate the legitimate role of subsidies in facilitating economic efficiency where there are market failures.

– Many emerging and 'developing' countries are significant exporters as well as importers of sustainable energy technologies. As a result, the political economy of the patterns of interests and influence in the international negotiations for a SETA are changing. Developing countries' increasing interests as technology exporters create incentives to participate in agreements that would reduce barriers to international diffusion of sustainable energy technologies. At the same time, those countries' expanding role in the world economy enhances their influence in international negotiations.

– The evolving energy technology revolution is also changing the international political economy of sustainable energy technology diffusion. New and evolving technologies are changing international trade, investment and technology diffusion patterns. As such patterns change, it is important that a SETA agenda be flexible so that it can expand to include new technologies.

– In sum, what is needed is a new international institutional architecture. While some opportunities for increased international cooperation can be exploited in existing institutional venues, others may require the creation of new international institutional arrangements. There is no single institutional setting, nor even only one type of institutional architecture that can fully exploit the gains from increasing the international diffusion of sustainable energy technologies.

## 1 Purpose, Scope and Structure

Facilitating the international diffusion of climate-friendly technologies is a central objective of proposals to develop a SETA. The purpose of this chapter is to assess the issues and options for enhancing climate-friendly technology diffusion in the development of such an agreement. The chapter explicitly considers a wide range of related topics that need to be addressed in designing and negotiating a SETA.

Throughout the chapter, the term 'trade' refers not only to international trade in services as well as goods, but also to international direct investment and international licensing. Similarly, the term 'technology' is used in its broad sense to refer to know-how and thus trade, investment and licensing in services, as well as hardware. The approach of the chapter is multi-level, in that it considers issues at the micro levels of technologies and firms, in addition to the macro levels of countries and groups of countries.

### *Country Groups*

As a result of discussions before, during and after recent climate change conferences, the terminology of country groups in international diplomacy has been changing. The changes were perhaps most significant during the Durban COP-17, when references to 'Annex I' and 'non-Annex I' countries and references to 'developing' and 'developed' countries and 'economies in transition' were not included in the Durban Platform for Enhanced Action.[1]

Although such references nevertheless continue to be embedded in some official documents and in much common discourse, it is often more precisely descriptive of current and prospective realities to use different terms. This is particularly so when referring to international technology issues, because the key features of countries are whether they are *sources* or *recipien*ts in specific international technology diffusion processes.

---

[1] Aldy and Stavins (2010: 5) have noted that 'approximately fifty non-Annex I countries – that is, developing countries and some others – now have higher incomes than the poorest of the Annex I countries with commitments under the Kyoto Protocol. Likewise, forty non-Annex I countries ranked higher on the Human Development K Index in 2007 than the lowest ranked Annex I country.'

At the aggregate level, it is especially important to note that many 'developing' countries are *sources* as well as recipients in such diffusion processes; it is also important to note that 'developed' countries are *recipients* as well as sources. The policy-relevant empirical issues are to what extent and in which technologies an individual country is a source or recipient; there are thus significant empirical questions in addition to the definitional issues about country groupings. Issues about the *scope* of a country's capabilities as a technology exporter and the *level* of its current and prospective capabilities are also often pertinent. The chapter considers empirical issues further below in relation to the *changing techno-economic geography* of international technology diffusion.

## Structure of the Chapter

The second section of the chapter puts the focal issues about international diffusion of sustainable energy technologies in the broader context of the interactions between sustainable development, technological change and trade. It includes definitions of key terms, as well as discussions of the scale of deployment of the sustainable energy technologies needed and the modes and geographic patterns of international technology diffusion. That section is thus about technology and economics.

The third section focuses on the government policies – domestic as well as international – that need to be taken into account in SETA design and policy making processes. It also briefly describes the panoply of international institutional arrangements concerning energy, climate, trade and development where SETA issues are or will be on the agenda.

The final section considers the way forward. It discusses the issues and options for change – including not only the issues and options that are SETA-specific – but also more broadly the political economy of participation and compliance issues.

## 2 Context: Sustainable Energy Technologies, Low-Carbon Economies and Trade

### Sustainable Energy in Low-Carbon Economies: Technologies and Terms

Sustainable development, energy technology diffusion and trade are inevitably integrated in economic and technological processes. Furthermore, technology spillovers – including sustainable technologies in the energy sector that is central to economic development – are important benefits of international technology diffusion, including developing countries in particular. The significance and role of technology spillovers are discussed, for instance, in Grossman and Helpman (1991) and Coe and Helpman (1995).

In order to understand better international technology diffusion processes, it is helpful to clarify key terms. The dialogue on technology transfer issues from the 1992 to the 2012 Rio Conference is discussed in Abdel Latif (2012b). Sampath and Roffe (2012) have also reviewed fifty years of technology transfer negotiations, as well as the associated terminology.

The accepted notions of *technology* have been expanded and refined as the technology diffusion dialogue has progressed in technology transfer and climate change mitigation policy making circles. In particular, it refers to 'know-how' and thus intangible, as well as tangible goods. *Technology diffusion* refers to a process that includes transfer but with the additional connotation of absorption of the technology into the recipient economy.

*Renewable* energy (RE) is 'any form of energy from solar, geophysical or biological sources that is replenished by natural processes at a rate that equals or exceeds its rate of use. Renewable energy is obtained from the continuing or repetitive flows of energy occurring in the natural environment and includes low-carbon technologies such as solar energy, hydropower, wind, tide and waves and ocean thermal energy, as well as renewable fuels such as biomass' (IPCC 2011, Annex 1, 9). *Sustainable energy* is 'inexhaustible and [does] not damage the delivery of environmental goods and services including the climate system' (IPCC 2011, Chapter 1: 18).

### *International Diffusion*

The international diffusion of sustainable energy technologies is not only important for climate change mitigation, but also for low-carbon growth and trade, including in developing countries. Since Solow (1956), economists have emphasised the importance of technological change in increasing productivity and economic growth. In order to improve understanding of these opportunities, ICTSD has completed technology mapping studies of the energy supply sector and climate-friendly technologies in two key end-use sectors: transport and buildings. Those studies identify products, including components, associated with these technologies. They also provide bases for analyses of the market access conditions and opportunities that arise from the deployment of such technologies, including for producers in developing countries. (See especially Goswami et al. 2009; Lako 2008; Kejun 2010; Vossenaar 2010; Vossenaar and Jha 2010; an extensive list of relevant ICTSD studies appears at the end of the chapter.)

The mapping studies also illustrate the large benefits that can be achieved through deploying climate-friendly technologies. These include improving technology efficiencies, as well as replacing carbon-intensive sources and introducing carbon capture and storage (CCS). Other possibilities include enhancing the role of nuclear power, introducing new technology configurations and systems (such as hydrogen as a carbon-free carrier to complement electricity, fuel cells and new storage technologies), as well as reducing GHG and $CO_2$ emissions from agriculture and land use.

Although technology is sometimes equated with hardware and thus as *goods* in international trade, a broader notion of technology as 'know-how' – and thus *services* – is now more widely accepted. Engineering, consulting and construction services are obvious examples (see, in particular, Kim 2011). As many as fifteen categories of services – from site analysis to installation of equipment and, finally, the transmission, distribution and sale of electricity – may be required to build and operate an off-shore wind farm.

In order to understand the diffusion of sustainable energy technologies, it is important to recognise that international technology transfers are often undertaken by firms as elements in complex 'bundles' of transactions involving services, the movement of personnel and the movement of financial capital, as well as goods. It is typically *firms' direct investment projects* that drive the composition of these bundles. Moreover, direct investment projects are often central to the key strategic choices of firms as they expand internationally within particular regions. For instance, in the wind industry, transportation economics create pressures to locate production of large towers, blades and turbines near wind farm installation sites. These pressures, in combination with economies of scale that create pressures to centralise production, lead to international regional hubs being strategically important for serving multiple international markets. A key ingredient of the hub is an international direct investment project, the success of which depends on market access to the countries of the region.

In many cases, technology can be transferred internationally through joint ventures involving direct investment by a foreign firm, as well as a local partner. For example, Japanese firms have partnered with Indian companies in joint ventures involving high-efficiency and low-emission coal technologies (IEA 2010: 575).

Access to technology can also be facilitated by international licensing. For example, one of China's largest wind technology manufacturers, Goldwind, gained access to wind technology by purchasing licences from the German wind turbine maker Vensys. Chinese firms have also acquired licences to produce boilers, turbines and generators for advanced technology coal-fired power plants.

## The Importance of Sustainable Energy Technologies and Energy Efficiency

Two fundamental facts about the significance of energy production and consumption as contributors to GHGs need to be recognised in order to understand the potential contributions of sustainable energy technologies to climate change mitigation (IEA 2010):
- Nearly two-thirds (65 per cent) of all GHGs worldwide can be attributed to energy supply and energy use.
- More than four-fifths (84 per cent) of global $CO_2$ emissions in particular are energy-related.

Addressing energy production and consumption issues is thus essential to mitigate climate change. According to the International Energy Agency (IEA), there is currently no single technology solution that can lead to a sustainable energy future (IEA 2010: 75; see also Appendix I in this chapter).

But, of course, not all energy technologies are equally cost-effective; nor are they similarly suitable for all locations. Yet, there are some consistent patterns in their relative attractiveness. One of those is that energy efficiency has been consistently found to be the most cost-effective approach to reducing the GHG emissions associated with energy production and consumption.

The relative cost-effectiveness of energy efficiency technologies – both at the global level and in some particular countries – has been a recurrent theme of analyses conducted by the IEA The following are indicative of the theme:
- End-use [energy] efficiency accounts for 36 per cent of all savings in the BLUE scenario. That scenario assumes that global energy-related $CO_2$ emissions are reduced to one-half their current levels by 2050 (IEA 2008).
- In the BLUE scenario for India, 'improved energy efficiency across both supply and end-use sectors is the single largest source of $CO_2$ reductions'.
- In China, 'higher rates of energy efficiency [are likely to] result in a 27 percent reduction in energy-related emissions in 2050 compared to Baseline levels' (IEA 2010: 382).
- In its study of paths for reaching a global GHG concentration level of 450ppm by 2100, the IEA (2010) found that the most cost-effective opportunities for energy-related abatement of $CO_2$ were particularly pronounced in energy efficiency measures in the early years. Thus, in 2020 the share of abatement from energy efficiency was projected to be 72 per cent in that path, and 44 per cent in 2035. Energy efficiency's share of the *cumulative abatement* during the period from 2010 to 2035

Table 8.1 *Energy Efficiency and Other Sources of Energy-Related $CO_2$ Abatement in an IEA 450 Scenario*[a]

| Source | Shares of Abatement | |
|---|---|---|
| | In 2020 per cent | In 2035 per cent |
| Energy efficiency[b] | 72 | 44 |
| Renewables, incl. biofuels | 19 | 25 |
| Nuclear | 5 | 9 |
| CCS | 3 | 22 |
| Amounts of abatement (Gt $CO_2$) | 2.5 Gt | 14.8 Gt |

*Notes:*
a The IEA 450 Scenario refers to the scenario in which a mixture of energy technologies could achieve a $CO_2$ concentration level of 450 ppm. The abatement represented here is the difference between the emissions in the 450 Scenario and the New Policies Scenario, in which recent and prospective policies are adopted beyond the status quo represented by the Baseline scenario and the more ambitious 450 scenario.
b Energy efficiency's share of the cumulative abatement during the period 2010–2035 is 50 per cent.
*Source:* Compiled by the author from IEA (2010: 214, Figure 6.4)

was 50 per cent. The relative contributions by renewable energy technologies and other sources are detailed in Table 8.1.

- 'The average energy efficiency in 2050 needs to be twice the level of today, a significant acceleration compared to the developments in the last 25 years' (IEA 2008).

As the building and transport sectors hold particularly significant potential gains in reducing GHG emissions from increased energy efficiency, much technical and policy research has been conducted in these sectors over the past several years. The significant potential for energy efficiency gains in the industrial sector has also been widely noted and researched, partly because firms have a clear incentive to reduce costs.

The *IPCC Special Report on Renewable Energies and Climate Change Mitigation* (IPCC 2011) is a useful source for assessing the relative importance and prospects of various types of sustainable energies. It lists six broad categories of technologies at the industry level: bioenergy, geothermal, hydro, ocean, solar and wind. Pacala and Socolow (2004) also have identified a wide range of sustainable energy technologies – including energy efficiency measures – that could significantly reduce GHG emissions. The IEA (2010) has identified seventeen 'key groups'

Table 8.2 *'Key Groups' of Technologies in the IEA BLUE Scenario*

| Supply Side | Demand Side |
| --- | --- |
| Onshore and offshore wind | Energy efficiency in buildings and appliances |
| Photovoltaic (PV) systems | Heat pumps |
| Concentrating solar power | Solar space and water heating |
| 2nd generation biofuels | Energy efficiency in transport |
| Biomass IGCC & co-combustion | Electric and plug-in vehicles |
| Nuclear power plants | Hydrogen fuel cell vehicles |
| CCS fossil-fuel power generation | CCS industry, hydrogen and fuel transformation |
| Coal: ultra-supercritical | Industrial motor systems |
| Coal: integrated-gasification combined cycle | |

*Source:* IEA (2010); the sequence of the entries has been changed by the author.

of technologies that could reduce GHG emissions 50 per cent by 2050, compared to current levels (see Table 8.2).

While lists like these are useful for appreciating the breadth of the technologies and sectors that can contribute to climate change mitigation, it is also useful to gain a more in-depth understanding of the technologies available in a particular sector. Sachs et al. (2009) have identified seventeen emerging technologies for heating, ventilating and air conditioning in the building sector, including *inter alia* advanced rooftop packaged air conditioners, commercial energy recovery ventilation systems, smart premium (robust) residential air conditioners, residential boiler controls and dehumidification enhancements for air conditioners in hot-humid climates.

The illustrative list in Appendix II of this chapter presents the results of a scanning of these and other such lists, in a preliminary effort to identify the universe of energy efficiency technologies in greater breadth and depth.

### *The Challenges of Scale and Price in the Evolving Energy Technology Revolution*

It is well known that the estimated technical potentials of world sustainable energy production are more than adequate to meet increasing energy demands for many decades, as the data of Table 8.3 indicate in detail.[2] The global technical potential of solar energy in particular has been estimated to be between three and 101 times recent world energy demand

---

[2] 'Technical potential' refers to the amount of [renewable energy] output obtainable by full implementation of demonstrated technologies or practices' (IPCC 2011: 6).

Table 8.3 *Approximate Ranges of Technical Potentials of World Sustainable Energy Production Relative to Recent Production*[a]

| Technology | Estimated Technical Potential Supply (EJ)[b] | Ratio to Recent World Total Demand | Recent World Demand (EJ) |
|---|---|---|---|
| Geothermal electricity | 118–1109 | 1.3–18.2 | Total electricity = 61 |
| Hydro electricity | 50–52 | 0.8–0.9 | |
| Ocean electricity | 7–331 | 0.1–5.4 | |
| Wind electricity | 85–580 | 1.4–9.5 | |
| Geothermal heat | 10–312 | 0.1–1.9 | Total heat = 164 |
| Bio primary energy | 50–500 | 0.1–1.0 | Total primary = 492 |
| Direct solar primary energy | 1575–49,837 | 3.2–101.3 | |

Notes:

a  The technical potentials reported here represent total worldwide potentials for annual RE supply and do not deduct any potential that is already being utilised. Ranges are based on various methods and apply to different future years; consequently, they are not strictly comparable across technologies.

b  EJ refers to exajoule, which is $10^{18}$ joules. For comparison, the energy consumption of the United States is approximately 100 EJ per year; in Table 8.3, total world primary energy consumption was 492 EJ.

Source: Compiled by the author from data in IPCC (2011, Technical Summary: 13).

(IPCC 2011: 10). Although the technical potentials of other energy sources are not so enormous, they are individually and collectively substantial. Yet, the technical potential has to be physically harnessed and distributed – at significant economic cost.

The *physical* challenge is evident in an IEA estimate: 'the average year-by-year investments between 2010 and 2050 needed to achieve a virtual decarbonisation of the power sector include, amongst others, 55 fossil-fuelled power plants with [carbon capture and storage], 32 nuclear plants, 17, 500 large wind turbines, and 215 million square metres of solar panels. [The BLUE scenario] also requires widespread adoption of near-zero emission buildings and, on one set of assumptions, deployment of nearly a billion electric or hydrogen fuel cell vehicles' (IEA 2008).

As for the *economic* challenge and the importance of technological innovation, 'using technologies available in 2005, the present value cost of achieving stabilisation at 550 ppm $CO_2$ would be over USD 20 trillion greater than with expected developments in energy efficiency, hydrogen energy technologies, advanced bioenergy, and wind and solar technologies' (IPCC 2011). The trends in comparative costs are thus centrally important in the evolving energy revolution. The IPCC (2011: 10) presents a summary of the 'levelized costs' of a variety of sustainable energy sources.[3] It concludes that 'Some [renewable energy] technologies are broadly competitive with existing market energy prices. ... [However, in] most regions of the world, policy measures are still required to ensure rapid deployment of many [renewable energy] sources.'

## *International Technology Flows: Changing Geographic Patterns*

The precise patterns and trends in the changing techno-economic geography of sustainable development vary according to the particular indicators used from among a wide range of possibilities. One is patents. Previous ICTSD studies (Abdel Latif 2012b; Abdel Latif et al. 2011) have found that firms in the Organisation for Economic Cooperation and Development (OECD) countries continue to dominate the issuance of patents.[4] Although developed countries remain predominant in patent applications and the granting of licenses, there is a trend of increasing innovation and sourcing of production in developing countries.[5] In some key industries and technologies, firms in a few large developing countries are world leaders in climate-friendly sustainable energy technologies. Moreover, some large multinational firms based in developed countries have significant global R&D centres in developing countries (Brewer 2008a, 2008b, 2008c, 2009).

The increasing importance of developing countries as suppliers of new technologies in general and climate-friendly technologies in particular is evident, for instance, in China for solar and in India for wind (IEA 2010: 573). Data computed for the twenty-one economies in Asia-Pacific Economic Cooperation forum (APEC) (Kuriyama 2012: 2) revealed

---

[3] The 'levelized cost' of energy refers to 'the cost of an energy generating system over its lifetime; it is calculated as the per-unit price at which energy must be generated from a specific source over its lifetime to break even. It usually includes all private costs that accrue upstream in the value chain, but does not include the downstream cost of delivery to the final customer; the cost of integration, or external environmental or other costs. Subsidies and tax credits are also not included' (IPCC 2011: 10).

[4] An extreme example is Siemens, which averages approximately forty new patents per day.

[5] Porter and van der Linde (1995) observed that domestic firms' compliance with environmental regulations can trigger technological innovations, since such inventions lower firms' cost of compliance. The existence of spillovers in climate change technology (i.e. transfers of technological know-how from one country to another) provides one mechanism by which developing countries' efforts to combat climate change can benefit from innovations in OECD countries.

that over the period from 2002 to 2010, developing countries' shares went from one-third to two-thirds of APEC exports of climate-friendly technology products. The shift in imports was less substantial: industrialised countries' share declined from slightly more than a half to about four-tenths, while developing countries' share went from slightly less than a half to nearly six-tenths.

China is the largest exporter of static converters that change solar energy into electricity, solar batteries for energy storage in off-grid PV systems, the concentrators used to intensify solar power in solar energy systems and wind turbine towers (IEA 2010: 574). An APEC policy brief (Kuriyama 2012) presents abundant data on the patterns and trends in environmental goods and services trade for the twenty-one APEC economies. The data include specific categories concerned with sustainable energy – including renewable energy plants, power generation, heat energy and management, resource efficient technologies and environmental monitoring, analysis and assessment equipment. 'Developing economies' in APEC have become significant technology exporters as well as importers, and the trend is for them to become increasingly net exporters. In these data, of course, the 'developing country' patterns and trends reflect particularly China's emergence as a major exporting economy.

Trends in climate-friendly products have been especially notable: '[W]orld trade for the climate-friendly technology products increased at an average annual rate of 16 percent during the period 2002–2010, reaching USD 224.4 billion in 2010. Over the same period, APEC exports of climate-friendly technology products to the world increased at a faster pace of 17.2 percent per year, totalling USD 123.7 billion. At the same time, APEC imports grew at a slower rate of 13.6 percent per year, reaching USD 100 billion.' Furthermore, 'intra-APEC trade of climate-friendly technology products nearly tripled from USD 23 billion to USD 63 billion between 2002 and 2010. Intra-APEC trade in 2010 was equivalent to 28.4 percent of the world total' (Kuriyama 2012: 2).

Data compiled by ICTSD (2011a) also reveal the increasing importance of many developing countries as exporters of sustainable energy products. Furthermore, data on international direct investment flows reveal increasing outward flows from developing countries (Hanni et al. 2011; also see UNCTAD 2010). Firm-level cases illustrate the nature of these flows in greater detail than is possible with the aggregated data. The Indian wind turbine manufacturer Suzlon is a particularly compelling example. Suzlon is the fifth largest turbine manufacturer in the world, with sales in nearly all regions. Its

> **Box 8.1. Internationalisation of Renewable Energy Industries**
>
> Over the last few years, RE has become as internationally distributed as any other large established industry.
>
> For instance, the USD 1.7 billion Thornton Bank wind project off the coast of Belgium will use turbines from German-based Repower, which is owned by Suzlon of India, has French and German utilities as shareholders; German and Danish export credit agencies as risk-guarantors for some of the debt; French, Dutch and German commercial banks and the European Investment Bank (EIB) as lenders; and a Danish company as blade supplier.
>
> The 212 MW Olkaria geothermal complex in Kenya has Israeli and Kenyan plant operators, and is ordering drilling equipment from Chinese companies with the help of finance from Chinese and French development banks. It also has a loan from Japan to pay for a transmission line, and from the German government towards the expansion of the project. It has pre-qualified four engineering companies, three from Japan and one from France, as bidders for the work to increase capacity by 280 MW.
>
> In Latin America, multinational involvement is also the norm. Two small hydro projects in Rio Grande do Sul province in Brazil, for instance, will use turbines made by French-owned Alstom and developed by a Canadian-owned company. Two 40 MW hydro projects in Chile, developed by a local company and backed by funds from a New York investor, secured loans from Spanish, French and German banks.
>
> Source: UNEP, Bloomberg New Energy Finance and Frankfurt School, UNEP Collaborating Centre for Climate and Sustainable Energy Finance (2011: 28).

expansion has been achieved through a combination of increasing exports from India and direct investment projects in foreign countries (see further details in Box 8.1).

## 3 Government Policies and International Institutional Arrangements

Within the broad array of governments' sustainable energy and trade policies, subsidies and other support measures are especially problematic in terms of their trade implications.

## Government Support Policies for Sustainable Energy and Trade Policies

There are two categories of market failures that justify government policy support for sustainable energy on the grounds of increasing economic efficiency – namely the market failures represented by (1) the negative externalities associated with the widespread use of fossil fuels and other activities that release GHGs, and (2) the positive externalities associated with investment in energy efficiency and RE or carbon sequestration projects with significant 'public goods' payoffs. Negative externalities are thus inherent in the cause of the problem, and positive externalities are inherent in the constraints on the solutions. There can therefore be economic efficiency rationales for governments to undertake subsidies to incentivise investment in and use of sustainable energy sources and practices (see e.g. Brewer 2014, Ch. 1).

Extensive use of support policies is common across all country groups. Feed-in tariffs (FITs) and a variety of types of tax reductions are especially frequent. In some countries – the United States in particular – several types of support policies are in effect only at the sub-national level.

From a trade policy perspective, of course, sustainable energy support policies prompt concerns about the possibility of disguised protectionism. Such concerns and possibilities have already been involved in trade disputes – some of which have reached the WTO dispute settlement process and some of which festered without being formally submitted to the WTO for resolution. Some are the subject of on-going dispute panel cases in mid-2012 as this chapter is being written. See Appendix III for summaries of these and other instances in which climate change and trade issues have collided.

Extensive analyses of subsidies issues are available in the ICTSD studies by Ghosh and Gangania (2012) and Howse (2010).

### Standards and Testing

Standards and testing procedures – like support policies – are widespread and diverse; and, like support policies, they are potentially problematic when they become enmeshed in trade relations because they can be easily perceived as disguised trade protectionism, as indeed they sometimes are. Yet, they can also be legitimate sustainable energy policies, which can be justified on economic efficiency grounds since they can be used to address market failures.

Their extent and diversity are evident in the results of a 'Survey of Market Compliance Mechanisms for Energy Efficiency Programs', undertaken by the APEC Expert Group on Energy Efficiency & Conservation (APEC 2012), which reported that 'Within the APEC region there are a total of 32 energy labelling and 16 minimum energy efficiency standards programs operated by 18 economies. These include programs that have been running since 1978 [and] those that are in their infancy; programs covering up to 50 product types [and] those spanning only one or two.'

In the EU, there are several Directives and many programmes that address sustainable energy standards and testing issues. Of particular importance is the Renewable Energy Directive (2009/28) which has spawned a variety of activities. For instance, the European Committee for Standardization (CEN) has been developing a set of standards for sustainably produced biomass for energy applications, which was made public in late 2012.

The usual trade policy problems concerning technical standards are significantly exacerbated for sustainable energy technologies because many of these technologies are relatively new and/or in a state of flux (see, for instance, ICTSD 2012a; IEA 2009; REN21 2011; Storm 2012; UNCTAD 2011).

### *International Venues: Who Does What*

The expansion of the development–energy–climate–trade agendas in recent years is reflected in the wide range of international institutional venues where those issues have been under consideration (for context, see Brewer 2008a, 2008b, 2008c, 2009). Much of the climate change diplomatic dialogue has been in the United Nations Framework Convention on Climate Change (UNFCCC) process (see in particular Aldy and Stavins 2007, 2009, 2010).

#### *The UNFCCC's Technology Mechanism, Centre and Network, and Executive Committee*

Since its beginning in 1992, the UN Framework Convention on Climate Change (UNFCCC) has included international technology diffusion as one of its central objectives. As a result of a series of Conferences of the Parties (COPs) in Bali, Cancún and Durban a new institutionalisation is being put into place in the form of a Technology Mechanism, which consists of a policy body, the Technology Executive Committee (TEC) and a Climate Technology Centre and Network (CTCN). Although the latter is not yet operational, the priorities of the Mechanism have been established by the various COP decisions and follow-up meetings. The priorities include the promotion of public–private partnerships (PPPs) and strengthening international research and development (R&D) cooperation, as well as the development of national innovations systems and technology action plans. Discussions of the tangible implications of these priorities and numerous suggestions for how the Mechanism can 'realise its potential' are available in an ICTSD report by Sampath et al. (2012).

### UNFCCC and WTO

The UNFCCC and the WTO – as the premier multilateral arrangements for climate and trade, respectively – have received the most attention from researchers, policy makers and other stakeholders, as well as the media. The attention has been mostly focused on legal–institutional issues, with a special concern about conflicts between trade and climate, but with a recognition of their compatibilities and complementarities (Cottier et al. 2009; Epps and Green 2010; Frankel 2010; WTO and United Nations Environment Programme 2009).

### Other Fora

The panoply of relevant international institutional arrangements, however, is much more extensive than just the UNFCCC and WTO. Among the specialised agencies of the UN system, the United Nations Environment Programme (UNEP) has been particularly active, including in particular the co-authorship with the WTO of a major study of issues at the intersection of trade and climate change (WTO and United Nations Environment Programme 2009). Two other UN agencies with industry-specific missions have become more salient in recent years, after many years of quiescence on climate–trade issues. The International Civil Aviation Organization (ICAO) has received much attention, particularly since the advent of the international controversy about the inclusion of aviation in the EU Emissions Trading Scheme (ETS), but it has done little on climate change issues as of this writing. The International Maritime Organisation (IMO) has also become more active in recent years.

Outside the UN system, several high-level fora for large economies have given periodic attention to climate and sustainable energy issues as well as trade and sustainable development, though typically not all of them at the same time. The annual G8 summit has sometimes focused on climate change issues, while the G20 has focused on sustainable energy; but in both cases international financial issues have become dominant in recent years. The Major Economies Forum on Energy and Climate (MEF), with seventeen members, was initiated in 2009 as a venue where major emitters of GHGs might make progress on negotiating a climate change agreement without the complications of nearly 200 participants in the UNFCCC negotiations. Although the MEF has stimulated and discussed major studies on a wide range of key issues, it has not taken any decisions of substance.

These and other international institutional arrangements for addressing the issues of this chapter can be described along several dimensions, including issue clusters, economic sectors and scope of membership. There are four directly relevant – and overlapping – issue clusters: energy, climate, trade and development. The economic sectors and sub-sectors can be summarised as energy production, energy distribution, buildings, transportation, industry and agriculture. The scope of membership extends from bilateral to regional to plurilateral to multilateral (i.e. nearly global).

Although it is not feasible to present a comprehensive listing of all the relevant international institutional arrangements and their key features in terms of issue clusters, economic sectors and membership, it is possible to consider a few to illustrate their extent and diversity.

It is easy to imagine at least five quite different venues where a SETA could be negotiated: a forum such as the G20 of large economies and large GHG emitters, a regional forum such as APEC or the Trans-Pacific Partnership (TPP) or perhaps a Trans-Atlantic Partnership, or bilateral free-trade agreements (FTAs) such as those being negotiated between the EU and several countries, particularly in Asia. Each, of course, presents its own distinctive issues as a result of the differences in trade patterns, GHG emissions trajectories and development paths, and it is not feasible to examine each one in detail. However, because SETA issues are already on the active agendas of APEC and EU bilateral FTA negotiations, they warrant special attention.

### *Sustainable Energy Issues in APEC*

A SETA-like agenda is already progressing in the APEC forum, whose twenty-one economies account for about 60 per cent of world energy demand.[6] Those economies' interest in sustainable energy issues is evident in official statements by leaders and ministers of the twenty-one APEC economies.[7] The November 2011 Leaders' Honolulu Declaration was particularly notable in its ambitions (APEC 2011):

---

[6] Compared with APEC, the Trans-Pacific Partnership (TPP) has an overlapping membership of only nine countries but a more wide-ranging agenda of issues.
[7] The Asia Pacific Partnership (APP) on Clean Development and Climate, which was specifically created to address issues of sustainable development, energy and climate, was disbanded in April 2011, with a transfer of its programmes to other venues.

In 2012, [our] economies will work to develop an APEC list of environmental goods ... on which we are resolved to reduce by the end of 2015 our applied tariff rates to 5 [per cent] or less, taking into account economies' economic circumstances, without prejudice to APEC economies' positions in the WTO.

They also committed to the elimination of NTBs, including local content requirements (LCRs) that distort environmental goods and services trade, as well phasing out inefficient fossil-fuel subsidies and incorporating low-emissions development strategies into their economic growth paths.

APEC trade ministers (APEC 2012) subsequently reaffirmed their commitment to promote trade and investment in environmental goods and services (EGS) in order to address environmental challenges. They also promised to 'strengthen regional cooperation on trade and environmental matters and to advance shared green growth objectives and to address both the region's economic and environmental challenges by accelerating the transition towards a global low-carbon economy, which will contribute to energy security and reduce APEC's aggregate energy intensity'.

Of course, the geographic scope of participation in APEC is limited by its regional focus, and thus major economies and GHG emitters such as Brazil, Argentina and European countries are excluded, as compared with the G20. Yet, at the same time, the inclusion of China and Russia, as well as a large number of southeast Asian countries, nevertheless gives it much diversity in terms of the structure of the economies represented in it.

*EU Bilateral FTAs*

The EU has engaged in a large number of bilateral FTA negotiations in recent years, several of them involving countries in east, south and southeast Asia (European Commission 2012). The first to enter into force was the FTA with South Korea (European Union 2011). The EU aims for each of the FTAs with Singapore, Malaysia and Vietnam to have an annex on sustainable energy technology issues, including provisions concerning GHG emissions, and NTBs to services trade and direct investments as well as tariffs on goods. Prospective EU FTAs with Japan and the United States are expected to include similar provisions. The FTA with India, however, is not expected to include such an annex.

## 4  Ways Forward: Issues and Options for Change

Carefully crafted trade policies could contribute to a more widespread and rapid deployment of efficient, sustainable technologies that promote sustainable growth. Within that context, trade can also provide incentives for innovation and investment in climate-friendly technologies. However, there are a variety of international institutional governance issues that need to be addressed in order to facilitate more effectively the contributions of trade to sustainable growth and climate change mitigation.

## *International Governance Issues*

Despite the wide array of existing international institutional arrangements, there are nevertheless some gaps in the institutional structure; and because of the number and diversity of the substantive development–energy–climate–trade issues involved, there are inevitably some inter-agency coordination issues, as well.

### *Assessing International Arrangements*

There are, of course, many ways to assess the adequacy of existing international institutional arrangements – political, economic, legal and administrative.[8] Each is a legitimate approach, with its own concepts and criteria, and each yields its own distinctive descriptive and prescriptive results. *Politically*, at the most fundamental level, the memberships of current institutions differ – and thus the patterns of interests and influence within them differ. A challenge for the development of a SETA, therefore, is to find a combination of countries that can form a politically viable arrangement; it seems unlikely that that will happen in a multilateral climate or trade venue, where the impasses between developing and industrial countries are so formidable. *Economically*, the current configurations obviously differ enormously in the sizes of the economies, individually and collectively – and in the quantities of their GHG emissions. A challenge for developing a SETA therefore is to find a venue that is large enough in economic terms and emissions terms to be significant, but small enough to be politically feasible. *Legally*, at a basic level, there are differences in the places of the institutions in international law – for instance, older international agreements have privileged positions relative to more recent ones; yet another challenge for a SETA is thus to fit it into an existing array of institutions and agreements.

[8] A less conventional – but potentially fruitful – alternative approach to assessing current or proposed arrangements from a technological or engineering perspective is to regard the arrangements as problem-solving and learning networks that overlap and adapt to changing circumstances. Such an approach is more akin to organisational behaviour and open systems approaches, or to a cybernetics approach emphasising communications and control. The three key questions of this approach are: What are the hubs and nodes of the networks for specific problems? Are there problems that are being ignored? Are there networks for specific issues that need more resources?

Finally, in simple *administrative* terms, there are already many multilateral institutions with overlapping missions and programme responsibilities, and therefore coordination and redundancy issues are inevitable. Coordination issues have been especially noteworthy in UNFCCC–WTO relations.

### UNFCCC and WTO at the Multilateral Level

The UNFCCC and the Kyoto Protocol have specific provisions that recognise the need to encourage international technology transfer. These include financial mechanisms such as specialised funds and market-based initiatives. The restructuring and expansion of the technology innovation and transfer programs as a result of Durban COP-17 decisions is in progress. Thus, the institutionalisation of issues concerning climate-friendly technologies within the UNFCCC framework continues to evolve. The institutionalisation of intellectual property (IP) rights issues within the WTO is well established.

How to coordinate the two is an open question. The continuing exchanges of information between the two secretariats and periodic attendance at one another's meetings are useful steps in the right direction. A potentially useful further institutionalisation of relations at the multilateral level of UNFCCC–WTO interactions would be formalised *mutual notification processes* whereby (a) climate-friendly policies such as sustainable energy subsidies would be notified to the WTO and (b) WTO dispute cases and the results of WTO Policy Reviews would be notified to the UNFCCC.

IP rights issues pose yet other questions about international governance, including the roles of the UNFCCC, the WTO and the World Intellectual Property Organization (WIPO). It has been suggested, for example, that 'multilateral discussions on sustainable development and climate change are not the appropriate venues for proposals that would modify international IP norms. Efforts to enhance the transfer and diffusion of green technologies in these discussions should take place in the framework of existing international rules established by the Trade-Related Aspects of Intellectual Property Rights (TRIPS) Agreement, and flexibilities are an integral part of the balance of rights and obligations in these rules' (Abdel Latif 2012a).

### *Elements of a SETA*

A previous ICTSD report on SETA-related issues (2011a: esp. 62–67) identifies a wide range of diverse questions about the potential elements of a SETA, and all of them are directly or indirectly relevant to the specific focus of the present chapter on technology diffusion issues. The current discussion builds on that report by highlighting some items in the previous report that are especially important for technology transfer issues and by supplementing them with a few additional observations.

A SETA should have a *broad scope* in terms of its coverage of industries and technologies. These should definitely include energy efficiency technologies – perhaps through a second track in addition to first-track negotiations concerning energy supply technologies. See ICTSD (2011a: 62) concerning the possibility of two-phase negotiations.

Policies that act as either *barriers* or *facilitators* of sustainable energy technology diffusion need to be addressed in a SETA. As for barriers, a SETA that would eliminate tariff barriers and NTBs on climate-friendly energy technologies could increase their traded volume by an average of 14 per cent in eighteen developing countries that emit high levels of GHGs, according to estimates in a World Bank study (2009). Reducing NTBs on international services and direct investments could have much bigger effects on technology transfers in the form of know-how.

As for *facilitation* policies, the issues are more complex. Government policies that facilitate innovation and investment in sustainable energy technologies are important because of the market failures associated with technology research, development and diffusion. Market forces alone are insufficient to bring about a rapid deployment of sustainable energy technologies; government regulations and incentives are also important. How to resolve the conflicts between sustainable energy subsidies and trade liberalisation is thus an urgent challenge. There are two approaches to the challenge: *formulating principles*, which has the potential advantage of creating clarity and reducing uncertainty, and *resolving dispute cases*, which has the advantage of pragmatic adaptation to the circumstances. Perhaps a mix of the two can be devised.

In any case, given the potential for conflicts between trade and sustainable energy/climate change issues, it is important that a SETA include provisions for *dispute resolution* – through either existing WTO or other arrangements.

There should be provisions concerning *government procurement* (at the sub-national as well as national levels) – given its importance in sustainable energy technology markets and the inadequacy of the coverage of the current WTO Agreement on Government Procurement.

*Competition policy* – which is not included in the WTO, but which has been broached as a possible agenda item in international negotiations in the past – should be on the agenda as a possible element of a SETA.

Anti-competitive industry structure and/or firm behaviour are issues in sustainable energy industries, as they are in any industry, where there are frequent international mergers and acquisitions. A SETA that is sensitive to technology diffusion issues should also reflect the following: 'Sustainable energy is a rapidly changing field, characterised by new technologies, innovation and market developments. Therefore, an open-ended SETA, where new issues could be added ... would enable the agreement to be dynamic and responsive to changing needs and circumstances' (ICTSD 2011a: 64).

### Key Political Economy Issues

Two key issues about the political economy of a SETA concern: (1) how the patterns and trends in countries' technology exports and imports affect their interests at stake in any SETA negotiations, and (2) the ways and extent to which a SETA may be able to circumvent the global public goods, free-rider problem of existing climate change agreements by creating club goods with incentives for countries to participate in a trade agreement and comply with its norms.[9]

### Changing Import/Export Patterns

The political economy of trade liberalisation has changed substantially in recent years, including for sustainable energy trade technologies in particular, as noted above. The IEA (2010: 586) has observed that 'The removal of [import trade] barriers is important for developing countries. China, Hong Kong (China), Mexico, Singapore and Thailand are among the top ten exporters of renewable energy technologies and, therefore, have significant export interests in trade liberalisation in the sector'. An APEC policy brief (Kuriyama 2012) concludes that 'Given the growing importance of [the EGS] market, removing barriers would be beneficial to both industrialised and developing APEC member economies. Barriers are not limited to tariffs and cover a wide array of issues, such as over-stringent technical regulations that go beyond what is reasonable to allow the commercialisation of a product; cumbersome certification procedures; quantitative restrictions to imports and exports; local content requirements; low enforcement to prevent infringement of intellectual property rights; subsidies to goods and services with higher

---

[9] Yet a third, related set of political economy issues concerns the ways in which varying arrangements for technology innovation and diffusion can address public goods and club goods issues – a set of issues under consideration for further research.

levels of carbon footprint; and sectoral restrictions to foreign investment, among others.'

As is generally true in international trade issues, sustainable energy technology importing countries and exporting countries have different interests. Though there are many examples of the interests of corporate and individual consumers of imports having sufficient influence to overcome the political pressures of protectionist firms and industry associations, it is nevertheless true that technology-importing countries – especially those with producers and/ or aspirations to become homes to producers – are more prone to resist liberalisation of trade barriers while exporting countries are likely to support them. The patterns and trends in imports and exports are therefore obviously significant determinants of countries' negotiating positions.

As noted above, services trade and direct investment are both centrally involved in international technology diffusion processes, in addition to goods trade. Thus, although patterns in goods trade are certainly indicative of some interests in international technology diffusion issues, they are not necessarily conclusive. In fact, the interests of economies are much more complex. The implications for negotiating a SETA are significant: governments' calculations of economic incentives and domestic economic groups' political pressures on governments tend to be more conflicted and their positions on issues in negotiations less clear cut. Consider, for instance, wind power issues for India. While India may be inclined to impede imports of wind turbines through tariffs and/or NTBs in order to protect its domestic producer Suzlon, it also true that Suzlon has an interest in access to foreign markets for its exports from India and for establishing foreign production and servicing facilities in foreign markets, and access to foreign technology through outward direct investment. Thus, like multinational firms with geographically diverse interests, Suzlon has mixed interests on trade issues, as does its home country.

A SETA, then, does *not* pose a simple conflict between the interests of industrialised countries as sustainable energy technology exporters and developing country importers. Indeed, many developing countries are already significant exporters of sustainable energy technology goods and services – and source countries of international direct investments. Therefore, they have incentives to support trade and investment liberalisation agreements.

### Club Goods – And Incentives for Participation and Compliance

Among the central challenges of developing any international environmental or trade agreement is to create an arrangement with sufficient incentives for countries – and perhaps other entities – to participate in its activities and to comply with its norms. Indeed, multilateral climate change agreements to date have been weak in precisely these two respects. The underlying problem, of course, is that there are incentives for countries to be 'free riders' in the presence of global 'public goods' – an endemic problem with international environmental agreements.

International trade agreements, on the other hand, can create 'club goods' such that only participants derive the benefits of the arrangement and such that non-compliance entails penalties. See especially, Victor (2011: 254–259), who emphasises how a 'club approach', as used in the WTO, can be adopted in climate change negotiations; also see Hoekman and Kostecki (2009: 158). Linking climate and trade issues is possible, of course, as when the EU made its support of Russian accession to the WTO conditional on the latter's participation in the Kyoto Protocol. Such a linkage made Russia's access to the WTO's 'club goods' dependent on its willingness not to be a 'free rider' on the Kyoto Protocol.

As a trade agreement, a SETA can create 'club goods' and thereby incentives for countries to participate in it and comply with its norms. By focusing directly on trade in sustainable energy goods and services (SEGS) – and thus indirectly on climate change mitigation – a SETA can incorporate the club goods features of a trade agreement, and circumvent (at least partially) the free-rider, public goods problem of climate change agreements. Also, a SETA could include provisions on capacity-building and technical assistance, and could refer to existing agreements on technology cooperation. (For example, see Africa–EU Renewable Energy Cooperation Programme 2010; NREL 2010; DEA 2012.)

There are yet other ways to increase the incentives for developing countries to participate in a SETA, particularly through linking SETA negotiations to other closely related issues. For instance, bilateral SETAs could be linked to existing bilateral international technology cooperation arrangements, in which developing countries already have a stake. SETAs can also be linked to capacity-building and to related development financing programmes. The many ways in which sustainable energy technologies can play crucial roles in sustainable development are discussed, for instance, in a World Bank (2012) report on *Inclusive Green Growth*.

### Prospects and Importance of a SETA

Proposals for a SETA are appearing at a critical but potentially propitious time in the evolution of both the international climate change regime and the international trade regime: both are undergoing strong fragmentation tendencies that are producing highly pluralistic regimes with diverse institutional manifestations at all levels, with multilateral institutions at their core – i.e. the UNFCCC and the WTO. A SETA that could be integrated into these expanding climate and trade institutional landscapes could make significant contributions to both.

Climate and trade concerns converge in the sustainable energy issues that are at the core of a SETA. It has been said (Schmalensee 2009: 897) that 'The most important and difficult [climate change] task is to move toward a policy architecture that can induce the world's poor nations to travel a much more climate-friendly path to prosperity than the one today's rich nations have travelled'. Such a climate-friendly alternative path could lead to a better quality of life, compared with a traditional carbon-intensive path. The development of a SETA – by whatever name – offers the prospect of being a milestone along that path. Indeed it may be an international institutional design that offers the rare combination of being a 'scientifically sound, economically rational, and politically pragmatic' contribution (Aldy and Stavins 2010: 2) to the array of international climate change agreements, as well as a contribution to the trade and development agenda.

A SETA offers developing countries, in particular, opportunities to participate more fully in the 'third industrial revolution' (Rifkin 2011) and in the 'remaking of the modern world' (Yergin 2011).

Existing international institutional arrangements for trade, energy, sustainable development and climate change are arguably inadequate to the challenges needing urgent attention. These are historically significant times needing historically significant diplomatic creativity. It is time for a burst of such creativity akin to that half a century ago when new international institutions were put into place to address the development and trade issues of that era. Now, new problems in new circumstances involving a broader and more diverse array of 'players' require new solutions based on new paradigms.

*How to Begin*

As negotiators sit down for their initial meetings to address the sustainable energy agenda, they should be articulating and acting upon the following key points about the issues and the context in which they will be considered.

As for the specific content of a SETA, the negotiators need to be direct about the challenge of addressing four especially important but difficult elements: subsidies, government procurement, standards and IP rights. They all need a fresh look. For instance, negotiators should take a pragmatic approach to intellectual property rights (IPR) issues – an approach that reflects the following observation: ' it is not mutually exclusive to recognise and encourage the important role of IPRs in promoting green innovation, while at the same time considering more closely their impact on technology transfer and dissemination, against the factor of "affordability". In this regard, it is difficult to reach categorical generalisations as the impact of IPRs varies according to many factors such as the technology in question, the sector, and country circumstances. Thus an examination of the role of IPRs in relation to diffusion of green technologies needs to be conducted on a case-by-case basis and in light of specific evidence' (Abdel Latif 2012b).

Negotiators' fresh look should reflect three new paradigms:

- A new paradigm of *international technology diffusion* that includes South–North and South–South transfers and other new techno-economic geographic realities at the macro level and the diverse modes of firms' technology transfers at the micro level. There must be realism about the new techno-economic geography: many economically significant 'developing' countries are also significant sources of sustainable energy technologies, not only in manufacturing goods but also in R&D and in services.
- A new paradigm of *economic growth* that incorporates the roles of sustainable energy and energy efficiency and that also recognises the potential contributions of private international technology transfers to climate change mitigation and sustainable development – transfers that occur through trade in services and direct investments as well as licensing and trade in goods.
- A new paradigm of the *role of government in economies* that recognises the market failures that are at the core of the climate change problem and that are embedded in technological solutions.

Finally, the negotiators should emphasise that the negotiations are not only about reaching agreements that take into account differences in countries' and industries' interests; they are also about the common interests of the international community in addressing effectively the global commons problems associated with climate change and about the global benefits of the widespread diffusion of sustainable energy technologies.

## Appendix I

### *Variations in Energy Scenarios Among Countries: China, Europe, India and the United States*

The IEA (2010) undertook a more detailed analysis of $CO_2$ trends and abatement options for four countries or regions that will have a major role in reducing global emissions: OECD Europe, the United States, China and India. Each faces unique challenges, reflecting current and future levels of economic development and diverse endowments of natural resources (represented in their energy mixes). Thus, each will have a very different starting point and future trajectory in terms of their $CO_2$ emissions, and develop in different ways in both the Baseline and the BLUE Map scenarios. Although many of the same technology options are needed to reduce emissions, the policy options associated with their application may be dramatically different.

In the Baseline scenario, $CO_2$ emissions in *India* show the largest relative increase, rising almost fivefold by 2050. *China* also shows a substantial rise, with emissions almost tripling between 2007 and 2050. *The United States* shows a much more modest rise, of 1 per cent, and emissions in OECD Europe decline by 8 per cent. In the BLUE Map scenario, all countries show considerable reductions from the Baseline scenario: emissions in 2050 (compared to 2007) are 81 per cent lower for the United States, 74 per cent lower for OECD Europe and 30 per cent lower in China, while India's emissions rise by 10 per cent.

The BLUE Map scenario also brings significant security of supply benefits to all four countries or regions, particularly through reduced oil use. In the United States and OECD Europe, oil demand in 2050 is between 62 per cent and 51 per cent lower than 2007 levels (gas demand shows similar declines). In China and India, oil demand still grows in the BLUE Map scenario, but is between 51 per cent and 56 per cent lower by 2050 than in the Baseline scenario.

In *OECD Europe*, the electricity sector will need to be almost completely decarbonised by 2050. More than 50 per cent of electricity generation is from RE, with most of the remainder from nuclear and fossil fuels using CCS (the precise energy mix varies widely among individual countries, reflecting local conditions and opportunities). In industry, energy efficiency and CCS offer the main measures for reducing emissions. In buildings, efficiency improvements in space heating can provide the most significant energy savings and more than half of the sector's emissions reductions in the BLUE Map scenario. Other mitigation measures include solar thermal heating, heat pumps, CHP/district heating and efficiency improvements for appliances. Transport volumes in OECD Europe are expected to remain relatively constant. Deep $CO_2$ emissions reductions in transport can be achieved through more efficient vehicles, a shift towards electricity and biofuels, and progressive adoption of natural gas followed by a transition to biogas and bio-syngas.

For the *United States*, energy efficiency and fuel switching will be important measures in reducing $CO_2$ emissions across all end-use sectors. Infrastructure investments will be vital to supporting the transition to a low-carbon economy, particularly in the national electricity grid and transportation networks. Most of the existing generation assets will be replaced by 2050 and low-carbon technologies such as wind, solar, biomass and nuclear offer substantial abatement opportunities. Many energy-intensive industries have substantial scope to increase energy efficiency through technological improvements. Similarly, the average energy intensity of Light Duty Vehicles (LDVs) is relatively high; doubling the fuel efficiency of new LDVs by 2030 can help reduce emissions. Advanced vehicle technologies can also play an important role in the LDV and commercial light- and medium-duty truck sectors. In buildings, improving the efficiency of space cooling, together with more efficient appliances, offers the largest opportunity to reduce $CO_2$ emissions.

Given the dominance of coal, *China* must invest heavily in cleaner coal technologies (such as CCS) and improve the efficiency of coal use in power generation and industry (which accounts for the largest share of China's energy use and $CO_2$ emissions). Priority should also be given to measures to improve energy efficiency and reduce $CO_2$ emissions in energy-intensive sectors such as iron and steel, cement and chemicals. The Chinese transport sector is evolving very rapidly, in terms of vehicle sales, infrastructure construction and the introduction of new technologies. The BLUE Map scenario shows that significant emissions reductions will depend on the electrification of transport modes and substantial decarbonisation of the electricity sector.

For *India*, the challenge will be to achieve rapid economic development – which implies a significant increase in energy demand for a growing population – with only a very small increase in $CO_2$ emissions. Electricity demand will grow strongly and the need for huge additional capacity creates a unique opportunity to build a low-carbon electricity system. While India has some of the most efficient industrial plants in the world, it also has a large share of small-scale and inefficient plants. Thus, improving overall industrial efficiency will be a significant challenge. Rising incomes and increased industrial production will spur greater demand for transport in India, making it imperative to promote public transport and new, low-carbon vehicle technologies. The buildings sector will also see strong growth in energy demand: efficiency improvements in space cooling and appliances will be critical to restraining growth in energy consumption and emissions.

*Source*: IEA (2010: 13–14); italics added by the author.

# Appendix II

## *Illustrative List of Energy Efficiency Technologies*[a]

### Building Sector

- Mineral wools (slag wool, rock wool)[b]
- Mineral insulating materials and articles[b]
- Multiple-walled insulating units of glass[b]
- Glass-filas insulation products[b]
- Phenolic resins, in primary forms[b]
- Plates, sheet, etc., non-cellular and not reinforced, of polymers of styrene[b]
- Plates, sheet, etc., cellular of polystyrene[b]
- Plates, sheet, etc., cellular of polyurethane[b]
- Heat pumps[b]
- Heat exchange units[b]
- Automatic regulating thermostats[b]
- Compact fluorescent lamps (CFLs)[b]
- Wood pellet burning stoves
- New plastic recycling methods for appliances
- Integrated building heating-air conditioning-lighting management systems

### Transport Sector

- Urban traffic control systems

### Motor Vehicles

- More efficient transmissions
- Recovery of energy from exhaust gases with turbo-charged engines
- More-efficient air conditioners, lighting
- High-strength steel; stainless steel; other advanced steels; aluminum; magnesium; plastic and plastic composites for lighter weight
- Efficient pumps, fans, air compressors, and heating, air conditioning and power-steering systems

### Rail

- Fuel efficiency monitoring systems
- More efficient engines
- Use of aluminum instead of conventional steels
- More efficient air conditioning

*Air*
- Improved aerodynamics
- Open rotor engines for improved propulsion
- Lightweight materials: carbon-fire reinforced plastic airframes
- Lightweight materials: composite materials with high- temperature tolerances for engines
- More efficient air traffic control

*Maritime Shipping*
- Flettner rotor (a spinning vertical rotor that converts wind into ship propulsion energy)
- Light weight alternatives for replacing steel
- Waste heat recovery

*Industry Sector*
- Power semiconductor modules that turn electricity on and off

*Agriculture, Forestry and Fishing Sector*
- Grassland management practices
- Rice production management practices

*Water and Water Treatment*
- Rehabilitating and retrofitting water treatment systems for greater energy efficiency
- Systems to reuse waste water from natural gas hydraulic fracturing drilling
- New materials for more efficient AC and DC inverters

*Notes:*
a  This is not intended to be an exhaustive, comprehensive list.
b  See ICTSD (2011a: Appendix A) for HS Codes.
*Sources: ICTSD (2010a, 2011b); McKinsey (2012); Mitsubishi (2012).*

## Appendix III

### Trade Conflicts Involving Sustainable Energy Technology and Climate Change Mitigation Issues

This Appendix provides an overview of cases in which sustainable energy, trade, climate change and technology diffusion issues have intersected publicly. Some of the cases have become subjects of formal disputes in the

WTO, while others have been addressed outside the WTO in bilateral negotiations.[10] See www.wto.org for updates.

## Chinese Investigation of US Solar Panel Polysilicon Dumping

In July 2012, the Chinese government announced that its Ministry of Commerce would investigate US exports of solar-grade polysilicon with a view of potentially imposing anti-dumping duties. A decision was not expected until July 2013.

## US Complaint in the WTO about Chinese Restrictions of Exports of Rare Earths

In 2012, the United States launched a formal complaint at the WTO (Dispute Settlement Case DS 440) about China's policy to limit exports of rare earths. Although this is not explicitly a sustainable energy case, it has direct implications for sustainable energy industries. In particular, some of the seventeen rare earth minerals are used in batteries and fluorescent lamps. The case was pending as of mid 2012.

## US Complaint in the WTO about Chinese Solar Panel Manufacturing Subsidies and Dumping

On November 9, 2011, the USITC began an investigation into issues concerning subsidies and dumping of Chinese-made solar cells. The subsidies are purported to include favoured treatment in taxes, raw materials prices, land prices, electricity prices and water prices, as well as loans, insurance and other assistance for exports. The investigation was instituted at the behest of a group of seven solar cell manufacturers; among them was SolarWorld Industries Americas, an affiliate of its parent firm, SolarWorld of Germany. The identities of the six other firms were not made public. The petition filed by the seven firms on October 19, 2011 contended that solar cells being manufactured in China were being subsidised by the Chinese government and dumped internationally below production and transportation costs in violation of WTO rules. They asked that tariffs up to 100 per cent be imposed on imports into the United States. The investigation and possible further US actions against the Chinese government and firms were opposed by a group of twenty-five US firms that buy and install solar panels and by some US-based environmental groups. Many of the US importers of solar panels are US subsidiaries of Chinese firms. The US ITO is required to make preliminary decisions about the merits of the case during 2012.

---

[10] E.g. the opposition of the automotive industry in the United States to steel import restrictions, and the opposition of retailers of compact fluorescent lights (CFLs) in the EU to restrictions on imports from Asia. In both instances, it should be noted, however, that the restrictions were in place for many years before they were removed under pressure from the affected domestic corporate consumers.

Additional background information includes the following: by 2011 China had more than half the world solar panel market, including more than half of the US market. Chinese imports into the United States in 2009 were USD 640 million and USD 1.5 billion in 2010. Three US manufacturers went bankrupt in 2011 – one of them being Solyndra, which had received a US government loan guarantee for approximately USD 500 million. Wholesale prices per watt of capacity declined from USD 3.30 in 2008 to USD 1.80 in early 2011 to USD 1.00–1.20 by November 2011.

**US Complaint in the WTO about Chinese Wind Power Subsidies**
The United States filed a request for WTO consultations in December 2010 about Chinese government RE subsidies, including in particular its subsidies in the wind power industry and domestic content requirements (WTO Dispute Settlement Case DS 419). At stake were the quantity and mode of international technology transfer into China, as well as the competitive implications for firms inside and outside China. During the consultation phase of the dispute settlement process in the WTO, before the case reached the phase involving the establishment of a dispute settlement panel, the parties reached an agreement according to which the Chinese government would end the subsidies.

**Japanese Complaint in the WTO about Canadian Domestic Content Requirements**
After requesting consultations with Canada in September 2010, Japan filed a formal WTO complaint in June 2011 against Canadian treatment of imported equipment for RE generation in its FIT programme due to LCRs. At issue specifically were the policies of the province of Ontario (the most populous in Canada). In view of the widespread interest in FITs in many countries, the outcome of the case was potentially significant for several renewable industries. The case is pending as WTO Dispute Settlement Case DS 412.

**EU Tariffs on Compact Florescent Lights from China and Other Asian Countries**
Although this case did not become a WTO dispute case, it is another example of how trade policy can limit international transfers of low-carbon technologies. In this case, the EU imposed special tariffs on CFLs from China unilaterally on the grounds that Chinese exporters were engaged in dumping. The conflict progressed through several stages, dragged on for a decade, and involved several exporting countries in Asia. At stake in China – and in other Asian countries to which some manufacturing was relocated from China – was the fate of hundreds of manufacturers/exporters of CFLs; at stake in Europe

was the energy efficiency of lighting in the commercial and residential sectors of twenty-seven countries. After several years the tariffs were reduced, but they were not eliminated altogether until European retail firms complained to the Commission that their desire to sell more low-cost, energy efficient CFLs was being thwarted by the EU tariffs. The issue of whether there was dumping was not resolved by an independent process because the conflict did not become a WTO dispute.

### US Limits on Rebates for Purchasers of Hybrid Automobiles from Certain Manufacturers

This case was unusual in that unnamed firms (which were in fact Honda and Toyota and which of course happened to be Japanese) encountered volume-based caps put on US government rebates to US purchasers of hybrid cars in the United States. Thus, the Japanese-based firms, which were leaders in the increasingly popular hybrid technology, reached their caps before their US-based rivals, so the US customers of the Japanese firms could no longer receive the US government tax credits. In response, the Japanese firms shifted production of hybrids from Japanese plants to their facilities in the United States in order to placate the US government. Over time, the effects of the rebates on international transfer of climate-friendly technology shifted: initially there was less transfer in the form of the *products* – i.e. hybrid vehicles – but subsequently there was a transfer of the *production process* technology from the plants in Japan to the Japanese firms' plants in the United States. Hybrid vehicles continue to be produced in the United States by Japanese firms. The US government rebate programme eventually expired and was not renewed. The Japanese government did not file a complaint in the WTO.

### EU GHG Emissions Standards on Imported Palm Oil-Based Biodiesel Fuel from South East Asia

This case is different from those above in that the issue was the imposition of barriers to the importation of biofuels produced in GHG-intensive production processes involving the clearing of rain forests (i.e. carbon sinks) in order to expand palm plantations. The trade barrier took the form of environmental standards because the imports were subjected to life cycle analyses in the EU.

### US Tariffs on Ethanol Imports from Brazil

As has been typical with agricultural goods, US tariffs on imports were causally connected to domestic subsidies – in this instance, subsidies for corn production and corn-based ethanol refineries. The special tariff of 57 cents per gallon on sugarcane-based ethanol from Brazil in addition to the regular 2½ per cent was a particularly troublesome issue in

Brazilian–US relations over many years, and it intruded into the WTO negotiating processes on trade in agricultural goods and trade in environmental goods. There was a surprising and unusual turn of events in this case in 2011, however, when the US Congress decided to end the tariffs as well as the domestic subsidies, as part of a broader movement against government spending in general and agricultural and energy subsidies in particular. As of 1 January 2012, both the special tariff and the subsidy programme were no longer in effect.

### EU Objections to US Exporters' Use of a 'Splash and Dash' Procedure to Subsidise US Exports of Biodiesel Fuel

The exploitation of a loophole in US legislation made it possible for US firms to import biodiesel fuel from southeast Asia, add a small portion of US petro diesel to it in order to qualify for a domestic US subsidy of the biodiesel portion of blended fuel, and then export it to Europe (Germany in particular, where it qualified for additional subsidies). The US firms were able to get US government subsidies for a product that was approximately 99 per cent based on imports with only 1 per cent US content and then export it without its entering the US market, even though the purpose of the subsidy programme was to expand the US biodiesel industry and increase the use of biodiesel in the US market. The practice ended after coming under pressure from European diplomats and business leaders and after the US industry association acknowledged that the practice was inconsistent with the intention of the legislation.

### Reactions to the EU's Inclusion of the Aviation Industry in the Emissions Trading Scheme

The inclusion in the ETS of commercial airline flights into and out of EU airspace became operational on 1 January 2012. When the EU decided to include the aviation sector in the ETS, in order to create a 'level playing field for all carriers in the aviation industry', it covered international flights into and out of EU countries – not only the flights of EU-based carriers but also those of carriers based outside the EU. US airlines and the US Air Transport Association complained to the European Court of Justice (ECJ) and also threatened legal action in the International Court of Justice (ICJ) on the grounds that the Chicago Convention concerning international cooperation in civil aviation was being violated. The ECJ ruled against the US complaint. Whether there would be any further actions by the US and/or other governments was unclear as of mid-2012. The Chinese government also objected to its airlines' flights into EU airspace being covered by the EU ETS. It put pressure on the EU by preventing a Chinese airline from entering into an agreement with the European manufacturer Airbus to buy nearly USD4 billion worth of new

A380 superjumbo planes. The case involved the use of a trade sanction by China on manufactured goods (airplanes produced in Europe), as well as the imposition of a restriction by the EU on international trade in services (commercial air transport services provided by Chinese airlines). At stake was not only the multi-billion dollar international trade transaction, but also the effectiveness of the EU ETS as a stimulator of technological innovation and diffusion. Indeed, a rationale for the establishment of a cap-and-trade system and thus the creation of a price for greenhouse gas emissions is, of course, precisely to foster technological innovation and diffusion based on price incentives. See www.icao.org.

Source: Compiled by the author from *Bridges Weekly, Climate Wire* and *Financial Times* (London); also see www.TradeAndClimate.net for additional information, including updates.

## References

Abdel Latif, Ahmed (2012a) *Intellectual Property Rights and Green Technologies from Rio to Rio*. ICTSD, Geneva

  (2012b) The UNEP–EPO–ICTSD 'Study on Patents and Clean Energy: Key Findings and Policy Implications', in David Ockwell and Alexandra Mallett (eds.), *Low-Carbon Technology Transfer – From Rhetoric to Reality*. Routledge, London and New York

Abdel Latif, Ahmed, Keith Maskus, Ruth Okediji, Jerome Reichman and Pedro Roffe (2011) *Overcoming the Impasse on Intellectual Property and Climate Change at the UNFCCC: A Way Forward*. ICTSD, Geneva

Africa–EU Energy Partnership (2010) *Africa–EU Renewable Energy Cooperation Programme*. African Union, Addis Ababa

Aldy, Joseph and R.N. Stavins (eds.) (2007) *Architectures for Agreement: Addressing Global Climate Change in the Post-Kyoto World*. Cambridge University Press

  (2009) *Post-Kyoto International Climate Policy: Summary for Policymakers*. Cambridge University Press

  (2010) *Post-Kyoto International Climate Policy: Implementing Architectures for Agreement*. Cambridge University Press

APEC (2011) *The Honolulu Declaration: Toward a Seamless Regional Economy*. APEC, Singapore

  (2012) Meeting of APEC ministers responsible for trade. APEC, Singapore

Barton, John (2007) *Intellectual Property and Access to Clean Energy Technologies in Developing Countries: An Analysis of Solar Photovoltaic*. ICSTD, Geneva

Brewer, Thomas (2008a) 'Climate change technology transfer: a new paradigm and policy agenda', *Climate Policy*, 8: 516–526

  (2008b) *International Energy Technology Transfers for Climate Change Mitigation*. Paper prepared for the CESifo Venice Summer Institute Workshop: Europe and Global Environmental Issues, Venice, 14–15 July; CESifo Working Paper, No. 2408

(2008c) 'The technology agenda for international climate change policy: a taxonomy for structuring analyses and negotiations', in Christian Egenhofer (ed.), *Beyond Bali: Strategic Issues for the Post-2012 Climate Change Regime*. Centre for European Policy Studies, Brussels: 134–145

(2009) 'Technology transfer and climate change: international flows, barriers and frameworks', in Lael Brainard (ed.), *Climate Change, Trade and Competitiveness*. The Brookings Institution, Washington, DC

(2014) *The United States in a Warming World: The Political Economy of Government, Business and Public Responses to Climate Change*. Cambridge University Press

Coe, David and Elhanan Helpman (1995) 'International R&D spillovers', *European Economic Review*, 39 (5): 859–887

Cottier, Thomas, Olga Nartova and Sadeq Bigdeli (eds.) (2009) *International Trade Regulation and the Mitigation of Climate Change*. Cambridge University Press

Danish Energy Agency (DEA) (2012) *Renewable Energy Cooperation with China*. Copenhagen

Epps, Tracey and Andrew Green (2010) *Reconciling Trade and Climate: How the WTO Can Help Address Climate Change*. Edward Elgar, Cheltenham

European Commission (2012) International Affairs: Free Trade Agreements. Brussels

European Union (2011) Official Journal, International Agreements, 2011/265/ EU, Council Decision of 16 September 2010 on the signing, on behalf of the European Union, and provisional application of the Free Trade Agreement between the European Union and its Member States, of the one part, and the Republic of Korea, of the other part

Frankel, Jeffrey (2010) 'Global environment and trade policy', in Joseph Aldy and Robert Stavins (eds.), *Post-Kyoto International Climate Policy: Implementing Architectures for Agreement*. Cambridge University Press

Ghosh, Arunabha and Himani Gangania (2012) *Governing Clean Energy Subsidies: What, Why, and How Legal?* ICTSD, Geneva

Goswami, Anandajit, Mitali Dasgupta and Nitya Nanda (2009) *Mapping Climate Mitigation Technologies and Associated Goods within the Buildings Sector*. ICTSD, Geneva

Grossman, G.M. and Helpman, E. (1991) *Innovation and Growth in the Global Economy*. MIT Press, Cambridge, MA

Hanni, Michael et al. (2011) 'Foreign direct investment in renewable energy: trends, drivers and determinants', *Transnational Corporations*, 20 (2): 29–65

Hoekman, Bernard and Michel Kostecki (2009) *The Political Economy of the World Trading System: The WTO and Beyond*, 3rd edn. Oxford University Press

Howse, Robert (2010) *Climate Mitigation Subsidies and the WTO Legal Framework: A Policy Analysis*. ICTSD, Geneva

Intergovernmental Panel on Climate Change (2011) *IPCC Special Report on Renewable Energy Sources and Climate Change Mitigation*. Cambridge University Press, Cambridge and New York

International Centre for Trade and Sustainable Development (2010a) *Climate-Related Single-Use Environmental Goods*. ICTSD, Geneva
  (2010b) 'Japan challenges Canadian sustainable energy incentives at WTO', *Bridges Trade BioRes*, 10 (17), 24 September: 3–5
  (2011a) *Fostering Low Carbon Growth: The Case for a Sustainable Energy Trade Agreement*. ICTSD, Geneva
  (2011b) *The Climate Technology Mechanism: Issues and Challenges*, Information Note, No. 18
International Energy Agency (IEA) (2008) *Energy Technology Perspectives 2008: Fact Sheet – The BLUE Scenario*. IEA, Paris
  (2009) *Technology Roadmap*. IEA, Paris
  (2010) *Energy Technology Perspectives, 2010*. IEA, Paris
Kejun, Jiang (2010) *Mapping Climate Mitigation Technologies and Associated Goods within the Transport Sector*. ICTSD, Geneva
Kim, Joy (2011) *Facilitating Trade in Services Complementary to Climate-Friendly Technologies*. ICTSD Global Platform on Climate Change, Trade and Sustainable Energy, Geneva
Kuriyama, Carlos (2012) *A Snapshot of Current Trends in Potential Environmental Goods and Services*. APEC Policy Support Unit, Policy Brief, No. 3, Singapore
Lako, Paul (2008) *Mapping Climate Mitigation Technologies and Associated Goods within the Renewable Energy Supply Sector*. ICTSD, Geneva
McKinsey (2012) Website, www.mckinsey.com, accessed on 1 June 2012
Mitsubishi (2012) Website, www.mitsubishi.com
National Renewable Energy Laboratory (NREL) (2010) *Strengthening Clean Energy Technology Cooperation under the UNFCCC: Steps toward Implementation*. NREL, Golden, CO
Pacala, Stephen and Robert Socolow (2004) 'Stabilization wedges: solving the climate problem for the next 50 years with current technologies', *Science*, 305, August 13: 968–972
Porter, Michael E. and Claas van der Linde (1995) 'Toward a new conception of the environment–competitiveness relationship', *Journal of Economic Perspectives*, 9 (4): 97–118
REN21 (2011) *Global Status Report*. REN21, Paris
Rifkin, Jerome (2011) *The Third Industrial Revolution*. Palgrave Macmillan, New York
Sampath, Padmashree, John Mugabe and John Barton (2012) *Realizing the Potential of the UNFCCC Technology Mechanism: Perspectives on the Way Forward*. ICTSD Global Platform on Climate Change, Trade and Sustainable Energy, Geneva
Sampath, Padmashree and Pedro Roffe (2012) *Unpacking the International Technology Controversy: Fifty Years and Beyond*. ICTSD, Geneva
Schmalensee, Richard (2009) 'Epilogue', in Joseph Aldy and Robert Stavins (eds.), *Post-Kyoto International Climate Policy: Implementing Architectures for Agreement*. Cambridge University Press
Solow, Robert (1956) 'A contribution to the theory of economic growth', *Quarterly Journal of Economics*, 70 (1): 65–94

Storm, Laura (ed.) (2012) 'Sustainia100'. Sustainia, Copenhagen
UNCTAD (2010) *World Investment Report*. United Nations, Geneva
  (2011) *Technology and Innovation Report*. United Nations, Geneva
UNEP, Bloomberg New Energy Finance and Frankfurt School, UNEP Collaborating Centre for Climate and Sustainable Energy Finance (2011) *Global Trends in Renewable Energy Investment 2011*. UNEP, Geneva
Victor, D. G. (2011) *Global Warming Gridlock: Creating More Effective Strategies for Protecting the Planet*. Cambridge University Press
Vossenaar, René (2010) *Deploying Climate-Related Technologies in the Transport Sector: Exploring Trade Links*. ICTSD, Geneva
Vossenaar, René and Veena Jha (2010) *Technology Mapping of the Renewable Energy, Buildings, and Transport Sectors: Policy Drivers and International Trade Aspects*. ICTSD, Geneva
World Bank (2009) *International Trade and Climate Change*. World Bank, Washington, DC
  (2012) *Inclusive Green Growth: The Pathway to Sustainable Development*. World Bank, Washington, DC
World Trade Organization and United Nations Environment Programme (2009) *Trade and Climate Change*. WTO and UNEP, Geneva
Yergin, Daniel (2011) *The Quest: Energy, Security, and the Remaking of the Modern World*. Penguin Press, New York

## Further Reading

Brewer, Thomas and Andreas Falke (2012) 'International transfers of climate-friendly technologies: how the world trade system matters', in David Ockwell and Alexandra Mallett (eds.), *Low-Carbon Technology Transfer – From Rhetoric to Reality*. Routledge, London and New York
de Coninck, Heleen (2009) *Technology Rules! Can Technology-Oriented Agreements Help Address Climate Change?* Vrije Universiteit, Amsterdam
Karp, Larry and Jinhua Zhao (2010) 'A proposal for the design of the successor to the Kyoto Protocol', in Joseph Aldy and Robert Stavins (eds.), *Post-Kyoto International Climate Policy: Implementing Architectures for Agreement*. Cambridge University Press
Keeler, Andrew and Alexander Thompson (2010) 'Mitigation through resource transfers to developing countries: expanding greenhouse gas offsets', in Joseph Aldy and Robert Stavins (eds.), *Post-Kyoto International Climate Policy: Implementing Architectures for Agreement*. Cambridge University Press
Lewis, Joanna (2007) 'Technology acquisition and innovation in the developing world: wind turbine development in China and India', *Studies in Comparative International Development*, 42: 3–4
MacGregor, James (2010) *Carbon Concerns: How Standards and Labelling Initiatives Must Not Limit Agricultural Trade from Developing Countries*. ICTSD, Geneva

Newell, Richard (2010) 'International climate technology strategies', in Joseph Aldy and Robert Stavins (eds.), *Post-Kyoto International Climate Policy: Implementing Architectures for Agreement*. Cambridge University Press

Ockwell, David and Alexandra Mallett (eds.) (2012a) *Low-Carbon Technology Transfer – From Rhetoric to Reality*. Routledge, London and New York

Sachs, Harvey, Amanda Lowenberger and Win Lin (2009) *Emerging Energy-Saving HVAC Technologies and Practices for the Buildings Sector*. ACEEE Research Report, A092. American Council for an Energy Efficient Economy, Washington DC

Sampath, Padmashree, John Mugabe and John Barton (2012) *Realizing the Potential of the UNFCCC Technology Mechanism: Perspectives on the Way Forward*. ICTSD, Geneva

Teng, Fey, Wenying Chen and Jiankun He (2010) 'Possible development of a technology clean development mechanism in a post-2012 regime', in Joseph Aldy and Robert Stavins (eds.), *Post-Kyoto International Climate Policy: Implementing Architectures for Agreement*. Cambridge University Press

US–China Clean Energy Research Center (2011) *Annual Report*. CERC, Cambridge

## *Related ICTSD Publications*

Abbott, Frederick (2009) *Innovation and Technology Transfer to Address Climate Change: Lessons from the Global Debate on Intellectual Property and Public Health*. Geneva

Abdel Latif, Ahmed et al. (2011) *Overcoming the Impasse on Intellectual Property and Climate Change at the UNFCCC: A Way Forward*. Geneva

Ancharaz, Vinaye and Riad Sultan (2010) *Aid for Trade and Climate Change Financing Mechanisms: Best Practices and Lessons Learned for LDCs and SVEs in Africa*. Geneva

Babcock, Bruce (2011) *The Impact of US Biofuel Policies on Agricultural Price Levels and Volatility*. Geneva

Bartels, Lorand (2012) *The Inclusion of Aviation in the EU ETS: WTO Law Considerations*. Geneva

Barton John (2005) *New Trends in Technology Transfer: Implications for National and International Policy*. Geneva

Cannady, Cynthia (2009) *Access to Climate Change Technology by Developing Countries: A Practical Strategy*. Geneva

Correa, Carlos (2009) *Fostering the Development and Diffusion of Technologies for Climate Change: Lessons from the CGIAR Model*. Geneva

Echols, Marsha (2009) *Biofuels Certification and the Law of the World Trade Organization*. Geneva

Faber, Jasper and Linda Brinke (2011) *The Inclusion of Aviation in the EU Emissions Trading System: An Economic and Environmental Assessment*. Geneva

Guey, Kamal et al. (2007) *Trade, Climate Change and Global Competitiveness: Opportunities and Challenges for Sustainable Development in China and Beyond*. Geneva

Harmer, Toni (2009) *Biofuels Subsidies and the Law of the World Trade Organization*. Geneva

Hervé, Alan and David Luff (2012) *Trade Law Implications of Procurement Practices in Sustainable Energy Goods and Services*. Geneva

International Centre for Trade and Sustainable Development (2007) 'Environmental groups target shipping emissions', *Bridges: Trade BioRes*, 1 (2), 1 December

(2008a) *Climate Change, Technology Transfer and Intellectual Property Rights*. Background Paper for the Trade and Climate Change Seminar, 18–20 June, Copenhagen

(2008b) *Liberalization of Trade in Environmental Goods for Climate Change Mitigation: The Sustainable Development Context*. Background Paper for the Trade and Climate Change Seminar, 18–20 June, Copenhagen

(2011a) 'Tensions build between EU, China over rare earths in aftermath of raw materials decision', *Bridges Weekly Trade News Digest*, 15 (27), 20 July: 7–8

(2011b) 'WTO Panel rules against China's export restrictions on raw materials', *Bridges Weekly Trade News Digest*, 15 (25), 6 July: 1–3

(2011c) 'US proclaims victory in wind power case; China ends challenged subsidies', *Bridges Weekly Trade News Digest*, 15 (21), 8 June: 3–4

(2011d) 'Beijing, Washington lock horns over Chinese wind power fund', *Bridges Trade BioRes*, 11 (1), 4 January: 1–3

Janssen, R. (2010) *Harmonising Energy Efficiency Requirements – Building Foundations for Cooperative Action*. Geneva

Jha, Veena (2008) *Environmental Priorities and Trade Policy for Environmental Goods: A Reality Check*. Geneva

(2009) *Trade Flows, Barriers and Market Drivers in Sustainable Energy Supply Goods: The Need to Level the Playing Field*. Geneva

Maskus, Keith (2004) *Encouraging International Technology*. Geneva

Maskus, Keith and Ruth Okediji (2010) *Intellectual Property Rights and International Technology Transfer to Address Climate Change: Risks, Opportunities, and Policy Options*. Geneva

Mytelka, Lynn (2007) *Technology Transfer Issues in Environmental Goods and Services: An Illustrative Analysis of Sectors Relevant to Air-Pollution and Renewable Energy*. Geneva

Nelson, Gerald et al. (2009) *The Role of International Trade in Climate Change Adaptation*. ICTSD-IPC Platform on Climate Change, Agriculture and Trade, Issue Paper, No. 4

Rai, Sunny and Tetyana Payasova. (2013a) *Selling the Sun Safely and Effectively: Solar Photovoltaic (PV) Standards, Certification, Testing and Implications for Trade Policy*. Geneva

Read, Robert (2010) *Trade, Economic Vulnerability, Resilience and the Implications of Climate Change in Small Island and Littoral Developing Economies*. Geneva

Selivanova, Yulia (2007) *The WTO and Energy: WTO Rules and Agreements of Relevance to the Energy Sector*. Geneva

Sell, Malena et al. (2006) *Linking Trade, Climate Change and Energy*. Geneva

UNEP, EPO and ICTSD (2010) *Patents and Clean Energy: Bridging the Gap between Evidence and Policy, Final Report*. Geneva

Vossenaar, René (2010a) *Climate-Related Single-Use Environmental Goods*. Geneva

(2010b) *Deploying Climate-Related Technologies in the Transport Sector*. Geneva

Wind, Izaak (2008) *HS Codes and the Renewable Energy Sector*. Geneva

(2009) *HS Codes and the Residential and Commercial Buildings Sector*. Geneva

(2010) *HS Codes and the Transport Sector*. Geneva

## *WTO Publications*

World Trade Organization (1998a) Committee on Trade and Environment. *United Nations Framework Convention on Climate Change*, Note by the Secretariat. Geneva

(1998b) Council for Trade in Services. *Environmental Services*, Background Note by the Secretariat. Geneva

(1999) *An Introduction to the GATS*. Revised version, October 1999, prepared by the Secretariat, Trade in Services Division. Geneva

(2001) Committee on Trade and Environment. *Matrix on Trade Measures Pursuant to Selected MEAs*, Note by the Secretariat. Geneva

(2005) Committee on Trade and the Environment. *Synthesis of Submissions on Environmental Goods*, Informal Note by the Secretariat. Geneva

See also www.TradeAndClimate.net

# 9 Legal Options for a Sustainable Energy Trade Agreement: July 2012

## Matthew Kennedy

This chapter looks at the legal options available to a group of countries interested in concluding a new international agreement dedicated to the interface between trade policy and climate change, which could be titled the Sustainable Energy Trade Agreement (SETA). Sustainable energy includes solar, wind, small-scale hydro and biomass-related fuels, as well as technologies and services that have the potential to mitigate greenhouse gas (GHG) emissions. The primary objective of a SETA would be to promote the scale-up and deployment of sustainable energy sources through the use of trade policy measures in order to reduce the emissions responsible for global warming.

While a SETA could be modelled either on the plurilateral Agreement on Government Procurement (GPA) or the Information Technology Agreement (ITA), it could also be a stand-alone agreement completely outside the World Trade Organization (WTO) framework. As regards the scope of issues and market barriers to be covered, it has been suggested that a SETA could be undertaken in a two-phased approach (ICTSD, 2011).[1] Phase one would address clean energy supply, i.e. goods and services relevant to sustainable energy generation in the areas of solar, wind, hydro and biomass as a starting point, although this could be extended to biofuels used for transportation, such as ethanol and biodiesel and other technologies.[2] Phase two could address the wider scope of energy efficiency products and standards, particularly those identified by the Intergovernmental Panel on Climate Change (IPCC) for GHG mitigation (buildings and construction, transportation, manufacturing and agriculture).

Different approaches have been negotiated for a SETA. On one approach, the agreement could initially focus on key trade-related issues as a cluster, comprising tariffs, non-tariff barriers (NTBs), subsidies,

---

[1] ICTSD (2011).
[2] For the full set that ICTSD has used see: Lako (2008) and Wind (2008). See also the EPO–UNEP–ICTSD joint 2010 report on Clean Energy Technologies, www.epo.org/news-issues/issues/clean-energy/study.html.

procurement and services. Depending on the ambitions of the parties, it could also proceed incrementally on an issue-by-issue agenda. It could, for instance, address issues related to domestic energy regulation, such as fossil-fuel subsidies, investment and competition policy, as well as trade facilitation and transit issues related to sustainable energy. The implementation of any eventual agreement or its scope and modalities would remain for Members and non-Member governments to decide.

The SETA would not be part of the Doha Round, although its subject matter would overlap with those negotiations, notably on environmental goods and services (EGS). In view of the lack of substantial progress in the multilateral negotiations and a tendency to converge on the lowest common denominator within a very large and diverse membership, a group of interested WTO Members and non-Members might wish to pursue negotiations on a SETA in parallel. If the negotiations lead to a successful conclusion, the parties could implement the agreement within the WTO framework, or outside it. This chapter addresses the institutional and formal issues to be taken into consideration in each scenario. Section 1 considers how the negotiations for SETA could be launched, conducted and implemented within the WTO framework, paying special attention to how non-participants would be protected. Section 2 considers how a SETA could be established outside the WTO framework, and what action could be taken to avoid conflicts between a SETA and the WTO. Each section ends with a summary of the outlook for the relevant scenario. Finally, the chapter presents conclusions and recommendations on ways forward.

## 1 Implementation of a SETA within the WTO Framework

Given the scope of issues and market barriers that a SETA might cover, the new agreement could be negotiated and concluded within the WTO. This would take advantage of the institutional framework that the WTO already provides for the conduct of trade relations among its Members and, potentially, allow the SETA to be enforced through the WTO dispute settlement mechanism. This would not only avoid a duplication of institutions, but also help prevent conflicts of jurisdiction. This section considers the available legal bases for such an exercise and some important issues to take into consideration. The options presented are available under the WTO's current institutional rules and procedures, but reform of the latter themselves can also be envisaged.[3]

---

[3] See, for example, Steger (2010); Cottier and Elsig (2011).

## Scope for a SETA to be a New WTO Agreement

The WTO framework is not set in stone. New agreements can be added at any time even though the WTO Agreement was originally agreed as a package deal. The Agreement in Article II:1 defines the scope of the WTO in terms of the Annexes to that Agreement, which can and do change over time. Three WTO Agreements have already terminated: the 1994 International Dairy Agreement and the 1994 International Bovine Meat Agreement were both deleted in 1997,[4] and the Agreement on Textiles and Clothing terminated at the end of 2004.[5] The results of other negotiations have been incorporated into the existing WTO Agreements, notably the ITA,[6] reviews of the Pharmaceutical Understanding,[7] a bilateral deal on tariffs on distilled spirits,[8] four General Agreement on Trade in Services (GATS) protocols on financial services, movement of natural persons and basic telecommunications[9] and a protocol to the Agreement on Trade in Civil Aircraft (TCA).[10] An amendment to the Trade-Related Aspects of Intellectual Property Rights (TRIPS) Agreement has been adopted, although it has not entered into force.[11] The WTO has also approved twenty-nine protocols with acceding Members. The WTO has yet to add a new formal agreement on trade facilitation to its framework, although one is under negotiation in the Doha Round.[12]

The single undertaking of the WTO does not prevent the addition of new agreements to the WTO framework, even without the successful conclusion of multilateral trade negotiations. The single undertaking

---

[4] General Council Decision on Deletion of the International Dairy Agreement from Annex 4 of the WTO Agreement, WT/L/251, 10 December 1997; General Council Decision on Deletion of the International Bovine Meat Agreement from Annex 4 of the WTO Agreement, WT/L/252, 10 December 1997.

[5] Agreement on Textiles and Clothing, Article 9.

[6] Ministerial Declaration on Trade in Information Technology Products, 13 December 1996, WT/MIN(96)/16 ('ITA').

[7] Reviews of the product coverage of GATT document L/7430, Trade in Pharmaceutical Products – Record of Discussion) at note 57 below), certified in WT/Let/251, 259, 270, 272, 361, 382, 405, 416, 442, 461 and 610.

[8] WT/Let/178 and 182.

[9] Second Protocol to GATS, S/L/11, text adopted by the Council for Trade in Services (CTS), S/L/13, 21 July 1995; Third Protocol to GATS, S/L/12, text adopted by the CTS, S/L/10, 21 July 1995; Fourth Protocol to GATS, S/L/20, text adopted by the CTS, S/L/19, 30 April 1996; Fifth Protocol to GATS, S/L/45, 3 December 1997, text adopted by the Committee on Trade in Financial Services, S/L/44, 14 November 1997.

[10] Protocol (2001) Amending the Annex to the Agreement on Civil Aircraft, done at Geneva on 6 June 2001 (TCA/4), and amended by Decision of the Committee on Trade in Civil Aircraft of 21 November 2001 extending the date for acceptance (TCA/7).

[11] Protocol Amending the TRIPS Agreement, done at Geneva on 6 December 2005, annexed to a WTO General Council Decision of the same date (WT/L/641).

[12] TN/TF/W/165/Rev.1, 2 March 2010. Many modifications and rectifications of concessions and commitments in individual Members' schedules have been certified.

Legal Options for a SETA 393

was an approach adopted for the organisation of two rounds of such negotiations, first the 1986 Punta del Este Declaration that launched the Uruguay Round[13] and later the 2001 Doha Declaration.[14] The single undertaking approach is reflected in the implementation of the results of the Uruguay Round insofar as members accepted the WTO Agreement and remain bound by it as a package, without the option to select *à la carte* among the agreements in Annexes 1, 2 and 3.[15] The preamble to the WTO Agreement recites the parties' resolution to develop 'an integrated, more viable and durable multilateral trading system' with respect to the results of negotiations *up to and including* the Uruguay Round.[16] The single undertaking approach was later adopted for the Doha Round by choice,[17] but that does not govern the organisation of other negotiations. Indeed, the WTO Agreement, Article X:9, expressly provides a procedure for the addition to Annex 4 of new agreements (and the deletion of old ones) to which some Members are parties but others are not.[18]

The WTO framework is open to agreements entered into by fewer than all 155 Members. The GPA is an example of one form of agreement. The ITA (formally the Ministerial Declaration on Trade in Information Technology Products) is an example of a different form of negotiating results concluded among a sub-set of Members and implemented within the WTO framework.[19]

There is a degree of mistrust among some developing countries regarding the GPA-type of optional agreement. Plurilateral agreements (i.e. optional agreements in Annex 4 to the WTO Agreement) are perceived as a throwback to the Tokyo Round of trade negotiations of 1973–1979.[20] The results of the Tokyo Round included nine agreements, or so-called 'codes' to which only some contracting parties to the GATT subscribed. One of the achievements of the Uruguay Round was to revise most of those codes and integrate them into the WTO framework. Five multilateral WTO Agreements were based on Tokyo Round Codes: the Agreement on Technical Barriers to Trade (TBT), the Anti-Dumping

[13] Ministerial Declaration on the Uruguay Round, GATT document MIN/DEC, adopted 20 September 1986, BISD 33S/19, para.B(ii).
[14] Doha Ministerial Declaration, WT/MIN(01)/DEC/1, adopted 14 November 2001, para 47.
[15] WTO Agreement, Articles II, XI, XII, XIII, XIV and XV, discussed in Appellate Body report in *Brazil – Coconut*, WT/DS22/AB/R: 12.
[16] WTO Agreement, Preamble, 4th recital.    [17] See n. 14.
[18] See further Kennedy (2011b): 77–120.
[19] For the purposes of this chapter, there is no need to consider in what sense the Ministerial Declaration setting out the ITA is an agreement or an instrument related to the WTO Agreement. See the Panel report in *EC – IT Products*, WT/DS377/R, paras 7.375–7.384.
[20] The use of the word 'code' to describe these agreements adds to that perception.

Agreement, the Customs Valuation Agreement, the Subsidies and Countervailing Measures (SCM) Agreement and the Agreement on Import Licensing Procedures. Four other WTO Agreements were based on, or identical to, a Tokyo Round Code but remained plurilateral. Two of these, the 1994 GPA and the 1979 Agreement on Trade in Civil Aircraft, remain in Annex 4 to the WTO Agreement.

An ITA-type of optional agreement among sub-sets of WTO Members does not arouse the same suspicions. The ITA provided for improvements in market access that were negotiated and concluded among only certain Members but given effect by means of certifications of modifications to participating Members' goods schedules,[21] which are integral parts of Part I of the GATT 1994.[22] The positive attitude toward this type of optional agreement lies in the fact that the market access improvements it provides accrue to all WTO Members. In reality, this may not be very different from the position of a SETA, regardless of its form, because it may still be applied on a most favoured nation (MFN) basis for reasons of substance. In other words, some or all of the benefits of a SETA may accrue to all WTO Members even if it is concluded as a GPA-type of optional agreement, for reasons discussed below.

A new optional Agreement, such as a SETA, could be concluded among a group of interested WTO Members and given effect according to either or both of these models, depending on what it contained. If a SETA were added to Annex 4 of the WTO Agreement as a GPA-type Agreement, the only substantive condition would be that it must be a trade agreement.[23] A SETA would clearly satisfy that condition, at least at an initial stage.[24] (There is also an important procedural condition, discussed below.) If a SETA were given effect through certifications of modifications to participating Members' schedules as an ITA-type Agreement, it could only provide for improvements in tariff concessions and improved commitments on non-tariff measures within the scope of the GATT 1994 and/or specific commitments on liberalisation of trade in services within the scope of the GATS. The improvements could make reference to an agreed set of rules that participants incorporate in their respective Schedules by reference, along the lines of the Understanding on Commitments in Financial Services[25] and the reference paper on

---

[21] Decision of the Contracting Parties of 26 March 1980, Procedures for Modification and Rectification of Schedules of Tariff Concessions, L/4962, BISD 27S/25, incorporated by Article 1(b)(iv) of GATT 1994.
[22] GATT 1994, Article II:7.   [23] WTO Agreement, Article X:9.
[24] ICTSD (2011): 63.
[25] Understanding on Commitments in Financial Services annexed to the Final Act Embodying the Results of the Uruguay Round.

Legal Options for a SETA 395

telecommunications services.[26] If these types of improvements were *all* that the new Agreement contained, it would not be necessary to add the SETA to Annex 4. However, schedules may only yield rights under the GATT 1994 and the Agreement on Agriculture and cannot diminish obligations under those agreements without some express authorisation in the text of those agreements. Therefore, a permanent derogation within the WTO framework would require a new agreement in Annex 4.[27] Other types of WTO instruments, such as an amendment, could also be considered.[28]

Either way, the WTO membership collectively retains control over which new agreements are added to the organisation's framework. Agreements among sub-sets of Members can be added to Annex 4 only by means of a decision of the Ministerial Conference (or the General Council conducting its functions between sessions) taken 'exclusively by consensus'.[29] This procedural condition could be a challenge to fulfil. Modifications of participating Members' goods or services schedules are certified only with the tacit consent of all Members to the proposed modification.[30]

A new optional Agreement among a group of interested WTO Members and open to all would not necessarily be a backward step. Transitional periods and other provisions granting special and differential treatment already create different sets of rights and obligations among different groups of WTO Members. The difficulty in obtaining agreement among all Members in the context of the Doha Round compels fresh consideration of alternative approaches to rule making in order for the WTO to remain relevant in light of contemporary developments.[31] In 2005, the Sutherland Report advised that possible plurilateral approaches to WTO negotiations should be re-examined – outside the context of the Doha Round. However, it outlined a series of

---

[26] The Reference Paper of 24 April 1996 contains a set of pro-competitive regulatory principles applicable to the telecommunication sector that certain Members incorporated in their services schedules, www.wto.org/english/tratop_e/serv_e/telecom_e/tel23_e.htm.

[27] See the GATT panel report in *US – Sugar*, adopted 22 June 1989, BISD 36S/331, para 5.7; Appellate Body report in *EC – Bananas III*, WT/DS27/AB/R, para 154; Appellate Body report in *EC – Sugar*, WT/DS283/AB/R, para 220.

[28] See Howse (2011): 17–24. Howse considers an authoritative interpretation of a WTO provision, revival of non-actionability or a waiver.

[29] WTO Agreement, Article X:9.

[30] 1980 Procedures for the Certification of Rectifications or Improvements to Schedules of Specific Commitments adopted by the Council for Trade in Services on 14 April 2000, (S/L/84).

[31] Harbinson (2009): 20. European Centre for International Political Economy, Working Paper, No. 10/2009 and Bacchus (2011). As regards the merits of optional agreements, see Low (2009).

problems to which attention should be paid, notably the rules under which any such negotiations take place.[32] In 2007, the Warwick Commission recommended that serious consideration be given to the re-introduction of the flexibility associated with what has come to be known as critical mass decision making, as opposed to the single undertaking approach.[33] Most of the successful negotiations in the WTO framework since its establishment have, in fact, been optional, sectoral exercises.[34]

There are legitimate concerns regarding WTO Agreements concluded among fewer members than the entire membership. These relate to the value and availability of the benefits of the Agreement; the inclusiveness of the negotiating process; the potential for pressure to participate after an Agreement has been concluded and potential conflicts between new Agreements and the existing WTO framework. Each of these concerns is discussed below.

### *Rights and Obligations Toward Non-Participants*

One of the most important issues regarding the conclusion of a SETA as a new Agreement within the WTO framework concerns the rights and obligations of participants towards WTO Members who do not participate in the SETA. Multilateralisation of benefits under critical mass agreements ensures that they do less damage to the integrated system that the WTO Agreement established in 1994 than might otherwise be the case.[35] The Warwick Commission recommended that the principle of non-discrimination should apply to all Members, regardless of whether they participate in critical mass agreements.[36] This could be a decisive issue in obtaining consent from the Ministerial Conference to add a SETA to Annex 4 of the WTO Agreement, if it were given legal effect as a GPA-type agreement.[37]

All improvements in tariff and other concessions effected through modifications to participating Members' goods schedules will accrue – in theory – to the benefit of all WTO Members through the application of the MFN obligation in the GATT 1994, Article 1. This is the position as regards the ITA and the Pharmaceuticals Understanding. Improvements in market access for services also apply on an MFN basis, subject to any relevant exemptions that participating Members may have entered under

---

[32] Sutherland et al. (2005), paras 291–300.
[33] Pettigrew et al. (2007): 30–32. See also Manfred Elsig, 'WTO decision-making: can we get a little help from the Secretariat and the critical mass?, in Steger(2010): 67–90.
[34] The Protocol Amending the TRIPS Agreement is a notable exception.
[35] Rodríguez (2011).   [36] Pettigrew (2007).   [37] Faizal and Vickers (2011): 472.

the GATS, Article II.[38] The application of improvements on an MFN basis is one of the attractions of this type of optional Agreement as regards non-participating Members. However, in practice, the position is more nuanced, as only those Members with an actual or potential trade interest in the relevant goods and services can obtain a benefit.

The position may not be so different as regards Agreements in Annex 4 of the WTO Agreement. Article II:3 provides that these agreements 'do not create either obligations or rights for members that have not accepted them'. This restates the basic rule of public international law that *pacta tertiis nec nocent nec prosunt*, which has been codified in the Vienna Convention on the Law of Treaties of 1969, Article 34, as '[a] treaty does not create either obligations or rights for a third State without its consent'.[39] However, parties to the agreements in Annex 4 still owe MFN obligations to all WTO Members under the multilateral WTO Agreements.[40] To the extent that the subject matter of any agreement in Annex 4 falls within the scope of any of these MFN obligations, such as the GATT 1994, Article I:1 and the GATS, Article II:1, the benefits must normally be extended to all WTO Members, including non-parties to the Agreement in Annex 4. These obligations are not created by the Agreements in Annex 4; hence they are not addressed by the WTO Agreement, Article II:3.

The GPA is an agreement in Annex 4 of the WTO Agreement, and it has no MFN effect, but that does not mean that SETA would have no MFN effect either if it were added to Annex 4. Article III of the GPA obliges the parties to grant non-discriminatory treatment only to other parties to that Agreement but if, in practice, those advantages are not applied on a multilateral basis, it is because the subject of government procurement does not fall within the scope of the GATT 1994, Article I:1,[41] or the other MFN obligations in the multilateral WTO Agreements. That is an important difference from the SETA, which is likely to cover issues that fall within the scope of existing MFN obligations, such as improvements in market access and eventually possibly product standards. The benefits of these obligations would normally accrue to all Members, including non-parties, in accordance with the WTO Agreements, as do benefits

---

[38] For example, certain members maintained broad MFN exemptions based on reciprocity, or limited MFN exemptions, as regards financial services after the conclusion of the Second GATS Protocol, and to a lesser extent after the Fifth GATS Protocol.

[39] It has been stated that this 'general rule is so well established that there is no need to cite extensive authority for it'. See Jennings and Watts (1992): 1260.

[40] Nottage and Sebastian (2006): 9.989–1016 at 1101.

[41] Note that the scope of the GATT 1994, Article I:1, includes 'all matters referred to in paras 2 and 4 of Article III' whereas government procurement is carved out of Article III by para 8(a).

under the ITA, even if a SETA were implemented as a GPA-type agreement. On the other hand, obligations in other areas might fall outside the scope of existing MFN obligations, and the benefits would not accrue directly to non-parties.

There is an alternative view that the parties to plurilateral agreements do not have to apply them on an MFN basis. This view, which would have the merit of preventing free riding, was expressed at least as early as the first plurilateral concluded under the auspices of the GATT (i.e. the Kennedy Round Anti-Dumping Code[42]). In 1967, the European Communities maintained that the parties were under no obligation to apply the provisions of that Code to non-signatories.[43] In 1968, the GATT Director General was asked for a ruling on whether the parties had a legal obligation under the GATT 1947, Article I, to apply the provisions of the Anti-Dumping Code in their trade with all GATT contracting parties. The Director General replied that in his judgement they *did* have such an obligation, because (a) the text of the GATT 1947, Article I, covered many of the matters dealt with in the Anti-Dumping Code; and (b) the MFN obligation under the GATT, Article I, was clearly unconditional.[44]

At the conclusion of the Tokyo Round in 1979, the results of the negotiations included several plurilateral agreements. The contracting parties to the GATT expressly addressed the issue of MFN treatment by adopting a decision reaffirming 'their intention to ensure the unity and consistency of the GATT system' and expressly confirming that 'existing rights and benefits under the GATT of contracting parties not being parties to [the plurilateral agreements among the results of the Round], including those derived from Article I, are not affected by these Agreements'.[45] In other words, it confirmed that the benefits of the Tokyo Round plurilateral agreements were to accrue to all contracting parties to the GATT, even those that were not parties to the plurilaterals, insofar as the subject matter of those agreements was covered by the GATT 1947, Article I. However, this did not always occur in practice.[46]

---

[42] 1967 Agreement on Implementation of Article VI, which entered into force on 1 July 1968.

[43] See first meeting of Committee on Anti-Dumping Practices of 15 November 1968, GATT document COM.AD/1, para 11.

[44] Note by the GATT Director General dated 29 November 1968, GATT document L/3149.

[45] Decision on 'Action by the CONTRACTING PARTIES on the Multilateral Trade Negotiations' of 28 November 1979, GATT document L/4905, BISD 26S/201, paras 1–3.

[46] The United States was party to the GATT Subsidies Code but did not apply the requirement of injury to a domestic industry under its countervailing duty law on an MFN basis: compare 19 USC §1671(a) and (b) with 19 USC §1303 (1988), discussed in

When the WTO was established, no derogation from MFN treatment was granted with respect to the plurilateral agreements in Annex 4. Therefore, the parties are required to extend the benefits of those agreements to all WTO Members, to the extent that their subject matter falls within the scope of the WTO MFN obligations. There is an alternative view that the language of the second sentence of the WTO Agreement, Article II:3 (quoted above) grants such a derogation, but it is not obvious that that sentence does anything more than restate the *pacta tertiis* rule.[47] No decision or interpretation on this issue has been adopted in the WTO.

In the interest of certainty, any decision to add a SETA to Annex 4 should address MFN treatment specifically. If the Ministerial Conference does so, it can confirm that a SETA does not affect non-parties' rights under the multilateral WTO Agreements, including MFN obligations. The benefits of SETA obligations would then accrue to all Members, in principle, but only insofar as those obligations are covered by one of the WTO MFN provisions. In any event, free riding under a sectoral agreement such as a SETA is not possible for all other Members in practice, as some will have no relevant trade interest.[48] Therefore, a careful definition of the critical mass required for implementation of the new Agreement can ensure that any free riding will not undermine attainment of the Agreement's objective.

A further incentive for a group of non-parties to consent to the addition of the new agreement might be the inclusion of an optional MFN exception in the text in favour of least-developed countries (LDCs). The GPA already contains such a provision in the second sentence of Article V:12.

The alternative is to grant a waiver to exclude a SETA from MFN treatment. If agreed, this would create an incentive to participate; but the exclusion of non-parties from the benefits of a SETA is likely to be perceived as promoting a 'two-speed' WTO, which would probably reduce the prospects of obtaining the Ministerial Conference's consent to add it to Annex 4.

In sum, the benefits of a SETA would normally accrue to all WTO Members, even if it was implemented as a GPA-type agreement, to the extent that it covered subjects within the scope of the MFN obligations in

---

Lowenfeld (1994): 477–488 at 478, n. 2, and see the US statement in the meeting of the GATT Committee on Subsidies and Countervailing Measures on 8 May 1980, GATT doc. SCM/M/3, para 11.

[47] See, for example, Ehlermann and Ehring (2005): 51–75 at 56, n. 15.
[48] Some Members' trade share would always be too small for them ever to form part of a 'critical mass'. The small countries could enjoy the benefits of such Agreements but recurrent use of critical-mass decision making could lead to their disenfranchisement and disengagement: see Amrita Narlikar 'Adapting to new power balances: institutional reform in the WTO', in Cottier and Elsig (2011): 111–128 at 122.

the multilateral WTO Agreements. The Ministerial Conference would be well advised to confirm this point in any decision to add a SETA to Annex 4 of the WTO Agreement.

### SETA Negotiations within the WTO

#### Preparatory Stage

The procedure to add a new agreement to Annex 4 envisages that an agreement has *already* been concluded among certain Members (as parties to it) and does not address the prior processes of negotiation and conclusion, which can take place outside formal WTO meetings.[49] While, in principle, the parties are free to decide how to handle those questions, it should be borne in mind that the way in which the participants proceed can affect the attitude of Members that are not parties, whose consent will be required for the addition of any eventual agreement to Annex 4.

The 1980 certification procedures for modifications to Members' schedules, which have been used to give effect to the results of sectoral tariff negotiations, do not restrict the prior processes either.[50] Renegotiation procedures have not been considered necessary, because the negotiations only lead to improvements (i.e. reduction or elimination of tariffs), hence no compensation is owed. The certification procedures adopted by the Council for Trade in Services to give effect to new services commitments and improvements to existing commitments do not stipulate who must participate in the negotiations either, although they do allow other Members to object to proposed changes.[51]

The ITA provides an illustration of one means to launch negotiations among a group of interested WTO Members (and an acceding Member). The initiative for the ITA came from a coalition of industry actors and was taken up by the governments of the old 'Quad' (Canada, the European Community, Japan and the United States). The initiative was endorsed at a bilateral EU–US summit in 1995 and in 1996 by Quad ministers, who instructed their negotiators to move forward. The United States submitted a proposal for a multilateral ITA to the WTO Committee on Market Access[52] and later provided the WTO Council for Trade in Goods with a summary of the agreement that had been developed among interested

---

[49] WTO Agreement, Article X:9. The procedure might also imply that an agreement has entered into force insofar as it refers to 'parties' to a trade Agreement rather than to 'contracting parties', but this need not prevent the Ministerial Conference approving the addition of an Agreement that had not yet entered into force.
[50] See n. 29.   [51] *Ibid.*
[52] Information Technology Agreement, Communication from the United States dated 4 October 1996, G/MA/W/8.

delegations. By that time, the ITA was envisaged as a plurilateral agreement. The United States offered to consult with any delegation that wished to know more about the specifics of the initiative.[53] Three of the old Quad members took up the issue in the Asia-Pacific Economic Cooperation (APEC) forum, because certain Asian countries were becoming important exporters in the information technology (IT) sector. A breakthrough came in the APEC 1996 Leaders' Declaration, which called for the conclusion of an ITA by the Singapore Ministerial Conference of the WTO in order to substantially eliminate tariffs by the year 2000.[54] The following month, at that session of the Ministerial Conference, Ministers of thirteen WTO members plus one acceding Member (comprising the old Quad, seven other APEC members and Norway, Switzerland and Turkey) made a joint declaration on trade in IT products that was the ITA.[55] The text of the 1996 declaration set out modalities for tariff reductions and product coverage, to be finalised and implemented later, and invited other members and acceding members to participate.[56] The Ministerial Conference took note of the ITA and welcomed the initiative.[57]

When a group of interested Members has reached an agreement on market access improvements and is willing to apply those improvements on an MFN basis, they can implement through the certification procedures without a WTO declaration or decision, provided they are willing to move forward without the participation of any other Members. For example, the Pharmaceutical Understanding was implemented in this way in 1994 among twelve parties, including the EC-12 as one party.[58]

A group of interested Members may reach an agreement on market access among themselves but not yet have secured what they consider to be sufficient participation for implementation to be viable and beneficial to their interests. For example, the participants in the ITA at the time of the 1996 Declaration accounted for 'well over 80 percent' of world trade in IT products, but this was less than what they considered to be the

---

[53] See minutes of the Council for Trade in Goods meeting of 1 November 1996, G/C/M/15, para 2.1.
[54] APEC Leaders' Declaration made at Subic, the Philippines, on 25 November 1996, para 13, www.apec.org/Meeting-Papers/Leaders-Declarations/1996/1996_aelm.aspx.
[55] Seven other WTO Members, including five APEC Members, India and the Czech Republic, indicated at that time that they were considering joining the ITA; see Fliess and Sauvé (1997): 13.
[56] WT/MIN(96)/16 of 13 December 1996.
[57] Singapore Ministerial Declaration, WT/MIN(96)/DEC, adopted on 13 December 1996, para 18. The Declaration also welcomed the expansion of the Pharmaceutical Understanding, see n. 57.
[58] Trade in Pharmaceutical Products – Record of Discussion dated 25 March 1994, GATT document L/7430.

critical mass for implementation of their Agreement.[59] However, when agreement is reached without the critical mass yet being assembled, there is a risk that other Members will seek to reopen the terms of the agreement as a condition of their participation.

When a new GPA-type Agreement foresees not only market access, but new rules as well, it is bound to be treated with scepticism if little is known about it. Transparency is an important means of building trust with non-parties, whose consent will eventually be required to add any Agreement to Annex 4 of the WTO Agreement. The aborted Multilateral Agreement on Investment (MAI) illustrates the point. The MAI was negotiated among OECD Member States from 1995 to 1998, but the negotiations were discontinued[60] in the face of fierce opposition from civil society after information was leaked regarding the proposed rules under discussion. The WTO established a working group on the relationship between trade and investment in 1996[61] but there was a belief that this would start a process towards a MAI in the WTO.[62] WTO negotiations were never launched, and the working group was disbanded in 2004.[63] In contrast, a General Council decision was taken in 2004 to launch multilateral WTO negotiations on trade facilitation in the context of the Doha Round. Of course, the difference between this new initiative and the MAI can be attributed to substance as well as to procedure, as sufficient Members believed that new rules were of potential benefit to them, at least in the context of a Round.[64] The point is that multilateral consent to the addition of a SETA could also be forthcoming if the negotiating process is transparent from the outset and the potential impact, including on non-parties, is communicated effectively.

A preparatory or exploratory group of Members could be formed ad hoc or based on a pre-existing arrangement. The old Quad would not be appropriate as it no longer has the pre-eminent role in the world trading

---

[59] WT/MIN(96)/16, 1st recital and Annex, para 4.
[60] OECD, The Multilateral Agreement on Investment Draft Consolidated Text, DAFFE/MAI(98)7/REV1.
[61] Singapore Ministerial Declaration of 13 December 1996, WT/MIN(96)/DEC, para 20.
[62] 'WTO Should Not Take up Trade and Investment – Joint NGO Statement on the Investment issue in WTO', para 10.
[63] General Council Decision of 1 August 2004, WT/L/579, para 1(g), second indent. Two other WTO working groups were also established in 1996 without the same 'baggage' but were likewise disbanded in 2004 without the launch of negotiations in any forum: one group had worked on trade and competition policy and the other on transparency in government procurement.
[64] General Council Decision of 1 August 2004, WT/L/579, para 1(g), first para and Annex D. The negotiating group produced a draft text for an agreement in 2010 but its conclusion and implementation are linked to the single undertaking of the Doha Round: see Annex D, para 10.

system as it did during the Uruguay Round and shortly thereafter. A contemporary alternative might be the Group of Twenty (G20), which accounts for approximately 80 per cent of world trade and two-thirds of the world's population.[65] The G20 includes both developed and developing countries as well as transition economies from all the major regions of the world, and its members include the top ten countries in terms of sustainable energy capacity and investment.[66] There are indications that this group includes many, but not all, large traders in sustainable energy-related equipment, such as photovoltaic (PV) cells, modules and panels and wind power generating sets.[67] However, the G20 does not aim to be representative of the WTO Membership and does not include any small economies or LDCs. APEC could possibly play a role, as it did in the negotiation of the ITA. APEC's membership overlaps with that of the G20, but it also includes certain other economies that appear to be significant traders in sustainable energy-related equipment.[68]

If a preparatory process becomes identified too strongly with one particular group, it risks jeopardising wider participation and the launch of WTO negotiations, or implementation of negotiating results within the WTO framework. Moreover, if the goal of a preparatory process conducted among a limited group of Members is not clearly focused and communicated to non-participants, there is a risk that the exercise will be perceived as an attempt to introduce a senior officials' consultative body or WTO steering group, as has been proposed in literature on WTO reform:[69]

Preparatory work and negotiations outside formal WTO meetings can be conducted among a group of interested WTO Members on an ad hoc basis or in an existing forum, with meetings in other fora playing a positive role, but transparency is important 'when the participation of other members will eventually be sought, or where the wider WTO membership's consent to add an agreement to the WTO framework will be required'.

### Launch of Negotiations Within the WTO

The SETA negotiating process can respect basic principles of transparency and inclusiveness even though the negotiating objective in the case of an optional agreement is not full participation by all WTO Members.

---

[65] See the G20 website, www.g20.org/. This is not to be confused with the G20 group of developing countries in the WTO agriculture negotiations.
[66] See ICTSD (2011): 10, Tables 1 and 3, sourced from Pew Charitable Trusts, 2010.
[67] Source: COMTRADE using World Integrated Trade Solution software, in Vossenaar (2010): 43–47, Table A4.
[68] Ibid.
[69] Garcia Bercero (2001): 103–115 at 108; Jackson 2001: 67–78 at 75; Sutherland et al. (2005): 70–71.

A decision of a WTO body is required to launch negotiations in formal WTO meetings. For example, in 1994, ministerial decisions were adopted at the end of the Uruguay Round to establish a Negotiating Group on Movement of Natural Persons and a Negotiating Group on Basic Telecommunications, which led to three GATS protocols.[70] In 2004, the Trade Negotiations Committee agreed to establish the Negotiating Group on Trade Facilitation, at the direction of the General Council.[71] One exception concerns the WTO Committee of Participants on Expansion of Trade in Information Technology Products, commonly known as the ITA Committee, which was formed not to negotiate the ITA but to implement it, including the review of its product coverage (so-called 'ITA II'). The ITA Committee was formed by a decision of the participants (two of whom were not even yet WTO Members), who then informed the Council on Trade in Goods of their decision in March 1997.[72] Meetings of the participants 'under the auspices of the Council on Trade in Goods' had been foreseen in the ITA.[73] This exceptional procedure for the formation of a WTO body occurred in the early days of the WTO, and it seems unlikely that a similar procedure would be acceptable to the membership today.

Non-participants as well as other Members should carefully review any decision to launch negotiations as it may be taken into account later in the interpretation of provisions in the WTO Agreements. For example, the Appellate Body in *US – Shrimp* referred to the ministerial decision to establish the Committee on Trade and Environment in its interpretation of the GATT 1994, chapeau to Article XX.[74]

Any decision to establish a WTO negotiating group would be taken by consensus among all WTO Members present at the meeting where such a decision was considered, in accordance with customary practice.[75] If the decision is taken by a subsidiary body of the General Council, consensus is the only option.[76] Although the General Council is empowered to take decisions by voting, it would be futile to launch negotiations over the

---

[70] Decision on Negotiations on Movement of Natural Persons, para 2; Decision on Negotiations on Basic Telecommunications, para 3, both annexed to the Final Act of the Uruguay Round.
[71] See minutes of Trade Negotiations Committee meeting of 12 October 2004, TN/C/M/14, paras 1–4.
[72] Implementation of the Ministerial Declaration on Trade in Information Technology Products, communication dated 26 March 1997, G/L/160 (para 3) considered by the Council for Trade in Goods at its meeting on 14 April 1997, G/C/M/19, paras 2.1–2.8. Estonia and Chinese Taipei were still in the process of acceding to the WTO.
[73] WT/MIN(96)/16, Annex, para 4.
[74] Appellate Body report in *US – Shrimp*, WT/DS58/AB/R, para 154.
[75] WTO Agreement, Article IX:1, first sentence, and subsequent practice.
[76] See, for example, the Rules of Procedure for meetings of the Council for Trade in Goods, Rule 33, WT/L/79.

objections of a substantial section of the membership if the proponents' goal is to add an eventual agreement to Annex 4 of the WTO Agreement, as that would require a decision taken exclusively by consensus. Naturally, consensus on launching negotiations does not imply that all Members should participate in the negotiations, nor implement the results.

If consensus can be found to launch negotiations, the WTO could establish a negotiating group on sustainable energy (but with a title that distinguishes it from bodies formed by the existing Trade Negotiations Committee). As a WTO negotiating group, it would report to a WTO Council or Committee and receive Secretariat support which can, among other things, help ensure the openness and transparency of the process, as well as generate a drafting history that could be taken into account in the interpretation of the results. The Secretariat can also provide technical support, such as tariff data and statistical information, although it would probably provide this information to Members in any case. Another advantage of conducting negotiations in a WTO negotiating group, in theory, can be the opportunity for trade-offs with other sectors and issues during a round of multilateral trade negotiations, but that advantage is illusory at present, given the current state of the Doha Round. If consensus cannot be found to establish a Negotiating Group on Sustainable Energy, negotiations can proceed in informal meetings of interested WTO Members with a view to requesting the incorporation of the eventual results into the WTO framework:

- A consensus decision of a WTO body would be required to launch negotiations in formal WTO meetings.

*Participation in the Negotiations*

There is no legal requirement regarding the minimum number of Members that must participate in a WTO negotiating process. The successful negotiations on basic telecommunications were launched among thirty-one governments, including the then-twelve EC Member States, which had already announced their intention to participate, with their names set out in a list and no express reference to a collective trade share.[77] The participants in the ITA at the time of the 1996 Ministerial Declaration noted that they accounted for 83 per cent of world trade in

---

[77] Ministerial Decision on Negotiations on Basic Telecommunications, para 4. The 1994 Ministerial Decision on Negotiations on Maritime Transport Services adopted the same approach, with thirty-three governments, including the then-twelve EC member states: Decision on Negotiations on Maritime Transport Services, para 3.

IT products, but by that stage, they had already agreed on product coverage and modalities for tariff elimination.[78]

The rules for negotiations should be clear in advance and appropriate to the WTO as an institution.[79] Negotiations on an ITA-type Agreement can be conducted among those who intend to participate in the results. However, negotiations on an optional Agreement to be added to Annex 4 of the WTO Agreement would be a novelty because the participants would negotiate among themselves in a group in the shadow of the wider membership. Therefore, this process would have to operate on two planes. Exclusion of non-participants from the negotiating process would only increase the likelihood that consent would not be given to add any such Agreement among the participants to the WTO Agreement. The process needs the input of non-participants, including those who have no intention of acceding.

A WTO negotiating group would ordinarily be open to all WTO Members that wished to participate, on a voluntary basis, according to an opt-in procedure.[80] For example, the 1994 Ministerial Decision on Negotiations on Basic Telecommunications and the 1994 Ministerial Decision on Negotiations on Maritime Transport Services provided that negotiations would be 'entered into on a voluntary basis' and that the negotiating groups would be open to all governments that announced their intention to participate. A procedure was established in the decisions that required notifications of intention to be addressed to the WTO Treaty depositary (i.e. the WTO Director General).[81]

All Members should be clear as to what participation in the negotiating group entails. The voluntary nature of negotiations should imply that those who have chosen to participate have a right to opt out at any time. However, as a practical matter, there is a risk that some participants' input will not be properly taken into account if their eventual participation is not assured or if participation in the negotiations will give rise to expectations of participation in the results in any case. Moreover, some Members will participate in the group only with the intention of following proceedings or sharing their own perspectives on the issues under negotiation. This role is important to ensure that negotiators are aware of the effect that their proposals may have on the wider membership, especially if the SETA includes new rules in areas already covered by existing WTO

---

[78] ITA, Preamble, 1st recital. See Hoda (2011).   [79] Sutherland et al. (2005), para 298.
[80] The only exceptions in the GATT or the WTO were the negotiations that led to the Agreement on Agriculture and the Agreement on Textiles and Clothing, which were both sectoral agreements negotiated in the multilateral framework of the Uruguay Round.
[81] Decision on Negotiations on Basic Telecommunications, paras 1 and 4; Decision on Negotiations on Maritime Transport Services, paras 1 and 3.

Agreements and not only improved market access. Therefore, it might be better to agree at the outset of negotiations in the WTO on what participation (or opting-in) means, and set up an appropriate structure for the negotiating process.

A practical solution is for a negotiating group's rules of procedures to distinguish between participants and observers. Participants in the negotiations could be expected to participate in the results. Observers would be free not to participate in the results, and they would not participate in decision making in the negotiating group.[82] Conditions of participation and observership, if any, could be laid down in the decision establishing the negotiating group or, in the case of observership, decided by the group. For example, only parties to the ITA are 'participants' in the ITA Committee.[83] The Committee adopted a decision allowing all WTO Members, and governments with observer status in the Council for Trade in Goods, to follow its proceedings in an observer capacity.[84] Similarly, only parties to the GPA are members of the Committee on Government Procurement (GPA Committee),[85] but any WTO Member and actual or potential acceding Members may apply to observe its meetings if they fulfil certain transparency conditions in their procurement tenders[86] and are 'interested in initiating accession negotiations' to the GPA.[87] In fact, over half the observers to the GPA Committee are not currently negotiating accession to the GPA.[88] The revised GPA will entitle non-parties to become observers by submitting a written notice to the GPA Committee.[89] During negotiations on the review of the GPA, it was agreed that observers were entitled to attend negotiations on the text in Committee meetings, but that they could attend bilateral negotiations on issues of coverage and elimination of discriminatory measures

---

[82] Harbinson (2009): 15.
[83] WT/MIN(96)/16, para 4, but see Rules of Procedure for meetings of the Committee of participants on the expansion of trade in information technology products, approved by the Committee on 30 October 1997, G/IT/3, para (i).
[84] Decision on Participation of Observers in the Committee of participants on the expansion of trade in information technology products, attached to ITA Committee Rules of Procedure, para 1.
[85] 1994 Agreement on Government Procurement, Article XXI:1.
[86] 1994 Agreement on Government Procurement, Article XVII:2.
[87] Decisions on Procedural Matters under the Agreement on Government Procurement (1994) of 27 February 1996, GPA/1, Annex 1: Decision on Participation of Observers in the Committee on Government Procurement.
[88] An updated list of observers in the WTO Committee on Government Procurement is available at www.wto.org/english/tratop_e/gproc_e/memobs_e.htm.
[89] Protocol Amending the Agreement on Government Procurement, Annex, Article XXI:4, appended to the Decision of 30 March 2012 of the Parties to the WTO Agreement on Government Procurement on the Outcomes of the Negotiations under Article XXIV:7, GPA/113.

and practices only if they had submitted an offer with a view to participating in the revised Agreement.[90]

Informal methods are also important parts of the negotiating process.[91] Consultations within smaller groups of participants can be used to advance negotiations, but there may also be value in including observers even in these groups and consulting other members representative of the wider membership. The goal is to avoid a situation where an Agreement is concluded only to require renegotiation because consent to annex it is denied. The identity of the Chair can also be important. Chairs are usually drawn from the ranks of the participants' representatives, but the novel nature of negotiations on a SETA might justify casting the net wider.[92] For example, the first Chair of the ITA Committee was an acknowledged expert on tariff negotiations and, at the time, a WTO Deputy Director General.[93]

Decisions to launch negotiations on a SETA can set out a date for the first meeting of the negotiating group to give impetus to the negotiations. They can also include requirements to report to a regular WTO body and require the negotiating group to issue a final report with a date for implementation. Several WTO bodies could have relevant responsibilities, given that a SETA could cover market access in goods, both agricultural and non-agricultural, as well as services, and possibly rules in different areas of trade in goods, such as technical barriers to trade, subsidies and procurement. Multiple reporting requirements are not helpful, but non-participants may wish to ask questions in different WTO committees. A practical solution would be for the SETA negotiating group to report directly to the General Council at least annually, with a request that the reports be included in the agenda of subsidiary bodies, or circulated in them, or both.

There is the possibility of assigning different aspects of a SETA to separate negotiating groups to take account of their specialised roles, but this is not necessarily the best way forward. While it is likely that different government agencies would have responsibility for different aspects of a SETA in many countries, separate processes will make it more difficult to achieve a coherent outcome on trade in sustainable energy goods and services (SEGS). Negotiations on issues relevant to market access for such goods and services have been scattered across various bodies in the Doha Round, including the special session of the Committee on Trade and Environment, the Negotiating Group on

---

[90] Decision of the Committee on Government Procurement of 16 July 2004, 'Modalities for the Negotiations on Extension of Coverage and Elimination of Discriminatory Measures and Practices', GPA/79.
[91] Faizel and Vickers (2011): 461–485 at 477.   [92] See further Odell (2005): 425–428.
[93] Hoda (2011).

Non-Agricultural Market Access and the special sessions of the Committee on Agriculture and the Council for Trade in Services, without results.

An alternative is to adopt a centralised approach. For example, negotiations relating to regional trade agreements (RTAs) take place in a single group, which in 2006 produced an 'early harvest' result in the Doha Round. That decision is now applied on a provisional basis.[94]

A practical solution for SETA would be to establish a single negotiating group that can organise its business as it sees fit. For example, it could take up rules in the group as a whole, with focused sessions on particular issues, and delegate market access to bilateral meetings, similar to the processes of negotiations on WTO accessions and the revision of the GPA. A group could also delegate certain issues to 'friends of the Chair' or 'facilitators' from within the negotiating group, who can make progress through focused bilateral, small group and open-ended meetings, similar to the process in the negotiations on trade facilitation.

Negotiations on a SETA would be separate from the Doha Round as an institutional matter, but each process could, nevertheless, influence the other, because the same parties may negotiate market access for the same products in two or more negotiating groups simultaneously. Proposals in one process might be taken into account inappropriately in another. Participants may be reluctant to make different, calibrated proposals on matters such as product coverage and modalities of treatment, adopting instead a 'wait and see' approach based on progress in one particular group. Therefore, it could be useful for participants to confirm that each negotiation is conducted in a separate context and forms part of a different balance of concessions.

### Implementation by a Critical Mass

While the size of a group of interested WTO Members required to implement a SETA has legal implications, it still depends only on participants' respective assessments as to what is the critical mass necessary in economic and political terms to ensure the viability of any eventual agreement, in light of those obligations that are to be applied on an MFN basis and those that are to be applied on a reciprocal basis among the parties only. The definition of the critical mass in effect determines the point beyond which the participants will tolerate free riding. The critical

---

[94] Transparency Mechanism for Regional Trade Agreements, General Council Decision of 14 December 2006, WT/L/671, which was negotiated in the Negotiating Group on Rules.

mass for implementation of a SETA could be assessed in light of the characteristics of the market for equipment and services for the generation of sustainable energy or those countries most responsible for $CO_2$ emissions, or both, depending on what the agreement contained.

When the participants have assembled a critical mass during the negotiations, they may simply agree that entry into force of the results is subject to all the participants accepting them by a target date. The GATS Protocols adopted this approach. If that condition was not fulfilled, a residual clause allowed those participants that had accepted on time to make a decision on entry into force.[95]

Alternatively, the participants in the negotiations may define the critical mass for implementation in terms of a percentage share of world trade in the relevant products. The ITA adopted this approach. The 1996 Declaration defined the critical mass for implementation as approximately 90 per cent of trade in IT products[96] but it was only in 1997, after several additional participants had joined, that this criterion was met.[97] The definition of the critical mass in terms of a collective share of world trade was inspired by the GATT, which was not limited to products in a particular sector. The GATT 1947, Article XXVI:6, as amended at the 1954–1955 Review Session, provided that the GATT would enter into force definitively after acceptance by parties accounting for 85 per cent of the total external trade of the thirty-four contracting parties. The trade share of each contracting party, based on the most recent four-year average, was stipulated in advance in two alternative scenarios.[98] The provision was never implemented and application of the GATT 1947 remained provisional until it was terminated.

The method of calculation of the critical mass can raise technical problems where the product coverage excludes certain products, such as parts, defined more narrowly than the Harmonized System (HS) six-digit level ('ex-outs'), or products that must be covered wherever they are classified in the HS. In these situations, a calculation that covers all trade in the relevant HS sub-headings (ignoring ex-outs) will over-estimate the actual trade involved. A calculation that only takes into account the fully covered HS sub-headings (ignoring sub-headings with ex-outs) will under-estimate the actual trade involved. The note by the WTO Secretariat on the calculation of the share of world trade in IT products under the ITA does

---

[95] See, for example, Protocol (2001), see n. 9, paras 2 and 3.   [96] ITA, Annex, para 4.
[97] Implementation of the Ministerial Declaration on Trade in Information Technology Products – Informal Meeting of 26 March 1997, Note by the Secretariat, Revision, G/L/159/Rev.1.
[98] GATT 1947, Annex H. The two scenarios depended on whether the threshold for entry into force was met before or after the accession of Japan.

Legal Options for a SETA 411

not specify how the calculation was carried out.[99] The only way to ensure an accurate calculation in a future negotiation is to ensure that product coverage is defined in terms of full HS sub-headings.

A definition of the critical mass in terms of a collective share of world trade may not be sufficient to limit free riding in sectors where there are major producers with as yet unrealised export potential. In these circumstances, a critical mass could be defined in terms of a collective share of world production as well, or instead.

An alternative basis for a definition of the critical mass can be found in the Kyoto Protocol to the UN Framework Convention on Climate Change (UNFCCC), concluded in 1997. That Protocol entered into force in accordance with a formula that comprised a minimum number of parties (55) plus a certain proportion of listed countries, expressed in terms of a collective share (55 per cent) of those countries' GHG emissions.[100] The relevance of this additional condition lay in the fact that the listed countries were the parties that assumed commitments to reduce GHG emissions under the Protocol. The Protocol entered into force just over seven years later in 2005.

A more flexible approach is to postpone a decision on implementation. This was the approach adopted for the entry into force of the WTO Agreement. At the conclusion of the Uruguay Round, no decision was made on the number of parties that would be required for entry into force of the WTO Agreement. Rather, a target date was set of 1 January 1995, and a decision on the timing of the entry into force of the results was postponed under the Implementation Conference in December 1994,[101] where the decision was taken based on the understanding that Members were committed to bringing the WTO into force on the target date and would be making every effort to conclude their domestic ratification processes to that end.[102] This approach was also used for the implementation of the results of the Kennedy Round.[103] However, such flexibility may be more appropriate for the results of a trade round, which spans disparate industrial sectors, because it can be more difficult to define the

---

[99] See n. 96.
[100] Protocol to the United Nations Framework Convention on Climate Change done at Kyoto on 11 December 1997, Article 25. The formula was not less than 55 Parties to the UNFCCC, incorporating Parties included in Annex I to the UNFCCC which accounted in total for at least 55 per cent of the total $CO_2$ emissions for 1990 of the Parties included in Annex I.
[101] WTO Agreement, Article XIV:1.
[102] Minutes of the meeting of the Preparatory Committee for the WTO held on 8 December 1994, PC/M/10, paras 4–5; Notification of entry into force and notification of acceptances, WT/Let/1, 27 January 1995.
[103] Geneva (1967) Protocol to the GATT, 30 June 1967, 620 UNTS 294, para 6.

participation necessary for the effective operation of the agreements reached and the adequacy of the benefits. A flexible approach is inherently less predictable and might not be as well suited to the case of a sectoral agreement like a SETA.

The critical mass may be quite a small minority of WTO Members in some sectors. For example, the TCA has just thirty-one parties, twenty of which are EU members.[104] The ITA had twenty-five participants, including the EU-15, at the time of its initial implementation,[105] and it now has forty-six participants, including the EU-27, representing approximately 97 per cent of world trade in IT products.[106] On the other hand, the minimum price provisions of the International Dairy Agreement (IDA) were suspended in 1995 because the limited membership of nine, including the EC, and in particular the non-participation of some major dairy-exporting countries, made the operation untenable.[107]

The critical mass for either the launch of negotiations or the implementation of a new Agreement may require a degree of diversity in participation. A critical mass expressed in strictly numerical terms will ordinarily involve some diversity if the number is large enough. For example, the ITA included several developing country participants at the time of implementation (India, Indonesia, Korea, Malaysia, Singapore, Thailand and Turkey). The definition of the critical mass could also include a target group of countries, but such a list would rankle many if it gave the impression that the new agreement was creating two classes in the WTO membership.[108] Participants will make their own individual assessments as to which other countries' participation is required in any case.[109] Finally, any critical mass for implementation should not include a contracting party who only accepts the agreement conditionally, lest it not fulfil the condition and free ride after entry into force.[110]

---

[104] The parties to the TCA comprise the old Quad; twenty EU Member States in their own right plus Albania, Egypt, Georgia, Macau, China, Norway, Switzerland and Chinese Taipei.

[105] See Implementation document, G/L/160, Preamble.

[106] Number of participants as of 12 May 2011, see Status of Implementation, Note by the Secretariat, G/IT/1/Rev.45, para 1.

[107] Twelve parties signed the IDA, but not all ratified.

[108] Sutherland et al. (2005), para 298.

[109] For example, in the United States, 19 USC 3511(b) authorised the President to accept the Uruguay Round agreements when he determined that a sufficient number of foreign countries were accepting the obligations of those agreements to ensure the effective operation of, and adequate benefits for the United States under, those agreements.

[110] Brazil accepted the Second, Fourth and Fifth GATS Protocols and the IDA 'subject to ratification', but did not later ratify them: see WTO Status of Legal Instruments 2008: 107, 114, 119 and 134 and 2009 Trade Policy Review of Brazil, Report by the Secretariat, WT/TPR/S/212/Rev.1, para 5.

Legal Options for a SETA 413

*Accession to a SETA After Entry into Force*

Accession to a SETA should remain open to all WTO Members after it has entered into force within the WTO framework, as this is consistent with the multilateral nature of the institution. An accession clause should be stated expressly in the Agreement.[111]

Accession could be open on the same terms accepted by original parties, or on terms to be agreed through negotiations. Where the same terms apply to original and acceding parties, this may promote wider acceptance of the SETA among the WTO membership in the long run. However, this reduces the incentive to participate early, which is essential to achieve the critical mass for implementation. Where terms of accession are agreed through negotiations, an acceding party will have little incentive to agree if it already obtains the benefits of the Agreement on an MFN basis.

Both approaches can be combined by initially leaving open the possibility that accession will be on terms to be agreed but, after successful implementation of the agreement, allowing all members to join on the terms accepted by the original parties. In the case of the ITA, the original participants decided in 1997 that accession would be on terms to be agreed,[112] but in 1998 they reached an understanding that new participants would join on the same terms and conditions as the original participants. They only negotiate if they want to deviate from the terms of the ITA.[113]

A choice is not possible in some cases. Accession negotiations are inevitable as regards market access agreements, such as the GATT 1994, the GATS and the GPA, unless the product coverage and modalities are defined in advance in objective terms.

A decision of the Ministerial Conference – taken exclusively by consensus – is required to add any new Agreement among a sub-set of WTO Members to Annex 4 of the WTO Agreement, however many or few they may be. One of the concerns regarding such Agreements is that non-parties are pressured to join. No current WTO Member is susceptible to pressure through WTO accession negotiations, but some may still be concerned that preferential trade arrangement (PTA) negotiations outside the WTO or even future multilateral negotiations inside the WTO could link market access and other benefits to accession to an Annex 4 Agreement. Therefore, WTO members who do not wish to participate in

---

[111] The accession clause should not include a limited period for acceptance. The WTO Agreement, Articles X:5 and X:7, refers to an acceptance period specified by the Ministerial Conference for amendments to the multilateral agreements, not an agreement added to Annex 4 under Article X:9.
[112] Implementation document, see n. 72, para 5.
[113] See the minutes of the ITA Committee meeting of 30 October 1997, G/IT/M/2, section 4.

a SETA may be more willing to consent to its addition to the WTO framework if they are given some assurances that accession is voluntary and shall not be made a condition for the launch or conclusion of negotiations on other subjects. Such an assurance could be included in the WTO decision to add the Agreement to Annex 4.

### *Modification of Existing WTO Disciplines*

The implementation of a SETA as an ITA-type agreement through participants' WTO schedules would not require any modification of the existing WTO institutional framework, save the optional addition of a committee to monitor implementation. The SETA commitments would become part of the participating Members' goods and/or services schedules and integral parts of Part I of the GATT 1994 and of the GATS, respectively. Other provisions of those Agreements, and other WTO Agreements, would apply to them in the same way that they apply to existing concessions and commitments in Members' schedules. These would include the Dispute Settlement Understanding (DSU) and the general exceptions and procedures for modifications of schedules. However, the scope for implementation by means of schedules is limited (as discussed above).

Conversely, the implementation of a SETA as a GPA-type Agreement added to Annex 4 of the WTO Agreement would not be subject to the same limits on scope, but clarification would be required as to the way in which the disciplines in the new Agreement would relate to the WTO dispute settlement procedures and substantive WTO rules. These issues are considered in turn below.

#### *Dispute Settlement Procedures*

*Availability of DSU Procedures*  The dispute settlement rules and procedures of the DSU can apply to disputes among parties to optional WTO agreements. For example, the dispute in *Korea – Procurement* arose under the GPA and was conducted under the DSU procedures, as adapted by the GPA, from 1999 to 2000.[114] The disputes in *EC – IT Products* concerned concessions made by the European Communities pursuant to the ITA and were conducted under the standard DSU procedures from 2008 to 2011.[115]

---

[114] WT/DS163, complaint by the United States. The panel report was adopted without appeal and no further action was required.

[115] WT/DS375, WT/DS376 and WT/DS377, complaints by the United States, Japan and Chinese Taipei, respectively. Implementation of the DSB recommendation was notified by the respondent in May 2011.

The availability of the DSU is likely to be a large part of the attraction of negotiating a SETA within the WTO framework for sponsors of the initiative because that would provide a means of ensuring that it is implemented faithfully.[116] The WTO's dispute settlement function is more efficient than its rule making function. The DSU provides an avenue for an independent review of implementation of treaty obligations in a system of compulsory jurisdiction, backed by the possibility of trade sanctions. The potential for dispute settlement has an *ex ante* effect on Members' efforts to implement obligations and, when disputes arise, Dispute Settlement Body (DSB) recommendations are implemented in most cases, usually without recourse to sanctions.

During the negotiation of a new Agreement (other than an ITA-type Agreement), some participants may be reluctant to accept the applicability of the DSU. For example, the October 2011 draft consolidated negotiating text on the Trade Facilitation Agreement (TFA) contains bracketed provisions on the applicability of the DSU in which various issues are as yet unresolved.[117] However, it should be borne in mind that the trade facilitation negotiations are multilateral and form part of the single undertaking of the Doha Round. Participants in a negotiation on an optional Agreement, such as a SETA, may be more willing to agree to the availability of dispute settlement, as they have the option not to accept any obligations under the Agreement at all.

*Suitability of DSU Procedures* The suitability of the DSU procedures for disputes under a SETA could change over time. During an initial phase at least, the SETA would closely relate to existing WTO Agreements. This would necessarily be the case where it was implemented partly or wholly through modifications to participants' goods and services schedules. However, a SETA may develop in later phases of negotiations to cover other issues, such as the wider scope of energy efficiency products and standards.[118] The potential inclusion of these issues – which can be made subject to the consent of the wider WTO membership – also affects the assessment of the suitability of DSU procedures.

The appropriateness of introducing standards into the GATT system on 'new' issues also arose in relation to the inclusion of intellectual

---

[116] See Bacchus (2011): 4.
[117] Negotiating Group on Trade Facilitation, Draft Consolidated Negotiating Text, 7 October 2011, TN/TF/W/165/Rev.11. The final content contains an array of provisions on due dates for implementation and grace periods for dispute settlement (WT/L/940).
[118] See n. 23: 63.

property (IP) in the Uruguay Round in 1986.[119] The objectives of the negotiations on that subject included the reduction of distortions and impediments to international trade and ensuring that measures and procedures to enforce intellectual property rights (IPRs) did not themselves become barriers to legitimate trade.[120] That did not prevent the elaboration of a set of rules and disciplines on intellectual property in the TRIPS Agreement that is more comprehensive than anything that had been agreed in the pre-existing IP conventions. During the negotiations, one participant proposed dividing responsibility between two panels, mandating a World Intellectual Property Organization (WIPO) panel first to assess compliance with IP standards and a GATT panel to assess any 'trade-related effects'.[121] This proposal was not accepted, and the DSU applies to disputes under the TRIPS Agreement.[122] Admittedly, IP is a branch of law, not of science, unlike many climate change issues.

The WTO DSM already applies to disputes that concern standards on complex non-trade issues arising under its trade Agreements. The TBT Agreement applies to technical standards and refers to scientific and technical information and fundamental climatic or geographical factors or fundamental technological problems in certain obligations.[123] The Sanitary and Phytosanitary (SPS) Agreement also refers to scientific principles and evidence and sanitary standards.[124] The ordinary DSU rules and procedures apply to disputes under all of these agreements, together with special procedures on the formation of technical expert groups.[125]

The DSU is partially equipped to manage detailed technical issues. Delegations of parties to panel proceedings may include persons other than trade officials.[126] Panels can and do on occasion include a person nominated for their specific expertise in a technical area, although rarely more than one. The Appellate Body is less likely to have expertise in a specialised area because its composition is limited to seven individuals.[127] The WTO Secretariat may lack experience in issues covered by an amended SETA, but that could be addressed through recruitment if such an Agreement were

---

[119] Ministerial Declaration on the Uruguay Round, GATT document MIN.DEC, adopted 20 September 1986, BISD 33S/19, Section D, under the sub-heading 'Trade-related aspects of intellectual property rights, including trade in counterfeit goods'.
[120] TRIPS Agreement, Preamble, 1st recital.
[121] Communication from Chile, 22 January 1990, MTN.GNG/NG11/W/61, Section B.
[122] TRIPS Agreement, Article 64.   [123] TBT Agreement, Articles 2.2 and 2.4.
[124] SPS Agreement, Articles 2.2 and 3.3.
[125] SPS Agreement, Article 11.2, and TBT Agreement, Articles 14.2 and 14.3 and Annex 2, as listed in Appendix 2 to the DSU.
[126] The respondent's Minister of Environment, Ms Marina Silva, made the initial statement to the panel in *Brazil – Retreaded Tyres* (WT/DS332).
[127] DSU, Article 17.1.

concluded in the WTO. Expeditious arbitration, by mutual agreement of the parties, is also available as an alternative means of dispute settlement.[128]

Ordinary and special DSU procedures allow the possibility for panels to seek technical advice and establish expert review groups.[129] Panels are not obliged to consult experts if the evidence is adequate. The panel in *US – Tuna II (Mexico)* made its own assessment of detailed scientific evidence regarding the adverse effects on dolphins of certain fishing practices and risks to dolphins in different locations in a dispute under the TBT Agreement, without consulting further experts.[130] On the other hand, the panel in *US – Clove Cigarettes* was unable to compare a series of surveys addressing smoking patterns and failed to rely on them in its assessment of consumers' tastes and habits in another dispute under the TBT Agreement.[131] Expert consultation typically occurs in disputes under the SPS Agreement regarding scientific issues and health risks.[132] Since 2005, disputes involving expert review have become very lengthy. Part of the problem concerns the expert review procedures themselves, notably in relation to the identification and selection of the experts and the preparation of the questions to pose to them. In *US/Canada – Continued Suspension* (a sequel to the dispute concerning beef produced with growth hormones), the Appellate Body found that two experts should not have been selected, so the procedure proved fruitless.[133] Expert evidence is also complex and ultimately it is the panel's task to make an objective assessment of the matter before it.

The WTO dispute settlement system has no experience in resolving disputes that fundamentally concern a non-trade objective, such as climate change. Until now, non-trade issues have tended to define the limits of the trade issues, rather than the objective of the obligations at issue. The usual bodies in the WTO dispute settlement system are not equipped to make judgements on sustainability issues, such as net carbon savings, as opposed to assessments of trade policy instruments applied to achieve environmental goals. If the subject matter of a SETA were extended at

---

[128] DSU, Article 25.
[129] DSU, Article 13.2; for provisions see n. 125 above. See also the Customs Valuation Agreement, Article 19.4, as regards consultations with a Technical Committee, all listed in DSU, Appendix 4.
[130] Panel report in *US – Tuna II (Mexico)*, WT/DS381/R, paras 7.491–7.506, 7.517–7.531, 7.546–7.564. This report is currently on appeal.
[131] *US – Clove Cigarettes*, Panel report, WT/DS406/R, paras 7.209–7.210. This approach was rejected on appeal, see WT/DS406/AB/R, paras 150–151.
[132] Consider, for example, *EC – Biotech ('GMOs')*, WT/DS291 and *Australia – Apples*, WT/DS367 but also *EC – Asbestos*, WT/DS135 that was not a SPS dispute but did consider health risks.
[133] Appellate Body report in *US – Continued Suspension*, WT/DS320/AB/R, paras 433–484, 736(b).

some later stage there might be concerns that a corresponding extension of the subject matter of jurisdiction under the DSU could lead to misguided decisions on important environmental issues and also destabilise the WTO dispute settlement system.

One possible means to address such concerns would be the establishment of a separate dispute settlement system under the SETA with the specialised expertise required to make an assessment of parties' implementation. This would be a less than satisfactory outcome for the parties, as the subject matter of a SETA would overlap significantly with other WTO Agreements, fragmenting the dispute settlement system as occurred in the GATT, where a single matter could involve claims under both a plurilateral agreement and the GATT 1947 and require a complaining party to choose which dispute settlement provisions to invoke.[134] Such a situation persists under the TCA, but its procedures are a legacy of the GATT and not ideal.[135] Today, parties and non-parties alike might consider that implementation of a SETA with both a separate membership *and* a separate dispute settlement system would be incompatible with the integrated nature of the WTO framework.

An alternative solution is to divide responsibility between the usual bodies of the WTO dispute settlement system, i.e. the DSB, panels, the Appellate Body and arbitrators on the one hand, and groups of experts in sustainable energy on the other. That is basically the solution that the DSU already offers. However, there is a need for clarity on issues such as expert selection procedures and the respective roles of the panel and the experts in a given case to ensure that expert review is efficient.[136] Recourse to an expert selection procedure could be either mandatory or at the discretion of the panel. These issues could be addressed in the additional dispute settlement rules applicable under a SETA (discussed further below).

*Procedures to Apply the DSU* If DSU procedures are to apply to disputes under a SETA as a GPA-type Agreement, three procedures must be followed. These are (1) the participants negotiate a provision in the text of the new Agreement that provides for the application of the DSU, subject to any special or additional rules and procedures; (2) the

---

[134] See the discussion in the Appellate Body report in *Brazil – Coconut*, WT/DS22/AB/R, Section IV:B.

[135] The TCA was concluded in 1979 and still refers to the old GATT dispute settlement procedures: see Agreement on Trade in Civil Aircraft, Article 8.8, although no WTO disputes have actually cited the TCA. The GATT dispute settlement procedures are also applicable in some respects to situation complaints under Article 26.2 of the DSU and, optionally, to complaints brought by a developing country Member against a developed country Member under the DSU, Article 3.12.

[136] See Kennedy (2011a): 221–253 at 248–250.

# Legal Options for a SETA

Ministerial Conference adopts a decision amending the list of 'covered agreements' in Appendix 1 to the DSU;[137] and (3) the parties to a SETA (in practice, the SETA Committee at one of its first meetings[138]) adopt a decision setting out the dispute settlement provisions of the new Agreement and notify that decision to the DSB, in accordance with the last paragraph of Appendix 1 to the DSU.[139]

The applicable procedure for a DSU amendment is essentially the same as the one for adding an Agreement to Annex 4 of the WTO Agreement, that is to say it requires a consensus decision of the Ministerial Conference. The willingness of non-parties to join a consensus may depend on the way in which DSU procedures would apply to disputes under a SETA and how it would affect the interpretation of other WTO Agreements (discussed below). If consensus can be reached, the decision to amend the list of covered Agreements in Appendix 1 to the DSU could be included in the decision to add the SETA to the WTO Agreement.[140]

The amendment to the list of 'covered agreements' does not dispense with the requirement for a decision of the parties to be notified to the DSB, due to the wording of the last sentence of Appendix 1. The requirement for the parties to adopt a decision is fairly redundant but it could serve a purpose in the case of the TCA.[141] The decision of the parties to a new Agreement does not dispense with the requirement to amend Appendix 1 to the DSU either – this point did not arise in the case of the original Annex 4 agreements because they were already listed.

*Administration of Standard DSU Rules and Procedures* The text of the SETA may state that DSU rules and procedures apply to disputes under the Agreement. However, the DSU already provides that it is not administered in the usual way as regards optional agreements in Annex 4 to the WTO Agreement, in two respects.

First, the DSU, Article 2.1, provides that the 'parties' to a SETA would be substituted for 'Members' in the DSU procedures.[142] This

---

[137] DSU, Article 1.1; WTO Agreement, Article X:8.
[138] The Committee on Government Procurement only notified the DSB of its special or additional dispute settlement procedures, see letter from Chairman of the Committee on Government Procurement to the Chairman of the DSB dated 6 July 1996 (GPA/5), approved by the Committee at its meeting of 4 June 1996 (GPA/M/2, Item I).
[139] DSU, Appendix 1, last sentence.
[140] WTO Agreement, Articles X:8 and X:9 and Article IV:2.
[141] The TCA contains a dispute settlement provision from the GATT days that the parties considered updating to refer to the DSU. A draft Protocol concerning technical rectifications needed to bring the Aircraft Agreement into the WTO framework was sent to TCA parties in 1999 but not adopted: see the minutes of the Aircraft Committee meeting in July 1999 (TCA/M/8, paras 3–9) and meetings thereafter until 2007.
[142] DSU, Article 2.1, 3rd sentence.

would mean, for example, that the interests of Members that were not parties to the SETA would not have to be taken into account during the panel process, and that such Members would have no third-party rights to be heard or to receive any submissions during a panel proceeding.[143] This rule has been applied in two disputes under the GPA, which illustrate its negative implications for non-parties to that agreement. In 1997, the complaints in *US – Massachusetts* concerned a secondary boycott with a human rights objective. Two Members with trade or systemic interests but not parties to the GPA were not allowed to participate.[144] (The dispute was later settled.)[145] In 2000, the panel report in *Korea – Procurement* included an examination of the non-violation remedy, which raised a systemic issue of interpretation under the DSU.[146]

Second, the DSU, Article 2.1 also provides that only those members that are parties to the optional Agreement may participate in decisions or actions taken by the DSB with respect to those disputes.[147] This may make little difference in practice, as a consensus decision is not required to adopt key DSB decisions but only to block them. Non-parties to the GPA were allowed to speak at the time of establishment of the panel in *US – Massachusetts* and at the adoption of the panel report in *Korea – Procurement*, although one party to the GPA expressly reserved its position in this regard on the latter occasion.[148]

There is no formal requirement that panellists in disputes under an optional agreement be citizens of parties to that Agreement, although the parties may make such a request during the process of panel composition.[149]

SETA participants and other Members may wish to consider whether to allow non-parties to exercise third-party rights in disputes under a SETA, and to confirm that non-parties may express their views in the DSB on reports and actions taken by the DSB with respect to those disputes. This could be achieved through an amendment of the rule in the DSU, Article 2.1, or the inclusion of a special rule in a SETA that would take precedence.

---

[143] DSU, Article 10, read in light of Article 2.1, 3rd sentence.
[144] See minutes of the DSB meeting of 21 October 1998, WT/DSB/M/49, Item 2.
[145] See *National Foreign Trade Council* v. *Baker*, 26 F. Supp. 2d 287 (D. Mass., 1998); and lapse of authority for establishment of the panel, WTO documents WT/DS88/6 and WT/DS95/6.
[146] See minutes of DSB meeting of 19 June 2000, WT/DSB/M/84, paras 64–72.
[147] DSU, Article 2.1, 4th sentence.   [148] See n. 146.
[149] All panellists in *US – Massachusetts* (WT/DS88 and WT/DS95) and *Korea – Procurement* (WT/DS163) were from parties to the GPA.

Legal Options for a SETA 421

*Special and Additional Dispute Settlement Rules and Procedures* The SETA text may state that DSU rules and procedures apply subject to special or additional dispute settlement rules and procedures. For example, special rules and procedures can provide for different timeframes for proceedings,[150] special expert review procedures,[151] specific remedies,[152] and non-violation and situation complaints.[153] They can exclude cross-retaliation.[154] They could also vary the rules in the DSU, Article 2.1, that exclude non-parties from dispute settlement proceedings under optional Annex 4 agreements (discussed above).

There is no need to amend Appendix 2 to the DSU because it only lists the special rules and procedures in the multilateral agreements, not those in optional Annex 4 agreements. The wider WTO membership would have the opportunity to review and approve special or additional procedures in the text of the SETA before it was added to the WTO Agreement.

A novel situation could occur in which a dispute arises under both the SETA and multilateral WTO Agreements, each with different rules. There is an existing practice with respect to combinations of standard and special rules: shorter timeframes are ignored when more than one is applicable but special remedies remain available with respect to the relevant claims.[155] However, there is no practice regarding the different status of parties to a SETA and other WTO Members in such a dispute, and it would be useful to clarify the position. Members could agree to exclude the rules in the DSU, Article 2.1, regarding plurilateral trade agreements, so that all Members may participate in the usual way whenever a dispute relates to a multilateral WTO Agreement taken in combination with a plurilateral Agreement.[156] The TCA contains a similar provision as regards the different dispute settlement systems within the GATT, in Article 8.8, second sentence. Members could also allow – or require – panels in such disputes to issue separate reports on matters under a SETA and matters under multilateral WTO Agreements. The DSU already contains such a provision as regards situation

---

[150] GPA, Article XXII:6; SCM Agreement, Articles 4 and 7.
[151] SPS Agreement, Article 11.2; TBT Agreement, Article 14.2–14.4; Customs Valuation Agreement, Article 19.4.
[152] SCM Agreement, Articles 4 and 7.
[153] GATS, Article XXIII; TRIPS Agreement, Articles 64.2 and 64.3.
[154] GPA, Article XXII:7 which contracts out of DSU, Article 22.2. Note also DSU, Article 22.3(g)(i).
[155] See for example the reports in *US – Upland Cotton* (WT/DS267), which included claims under Parts II and III of the SCM Agreement and the Agreement on Agriculture subject to special and standard rules and procedures, respectively.
[156] See DSU, Article 2.1, last sentence.

complaints in Article 26.2(b). These results could be achieved through an amendment of the rules in the DSU, Article 2.1.

*Future Amendments to the Scope of Jurisdiction* A potential concern of non-parties to a SETA relates to the scope of future amendments of an optional Annex 4 Agreement and, in particular, jurisdiction in disputes related to the amended versions. Sustainable-energy technology and markets are rapidly changing fields; hence a SETA is likely to require revision in the future to keep pace with the development of new products and other changes. Indeed, the parties to the TCA updated and expanded its product coverage,[157] the parties to the GPA have agreed to revise its text and expand its coverage,[158] and the ITA also foresees expansion.[159]

The WTO Agreement, Article X:10, creates a loophole in that it provides that amendments to agreements in Annex 4 are governed by the provisions of those Agreements, rather than the multilateral amendment procedures. The amendment procedures in optional Agreements would normally exclude the participation of non-parties. For example, the 2001 protocol amending the Annex to the TCA was agreed among the signatories to that Agreement, without the involvement of the wider WTO membership.[160] The 2012 Protocol amending the GPA was approved by the parties to the GPA and adopted by the GPA Committee.[161]

The only substantive condition for an Agreement to be added to Annex 4 is that it is a 'trade' Agreement, which is not defined. That might by implication prevent future amendments transforming the added Agreement into something else, such as an environmental Agreement with trade provisions. Yet even a 'trade' agreement can be interpreted broadly in light of the objectives set out in the Preamble to the WTO Agreement which, within the realm of 'the parties' relations in the field of trade and economic endeavour', refer to the classic issue of 'expanding the production of and trade in goods and services' followed by the clauses 'while allowing optimal use of the world's resources in accordance with the objective of sustainable development', and 'seeking both to protect and preserve the environment'.[162] This might give considerable latitude to amend the SETA in a later phase of negotiation.

---

[157] TCA/4, see n. 10.   [158] GPA/113, see n. 89.   [159] ITA, Annex, para 3.
[160] Protocol (2001) Amending the Annex to the Agreement on Trade in Civil Aircraft, TCA/4.
[161] Decision of 30 March 2012 of the Parties to the WTO Agreement on Government Procurement, see n. 89, adopting *inter alia* the Decision of 30 March 2012 of the Committee on Government Procurement, GPA/112.
[162] WTO Agreement, Preamble, 1st recital. The Appellate Body referred to this recital in *US – Shrimp* (WT/DS58/AB/R), para 129. See further the references in n. 107 to that report: G. Handl, 'Sustainable development: general rules versus specific obligations',

There is a risk that parties to an Agreement among a sub-set of Members might introduce provisions to which substantial sections of the WTO membership object, and that the DSU might become applicable to these provisions without their consent. A practical solution to provide certainty would ensure that any decision that amends the list of covered Agreements in the DSU in order to add an optional Agreement also states expressly whether it covers future amendments of that Agreement, or whether a further consensus decision of the Ministerial Conference is required. Transitional provisions could usefully be added as well.

### Substantive Rules

*Adding to, and Diminishing, WTO Rights and Obligations*   A SETA would be a sectoral Agreement, which could govern different types of trade policy instruments.[163] Its precise coverage would require definition.[164] The SETA would overlap with other WTO Agreements, because the WTO's rules already apply to trade in SEGS. The position is different with regard to government procurement, for which some of the multilateral agreements contain express carve-outs.[165] Therefore, the relationship between any rules negotiated in a SETA and existing WTO substantive rules requires careful consideration. As an illustration, one can consider the TCA, where uncertainty regarding its relationship to the SCM Agreement, particularly the question of which Agreement takes precedence, is one of the reasons the parties to the TCA have never agreed to apply the DSU to disputes under that Agreement.[166] Moreover, a SETA would be added to the WTO Agreement at a later time than the other Agreements, which might be a source of uncertainty as regards interpretation, if not clearly addressed in the text.

WTO sectoral Agreements can create exceptions to other Agreements, adapt disciplines in them and add new disciplines. For example, the Agreement on Agriculture provides for market access for agricultural products, establishes a special safeguard mechanism, creates a new discipline on numerical reductions in domestic support and excludes certain

---

in Lang (1995): 5; World Commission on Environment and Development. (1997): 43, regarding the scope of the concept of sustainable development.

[163] See Cottier and Delimatsis (2011): 211–244.

[164] See n. 23. For experience regarding definitional issues in the Doha Round environmental goods and services negotiations, see Thomas Cottier and Donah Baracol-Pinhão (2009) 'Environmental goods and services: the Environmental Area Initiative approach and climate change', in Cottier et al. (2009).

[165] GATT 1994, Article III:8(a); TBT Agreement, Article 1.4; GATS, Article XIII.

[166] See n. 141.

products in certain Members from the SCM Agreement disciplines on export subsidies.[167] Meanwhile, the SPS combines elements of GATT 1994 and elaborates rules for the application of Article XX(b) to the use of SPS measures.[168] Agreements can also extend the scope of existing obligations, such as the non-sectoral Agreement on Trade-Related Investment Measures (TRIMS) Agreement, which extends the application of GATT 1994, Articles III and IV, to investment measures related to trade in goods.[169] A SETA could employ the same techniques but, as an optional rather than a multilateral agreement, certain additional considerations would apply.

New SETA rules are likely to clarify or *add* to the obligations of SETA parties. Any advantages that those parties extend to each other that fall within the scope of MFN obligations in the multilateral WTO Agreements would accrue to the benefit of all other WTO Members (as discussed above). Any obligations assumed by SETA parties would not be assumed by other WTO Members in view of the *pacta tertiis* rule, as expressed in the WTO Agreement, Article II:3.

However, some SETA rules might purport to *diminish* obligations in the WTO Agreements. For example, the creation of a new general exception-style provision would purport to entitle SETA parties to breach obligations, such as national treatment (NT) and the prohibition of quantitative restrictions, or at least to confirm liberal interpretations of existing obligations or exceptions. The implementation of a SETA as an ITA-style Agreement through modification of participants' schedules would not be an effective means to diminish obligations because scheduling concessions and commitments can only yield rights, absent some express authorisation in the text of the Agreements.[170] Therefore, a choice to implement the SETA as a GPA-style Agreement added to Annex 4 of the WTO Agreement may be motivated, at least partly, by the objective of diminishing certain of the parties' WTO obligations as between themselves.

The WTO Agreement allows an optional Agreement added to Annex 4 to add to the parties' WTO obligations as well as to diminish them. Article X:9 provides for the possibility of two or more Members modifying their rights and obligations between themselves within the WTO framework (i.e. entering into an *'inter se'* agreement). It recognises the possibility of adding a trade Agreement between two or more members but does not state that the trade Agreement must be consistent with the multilateral

---

[167] Agreement on Agriculture, Articles 4, 5, 6, 8 and 9; SCM Agreement, Article 3.1.
[168] Agreement on the Application of Sanitary and Phytosanitary Measures, Preamble, 8th recital.
[169] Agreement on Trade-Related Investment Measures, Articles 2 and 3. [170] See n. 27.

WTO Agreements. Also, it does not state that the multilateral agreements in Annexes 1, 2 and 3 to the WTO Agreement prevail over the optional agreements in Annex 4. These issues are within the discretion and authority of the Ministerial Conference in considering whether or not to consent to the addition of the agreement to Annex 4. This provides non-parties with the means to prevent certain modifications of the parties' WTO obligations as between themselves within the WTO framework, but the WTO Agreement itself does not set any substantive conditions.[171] Even if the Ministerial Conference's consent is forthcoming, the WTO Agreement, Article II:3, ensures that any modifications of multilateral WTO rules in an Annex 4 agreement will be ineffective as regards non-parties.

The Vienna Convention on the Law of Treaties (1969), in Article 41, provides for agreements to modify a multilateral treaty between certain of the parties only.[172] In cases where the possibility of such a modification is provided for by the multilateral treaty, as in the WTO Agreement, Article X:9, the Vienna Convention, Article 41(1), sets no substantive conditions. Article 41(2) sets out a procedural requirement that the parties 'notify the other parties of their intention to conclude the agreement and of the modification to the treaty for which it provides'. This would normally require the participants in the negotiations to notify non-participating parties to the multilateral treaty *before* they conclude their agreement between themselves. Article X:9 might contract out of this requirement because it envisages that the parties will have already concluded an agreement at the time they request that it be added to Annex 4 of the WTO Agreement. Naturally, nothing prevents the text of the Agreement, including drafts, being communicated to non-participants in the negotiations earlier.

*Affecting the Interpretation of WTO Agreements* The addition of an optional Agreement to Annex 4 could indirectly affect the rights of non-parties through the interpretation of the covered Agreements, if the issue is not addressed in the text of the new Agreement. The WTO Agreement, Article II:3, provides that optional agreements in Annex 4 are part of the WTO Agreement for those Members that have accepted them. Therefore, a panel or the Appellate Body might make reference to the terms of the optional agreement as context in interpreting the terms of

---

[171] The Vienna Convention on the Law of Treaties (1969), Article 41, adds certain conditions in cases where the multilateral treaty neither provides for nor prohibits *inter se* modifications. This is discussed in relation to modifications of WTO Agreements effected among two or more Members in a non-WTO Agreement in Section 3((a)0) below.

[172] The status of the rules in of the Vienna Convention, Article 41, is discussed below.

the multilateral trade Agreements – at least in a dispute between the parties to the optional Agreement, who are bound by both. This is more likely in the case of an Agreement that is closely related to the subject matter of the multilateral Agreements, such as a SETA, than one that refers to a separate subject, such as the GPA. Consequently, the same provisions in the multilateral Agreement will either have the same meaning in a dispute involving a non-party or non-parties to the optional Agreement – in which case their rights *are* affected – or else the same provisions will have different meanings as between different WTO Members, which is a threat to the integrity of the WTO Agreement.

The risk of diverging interpretations that affect the rights of non-parties is greatest in cases of potential conflicts among different WTO Agreements without a clause indicating an order of precedence. Conflicts are to be avoided between different WTO Agreements, as they are integral parts of a single Agreement.[173] One provision may be 'read down' to avoid conflict with another. What constitutes a conflict is a matter of debate: it may arise not only where it is impossible for a Member to comply with both Agreements simultaneously, but also where one Agreement explicitly authorises a measure that is prohibited by the other.[174]

Overlapping Agreements ideally should contain precedence clauses that provide how substantive rules in different Agreements interrelate. These may be conflict provisions in which an Agreement provides that, to the extent of any conflict, it prevails over others (as in the WTO Agreement and the Agreement on Agriculture[175]) or that the others prevail over it (as in the case of the GATT 1994).[176] However, conflict provisions only apply after the process of treaty interpretation has failed to reconcile competing rules, which means that the terms of one Agreement can affect the meaning of the other even without a conflict. Saving provisions also create an order of precedence but provide that rights and obligations under the multilateral Agreements are not affected by a new Agreement. In these cases, the interpretation of the former does not change after the new Agreement is concluded (as was the case with the IDA).[177] A saving provision can be tailored to an optional Agreement so that non-participants obtain the benefits of market access but are not

---

[173] WTO Agreement, Article II:2 and 3; and the principle of effective treaty interpretation, see Appellate Body report in *Korea – Dairy*, WT/DS98/AB/R, at para 81 (referring to Annexes 1, 2 and 3).

[174] See Vranes (2006): 395–418.

[175] WTO Agreement, Article XVI:3; Agreement on Agriculture, Article 21.1.

[176] General interpretative note to Annex 1A to the WTO Agreement regarding the GATT 1994 and the other agreements in that Annex.

[177] IDA, Article VIII:6. The Anti-Counterfeiting Trade Agreement (ACTA), which was concluded outside the WTO framework, provides that nothing shall derogate from any

Legal Options for a SETA 427

affected by any new rules. For example, the 2001 TCA Amending Protocol stated that, apart from duty-free treatment, nothing in that Protocol or the amended TCA changed or affected a party's rights and obligations under any of the WTO Agreements.[178] This might be a suitable solution for the SETA as regards non-parties. The way in which these issues are handled may affect the wider membership's willingness to consent to amend the DSU to apply to disputes under the SETA.

*Subsidies' Disciplines* The negotiating objectives for a SETA with respect to subsidies may need clarification as to whether the aim is to add to or diminish the disciplines of the SCM Agreement, or both.[179] This will impact the prospects for market access concessions, as new commitments not to challenge or countervail subsidy programmes would act as a significant disincentive for tariff reductions.[180] Moreover, given that the SETA would not be a multilateral Agreement, any new rules that purported to diminish disciplines would be subject to certain limitations.

The SETA could be designed to add to the WTO disciplines on subsidies. The SCM Agreement applies to trade in goods (although a subsidy may take the form of government provision of services) whereas the GATS as yet provides no multilateral disciplines to avoid the distortive effects of subsidies on trade in services.[181] Subsidies for the generation of sustainable energy, as opposed to the production and supply of sustainable energy equipment and services, are not covered in a systematic way by the existing SCM disciplines. A SETA could address these matters, including an appropriate classification of energy and energy-related services.[182]

Depending on its coverage, a SETA could add to the disciplines in the SCM Agreement. For example, it could expand or clarify the definition of a subsidy in the SCM Agreement, Article 1, as between the parties with respect to known subsidies in the sustainable energy sector,[183] as well as 'measures subsidies' such as the allocation of emissions permits. It could also clarify the assessment of when they confer a benefit. A SETA could also expand the category of prohibited subsidies beyond domestic content or import substitution subsidies.[184]

obligation between the parties under existing agreements, including the TRIPS Agreement: see Article 1.
[178] *Ibid.*, para 6.   [179] See Rubini (2012): 525–579.
[180] See Jha (2009): 39. See also Bagwell and Staiger (2006): 877–895.
[181] GATS, Article XV. See Poretti (2008): 466–489.
[182] See Cottier and Delimatsis (2011): 222–223.
[183] See, for example, the typology of subsidies in certain WTO members in Ghosh et al. (2012).
[184] SCM Agreement, Article 3.1(b). They are also inconsistent with the GATT 1994, Article III:4.

In this case, the SETA would not need to create its own system of countermeasures but could extend the one already found in Part II of the SCM Agreement.

The SETA might also be designed to diminish the SCM disciplines. One proposal is to revive the concept of non-actionable subsidies, which was the 'green box' of the SCM Agreement, in contrast to prohibited (red) and actionable (amber).[185] Non-actionable status protected subsidies from challenges on the grounds that they caused adverse effects to the interests of another Member and from countervailing action, subject to exceptions.[186] For example, the SETA could treat non-discriminatory feed-in tariffs (FITs) as non-actionable subsidies.

Adverse effects may consist of 'serious prejudice' to the interests of another Member, for example, through export displacement of a like product, or through significant price undercutting, significant price suppression, price depression or lost sales of a like product in the same market or through an increase in world market share of the subsidising Member following a consistent trend, on certain conditions.[187] For example, in 1998 in *Indonesia – Autos*, duty and sales tax exemptions under Indonesia's national car programme were found to have caused serious prejudice to like imports of EC vehicles through significant price undercutting.[188] In 2010–2011, in *EC – Airbus*, launch aid, certain equity infusions and provision of certain infrastructure by the EC, France, Germany, Spain and the United Kingdom were found to have caused serious prejudice to the interests of the United States through displacement of exports by Boeing and significant lost sales in the same market.[189] Adverse effects may also arise in other circumstances.[190] Non-actionable status could prevent parties from bringing similar challenges to the WTO with respect to subsidies for sustainable energy equipment.

Countervailing action is available where subsidised imports cause injury to a Member's domestic industry. Non-actionable status could, for example, prevent importing parties from imposing countervailing duties or even initiating countervailing investigations with respect to subsidies for sustainable energy equipment.

---

[185] Howse (2011): 20.    [186] SCM Agreement, Article 9.1; Article 10, n 35.
[187] SCM Agreement, Articles 5(c) and 6.3.
[188] Panel report in *Indonesia – Autos*, WT/DS54/R.
[189] Panel report in *EC and Certain Member States – Large Civil Aircraft*, WT/DS316/R, and Appellate Body report, WT/DS316/AB/R.
[190] 'Adverse effects' may consist of injury to a domestic industry caused by the subsidised imports or through nullification or impairment of benefits accruing under the GATT 1994, most typically where the improved market access expected to flow from a tariff concession is undercut by subsidisation: SCM Agreement, Article 5(a) and (b).

Legal Options for a SETA

The original classes of non-actionable subsidies were set out in the SCM Agreement, Article 8.2, but it expired in 1999. There has been no consensus to extend or revive that provision.[191] The design of any new category of non-actionable subsidies is a matter for negotiation. The final text of Article 8.2 and its warren of footnotes do not focus on sustainable energy and are heavily qualified so that they probably would be unsuitable to serve as an initial draft for any SETA negotiations. Non-actionable status could have a positive *ex ante* effect by encouraging parties to adopt or maintain measures promoting sustainable energy – the goal is not simply to allow a Member to escape sanctions at the end of a lengthy and costly WTO proceeding or to avoid the imposition of definitive duties at the end of a countervailing investigation, or to have duties revoked after judicial review. If the qualifying criteria were simple enough and, say, did not require detailed analysis of effects, they would probably generate fewer disputes regarding the consistency of measures.[192]

A key to the original category of non-actionable subsidies was a clause stating expressly that the provisions of Part III of the SCM Agreement (on actionable subsidies) and Part V (on countervailing measures) 'shall not be invoked' regarding measures considered non-actionable in accordance with its provisions.[193] If the new category is to grant similar immunity, the SETA must specifically state so, because the provision in the SCM Agreement has expired and will not attach to a new category created in another Agreement in any case. It can also provide that it prevails over other provisions of the multilateral WTO Agreements, although the original category of non-actionable subsidies did not.

There are limits to non-actionable status if it is only recognised in an optional agreement in Annex 4 to the WTO Agreement. For one thing, non-parties that are WTO Members would remain free to take countervailing action. This would not necessarily prevent any SETA provisions from being effective as the domestic industry in a non-party must still satisfy the standing requirements for a countervailing investigation and a case must still satisfy the conditions for a countervailing duty, including

---

[191] See, for example, the Doha Ministerial Decision of 14 November 2001 on Implementation-Related Issues and Concerns, WT/MIN(01)/17, para 10.2.

[192] Some disagreements are inevitable but simplified procedures can be agreed to determine which subsidies do in fact qualify. The SCM Agreement, Articles 8.3–8.5, contained notification and arbitration procedures and the SCM Committee adopted formats for notifications and updates and procedures for arbitration, which could serve as a model for the SETA Committee: see Format for notifications: PC/IPL/11, G/SCM/14; Format for updates of notifications: G/SCM/13; Procedures for arbitration: G/SCM/19. No Member ever availed itself of these notifications or procedures.

[193] See the SCM Agreement, n. 35, second sentence. The subsequent sentences backtracked in certain respects.

the existence of a subsidy, benefit and injury. Even where these conditions are all fulfilled, countervailing duties are only applied on a bilateral basis.

A more serious problem is that non-parties that are WTO Members would remain free to challenge actionable subsidies in the WTO. If a non-party Member could demonstrate adverse effects to its interests,[194] it could obtain a DSB recommendation that the SETA party remove the adverse effects or withdraw the subsidy.[195] Withdrawal is not bilateral, while removal of the adverse effects with respect to non-parties only could be impractical.

Non-actionability as regards the SCM Agreement cannot alter the status of measures under the Agreement on Agriculture. Certain measures, such as biofuel production-related payments, will still be subject to domestic support commitments unless they qualify for an exemption in that Agreement, such as the 'green box' criteria in Annex 2. Any attempt to exclude them from those commitments would be ineffective for reasons similar to those regarding prohibited subsidies. However, the prospects for obtaining a WTO remedy would depend on many other factors making up the Member's current total aggregate measure of support (AMS). In any case, there are no product-specific caps for domestic support under the Agreement on Agriculture.

A SETA could add to or clarify the disciplines on subsidies in the SCM Agreement. It might also diminish those disciplines by creating a new category of non-actionable subsidies, but this would be less effective when handled in an optional agreement.

*General Exception-Style Approach*   The general exceptions found in the GATT 1994, Article XX, and the GATS, Article XIV, could provide a model for a SETA. The GATT 1994, Article XX(b), in particular, sets out a general exception for health measures. In 2007, the Appellate Body in *Brazil – Retreaded Tyres* explicitly mentioned measures adopted in order to attenuate global warming and climate change in its interpretation of that provision.[196] This suggests that the GATT 1994, Article XX(b), could ground an authoritative interpretation under the WTO Agreement, Article IX:2, or provide the basis for an agreement elaborating rules for its application to measures to promote sustainable energy.[197] The GATS, Article XIV(b), contains an almost identically worded health exception as regards trade in services. The GATT 1994, Article XX(g), sets out a general exception for measures relating to the

---

[194] SCM Agreement, Articles 5(c) and 7.2.   [195] SCM Agreement, Article 7.8.
[196] Appellate Body report on *Brazil – Retreaded Tyres*, (WT/DS332/AB/R) at para 151.
[197] See Howse (2011): 17–19.

conservation of exhaustible natural resources, which can also apply to environmental measures.

The general exceptions are subject to significant limitations. By their own terms, they create exceptions to obligations in the same respective Agreements in which they are found. It is not clear whether and how a SETA would purport to create a general exception to other Agreements in Annex 1A of the WTO Agreement, notably the TBT and the SCM Agreements.[198] Further, every general exception in the GATT 1994 and the GATS contains conditions regarding the content of the measures to which they apply, and all general exceptions are subject to conditions regarding the manner in which measures are applied, which may exclude certain measures that the parties to a SETA may wish to allow.

However, the existing general exceptions are relevant to a *multilateral* approach to sustainable energy within the WTO because the general exceptions are found in the Agreements in Annex 1 to the WTO Agreement. (They are also relevant to any implementation of a SETA outside the WTO.) If a SETA were added to Annex 4 of the WTO Agreement as a GPA-type Agreement, there would be no need for it to conform to the requirements of the general exceptions. It could in any case prevail over the parties' obligations under the Agreements in Annex 1 to the WTO Agreement – including not only the GATT 1994 but also the SCM Agreement. The limitation of this approach is that it cannot bind non-parties. They retain their rights under the multilateral WTO Agreements, which could undermine derogations from certain obligations, as described in the previous section on subsidies' disciplines.

A general exception-style provision could form part of a SETA's sectoral approach. A SETA could be inspired by the structure and concepts in the general exceptions, setting out classes of policy objectives and qualifying criteria for each, with rules on the method of implementation such as non-discrimination. The general exception could cover environmental measures, broadly defined.[199] It could also interpret and extend the TBT Agreement.[200] Clear references to the existing WTO Agreements would have the advantage of clarifying the new Agreement's relationship to them. This approach could subsume the new category of non-actionable subsidies by providing not only that the general exception justifies certain environmental measures under the WTO multilateral trade Agreements, but also that other SETA parties shall not invoke certain provisions of the SCM Agreement.

---

[198] As regards the relationship between the general exceptions in the GATT 1994, Article XX, and subsidies covered by the SCM Agreement, see Rubini (2009): 195 and n. 179.
[199] Green (2006): 377–414 at 409, cited in Howse (2011): 17.
[200] See Hufbauer et al. (2009): 71.

The disputes that have arisen regarding the relationship between general exceptions and commitments in China's Protocol of Accession[201] would not arise if the SETA were implemented following the GPA model. General exceptions in the multilateral trade agreements in Annex 1 of the WTO Agreement do not apply to agreements in Annex 4, unless incorporated by the latter.

A SETA could be inspired by the structure and concepts of the WTO general exceptions but, as an optional Agreement in Annex 4 of the WTO Agreement, it would not be necessary to conform to the requirements of a general exception as between the parties.

## Outlook

There is scope for a SETA to be an optional agreement within the WTO framework. One model is an ITA-type Agreement to be implemented through modifications to participating Members' goods and services schedules. These could also include disciplines on non-tariff measures agreed among the participants and incorporated in their schedules by reference. Another model is a GPA-type Agreement that could be added to Annex 4 of the WTO Agreement, exclusively with the consent of the wider WTO membership. This type of agreement could vary the parties' WTO rights and obligations up or down.

Under either approach, the existing WTO rights of non-participants are guaranteed through the MFN obligations in the multilateral WTO Agreements. A GPA-type Agreement can cover subjects that fall outside the scope of those obligations, but SETA would overlap much more with existing Agreements.

Preparatory work on negotiations can be initiated among a group of interested members on an ad hoc basis or in existing groupings, such as the G20 and APEC. The issue can then be raised in the WTO when suitable, which may be earlier rather than later in cases where the interested parties will need the participation of other Members.

A WTO committee for the conduct of negotiations on a GPA-type Agreement can be formed by decision of a WTO body. The participants negotiate among themselves but in the shadow of the wider WTO membership, which must consent before the agreement can be added to the WTO framework. Participation should be voluntary and open to all Members, but Members' different roles can be reflected in their status as either participant or observer.

---

[201] See Appellate Body reports in *China – Audiovisual Services*, WT/DS363/AB/R, paras 205–233, and *China – Raw Materials*, WT/DS394/AB/R, paras 278–307.

The critical mass required for implementation represents the scale of participation that is required to render an Agreement viable, which depends on its product coverage, the nature of the obligations that it contains and which of them are applied on an MFN basis. If a critical mass has been assembled prior to the conclusion of the Agreement, no threshold is required other than acceptance by all participants in the negotiations.

DSU procedures are available to enforce a SETA if it is implemented within the WTO. If the SETA is a GPA-style Agreement, this will require dispute settlement provisions in the SETA text, a consensus decision by the Ministerial Conference to amend the list of Agreements covered by the DSU and a decision of the SETA Committee notified to the DSU.

Standard DSU rules and procedures can apply to disputes under the SETA but, if it is implemented as a GPA-style Agreement, special rules can be adopted to protect non-parties' interests, including granting the right to participate in dispute settlement procedures under the new Agreement. Special rules and procedures – particularly expert review procedures – can also be adopted to take account of any scientific or technical issues that may arise under a SETA. The expert review procedures should be more detailed than those that currently exist.

Depending on the level of ambition, the SETA can adopt a sectoral approach, like the Agreement on Agriculture, and combine different disciplines in a single Agreement. It can add to, and diminish, the rights and obligations of parties under the existing WTO Agreements, but the rights of non-parties are saved by the WTO Agreement, Article II:3. This would reduce the effectiveness of any new category of non-actionable subsidies in certain respects, but it would not prevent the creation of a general exception-style provision for sustainable energy among the parties.

## 2      Implementation of a SETA Outside the WTO Framework

A SETA can be implemented outside the WTO. In this scenario, the participants would have to find or establish another institutional framework including, if they wished, a dispute settlement mechanism. Although the SETA would not require multilateral approval in the WTO, it would still need to comply with certain WTO disciplines. Care would also be required to prevent conflicts with the WTO in terms of both jurisdiction and substantive norms. This section considers the relevant WTO disciplines and other international law, including options for dispute settlement.

434    Matthew Kennedy

Transparency in the process of negotiation of a trade Agreement is still important, even when there is no requirement to obtain approval from non-participants in the WTO. For example, the ACTA[202] was negotiated and concluded in 2011 among a group of interested WTO Members outside the WTO framework. The participants did not publish a draft text until the final year of negotiations in 2010, which attracted criticism from parts of civil society.

## Relevant WTO Disciplines

### Preferential Trade Agreements

The WTO Agreements expressly envisage the possibility that two or more WTO Members may enter into certain types of trade agreements among themselves (i.e. *inter se* agreements). The GATT 1994, Article XXIV, and the GATS, Article V, provide that they do not prevent the conclusion of certain types of preferential trade agreements (PTAs) while the Enabling Clause, which is incorporated in the GATT 1994[203], expressly provides that Members may enter into certain types of PTAs.[204] The term PTA is preferable to 'regional trade agreement', because an increasing number of these agreements are being negotiated between countries on different continents. The majority of PTAs are free-trade agreements (FTAs) but there are also partial scope agreements and customs unions.

The principal attraction of a PTA for the parties is that it may qualify for an exception to the MFN obligations in the GATT 1994 and the GATS. If it does, the benefits granted under the PTA need not be extended to any non-parties, which can increase the incentive to participate. There is no requirement that the parties to a PTA obtain the approval of the wider WTO membership, although there are certain transparency obligations. Instead, each PTA exception sets out separate conditions that must be met for an Agreement to qualify. The salient conditions in the three relevant PTA exceptions are set out below.

The GATT 1994, Article XXIV, is applicable to FTAs as regards trade in goods and customs unions. There is no requirement that the parties to a FTA or customs union be in the same geographical region or at the same or different levels of development. Among the qualifying criteria for both is a condition that the duties and other restrictive regulations of

---

[202] Text of the ACTA dated 3 December 2010, www.dfat.gov.au/trade/acta/Final-ACTA-text-following-legal-verification.pdf.
[203] GATT 1994, Article 1(b)(iv) of the incorporation text.
[204] Decision of the Contracting Parties to the GATT of 28 November 1979 on 'Differential and More Favourable Treatment, Reciprocity and Fuller Participation of Developing Countries', GATT document L/4903.

## Legal Options for a SETA

commerce are eliminated on 'substantially all the trade' between the parties in products originating in their territories.[205] While there is no precise definition of this condition, it is clear that a SETA would not qualify if it were limited to SEGS in the trade between any two parties.

Similarly, the GATS, Article V, is applicable to economic integration agreements. Among other things, these agreements must have 'substantial sectoral coverage', which is understood in terms of number of sectors, volume of trade affected and modes of supply. In order to meet this condition, agreements should not provide for an *a priori* exclusion of any mode of supply.[206] It is clear that a SETA would not qualify if it were limited to sustainable energy services, even though those services are scattered across different sectors, due to the volume of trade affected.

A SETA could qualify under the GATT 1994, Article XXIV, and the GATS, Article V, if it were negotiated as part of comprehensive FTAs, either existing ones or others to be negotiated in the future. The modalities for such negotiations would necessarily require substantial coverage of trade in goods or substantial sectoral coverage in trade in services (to the extent that the SETA covers services). Where the broader agreement satisfied the conditions in the GATT 1994, Article XXIV, and the GATS, Article V, the advantages granted under it, including those for SEGS, would be excepted from MFN treatment, and not accrue to non-parties.

The Enabling Clause is applicable to PTAs for the mutual reduction or elimination of tariffs among developing country members on products imported from one another.[207] The SETA would not qualify under the Enabling Clause if it included developed countries, unless those parties granted non-reciprocal preferences to the developing country members in accordance with the Generalized System of Preferences (GSP).[208]

The SETA, as a stand-alone agreement, is unlikely to satisfy the conditions of any of the PTA exceptions. However, that is not essential for the SETA to be viable outside the WTO framework. Failure to satisfy the conditions of a PTA exception does not invalidate an Agreement. It only means that commitments on improved market access and certain other trade advantages must be extended to all WTO Members in accordance with MFN treatment. That would not render an Agreement unviable if the critical mass for implementation were defined widely enough to reduce the risks of free riding (as discussed above). Further, failure to satisfy the conditions of a PTA exception would not affect the extension

---

[205] GATT 1994, Article XXIV:8(b).   [206] GATS, Article V:1(a) and n. 1.
[207] Enabling Clause, para 2(c), read together with the GATT 1994, Explanatory note 2(a) in the incorporation text.
[208] Enabling Clause, para 2(a).

of benefits under any SETA rules on subjects that are not covered by the MFN obligations in the WTO Agreements.

### Other Non-WTO Agreements

An advantage of implementing a SETA through an agreement outside the WTO framework is that non-WTO Members can join. The latter form a smaller group now that Russia has joined the global trade body, but two of the top twenty-five emitters of GHG in electricity and heat (Iran and Kazakhstan)[209] are still outside the club, as are two large destinations of greenfield investment in the manufacturing of environmental technology products (ETS) (Algeria and Libya).[210] However, even if the SETA is implemented within the WTO, participants can include acceding Members (as the ITA did) pending their accession, and non-Members could also apply the SETA on a *de facto* basis.

Two or more Members could agree to conclude the SETA on a stand-alone basis or in an existing non-WTO framework. With respect to trade in sustainable energy, two relevant intergovernmental frameworks already exist in the UNFCCC and the Energy Charter. If the new Agreement modified the parties' WTO obligations as between themselves, this would be an *inter se* modification but, unlike those discussed above, it would not be implemented within the WTO framework. Nevertheless, even outside the WTO framework, any SETA modifications of WTO obligations could still be valid as between the parties.

The WTO Agreement does not prohibit the conclusion of a trade Agreement among two or more of its Members outside the WTO, even where the trade Agreement does not qualify as a PTA. While the WTO Agreement, Article X:9, sets out a procedure to add trade agreements to Annex 4, it does not oblige the parties to an Agreement to request that it be added, and it does not prohibit such an Agreement where the parties do make a request, but the Ministerial Conference does not consent to it. Also, the other amendment procedures in Article X do not purport to exclude the operation of customary rules of public international law regarding *inter se* modifications.[211]

The Vienna Convention on the Law of Treaties (1969), in Article 41, provides that '[t]wo or more of the parties to a multilateral treaty may conclude an agreement to modify the treaty as between themselves alone'.

---

[209] ICTSD (2011): 5, Figure 4.  [210] *Ibid.*: 11
[211] An analogy might be seen *in US – Clove Cigarettes*, in which the Appellate Body found that the procedure for authoritative interpretations in the WTO Agreement, Article XI:2, did not exclude the customary rules of interpretation to which of the DSU, Article 3.2, refers: WT/DS406/AB/R, paras 257–259.

This basic rule is considered well established.[212] Article 41 adds conditions to the rule, which appeared innovatory when the International Law Commission (ILC) took up its study of the matter, but States have not subsequently called them into question.[213] Two substantive conditions in Article 41(1)(b) apply where a multilateral treaty neither provides for nor prohibits its modification by an agreement between certain of the parties only. The conditions are that (i) the modification in question does not affect the enjoyment by other parties (to the multilateral treaty) of their rights under that treaty or the performance of their obligations, and (ii) the modification in question does not relate to a provision, derogation from which is incompatible with the effective execution of the object and purpose of the multilateral treaty as a whole.

As regards condition (i), a SETA would only create obligations for its parties. Non-parties that are WTO Members would continue to enjoy their rights under the WTO Agreement, including MFN treatment, to the extent applicable. Non-parties' rights are in any case limited by the general exceptions in the GATT 1994 and the GATS, which could justify certain SETA implementing measures. If the SETA does not purport to modify the WTO Agreement *inter se*, the saving provision could confirm that it does not affect the WTO rights and obligations of any WTO Member.[214] In other cases, it would be useful to include a saving provision in the SETA text confirming that it does not affect the WTO rights and obligations of non-parties. A more specific clause could also reiterate the text of particular WTO provisions, depending on the circumstances. For example, if the SETA created a new category of non-actionable subsidies, the text could confirm that no Member should cause, through the use of any subsidy, adverse effects to the interests of other WTO Members not party to the SETA.

As regards condition (ii), it is difficult to discern a non-derogable provision in the WTO Agreement.[215] A SETA would in any case be designed to promote the attainment of the WTO's objectives, as set out in its Preamble, regarding expansion of 'production of and trade in goods and services, while allowing for the optimal use of the world's resources in

---

[212] 1928 Havana Convention on Treaties, Article 19, para 1; *Oscar Chinn* case, PCIJ (1934) Series A/B no. 63, 80ff., see the separate opinions of Judges van Eysinga and Schücking; Christian Feist, *Kündigung, Rücktritt und Suspendierung von multilateralen Verträgen* (Kiel: Duncker & Humblot, 2001): 197, cited in Villiger (2009): 98.

[213] Villiger (2009). See also Aust (2000): 222. The WTO Agreement, Article X:9, appears to contract out of the procedural condition in cases where the parties request that their agreement be annexed to the WTO Agreement.

[214] ACTA, see n. 176, contains such a provision.

[215] See Pauwelyn (2001): 535–577 [2001] at 547–550; (2003): 320; and (2003a): 907–951 at 914–915.

accordance with the objective of sustainable development'.[216] It would contribute to these objectives by the same means found in the WTO Agreement, i.e. through reciprocal and mutually advantageous arrangements directed to the substantial reduction of tariffs and other barriers to trade and the elimination of discriminatory treatment.[217] Therefore, the SETA should not run afoul of this condition. The Vienna Convention (1969), Article 41(2), also sets out a procedural requirement that the parties 'notify the other parties of their intention to conclude the agreement and of the modification to the treaty for which it provides'. It can be noted that this provides for notification *before* the parties conclude their agreement between themselves.[218]

## WTO Institutional and Dispute Settlement Provisions

The conclusion of a SETA would create a specialised system of rules tailored to address the issues unique to production of and trade in SEGS. Its implementation outside the WTO framework would grant the SETA system a degree of autonomy but its subject matter would still overlap with the WTO. This could lead to conflicts of substantive norms and of jurisdiction, which are considered below.

### Conflicts of Substantive Norms

Potential conflicts of norms between international agreements can lead to uncertainty regarding their compatibility, which require the parties to make efforts to ensure their mutual supportiveness. The SETA's relationship with the WTO Agreements might not be straightforward where it contains new disciplines, which could lead to proposals to amend it, or the WTO Agreements, or both. For example, there is a long-running discussion in multiple fora, including the WTO, regarding the relationship between the TRIPS Agreement, Article 27.3(b), which governs the patentability of plants and animals, and the Convention on Biological Diversity (CBD). This has led to proposals to amend the TRIPS Agreement to enhance the mutual supportiveness of the two.[219]

---

[216] WTO Agreement, Preamble, 1st recital.
[217] WTO Agreement, Preamble, 3rd recital.
[218] As regards notification of Agreements that the parties request be added to Annex 4 of the WTO Agreement, see p. 212.
[219] See Communication from Brazil, China, Colombia, Ecuador, India, Indonesia, Peru, Thailand, the ACP Group and the African Group dated 15 April 2011, setting out a draft Decision, TN/C/W/59.

Legal Options for a SETA                                                            439

A SETA could reduce uncertainty through the express clarification of its relationship to relevant WTO Agreements. For example, it could make reference to particular WTO Agreements or provisions in its Preamble and consistently use terminology and definitions found in WTO Agreements to the extent appropriate. This could also assist national administrations and courts in implementation of the Agreement.[220]

Actual conflicts of substantive norms may arise in different scenarios (as discussed previously). These could also arise between the WTO Agreements and the SETA if it is implemented outside the WTO framework. The SETA could contain a conflict clause providing that it prevailed to the extent of any inconsistency. However, WTO law would be applicable in any proceeding initiated in the WTO dispute settlement system.

The implementation of a SETA outside the WTO framework would not entirely insulate the WTO Agreements from its impact. It is clear that the WTO dispute settlement system has a limited mandate, which is to determine conformity with the 'covered Agreements' as listed in Appendix 1 to the DSU. It is also clear that the DSU, Article 3.2, directs dispute settlement panels and the Appellate Body to apply the customary rules of interpretation of public international law. In 1996, the Appellate Body in *US – Gasoline* considered that this direction reflected a measure of recognition that the GATT was 'not to be read in clinical isolation from public international law'.[221] The exact role in the WTO of external rules of public international law has become a subject of much debate. A 2003 suggestion by the Appellate Body in *US – Byrd Amendment* that a WTO panel might in an appropriate case find that a Member had not complied with a rule of customary international law – even one as fundamental as good faith – sparked controversy.[222]

The general rule of treaty interpretation itself leads to debate as to which instruments and rules of international law are relevant to the interpretative exercise in a given case.[223] For example, in 1998 the Appellate Body in *US – Shrimp* referred to certain multilateral environmental agreements and declarations adopted outside the WTO framework to support its interpretation of the term 'natural resources' in the GATT 1994, Article XX(g), because those instruments reflected contemporary concerns of the community of nations about the protection

---

[220] See Kuijper (2010): 19–20.
[221] Appellate Body report, *US – Gasoline*, WT/DS2/AB/R: 17.
[222] See statement of the United States on adoption of the reports in *US – Continued Dumping and Subsidy Offset Act of 2000 ('Byrd Amendment')* in the minutes of DSB meeting of 27 January 2003, WT/DSB/M/142, para 57.
[223] Vienna Convention on the Law of Treaties (1969), Article 31.

and conservation of the environment.[224] In 2011, the Appellate Body in *US – Anti-Dumping and Countervailing Duties (China)* made extensive reference to the ILC Articles on Responsibility of States for Internationally Wrongful Acts[225] regarding the meaning of 'public bodies' in the SCM Agreement, Article 1.1(a)(1), without resolving definitively the extent to which the ILC Articles, Article 5, reflects customary international law.[226]

### Conflicts of Jurisdiction

The implementation of a SETA dispute settlement system outside the WTO framework could also lead to conflicts of jurisdiction with the DSU, unless the SETA contained a clear jurisdiction clause.[227] Three examples that have arisen in practice illustrate the type of conflicts that can arise between a PTA or multilateral environmental Agreement dispute settlement system, on the one hand, and the DSU on the other.

Simultaneous proceedings might be initiated in the two dispute settlement systems, straining parties' resource and creating a risk of conflicting decisions. For example, in 2000 the EC initiated a dispute in the WTO in *Chile – Swordfish* regarding measures that prevented EC fishing vessels operating in the South-Eastern Pacific Ocean from unloading their swordfish in Chilean ports for warehousing or transshipment, alleging breaches of the GATT 1994, Articles V and XI, on freedom of transit and quantitative restrictions.[228] Meanwhile, Chile initiated a dispute before the International Tribunal for the Law of the Sea (ITLOS), asserting that the EC had not complied with its obligations under the UN Convention on the Law of the Sea to ensure the conservation of swordfish in the fishing activities of its vessels in the South-Eastern Pacific Ocean.[229] A WTO panel was established (but not composed) and a special chamber of the ITLOS was constituted. Fortunately, before either proceeding advanced further, the parties came to a provisional arrangement on bilateral cooperation on the swordfish stocks in the South East Pacific. In 2008, they agreed

---

[224] Appellate Body report, *US – Shrimp*, WT/DS58/AB/R, para 129.
[225] See *US – Anti-Dumping and Countervailing Duties (China)*, WT/DS379/AB/R, paras 35–41, regarding the ILC Articles, Articles 4, 5 and 8. The Appellate Body had referred to the ILC Draft Articles (as they then were), Article 51, in *US – Cotton Yarn*, WT/DS192/AB/R, para 120 and n. 90; and *US – Line Pipe*, WT/DS202/AB/R, para 279.
[226] WT/DS379/AB/R, paras 304–316. The ILC Articles, Article 5, sets out a rule of attribution regarding the conduct of persons or entities exercising elements of governmental authority.
[227] See Kuijper (2010): 25–38.
[228] *Chile – Measures Affecting the Transit and Importation of Swordfish*, WT/DS193.
[229] *Case concerning Conservation and Sustainable Exploitation of Swordfish Stocks in the South-Eastern Pacific Ocean (Chile/European Union)*, ITLOS Case No.7.

on a more structured understanding. The proceedings in ITLOS were discontinued at the request of the parties in 2009, and the parties agreed unconditionally not to exercise any procedural rights concerning the dispute under the DSU in 2010.[230] The point to note is that the conflict of jurisdiction was avoided by the parties' mutual agreement, not by application of the rules of either dispute settlement system.

A ruling in one dispute settlement system may not take account of issues arising under the law of the other system. For example, in the late 1990s Mexico initiated dispute settlement procedures under the North American Free Trade Agreement (NAFTA) regarding US quota restrictions on imports of Mexican sugar, but a NAFTA arbitral panel was not established at its request. In 1998, Mexico also imposed anti-dumping duties on US high-fructose corn syrup (also a sweetener) that were condemned in WTO dispute settlement proceedings[231] and later in a NAFTA panel proceeding. From 2002, Mexico imposed discriminatory taxes on drinks sweetened with high-fructose corn syrup, most of which were produced in the United States. In 2004, the United States initiated a dispute in the WTO regarding the taxes and other requirements in *Mexico – Soft Drinks* on the basis of the GATT 1994, Articles III:2 and III:4. The WTO panel and Appellate Body upheld the US claims under WTO law and found that they could not take the NAFTA issues into account (discussed below).[232] In 2006, after twelve years, the two sides reached a negotiated agreement to settle the broader dispute, including implementation of the WTO ruling.

Both dispute settlement systems may rule, leading to incoherent jurisprudence. For example, there were successive proceedings in MERCOSUR and the WTO regarding a Brazilian ban on imports of retreaded tyres, originally imposed in 2000. In response to a 2002 ruling of a MERCOSUR arbitral tribunal, Brazil exempted imports from MERCOSUR countries but imports from other countries remained subject to the ban. In 2005, the EC initiated a WTO dispute alleging that the measure was a quantitative restriction inconsistent with the GATT 1994, Article XI:1. Brazil sought to justify the ban on the basis of the general exception for health measures in the GATT 1994, Article XX(b), among other things, due to the facts that mosquitoes that transmit dengue, yellow fever and malaria use waste tyres as breeding grounds and tyre fires produce toxic emissions. The difficulty with this argument was that the same problems were presented by tyres from MERCOSUR countries that were exempt from the ban. The WTO panel considered that the

---

[230] WT/DS193/4.   [231] WT/DS132/R and RW.
[232] Appellate Body report in *Mexico – Soft Drinks*, WT/DS308/AB/R, paras 44–54.

MERCOSUR ruling provided a reasonable basis for the MERCOSUR exemption, the implication being that the resulting discrimination was not arbitrary. However, the Appellate Body ruled that the measure did *not* satisfy the general exception for health measures because the MERCOSUR exemption bore no relationship to the public health objective pursued by the measure. The Appellate Body expressly noted that the discrimination associated with the MERCOSUR exemption did not necessarily result from a conflict between provisions under MERCOSUR and the GATT 1994, because Brazil had not raised the public health exception available under MERCOSUR. It also suggested that the respondent might be able to justify its measure under the PTA exception in the GATT 1994, Article XXIV:8(b), as a further means of reconciling the two systems.[233]

In cases of conflict of jurisdiction, the WTO dispute settlement system may well prevail. The DSU provides for exclusive WTO jurisdiction where Members seek redress in disputes arising under the WTO covered Agreements. The DSU, Article 23, titled 'Strengthening of the Multilateral System' (in which the key word is 'Multilateral') establishes the WTO dispute settlement system as the exclusive forum for the resolution of such disputes and requires adherence to the rules of the DSU. It also prohibits certain unilateral action.[234] The DSU also provides for compulsory jurisdiction in cases where a Member initiates a dispute under its procedures. A recalcitrant respondent can delay but not block the establishment of a panel, the composition of a panel, the adoption of panel and Appellate Body reports and the authorisation of retaliation.[235] The DSU may also oblige the WTO itself to accept jurisdiction where a Member initiates a dispute under its procedures. The DSU, Article 3.3, entitles a Member to initiate a dispute and this, read together with Article 23 and the mandate and functions of panels, has been interpreted as preventing panels from declining jurisdiction. In *Mexico – Soft Drinks* the respondent, Mexico, asked a WTO panel to decline to exercise jurisdiction in favour of a NAFTA panel. The WTO panel did not consider that it had a choice to decline jurisdiction, and the Appellate Body agreed.[236]

The DSU probably excludes countermeasures affecting WTO obligations taken in response to a breach of a non-WTO Agreement such as

---

[233] Appellate Body report in *Brazil – Retreaded Tyres*, WT/DS332/AB/R, paras 217–234 and n. 245.
[234] Appellate Body reports in *US – Certain EC Products*, WT/DS165/AB/R, para 111 and *US – Continued Suspension*, WT/DS320/AB/R, para 371.
[235] DSU, Articles 6.1, 8.7, 16.4, 17.14 and 22.6.  [236] See n. 231.

a SETA.[237] The issue has never been decided by a WTO body, and commentators have expressed differing views as regards the countermeasures considered lawful under non-WTO norms of international law.[238] However, it seems implausible that the WTO would allow countermeasures as redress for a non-WTO violation when it has so narrowly confined its power to authorise suspension of obligations in response to a violation of its own rules.[239] Were the SETA to establish a DSM endowed with powers to authorise trade sanctions, any suspension of WTO obligations pursuant to such authorisation would most probably conflict with DSU rules and procedures. The mechanism could only effectively authorise suspension of any SETA obligations that lie outside the WTO framework. Cross-retaliation would not be possible.

Even if the WTO system prevails in a situation of competing and parallel procedures, this could still be damaging for both dispute settlement systems. If the WTO's exercise of jurisdiction hinders the fulfilment of the object and purpose of the non-WTO Agreement, this may in the long run affect the legitimacy of the DSU.

There exists the (as yet unrealised) possibility that a legal impediment could preclude a WTO panel from ruling on the merits of a claim before it in a case of competing jurisdiction. The Appellate Body highlighted in its report in *Mexico – Soft Drinks* that it did not express a view as to whether such a legal impediment could exist in different circumstances. It evoked the possibility of two proceedings, one in another forum and one in the WTO, with identical subject matter and identical parties' positions, a legal basis to raise the claims made in the other dispute in the WTO as well, a prior decision in the other dispute and invocation of a choice of forum, or so-called fork-in-the-road provision.[240] A WTO panel might one day rule a claim inadmissible in these circumstances.

Potential conflicts between a SETA and the WTO can be anticipated to some extent through various techniques, either in the WTO framework or in the text of the SETA, or both. The options of a waiver, a peace clause and a fork-in-the-road provision are considered below.

---

[237] The WTO Agreements only recognise authorisation to suspend concessions granted by the DSB, or under the UN Charter for the maintenance of international peace and security: see the GATT 1994, Article XXI(c), the GATS, Article XIVbis(c) and the TRIPS, Article 73(c).
[238] Pauwelyn (2003): 232; contrast Bianchi and Gradoni (2008) and Kuijper (2010): 26.
[239] DSU, Article 3.2.
[240] Appellate Body report in *Mexico – Soft Drinks*, WT/DS308/AB/R, para 54, discussed in Pauwelyn and Salles (2009): 77–118 at 90.

### Waiver

One option to reconcile any substantive provisions in a SETA that were inconsistent with the WTO Agreements would be to request a waiver. A waiver does not presume a hierarchy among international Agreements.[241] In practical terms, a request for a waiver is indicative only of a potential conflict between two systems and the existence of a functioning dispute settlement mechanism in one of them. Through a waiver, the WTO yields to the other system. A waiver could cover any SETA-implementing measures, but it would be subject to important limitations.

The WTO Agreements set out a waiver procedure that includes a series of requirements that begin with the submission of a request stating the existence of exceptional circumstances justifying the waiver decision. A request for a waiver from obligations under the GATT 1994 must describe the measures the requesting Members propose to take, the specific policy objectives that they seek to pursue and the reasons that prevent them from achieving those policy objectives by measures consistent with the GATT 1994. A waiver is granted subject to conditions that must include a termination date.[242] WTO waiver decisions are taken by consensus, in accordance with the decision making procedures agreed by the WTO General Council in 1995.[243]

One model for the SETA could be the 2003 waiver granted for measures implementing the Kimberley Process Certification Scheme for Rough Diamonds (the Kimberley process), which regulates the import and export of rough diamonds.[244] Eleven of the participants[245] submitted a request to the WTO for a waiver with respect to measures necessary to prohibit the export and import of rough diamonds to and from non-participants, many of whom are WTO Members. They cited the exceptional circumstances presented by the trade in conflict diamonds that fuels armed conflict, which has a devastating impact in affected countries. They also recalled that the UN Security Council had adopted a resolution supporting the Kimberley process.[246] The WTO General Council

---

[241] Pauwelyn (2003b): 1177–1207; Kuijper (2010): 12.
[242] WTO Agreement, Article IX:3 and IX:4; Understanding in Respect of Waivers of Obligations under the GATT 1994, para 1.
[243] WT/L/93. Consensus is the only option in certain cases concerning transition periods and staged implementation: see WTO Agreement, Article IX:3, n. 4.
[244] WTO General Council Decision of 15 May 2003 on Waiver Concerning Kimberley Process Certification Scheme for Rough Diamonds, WT/L/518.
[245] Australia, Brazil, Canada, Israel, Japan, Korea, Philippines, Sierra Leone, Thailand, United Arab Emirates and the United States. Eight other participants are listed in the 2006 extension.
[246] S/RES/1459(2003).

Legal Options for a SETA 445

granted the requesting members a waiver until 2006 from specific WTO obligations subject to the condition that the participants' measures were consistent with the Kimberley scheme together with other conditions regarding transparency. The waiver was later extended until the end of 2012.[247] As of 2009, seventy-five WTO Members participate in the Kimberley process; this includes all major rough diamond producing, exporting and importing countries in the world.[248]

This model has inspired a proposal to establish a mechanism within the UNFCCC to govern climate change subsidies. Subsidies notified and subject to the discipline of this mechanism could benefit from a WTO waiver from certain disciplines of the SCM Agreement.[249]

The SETA could fulfil certain of the requirements for a waiver. The unprecedented challenge to humanity presented by climate change can constitute 'exceptional circumstances' for the purposes of the WTO Agreement., Article IX:3. A request for a waiver should carefully list every specific WTO obligation with which the SETA might be inconsistent, as other obligations will not be covered simply by implication. The parties to the SETA could agree to transparency conditions, and those who benefit from the waiver could be listed, leaving open the possibility for parties acceding to the SETA later also to benefit from the waiver on notification to the WTO, as in the Kimberley process waiver. The waiver could be granted whether SETA was a stand-alone agreement or incorporated in another agreement, such as the UNFCCC, the Energy Charter, a PTA or a web of bilateral investment treaties.

However, there are limitations to what waivers can achieve. In 1997, the Appellate Body noted in *EC – Bananas III* that waivers have an exceptional nature, are subject to strict disciplines and should be interpreted with great care. In 2008, the Appellate Body in *EC – Bananas III (Article 21.5 – US)* reiterated those findings and emphasised that the purpose of waivers is 'not to modify existing provisions in the agreements, let alone create new law or add to or amend the obligations under a covered agreement or Schedule'.[250] A SETA may clearly be intended to modify existing provisions in the WTO Agreements on a permanent basis, something that a waiver cannot accomplish. In contrast, the Kimberley process waiver was granted and extended for

---

[247] General Council Decision of 15 December 2006 on Kimberley Process Certification Scheme for Rough Diamonds, WT/L/676.
[248] See www.kimberleyprocess.com/structure/participants_world_map_en.html, accessed on 27 January 2012.
[249] Howse (2011): 24.
[250] Appellate Body report in *EC – Bananas III (Article 21.5 – US)*, WT/DS27/AB/RW/R, paras 381–382, citing the Appellate Body report in *EC – Bananas III*, WT/DS27/AB/R, para 185.

the sake of legal certainty[251] and may not in fact be necessary under WTO rules at all due to the general exceptions in the GATT 1994, Article XX. The waiver does not even apply to most Kimberley participants, which indicates that the lack of WTO disputes regarding Kimberley-implementing measures is partly, if not wholly, due to other reasons.

Another major hurdle is that the wider WTO membership must normally consent to a waiver and to any extensions.[252] While the Kimberley process waiver was granted and renewed, it addresses a different policy objective from sustainable energy and covers a different product. Non-parties may block consensus on a SETA waiver. Although the waiver procedures provide for a decision by voting by a three-fourths majority, that is not applied in practice[253] and some Members, who may be parties to a SETA, would probably be unwilling to resort to a voting procedure for systemic reasons.

There is a risk that even a request for a waiver can be used against a Member in WTO dispute settlement, as in the Appellate Body report on *EC – Tariff Preferences*.[254] Therefore, if a waiver is only sought for the sake of legal certainty, the request and any decision should state so expressly and provide that both are without prejudice to the WTO consistency of the SETA-implementing measures.

### Peace Clause

One option that has been suggested is a 'peace clause' under which WTO Members agree that certain measures will be shielded from complaints under the DSU.[255] This option would also be subject to important limitations.

The original WTO peace clause, which was crucial to the conclusion of the Uruguay Round, basically provided that subsidies that conformed to the new disciplines of the Agreement on Agriculture would be exempt from certain types of actions under the SCM Agreement and the GATT 1994 until the end of 2003.[256] Alternatively, a peace clause could simply commit members to exercise due restraint in having recourse to the DSU regarding certain types of measures. Such a clause does not provide legal

---

[251] WT/L/518, 4th recital, WT/L/676, 5th recital.
[252] The TRIPS and Public Health waiver (General Council Decision 30 August 2003 on the Implementation of para 6 of the Doha Declaration on the TRIPS Agreement and Public Health, WT/L/540 and Corr.1) exceptionally expressed the termination date in terms of a condition subsequent.
[253] WTO Agreement, Article IX:3(a), and the decision making procedures at n. 243.
[254] Appellate Body report in *EC – Tariff Preferences*, WT/DS246/AB/R, para 186.
[255] See Hufbauer et al. (2009): 103.   [256] Agreement on Agriculture, Article 13.

Legal Options for a SETA 447

certainty but may be sufficient for Members to reach an agreement, particularly if it is temporary. A peace clause could be tied, for instance, to a timetable for further negotiations.

A SETA could also include a peace clause if it purported to authorise measures that were WTO-inconsistent. This would occur if it diminished WTO obligations among the parties (as discussed previously). A SETA peace clause could provide that subsidies that conformed to the SETA would be exempt from actions or countervailing measures under the SCM Agreement. Such a clause is worthwhile only if one or more Members consider that their measures may be vulnerable to challenge, which would not be the case if the SETA only clarified and added to WTO obligations without diminishing them.

Although the Uruguay Round peace clause was negotiated bilaterally between the EC and the United States to end the litigation surrounding EC agricultural policies,[257] the deal was incorporated in the text of a multilateral agreement (the Agreement on Agriculture, Article 13, with multiple cross-references from the SCM Agreement) and binding on all Members. If a SETA is implemented outside the WTO, a SETA peace clause would not be incorporated in the WTO Agreement and would not be binding on non-parties, in accordance with the *pacta tertiis* rule. Therefore, it would not prevent non-parties challenging the measures under the DSU, should they have grounds for action. Indeed, such a clause would not even prevent recourse to the DSU by the parties themselves, although there exists the possibility that a WTO panel might rule such a claim inadmissible in certain circumstances (as discussed above).

In practice, a SETA might be sufficiently effective to prevent disputes under the DSU where the critical mass of parties includes all the likely eventual complainants. It can be recalled that, despite the proliferation of FTAs throughout the world, no WTO Member has ever challenged an agreement under the DSU for failure to comply with the 'substantially all the trade' requirement in the GATT 1994, Article XXIV, due at least partly to most Members' mutual interest in justifying their own respective agreements under that Article.

However, the degree of comfort provided by the participation of the critical mass depends on the market for the relevant product and the nature of the relevant WTO obligations (as discussed previously in relation to the SCM Agreement). The longer an Agreement remains in effect,

---

[257] Under the peace clause, the United States renounced further litigation regarding EC agricultural policies after the GATT Panel reports on *EEC – Oilseeds I*, BISD 37S/86 and *EEC – Oilseeds II*, BISD 39S/91.

the more likely that trade flows will change and challenges will arise from unexpected quarters. For example, the Uruguay Round peace clause was negotiated bilaterally between the United States and the EC, but when it was eventually litigated in 2002 it was Brazil that brought an action, challenging US subsidies in *US – Upland Cotton*.

A WTO moratorium can, in effect, suspend recourse to the DSU regarding certain measures for a period of time, by all WTO Members. For example, a moratorium on the initiation of non-violation and situation complaints under the TRIPS Agreement was agreed at the Doha Ministerial Conference in 2001 in a Ministerial Decision and has been renewed at successive Ministerial Conferences by declaration or decision, most recently in 2011.[258] The moratorium is temporary each time, but free of conditions. Irrespective of the legal status of the moratoria granted in the declarations and decisions, the fact is that no such complaints have been filed.

A moratorium would not be suited to SETA-implementing measures because it would have to be subject to conditions. The minimum condition would be that the implementing measures were consistent with the SETA.[259] It should also be clear whether that condition encompasses not only measures that implement SETA obligations, but also voluntary measures that exercise SETA rights. As soon as conditions apply, it is not feasible to prevent the initiation of disputes under the DSU. Members would be free to resort to the DSU to claim that another member's measures did not comply with the relevant conditions – as Brazil did, successfully, in *US – Upland Cotton*. In that sense, the decision would not be a moratorium but would simply purport to add the terms of a SETA as conditions to WTO obligations without modifying the WTO Agreements. In any event, a Ministerial Declaration or decision (or a General Council decision) is adopted by consensus in accordance with customary practice, which could be extremely difficult to reach for a decision authorising a WTO-inconsistent agreement, particularly if the WTO membership had already declined to implement it within the WTO framework.

---

[258] Doha Ministerial Decision on Implementation-Related Concerns, WT/MIN(01)/17, para 11.1; Hong Kong Ministerial Declaration, WT/MIN(05)/DEC, para 45; Ministerial Conference Decisions of 2 December 2009 and 17 December 2011 on 'TRIPS Non-Violation and Situation Complaints', WT/L/783, WT/L/842.

[259] See, for example, Decision of 8 December 1994 of the Preparatory Committee for the WTO and the CONTRACTING PARTIES to the GATT on 'Transitional Co-Existence of the GATT 1947 and the WTO Agreement', PC/12, GATT document L/7583, para 2, and corresponding decision regarding the Anti-Dumping Code, PC/13, GATT document L/7584, para 2.

### 'Fork-in-the-Road' Provision

A 'fork in the road' provides that a complainant's first choice of forum to resolve a dispute is irrevocable where an international Agreement offers a choice of dispute settlement systems or procedures.[260] It is designed to anticipate the risks that the same matter may become the subject of multiple proceedings in different fora and that those proceedings may lead to different results. Once a complainant initiates a dispute settlement procedure in one forum, the fork-in-the-road provision excludes recourse to the other forum regarding the same dispute between the same parties. For example, many PTAs incorporate the WTO Agreements and offer an irrevocable choice between the WTO's dispute settlement procedures and the PTA's dispute settlement procedures.[261]

A SETA would contain substantive norms on many WTO subjects and might therefore incorporate WTO Agreements. In that scenario, it would be important to include a fork-in-the-road provision to anticipate conflicts of jurisdiction where a SETA is implemented outside the WTO framework. The WTO has not yet pronounced on the effectiveness of such a provision, but it is one circumstance among others that could potentially create a legal impediment that would preclude a WTO panel from ruling on a claim (as discussed above). Even if the provision is not legally effective, its deterrent effect might be sufficient for it to achieve its objective.

### Comity

Consideration of mutual respect and comity between judicial institutions can allow one tribunal to cede jurisdiction to another in the same issue. For example, in the *MOX Plant* case between Ireland and the United Kingdom (both EU Member States), a UN Convention on the Law of the Sea (UNCLOS) arbitral tribunal that had *prima facie* jurisdiction nevertheless suspended further proceedings to avoid a situation in which it and the European Court of Justice (ECJ) could both render final and binding decisions. The tribunal considered that 'a procedure that might result in two conflicting decisions on the same issue would not be helpful to the

---

[260] Bilateral investment treaties often give the investor an irrevocable choice between different arbitration procedures: McLachlan et al. (2007), para 4.75.

[261] For example, the NAFTA, Article 2005; the Olivos Protocol for the Settlement of Disputes in MERCOSUR, Article 1; the Agreement on Dispute Settlement Mechanism of the Framework Agreement on Comprehensive Economic Co-Operation Between the Association of Southeast Asian Nations and the People's Republic of China, Article 2(6).

resolution of the dispute between the Parties', although it remained seised of jurisdiction.[262]

A SETA panel or tribunal operating in a separate dispute settlement system could potentially suspend its proceedings in favour of a WTO panel in order to avoid a conflict. However, it seems unlikely that a WTO panel would cede jurisdiction or refrain from ruling on a claim for reasons of comity alone, particularly in view of the terms of the DSU as interpreted by the Appellate Body in *Mexico – Soft Drinks* (discussed above).

## Outlook

If a SETA is more than a market access Agreement and the WTO Ministerial Conference does not consent to add it to the WTO Agreement, participants may choose to implement it outside the WTO framework. As a stand-alone agreement it would be unlikely to satisfy all of the conditions in any of the PTA exceptions, but that would not prohibit its conclusion.

Conflicts of substantive norms with the WTO Agreements may arise. They could create challenges for those that are party to both, to ensure that they are mutually supporting. If the SETA creates its own DSM, conflicts of jurisdiction with the DSU may also arise. These can involve multiple procedures in different fora, applying different rules and leading to different results. The WTO system may well prevail in such a conflict, which could hinder the fulfilment of SETA's object and purpose but might also reflect poorly on the WTO.

A waiver is possible on a temporary basis to shield SETA-implementing measures from conflicts with the WTO and the DSU, but this is not a suitable vehicle to create a permanent exception from WTO obligations. A peace clause could achieve the same purpose, but it will not be binding on non-participants unless it is included in a multilateral agreement. A fork-in-the-road provision would avoid procedures in multiple fora in many cases, as it would require a party to make a single choice between dispute settlement systems. There is the possibility that a WTO panel might rule a claim inadmissible in a dispute under the DSU between SETA participants on the basis of such a provision in appropriate circumstances.

---

[262] Arbitral Tribunal Constituted Pursuant to Article 287, and Article 1 of Article VII, of the United Nations Convention on the Law of the Sea for the Dispute Concerning the MOX Plant, International Movements of Radioactive Materials and the Protection of the Marine Environment of the Irish Sea, the MOX Plant case, *Ireland* v. *United Kingdom*, Order No. 3 'Suspension of Proceedings on Jurisdiction and Merits, and Request for Further Provisional Measures', dated 24 June 2003, paras 14–30.

## 3 Conclusions and Recommendations

A SETA can promote the scale-up and deployment of sustainable energy sources through the use of trade policy measures in order to reduce the emissions responsible for global warming. The scope of the Doha Round is fixed and in any case the lack of progress in that exercise compels fresh thinking about ways to negotiate new agreements. The best way to proceed depends on further work on the substance of a new Agreement, but certain options can be identified now.

A SETA could be negotiated, concluded and implemented among a group of interested WTO Members (and even non-Members) separately from the Doha Round. It could be implemented either within the WTO framework or, on the other hand, outside it as a stand-alone agreement or within another existing framework. Interested Members can begin preparatory work in other fora or in informal meetings at the WTO, but transparency is important to build confidence in the process, particularly since other Members' participation or consent may eventually be required.

A SETA could be a 'critical mass' agreement concluded among a group of countries large enough to render its implementation viable, but less than the total WTO membership. The definition of the 'critical mass' would depend on the coverage of the new Agreement, the nature of its obligations and which of them would be applied on an MFN basis, but a sufficient collective share of world trade in the products covered would seem to be essential for any market access agreement. Irrespective of the form of the Agreement, the benefits of improved market access would still accrue to any WTO Members that did not join, in accordance with their rights to MFN treatment. Free riding could be reduced by defining the critical mass widely enough to include all Members with significant volumes of trade in the relevant products.

A SETA could be a vehicle for market access improvements in an ITA-type agreement, which could cover not only tariff concessions, but also agreed rules on non-tariff measures and services commitments through incorporation of a reference paper. However, certain types of rules, namely those outside the scope of the GATT 1994 and the GATS, and any rules that diminish the parties' WTO obligations between themselves, would require a GPA-type agreement. The major hurdle to implementation of such an Agreement within the WTO framework is that the Ministerial Conference (or General Council between sessions) must agree to it by consensus. No plurilateral agreement has been added through that procedure since the establishment of the WTO. Consent might nevertheless be forthcoming if (a) the negotiating process is transparent and takes into account the views of non-parties as observers; and (b) the new Agreement protects the interests of non-parties, including their rights to MFN treatment.

The application of the DSU to disputes under a SETA is likely to be an incentive for parties to implement it within the WTO framework. However, this would also require the Ministerial Conference to agree that the DSU should be applicable, by consensus. That consent might be obtained if the interests of non-parties were specifically addressed in the way that the dispute settlement rules and procedures are applied to the new Agreement and the way in which non-parties' rights and obligations under the existing substantive rules were preserved.

If the Ministerial Conference did not consent to add the SETA to the WTO framework, the parties could implement it outside. This would be a second-best outcome, as the Agreement could not benefit from the WTO's institutional structure, in particular, its dispute settlement system. It is unlikely that the SETA would qualify for a PTA exception; hence, the benefits would still accrue to all WTO Members in accordance with MFN obligations. This would not necessarily render the Agreement unviable, depending on how many parties there were and who they were. Conflicts of norms and jurisdiction with the WTO would have to be avoided through the negotiation of appropriate rules and, ideally, an effective choice-of-forum clause.

## References

### *ICTSD Publications*

Bacchus, James (2011) 'A way forward for the WTO', *Essay for the Trade and Development Symposium, Perspectives on the Multilateral Trading System*. ICTSD and Swiss Federal Department of Economic Affairs, Geneva

Bianchi, Andrea and Lorenzo Gradoni (2008) *Developing Countries, Countermeasures and WTO Law: Reinterpreting the DSU against the Background of International Law*. Geneva

Ghosh, Arunabha and Himani Gangania (2012) *Governing Clean Energy Subsidies: What, Why and How Legal?* Geneva

ICTSD (2011) *Fostering Low Carbon Growth – The Case for a Sustainable Energy Trade Agreement*. Geneva

Jha, Veena (2009) *Trade Flows, Barriers and Market Drivers in Renewable Energy Supply Goods: The Need to Level the Playing Field*. Geneva

Kuijper, Pieter Jan (2010) *Conflicting Rules and Clashing Courts: The Case of Multilateral Environmental Agreements, Free Trade Agreements and the WTO*. Geneva

Lako, Paul (2008) *Mapping Climate Mitigation Technologies and Associated Goods within the Renewable Energy Supply Sector.* Geneva

Rodríguez, Miguel (2011) 'Towards plurilateral plus agreements', Essay for the Trade and Development Symposium, *Perspectives on the Multilateral Trading System.* ICTSD and the Swiss Federal Department of Economic Affairs, Geneva

Vossenaar, René (2010) *Climate-Related Single-Use Environmental Goods.* Geneva

Wind, Izaak (2008) *HS Codes and the Renewable Energy Sector.* Geneva

## Books and Reports

Aust, Anthony (2000) *Modern Treaty Law and Practice.* Cambridge University Press

Cottier, Thomas and Manfred Elsig (eds.) (2011) *Governing the World Trade Organization: Past, Present and Beyond Doha*, including Amrita Narlikar, 'Adapting to new power balances: institutional reform in the WTO'. Cambridge University Press: 111–128

Hoda, Anwarul (2011) *Tariff Negotiations and Renegotiations under the GATT and the WTO: Procedures and Practices.* Cambridge University Press

Hufbauer, Gary, Steve Charnovitz and Jisun Kim. (2009) *Global Warming and the World Trading System.* Peterson Institute for International Economics, Washington, DC

Jennings, Robert and Arthur Watts (eds.) (1992) *Oppenheim's International Law.* Longman, London

Lang, W. (ed.) (1995) *Sustainable Development and International Law.* Graham & Trotman, London

McLachlan, Campbell, Laurence Shore and Matthew Weiniger (2007) *International Investment Arbitration: Substantive Principles.* Oxford University Press

Pauwelyn, Joost (2003) *Conflict of Norms in Public International Law: How WTO Law Relates to Other Rules of Public International Law.* Cambridge University Press

Pettigrew, Pierre et al. (2007) *The Multilateral Trade Regime: Which Way Forward?* The Report of the First Warwick Commission, University of Warwick

Rubini, Luca (2009) *The Definition of Subsidy and State Aid: WTO and EC Law in Comparative Perspective.* Oxford University Press

Steger, Debra (ed.) (2010) *Redesigning the World Trade Organization for the Twenty-First Century*, including Manfred Elsig, 'WTO decision-making: can we get a little help from the Secretariat and the critical mass?' Wilfrid Laurier University Press, Waterloo, Ontario: 67–90

Sutherland, Peter et al. (2005) *Addressing Institutional Challenges in the New Millennium.* Report by the Consultative Board to the former Director-General Supachai Panitchpakdi. WTO, Geneva

Villiger, Mark (2009) *Commentary on the 1969 Vienna Convention on the Law of Treaties.* Martinus Nijhoff, Boston, MA

World Commission on Environment and Development (1997) *Our Common Future*. Oxford University Press

## Articles, Discussion Papers and Book Chapters

Bagwell, Kyle and Robert Staiger (2006) 'Will international rules on subsidies disrupt the world trading system?', *The American Economic Review*, 96 (3): 877–895

Ehlermann, Claus-Dieter and Lothar Ehring (2005) 'Decision-making in the World Trade Organization: is the consensus practice of the World Trade Organization adequate for making, revising and implementing rules on international trade?', *Journal of International Economic Law*, 8 (1): 51–75

Faizel, Ismail and Brendan Vickers (2011) 'Fairer decision-making in the WTO negotiations', in Carolyn Deere Birkbeck (ed.), *Making Global Trade Governance Work for Development*, University of Oxford, Global Economic Governance Program: 461–485

Fliess, Barbara and Pierre Sauvé (1997) *Of Chips, Floppy Disks and Great Timing: Assessing the Information Technology Agreement*. Institut Français des Relations Internationales and the Tokyo Club Foundation of Global Studies, Paris

Garcia Bercero, Ignacio (2000) 'Functioning of the WTO system: elements for possible institutional reform', *International Trade Law and Regulation*, 6 (4): 103–115

Green, Andrew (2006) 'Trade rules and climate change subsidies', *World Trade Review*, 5 (3): 377–414

Harbinson, Stuart (2009) *The Doha Round: 'Death-Defying Agenda' or 'Don't Do It Again'?*, European Centre for International Political Economy, Working Paper, No. 10/2009

Howse, Robert (2011) *Climate Mitigation Subsidies and the WTO Legal Framework: A Policy Analysis*. IISD, Winnipeg

Jackson, John (2001) 'The WTO Constitution and proposed reforms: seven mantras revisited', *Journal of International Economic Law*, 4 (1): 67–78

Kennedy, Matthew (2011a) 'Why are WTO Panels taking longer? And what can be done about it?', *Journal of World Trade*, 45: 221–253

 (2011b) 'Two single undertakings: can the WTO implement the results of a Round?', *Journal of International Economic Law*, 14 (1): 77–120

Low, Patrick (2009) *WTO Decision-Making for the Future*. Background Paper prepared for the Inaugural Conference of Thinking Ahead on International Trade. Geneva, 17–18 September

Lowenfeld, Andreas (1994) 'Remedies along with rights: institutional reform in the new GATT', *American Journal of International Law*, 88 (3): 477–488

Odell, John (2005) 'Chairing a WTO negotiation', *Journal of International Economic Law*, 8 (2): 425–448

Pauwelyn, Joost (2001) 'The role of public international law in the WTO: how far can we go?', *American Journal of International Law*, 95: 535–577

(2003a) 'A typology of multilateral treaty obligations: are WTO obligations bilateral or collective in nature?', *European Journal of International Law*, 14 (5): 907–951

(2003b) 'WTO compassion or superiority complex?: What to make of the WTO waiver for "conflict Diamonds"', *Michigan Journal of International Law*, 24: 1177–1207

Pauwelyn, Joost and Luiz Eduardo Salles (2009) 'Forum shopping before international tribunals: (real) concerns, (im)possible solutions', *Cornell International Law Journal*, 42 (1): 77–118

Poretti, Pietro (2008) 'Waiting for Godot: subsidy disciplines in services trade', in Marion Panizzon et al. (eds.), *GATS and the Regulation of International Trade in Services*. Cambridge University Press: 466–489

Rubini, Luca (2012) 'Ain't wastin' time no more: subsidies for renewable energy, the SCM Agreement, policy space, and law reform', *Journal of International Economic Law*, 15 (2): 525–579

Vranes, Erich (2006) 'The definition of norm conflict in international law and legal theory', *European Journal of International Law*, 17 (2): 395–418

## Further Reading

Cottier, Thomas and Panagiotis Delimatsis (eds.) (2011) *The Prospects of International Trade Regulation: From Fragmentation to Coherence*, including Thomas Cottier et al., 'Energy in WTO law and policy'. Cambridge University Press: 211–244

Cottier, Thomas et al. (eds.) (2009) *International Trade Regulation and the Mitigation of Climate Change*, including Thomas Cottier and Donah Baracol-Pinhão, 'Environmental goods and services: the Environmental Area Initiative approach and climate change'. Cambridge University Press: 395–419

EPO–UNEP–ICTSD joint 2010 report on Clean Energy Technologies, www.epo.org/news-issues/issues/clean-energy/study.html

Lawrence, Robert (2006) 'Rulemaking amidst growing diversity: a club-of-clubs approach to WTO reform and new issue selection', *Journal of International Economic Law*, 9 (4): 823–835

Nottage, Hunter and Thomas Sebastian (2006) 'Giving legal effect to the results of WTO trade negotiations: an analysis of the methods of changing WTO law', *Journal of International Economic Law*, 9 (4): 989–1016

Wilke, Marie (2011) *Feed-in Tariffs for Renewable Energy and WTO Subsidy Rules: An Initial Legal Analysis*. ICTSD, Geneva

# Index

Agreement on Agriculture, 395, 406, 423, 426, 430, 433, 446
Agreement on Basic Telecommunications Services (BTA), 76, 406
  Negotiating Group on Basic Telecommunications, 404–405
Agreement on Government Procurement (GPA), 12–13, 16, 73, 158, 196, 199, 201, 203–206, 208–209, 213, 224–231, 234, 350, 369, 390, 393, 397, 399, 407, 409, 414, 420, 422
  GPA Committee, 407, 422
Agreement on Subsidies and Countervailing Measures (SCM), 86, 160, 178–179, 182–184, 226, 327, 394, 423, 427–431, 440, 445, 446
  SCM Committee, 181
Agreement on Technical Barriers to Trade (TBT), 69, 230–231, 283–302, 416–417, 431
  Committee, 70, 292–293, 295–296
Agreement on Trade Related Investment Measures (TRIMS), 86, 183, 226–227, 327–328, 424
Airbus, 382
Algeria, 436
American National Standards Institute (ANSI), 251, 257, 269, 297
American Society for Testing and Materials (ASTM), 250
Andean Community, 208
Argentina, 75, 153
  Renewable Energy for Rural Markets Projects (PERMER), 154
Asian Development Bank (ADB), 155
Asia-Pacific Economic Cooperation (APEC), 8, 21, 28, 277, 335, 363, 365, 401
  Government Procurement Expert Group (GPEG), 199
  Honolulu Declaration, 335, 365
  list of environmental goods, 8–11, 18, 28, 74

Australia, 21, 30, 75, 101, 118, 126, 130, 263, 267
Austria, 214

Belgium
  Thornton Bank wind project, 361
Bloomberg New Energy Finance, 5, 8
Bolivia, 143, 153, 154
Brazil, 30, 69, 75, 100–101, 103, 131, 133, 141, 148, 163, 165, 176, 182, 258–259, 366, 448
  cane-based ethanol, 156, 381
  Eletrobrás, 165
  Luz para Todos Electrification Programme, 165
  Programme of Incentives for Alternative Electricity Sources (PROINFA), 165–166
*Brazil– Retreaded Tyres*, 430–442
Business 20 (B20) Summit, 7

Canada, 21, 72, 92, 103, 110, 129–130, 176, 204, 207, 250, 257, 276, 297, 326, 329, 335
Canadian Electrical Code (CEC), 252
Canadian Solar, 7
Canadian Standards Association (CSA), 251
  feed-in-tariff (FIT) program, 66, 74, 225–227, 257, 327, 380
  Green Energy and Green Economy Act, 66, 179, 326
  local content requirements (LCRs), 15, 179, 225, 380
  Ontario, 179, 225–226, 257, 326–327, 329, 380
  Quebec, 328, 329
  Standards Council of Canada (SCC), 251, 253
  wind energy, 325
CanmetENERGY, 276
carbon capture and storage (CCS), 93, 95, 220

# Index

Center for Evaluation of Clean Energy Technology (CECET), 249
Chile, 29, 75, 361, 440
*Chile– Swordfish*, 440
China, 8, 29, 30, 43, 63, 65, 68–69, 71, 73, 75, 78, 113, 116, 124, 127, 129, 133, 141–143, 145–146, 148, 153–154, 157, 163, 167, 180–182, 184, 200, 210, 217–219, 263–264, 288, 355, 360, 366, 370, 375–376, 379–380, 383, 432
  11th Five-Year Plan for National and Economic Development, 215
  12th Five-Year Plan, 145, 168
  Bidding Law, 216
  Chamber of Commerce for Import and Export of Machinery and Electronic Products, 181
  Circular Economy Law, 216
  Clean Production Promotion Law, 216
  Golden Sun programme, 166
  Government Procurement Law, 216
  Law on Energy Conservation, 217–218, 223
  local content requirements (LCRs), 66, 182
  National Development and Reform Commission (NDRC), 66, 147, 167, 181, 218
  Renewable Energy Industries Association, 181
Chu, Steven, 146
Clean Development Mechanism (CDM), 75
clean energy subsidies, 13, 17, 140, 142, 145, 149, 151, 153–154, 163, 178, 183, 185–191
climate change, 2, 6–7, 10, 13, 15, 19, 77, 94, 117–118, 133, 140, 151, 183–185, 188, 282, 317, 333, 362–374, 378, 390, 416, 430, 445
  mitigation, 111, 124, 142, 333, 349, 353, 355, 357, 367, 372, 374
Colombia, 30, 72, 75, 208
Committee on Trade and Environment Special Session (CTESS), 22
Concordia Buses, 233

Danfoss, 7
Denmark, 5, 8
Doha Development Round, 9, 13, 74, 87, 91, 103, 117, 128, 131–132, 207, 391–393, 395, 402, 405, 408, 415, 451
  Committee on Agriculture, 409
  Committee on Trade and Environment, 408
  Council for Trade in Services, 409
  Doha Ministerial Declaration, 20, 282, 393, 448
  Negotiating Group on Non-Agricultural Market Access, 409

*EC– Airbus*, 428
*EC– Asbestos*, 286
*EC– Bananas III*, 445
*EC– ITProducts*, 414
*EC– Sardines*, 286
Egypt, 112–113, 261
Engel, Ditlev, 77
Environmental Goods Agreement (EGA), 9, 18
European Commission (EC), 8, 182, 197, 211, 214
  Buying Green, 211
  Foreign Affairs Council, 8
  Trade Policy Committee, 8
European Economic Community (EEC), 9, 205–206, 398, 400, 441
  Statistical Office of the European Community (EuroStat), 92
European Investment Bank, 361
European Producers Union of Renewable Ethanol Association (ePURE), 182
European Union (EU), 8, 15, 21–22, 43, 63, 66, 72, 75, 112, 128–129, 204, 207, 210, 215, 327, 412
  Court of Justice, 232–234, 382, 449
  Emissions Trading Scheme (ETS), 364, 382, 383
  European Committee for Standardization (CEN), 363
  Public Sector Directive, 210
  Utilities Directive, 211
Evergreen Solar, 180
Export-Import Bank of the United States (EXIM Bank), 330

Finland
  Helsinki City Council, 233
France, 69, 255, 259, 428

General Agreement on Tariffs and Trade (GATT), 437, 439–442, 444, 446–447, 451
  Article XXIV, 435, 442, 447
  Peace Clause, 446–448
  Uruguay Round, 72–73, 90–91, 101, 113, 199, 319, 393, 403–404, 411, 416

458   Index

General Agreement on Trade in Services
    (GATS), 12, 16, 20, 300, 392, 394,
    404, 410, 413, 427, 431, 434,
    437, 451
  Article II, 397
  Article V, 434–435
  Article VI, 300, 302
  Article XIV, 430–431
General Electric Company (GE), 77, 95
Germany, 69, 116, 122, 148, 154, 163, 168,
    170, 176, 182, 254, 256, 259–260,
    265, 382, 428
  DIN EN 13501-1, 270
  Renewable Energies Heat Act
    (EEWärmeG), 170
  Renewable Energy Sources Act
    (EEG), 168
Global Green Growth Institute (GGGI),
    5, 19
Global Partnership for Output Based Aid
    (GPOBA), 154
Global Trade Alert, 189
global warming, 16, 390, 430, 451
green energy, 218, 319, 327, 334–335
Green Growth Action Alliance (G2A2), 7
Green Growth Task Force, 7
Green Public Procurement in Europe 2006
    Report, 215
Grundfos, 7

Hanwha Solar, 7, 266
Honda, 381

IBM, 76
India, 4, 30, 73–75, 78, 101, 103, 112–113,
    116, 124, 127, 129, 131, 133,
    141–142, 147, 154–156, 159, 163,
    171–172, 176, 182, 184, 264–265,
    329–330, 371, 375–376, 412
  Indian Renewable Energy Development
    Agency (IREDA), 155, 170
  local content requirements (LCRs),
    331–332
  Ministry of New and Renewable Energy
    (MNRE), 170, 264
  National Solar Mission, 155, 160–161,
    264, 330, 332
  thin film, 330–331
Indonesia, 30, 75, 143, 231, 412, 428
*Indonesia– Autos*, 428
Information Technology Agreement (ITA),
    83, 87, 127, 390, 392–394, 397–398,
    400–402, 404, 410–415, 422, 424,
    432, 451
  ITA Committee, 404, 407

Institute for Sustainable Power Quality
    (ISPQ), 248
Institute of Electrical and Electronics
    Engineers (IEEE)
  Standards Association (IEEE-SA), 250
Inter-American Development Bank
    (IADB), 7
International Centre for Trade and
    Sustainable Development (ICTSD),
    10–11, 88–89, 97, 353
International Chamber of Commerce, 336
International Civil Aviation Organization
    (ICAO), 364
International Council for Local
    Environmental Initiatives
    (ICLEI), 214
International Court of Justice (ICJ), 382
International Electrotechnical Commission
    IECEE, 253
International Electrotechnical Commission
    (IEC), 70, 250–251, 253, 257, 260–262,
    265–267, 269, 271, 273–274, 276–277,
    280, 283, 293, 298, 301
  61215, 246, 266, 274
  61646, 70, 246
  61730-2, 266
  IEC standards, 70, 250–253, 255, 257,
    260–262, 265–269, 271–272, 276, 298
  IECEE, 253, 301
  TC 82, 251
International Energy Agency (IEA), 164,
    355–356, 374
International Maritime Organisation
    (IMO), 364
International Organisation for Cooperation
    and Development (OECD), 198
International Organization for
    Standardization (ISO), 70, 253,
    293, 302
International Renewable Energy Agency
    (IRENA), 7, 281
International Services Agreement (ISA), 71
International Trade Administration
    (ITA), 68
International Trade Commission (ITC), 68
Interstate Renewable Energy Council
    (IREC), 248
Intertek, 249, 257, 277
Iran, 436
Ireland, 449
Italy, 261

Japan, 15, 29–30, 43, 63, 66, 74–75, 112,
    129, 179, 205–206, 225, 256, 263,
    265–266, 327, 361, 366, 380–381, 400

# Index

Japan Electrical Safety and Environment Technology Laboratories (JET), 266
Japanese Industrial Standardization (JIS), 266
  Ministry of Transport, 205
Jordan, 200, 261
  Arab Gas Pipeline, 261

Kazakhstan, 436
Kenya, 157, 361
*Korea– Procurement*, 226, 414, 420
Korea, Republic of, 5, 30, 63, 70, 75, 100, 103, 148, 206–207, 266, 282–283, 288, 325, 366, 412
Korea Energy Management Corporation (KEMCO), 70
Korean Airport Construction Authority (KACA), 206
Korean Management Energy Cooperation, 283
Krenicki, John, 77

Laos, 153
LG, 266
Libya, 436
Liebreich, Michael, 5
Local Authority Environmental Management Systems and Procurement (LEAP), 214

Major Economies Forum on Energy and Climate (MEF), 364–365
Malaysia, 112–113, 366, 412
MERCOSUR, 441
Mesa Power Group, 184
Mexico, 207, 325, 370, 441
*Mexico– Soft Drinks*, 441, 442, 450
MIT Energy Initiative, 8
Mitsubishi, 97
Morocco, 208, 222
  National Electricity Office (ONE), 222
  TEMASOL, 222
Myanmar, Burma, 206

Nepal, 153
New Zealand, 30, 75
North American Free Trade Agreement (NAFTA), 29–30, 43, 63, 75, 184, 207
Norway, 401

Obama, Barack, 8–9, 65, 219
Organisation for Economic Co-operation and Development (OECD), 7, 189, 359, 375, 402
Overseas Private Investment Corporation (OPIC), 330

Papua New Guinea, 153
Peru, 30, 75, 208
Peterson Institute for International Economics (PIIE), 5
Philippines, 143, 153, 156
Porges, Amy, 69
Protocol on Energy Efficiency and Related Environmental Aspects (PEEREA), 184

Qi, Liu, 8

Renault, 66
renewables, 3, 95–96, 172, 258
Russian Federation, 100, 116, 129, 133, 366, 372, 436

Samsung, 266
Sanyo, 258
Saudi Arabia, 261
Siemens AG, 7, 97
Singapore, 30, 115, 366, 370, 412
Sinovel, 167
Solar Clarity, 7
SolarWorld, 180, 379
Solyndra, 146, 180, 380
South Africa, 30, 75, 159, 262
Spain, 147, 163, 172–174, 259–260, 428
  Programme for the Promotion of Solar Thermal Energy Installations (PROSOL), 172
  Renewable Energy Plan (REP), 173
  Sustainable Economy Law, 172
Spectra Watt, 180
Sri Lanka, 143, 156, 159
SunPower, 258, 264
Suntech, 181
sustainable energy goods and services (SEGS), 4–9, 13–17, 75, 86–89, 196–197, 203–204, 207–211, 214–215, 218–236, 337, 372, 408, 423, 435, 438
Sustainable Energy Trade Agreement (SETA), 5–19, 21–22, 28–30, 43, 63, 68–78, 83, 85–89, 95, 124–128, 131, 133–135, 187–190, 196–197, 209, 234–236, 242, 283, 300–302, 336–337, 349–352, 365, 367–373, 390–391, 394, 396–400, 402–403, 406–410, 412–440, 443–452
Suzlon, 7, 360–361, 371
Switzerland, 92, 401

Tanzania, 153, 157
Thailand, 370, 412

# Index

Toshiba, 76
Toyota, 381
Trade in Services Agreement (TISA), 13, 117, 125–126, 132–133, 134
Trans-Pacific Partnership (TPP), 127, 365
Turkey, 30, 43, 75, 115, 262, 401, 412

Ukraine, 200
United Arab Emirates, 261
United Kingdom (UK), 8, 116, 145, 147, 161–162, 220–221, 428, 449
  hydrogen fuel-cell and carbon abatement technology fund, 221
  Sustainable Procurement Action Plan, 214
United Nations Commission on International Trade Law (UNCITRAL), 197–198
  Model Law on Public Procurement, 198, 224
United Nations Conference on Sustainable Development (Rio+20), 18–19, 190
United Nations Convention Against Corruption, 199
United Nations Convention on the Law of the Sea (UNCLOS), 449
United Nations Economic Commission for Latin America and the Caribbean (UN ECLAC), 7
United Nations Environment Programme (UNEP), 19, 164, 364
United Nations Framework Convention on Climate Change (UNFCCC), 5–7, 18, 363–365, 368, 373, 436, 445
  COP 17, 351, 368
  COP 21, 19
  Kyoto Protocol, 368, 372, 411
  Technology Mechanism, 158, 363
United Nations Industrial Development Organization (UNIDO), 7, 191
United Nations Security Council, 444
United Nations Sustainable Energy for All Initiative (SE4All), 6, 124, 189
United States (US), 21–22, 29, 65–66, 68–71, 75, 92, 112, 116, 129, 144, 146–148, 156, 159, 163, 175, 206–207, 257, 325, 330, 375, 401, 441
  Air-Conditioning, Heating, and Refrigeration Institute (AHRI), 247
  American Recovery and Reinvestment Act (ARRA), 176
  California Energy Commission (CEC), 275
  California Solar Initiative (CSI), 177, 275
  Coalition for Affordable Solar Energy (CASE), 182
  Department of Energy (DOE), 276
  Energy Information Administration (EIA), 151, 164
  ethanol, 156
  Family Smoking Prevention Tobacco Control Act of 2009, 231
  Florida Solar Energy Center (FSEC), 275
  General Services Administration (GSA), 158
  National Electric Code (NEC), 245–246, 252, 268
  Occupational Safety and Health Administration (OSHA), 253–254, 257
  Renewable Energy Production Incentive (REPI), 175
  Renewable Portfolio Standard (RPS), 177, 220
  Solar ABCs, 276, 280
  Volumetric Ethanol Excise Tax Credit (VEETC), 176
*US– Anti-dumping and countervailing duties (China)*, 440
*US– Byrd Amendment*, 439
*US– Clove Cigarettes*, 231, 286, 288–291, 417
*US– COOL*, 284, 290–291
*US– Gasoline*, 439
*US– Massachusetts*, 420
*US– Tuna*, 290
*US– Tuna II (Mexico)*, 284, 286–287, 290–292, 294, 417
*US– Upland Cotton*, 448
*US/Canada– Continued Suspension*, 417

Vietnam, 366

World Bank, 29, 86, 154–155, 189
World Business Council for Sustainable Development (WBCSD), 77
World Economic Forum, 8
  Global Agenda Council on Sustainable Energy, 5
World Intellectual Property Organization (WIPO), 368, 416

Xinjiang Goldwind, 167, 354

Zambia, 153

Lightning Source UK Ltd.
Milton Keynes UK
UKOW06n0249090916

282578UK00004B/4/P